SHOCK WAVES, EXPLOSIONS, AND DETONATIONS

Edited by
J.R. Bowen
University of Washington
Seattle, Washington

N. Manson
Université de Poitiers
Poitiers, France

A.K. Oppenheim
University of California
Berkeley, California

R.I. Soloukhin
Institute of Heat and Mass Transfer
BSSR Academy of Sciences
Minsk, USSR

Volume 87
PROGRESS IN
ASTRONAUTICS AND AERONAUTICS

Martin Summerfield, Series Editor-in-Chief
Princeton Combustion Research Laboratories, Inc.
Princeton, New Jersey

Technical papers selected from the Eighth International Colloquium on Gasdynamics of Explosions and Reactive Systems, Minsk, USSR, August 1981, and subsequently revised for this volume.

Published by the American Institute of Aeronautics and Astronautics, Inc.
1633 Broadway, New York, NY 10019

American Institute of Aeronautics and Astronautics, Inc.
New York, New York

Library of Congress Cataloging in Publication Data
Main entry under title:

International Colloquium on Gasdynamics of Explosions
 and Reactive Systems (8th: 1981: Minsk, Byelorussian S.S.R.)
 Shock waves, explosions, and detonations.

 (Progress in astronautics and aeronautics: v. 87)
 Includes index.
 1. Explosions—Congresses.   2. Shock waves—Congresses.
I. Bowen, J.R. (J. Ray)   II. Title.   III. Series.
TL507.P75 vol. 87   [QD516]   629.1s   [541.3'6]   83-15454
ISBN 0-915928-76-0;   Set: 0-915928-79-5

Copyright © 1983 by
American Institute of Aeronautics and Astronautics, Inc.

All rights reserved. No part of this book may be reproduced in any form or by any means, electronic or mechanical, including photocopying, recording, or by any information storage and retrieval system, without permission in writing from the publisher.

Progress in Astronautics and Aeronautics

Series Editor-in-Chief

Martin Summerfield
Princeton Combustion Research Laboratories, Inc.

Series Associate Editors

Burton I. Edelson
*National Aeronautics
and Space Administration*

Leroy S. Fletcher
Texas A&M University

Allen E. Fuhs
Naval Postgraduate School

J. Leith Potter
Vanderbilt University

Norma J. Brennan
Director, Editorial Department
AIAA

Camille S. Koorey
Series Managing Editor
AIAA

Table of Contents

Preface xi

List of Series Volumes xv

Introduction 1
    Y.B. Zel'dovich, *Academy of Sciences, Moscow, USSR*

Chapter I.   Shock Waves Interactions 7

**The Study of Shock-Induced Signals and Coherent Effects
in Solids by Molecular Dynamics** 9
    A.M. Karo, F.E. Walker, and W.G. Cunningham, *Lawrence Livermore
National Laboratory, Livermore, Calif.,* and J.R. Hardy,
University of Nebraska, Lincoln, Neb.

Oblique Shock Waves in Two-Phase Flow 22
    K. Hayashi, *Princeton University, Princeton, N.J.,* and M.C. Branch,
University of Colorado, Boulder, Colo.

Equilibrium Shock Wave Properties in Dusty and Clean Air 41
    A.L. Kuhl, *R&D Associates, Marina del Rey, Calif.*

Shock Waves in Water Induced by Focused Laser Radiation 64
    O.G. Martynenko, N.N. Stolovich, G.I. Rudin, and S.A. Levchenko,
Academy of Sciences, Minsk, USSR

Ignition of Small Particles Behind Shock Waves 71
    V.M. Boiko, A.V. Fedorov, V.M. Fomin, A.N. Papyrin,
and R.I. Soloukhin, *Academy of Sciences, Novosibirsk, USSR*

Reflection of Shock Waves at Rigid Walls in Two-Phase Media 88
    A.A. Borisov, B.E. Gel'fand, R.I. Nigmatulin, K.A. Rakhmatulin,
and E.I. Timofeev, *Academy of Sciences, Moscow, USSR*

Relaxation Phenomena in a Foamy Structure 96
    V.M. Kudinov, B.I. Palamarchuk, V.A. Vakhnenko, A.V. Cherkashin,
S.G. Lebed, and A.T. Malakhov, *Academy of Sciences, Kiev, USSR*

Chapter II.   Blast Waves 119

**Self-Similar Blast Waves Incorporating Deflagrations
of Variable Speed** 121
    R.H. Guirguis, M.M. Kamel, and A.K. Oppenheim, *University of
California, Berkeley, Calif.*

**Analysis of Reactive Blast Waves Propagating Through Gaseous
Mixtures with Spatially Distributed Chemical Energy** 157
    S. Ohyagi and A. Ohsawa, *Saitama University, Saitama, Japan*

**On the Use of General Equations of State in Similarity
Analysis of Flame-Driven Blast Waves** 175
  A.H. Kuhl, *R&D Associates, Marina del Rey, Calif*

Optical Interferometry of Spherical Shock Waves 196
  V.F. Klimkin and V.V. Pickalov, *Academy of Sciences, Novosibirsk,
  USSR,* and R.I. Soloukhin, *Academy of Sciences, Minsk, USSR*

A Study of Explosive Shock Tubes 205
  G.A. Shvetsov, V.M. Titov, V.P. Chistyakov, and I.A. Stadnichenko,
  Academy of Sciences, Novosibirsk, USSR

**The Taylor Instability of Contact Boundary Between Expanding
Detonation Products and a Surrounding Gas** 218
  S.I. Anisimov, Y.B. Zel'dovich, N.A. Inogamov, and M.F. Ivanov,
  L.D. Landau Institute for Theoretical Physics, Moscow, USSR

Chapter III.   Gaseous Detonations...................... 229

**Properties of Detonation Waves in Hydrocarbon-Oxygen-Nitrogen
Mixtures at High Initial Pressures**........................ 231
  P. Bauer and C. Brochet, *Laboratoire d'Energétique et de Détonique,
  Poitiers, France*

Overdriven Gaseous Detonations 244
  T.P. Gavrilenko and E.S. Prokhorov, *Academy of Sciences,
  Novosibirsk, USSR*

Motion of Solid Bodies in Combustible Gas Mixtures 251
  M.M. Gilinsky, L.I. Zak, and T.S. Novikova, *Moscow State
  University, Moscow, USSR*

Direct Initiation of Detonation in LNG/Air Clouds............ 262
  J. Kurylo, J.M. Thomsen, and F.M. Sauer, *Physics International
  Company, San Leandro, Calif.*

**Influence of Walls on Pressure Behind Self-Sustained Expanding
Cylindrical and Plane Detonations in Gases** 302
  D. Desbordes and N. Manson, *Université de Poitiers, Poitiers, France,*
  and J. Brossard, *Université d'Orléans, Bourges, France*

Kinetic Modelling of Ethane/Air Detonability 318
  R. Atkinson and D.C. Bull, *Shell Research Ltd., Thornton
  Research Center, Chester, England*

Chapter IV.   Heterogeneous Detonations 333

**Detonations Supported by Physical Explosions
of Liquefied Gases**...................................... 335
  S. Tsugé and S. Kadowaki, *University of Tsukuba, Ibaraki, Japan*

Effect of Liquid Films on Detonation in a Gaseous Mixture 352
    J.P. Saint-Cloud, C. Guerraud, and N. Manson, *Université de Poitiers, Poitiers, France*

Ignition of Aluminum Particles in a Gaseous Detonation 362
    B. Veyssiere, *Université de Poitiers, France*

A Model of Blast Waves Propagating in Coal Mines 376
    V.P. Korobeinikov and I.S. Men'shov, *Academy of Sciences, Moscow, USSR*

Multifront Combustion of Two-Phase Media 394
    L.A. Afanasiva, V.A. Levin, and Y.V. Tunik, *Moscow State University, Moscow, USSR*

Flame Propagation in Dust-Air Mixtures at Minimum Explosive Concentration 414
    W. Buksowicz, *Fire Protection Research and Development Center, Józefów, Poland*, and P. Wolánski, *Technical University of Warsaw, Warsaw, Poland*

Chemical Kinetics of Detonation in Some Organic Liquids 426
    B.N. Kondrikov, *Mendeleev Institute of Chemical Technology, Moscow, USSR*

Chapter V. Explosions in Solids 443

Shock Induced Hot-Spot Formation and Subsequent Decomposition in Granular, Porous HNS Explosive 445
    D.B. Hayes, *Sandia National Laboratories, Albuquerque, N. Mex.*

Initiation of Detonations 468
    C.L. Mader, *Los Alamos National Laboratory, Los Alamos, N. Mex.*

Shock Wave Predetonation Processes in Porous High Explosives 492
    B.A. Khasainov, A.A. Borisov, and B.S. Ermolayev, *Academy of Sciences, Moscow, USSR*

Author Index for Volume 87 505

Table of Contents for Companion Volume 88

Chapter I. Laminar Flames 1

  Resonant Response of a Flat Flame Near a Flame-Holder 3
    A.C. McIntosh and J.F. Clarke, *Cranfield Institute of Technology, Bedford, England*

  Temperature and Pressure Effect in Cool and Blue Flames 38
    Y. Ohta and H. Takahashi, *Nagoya Institute of Technology, Nagoya, Japan*

  An Excess Enthalpy Flame Stabilized in Ceramic Tubes 57
    T. Hashimoto and S. Yamasaki, *Hitachi, Ltd., Tsuchiura, Japan*, and T. Takeno, *The University of Tokyo, Tokyo, Japan*

Chapter II. Turbulent Flames 79

  Differential Diffusion Effects on Measurements in Turbulent Diffusion Flames
  by the Mie Scattering Technique 81
    S.H. Stårner and R.W. Bilger, *The University of Sydney, Sydney, Australia*

  Concentration and Velocity Measurements in a Turbulent Reacting Mixing Layer 105
    J.L. Bousgarbies and J. Nérault, *Laboratoire d' Études Aérodynamiques et Thermiques, Poitiers, France*

  Turbulent Combustion Zone in a Tubular Reactor 119
    Y. Chauveau, P. Cambray, E. Gengembre, M. Champion, and J.C. Bellet, *Université de Poitiers, Poitiers, France*

  Premixed Turbulent Flames—Interplay of Hydrodynamic and Chemical Phenomena 133
    A.M. Klimov, *Academy of Sciences, Moscow, USSR*

  Modification of Turbulent Flowfield by an Oblique Premixed Hydrogen-Air Flame 147
    D. Escudie, M. Trinite, and P. Paranthoen, *Laboratoire de Thermodynamique, Mont Saint Aignan, France*

Chapter III. Combustion of Solids 165

  Combustion of Lithium Perchlorate, Ammonium Chloride-Ammonium
  Perchlorate Solid Mixtures 167
    M.S. Al Fakir, P. Joulain, and J.M. Most, *Université de Poitiers, Poitiers, France*

  Radiation from Polyurethane Pile Fires 182
    J.M. Souil, H. Azov, and P. Joulain, *Université de Poitiers, St. Julien l'Ars, France*, and S. Galant, *Bertin et Cie, Plaisir, France*

  Study of Condensed System Flames by Molecular Beam Mass Spectrometry 197
    O.P. Korobeinichev and L.V. Kuibida, *Academy of Sciences, Novosibirsk, USSR*

  Unsteady Burning of Double-Base Propellants 208
    V.E. Zarko, V.N. Simonenko, and A.B. Kiskin, *Academy of Sciences, Novosibirsk, USSR*

  Surface Layer Destruction during Combustion of Homogeneous Powders 220
    V.E. Zarko and V.Y. Zyryanov, *Academy of Sciences, Novosibirsk, USSR*

  Heat Transfer Ahead of Flame Spreading over a Cured Epoxy Resin Surface
  in an Opposed Flow 228
    B.Y. Kolesnikov, V.L. Efremov, N.S. Umarbekov, and A.B. Kolesnikov *Kazakh State University, Alma-Ata, USSR*

Chapter IV. Ignition and Extinction 237

  Self-Ignition of Atomized Liquid Fuel in Gaseous Medium 239
    A.A. Borisov, B.E. Gel'fand, E.I. Timofeev, S.A. Tsyganov, and S.V. Khomik, *Academy of Sciences, Moscow, USSR*

Hydrocarbon Induced Acceleration of Ignition of Methane-Air Ignition 252
    R. Zellner and K.J. Niemitz, *Institut für Physikalische Chemie der Universität Göttingen, Göttingen,*
    W. Germany, and J. Warnatz, *Institut für Physikalische Technischen Hochschule Darmstadt, Darmstadt,*
    W. Germany, and W.C. Gardiner Jr., C.S. Eubank, and J.M. Simmie, *University of Texas, Austin, Texas*

Extinction of In-Flight Engine Fuel-Leak Fires with Dry Chemicals 273
    R.L. Altman, *NASA Ames Research Center, Moffett Field, Calif.*

Chapter V. Nonequilibrium Systems 291

**Linearized Kinetic Models for Polyatomic Gases and Mixtures of Gases: Application
to Vibrationally Relaxing Flows** 293
    R. Brun, G. Duran, P.C. Philippi, M.F. Dourieu, and R. Tosello, *Université de Provence, Marseille, France*

Theoretical Model for Sound Output from a Pulsating Arc Discharge 305
    N.I. Kidin and V.B. Librovich, *Academy of Sciences, Moscow, USSR,* and M.L. Vuillermoz and J.P. Roberts,
    Polytechnic of the South Bank, London, England

Chapter VI. Lasers 317

**Gain Coefficient and Vibrational Temperature Measurements in Shock Tube Driven CO_2-GDL,
$CO_2 + N_2 + He$, and $CO_2 + CO(N_2) + H_2$ Mixtures** 319
    S.S. Novikov, V.M. Doroshenko, and N.N. Kudryavtsev, *Academy of Sciences, Moscow, USSR*

Time-Dependent Nozzle and Base Flow/Cavity Model for CW Chemical Laser Flowfields 336
    N.L. Rapagnani and D.W. Lankford, *Air Force Weapons Laboratory, Kirtland Air Force Base, N. Mex.*

Modeling of a Chemically Driven H_2-HCl Transfer Laser 369
    V.K. Baev and V.I. Golovichev, *Academy of Sciences, Novosibirsk, USSR,* H. Guenoche and C. Sedes,
    Université de Provence, Marseille, France

A Gasdynamic Laser Using Products of Acetylene Explosions 391
    A.B. Britan, A.N. Khmelevskii, V.A. Levin, S.A. Losev, V.V. Lugovskoi, G.D. Smekhov, and A.M. Starik,
    Moscow University, Moscow, USSR

Influence of Flow Structure on Optical Gain in Gasdynamic Lasers 411
    M.G. Ktalkherman, V.M. Malkow, and N.A. Ruban, *Academy of Sciences, Novosibirsk, USSR*

Operation of Arc-Heated Gasdynamic CO_2 Laser at 16.4-18.6 μm 425
    A.I. Demin, E.M. Kidriavtsev, and A.Y. Volkov, *Academy of Sciences, Moscow, USSR,* and
    D.G. Bakanov, A.I. Fedoseev, and A.I. Odintsov, *Moscow State University, Moscow, USSR,* and
    V.F. Sharkov, *Kurchatov Institute of Atomic Energy, Moscow, USSR*

Preface

This and a companion volume include revised and edited papers that were presented at the Eighth International Colloquium on the Gasdynamics of Explosions and Reactive Systems held in Minsk, USSR, in August 1981. The International Gasdynamic Colloquia had their origin in 1966 as a consequence of revolutionary advances in the understanding of detonation wave structure. Leading researchers in this field concluded that a regular forum should be available for discussion of important findings in the gasdynamics of flows associated with the exothermic process—the essential feature of detonation waves. However, it was felt that a much broader scope of applications should be included.

The gasdynamics of explosions is a subject concerned principally with the interrelationship between the rate processes of energy deposition in a compressible medium and its concurrent nonsteady flow as it occurs typically in explosion phenomena. Gasdynamics of reactive systems is a broader term referring to the processes of coupling between the dynamics of fluid flow and molecular transformations in reactive media occurring in any combustion system. In this connection, in addition to the usual topics of explosions, detonations, shock phenomena, and reactive flow, the Eighth Colloquium included papers that dealt especially with the gasdynamic aspects of nonsteady flow in combustion systems, the fluid mechanic aspects of combustion with particular emphasis on the effects of turbulence, and the diagnostic techniques for the study of combustion phenomena. Of special interest were papers dealing with the radiative heat-transfer effects on the fluid dynamic features of reactive systems such as luminous flames, intense fires, and gasdynamic lasers.

The contributions have been assembled into two volumes: *Shock Waves, Explosions, and Detonations* and *Flames, Lasers, and Reactive Systems*. In this volume, the papers have been arranged into chapters on shock interactions, blast waves, gaseous detonations, heterogeneous detonations, and explosions in solids. The material should be particularly useful to the reader who wishes to be informed about the current directions and important findings of Soviet scientists. About half of the papers are of Russian origin and provide a valuable insight into the current status of Russian research. The Introduction by Academician *Y.B. Zel'dovich* provides a historical perspective of the subject matter and a critical assessment of its progress.

Many of the 55 papers in these two volumes provoked interesting discussions during the Colloquium. While the brevity of this Preface does not permit the editors to do justice to all of the stimulating papers, the more noteworthy contributions among them will be highlighted in the following.

The chapter titles indicate the diversity of topics treated in this volume. In Chap. I, Shock Wave Interactions, *Karo, Walker, Cunningham,* and *Hardy* report on their molecular dynamic calculations that reveal the extent to which microscopic structure may be deduced from macroscopic observations of shock wave propagation in condensed systems. *Boiko, Fedorov, Fomin, Papyrin,* and *Soloukhin* report on the application of laser diagnostics to determine the ignition characteristics of small particles behind a shock wave.

In Chap. II, Blast Waves, *Anisimov, Zel'dovich, Inogamov,* and *Ivanov* report an interesting analysis of the Taylor instability of the contact surface between expanding detonation products and the surrounding gas.

In Chap. III, Gaseous Detonations, *Gavrilenko* and *Prokhorov* report experimental results on the overdriven detonation waves that are generated when a Chapman-Jouguet detonation wave encounters a change in confinement (i.e., a wedge in a flat channel or a conical insert in a round tube). The results provide additional insight into transient detonation response. *Kurlyo, Thomsen,* and *Sauer* report on a numerical simulation of the direct initiation of LNG/air mixtures by subcritical spherical energy sources. The simulation includes detailed treatment of the reaction kinetics associated with gasdynamics modeling. Their results indicate three distinct modes of propagation of the shock and reaction fronts: strongly coupled, weakly coupled, and uncoupled. *Atkinson* and *Bull* describe the development of a method for predicting the detonability of ethane/air mixtures. They used a detailed kinetic model of ethane oxidation combined with the Zel'dovich criterion to derive predicted detonability limits yielding good agreement with available experimental data.

In Chap. IV, Heterogeneous Detonations, *Tsugé* and *Kadowaki* present an interesting treatment of detonations supported by endothermic phase transitions that involve large changes in volumes. Their analysis suggests the existence of a critical temperature for each liquid below which physical explosions are impossible. *Korobeinikov* and *Men'shov* present a theoretical model of combustion and detonation of coal dust in methane/air mixtures and report solutions for several cases.

In Chap. V, Explosions in Solids, *Mader* presents results of two computational methods for the analysis of detonation initiation: 1) a two-dimensional Lagrangian code combined with forest fire explosive decomposition rates and 2) a three-dimensional, reactive,

Eulerian hydrodynamic code. Shock-induced hot-spot formation and explosive decomposition of granular porous hexanitrostilbene is the subject of an investigation by *Hayes*. This work provides useful information about the initiation of explosions of granular materials.

The companion volume includes papers on laminar and turbulent flames, the combustion of solids, ignition and extinction, nonequilibrium systems, and lasers (Vol. 88 in the *AIAA Progress in Astronautics and Aeronautics* series).

The first Colloquium was held in 1967 in Brussels, and Colloquia have been held on a biennial basis since then (1969 in Novosibirsk, 1971 in Marseilles, 1973 in La Jolla, 1975 in Bourges, 1977 in Stockholm, 1979 in Göttingen, and 1981 in Minsk). They have now achieved the status of a prime international meeting on these topics and attract contributions from scientists and engineers throughout the world. The Proceedings of the First through the Sixth Colloquia have appeared as part of the journal, *Acta Astronautica*, or its predecessor, *Astronautica Acta*. With the publication of the Seventh Colloquium the Proceedings now appear as part of the AIAA *Progress in Astronautics and Aeronautics* series.

Acknowledgments

The Eighth Colloquium was held under the auspices of the Institute of Heat and Mass Transfer, Soviet Academy of Sciences, Minsk, BSSR, Aug. 24-28, 1981. Arrangements in Minsk were made by Prof. R.I. Soloukhin. The publication of the Proceedings has been made possible by grants from the National Science Foundation (USA) and the Army Research Office (USA).

Preparations for the Ninth Colloquium are under way. The meeting is scheduled to take place in July 1983 at the Ecole Nationale Supérieure de Mécanique et d'Aérotechnique in Poitiers, France.

J. Ray Bowen
Numa Manson
Antoni K. Oppenheim
R.I. Soloukhin
April 1983

R.I. Soloukhin and Y.B. Zel'dovich in discussion.

Participants in session.

Progress in Astronautics and Aeronautics

Volume Titles

Volume Editors

*1. Solid Propellant Rocket Research. 1960

Martin Summerfield
Princeton University

*2. Liquid Rockets and Propellants. 1960

Loren E. Bollinger
The Ohio State University
Martin Goldsmith
The Rand Corporation
Alexis W. Lemmon Jr.
Battelle Memorial Institute

*3. Energy Conversion for Space Power. 1961

Nathan W. Snyder
Institute for Defense Analyses

*4. Space Power Systems. 1961

Nathan W. Snyder
Institute for Defense Analyses

*5. Electrostatic Propulsion. 1961

David B. Langmuir
Space Technology Laboratories, Inc.
Ernst Stuhlinger
NASA George C. Marshall Space Flight Center
J. M. Sellen Jr.
Space Technology Laboratories, Inc.

*6. Detonation and Two-Phase Flow. 1962

S. S. Penner
California Institute of Technology
F. A. Williams
Harvard University

*7. Hypersonic Flow Research. 1962

Frederick R. Riddell
AVCO Corporation

*8. Guidance and Control. 1962

Robert E. Roberson
Consultant
James S. Farrior
Lockheed Missiles and Space Company

---

*Now out of print.

*9. Electric Propulsion
Development. 1963
Ernst Stuhlinger
NASA George C. Marshall Space Flight Center

*10. Technology of Lunar
Exploration. 1963
Clifford I. Cummings and
Harold R. Lawrence
Jet Propulsion Laboratory

*11. Power Systems for Space
Flight. 1963
Morris A. Zipkin and
Russell N. Edwards
General Electric Company

*12. Ionization in High-
Temperature Gases. 1963
Kurt E. Shuler, Editor
National Bureau of Standards
John B. Fenn, Associate Editor
Princeton University

*13. Guidance and Control—II.
1964
Robert C. Langford
General Precision Inc.
Charles J. Mundo
Institute of Naval Studies

*14. Celestial Mechanics and
Astrodynamics. 1964
Victor G. Szebehely
Yale University Observatory

*15. Heterogeneous Combustion.
1964
Hans G. Wolfhard
Institute for Defense Analyses
Irvin Glassman
Princeton University
Leon Green Jr.
Air Force Systems Command

*16. Space Power Systems
Engineering. 1966
George C. Szego
Institute for Defense Analyses
J. Edward Taylor
TRW Inc.

*17. Methods in Astrodynamics
and Celestial Mechanics. 1966
Raynor L. Duncombe
U. S. Naval Observatory
Victor G. Szebehely
Yale University Observatory

*18. Thermophysics and
Temperature Control of
Spacecraft and Entry
Vehicles. 1966
Gerhard B. Heller
NASA George C. Marshall Space Flight Center

*19. Communication Satellite Richard B. Marsten
    Systems Technology. 1966 *Radio Corporation of America*

*20. Thermophysics of Spacecraft Gerhard B. Heller
    and Planetary Bodies: *NASA George C. Marshall Space*
    Radiation Properties of *Flight Center*
    Solids and the
    Electromagnetic Radiation
    Environment in Space. 1967

*21. Thermal Design Principles of Jerry T. Bevans
    Spacecraft and Entry Bodies. *TRW Systems*
    1969

*22. Stratospheric Circulation. Willis L. Webb
    1969 *Atmospheric Sciences Laboratory,*
    White Sands, and University of
    Texas at El Paso

*23. Thermophysics: Applications Jerry T. Bevans
    to Thermal Design of *TRW Systems*
    Spacecraft. 1970

24. Heat Transfer and Spacecraft John W. Lucas
    Thermal Control. 1971 *Jet Propulsion Laboratory*

25. Communications Satellites for Nathaniel E. Feldman
    the 70's: Technology. 1971 *The Rand Corporation*
    Charles M. Kelly
    The Aerospace Corporation

26. Communications Satellites for Nathaniel E. Feldman
    the 70's: Systems. 1971 *The Rand Corporation*
    Charles M. Kelly
    The Aerospace Corporation

27. Thermospheric Circulation. Willis L. Webb
    1972 *Atmospheric Sciences Laboratory,*
    White Sands, and University of
    Texas at El Paso

28. Thermal Characteristics John W. Lucas
    of the Moon. 1972 *Jet Propulsion Laboratory*

29. Fundamentals of Spacecraft John W. Lucas
    Thermal Design. 1972 *Jet Propulsion Laboratory*

30. Solar Activity Observations and Predictions. 1972 — Patrick S. McIntosh and Murray Dryer, *Environmental Research Laboratories, National Oceanic and Atmospheric Administration*

31. Thermal Control and Radiation. 1973 — Chang-Lin Tien, *University of California, Berkeley*

32. Communications Satellite Systems. 1974 — P. L. Bargellini, *COMSAT Laboratories*

33. Communications Satellite Technology. 1974 — P. L. Bargellini, *COMSAT Laboratories*

34. Instrumentation for Airbreathing Propulsion. 1974 — Allen E. Fuhs, *Naval Postgraduate School*; Marshall Kingery, *Arnold Engineering Development Center*

35. Thermophysics and Spacecraft Thermal Control. 1974 — Robert G. Hering, *University of Iowa*

36. Thermal Pollution Analysis. 1975 — Joseph A. Schetz, *Virginia Polytechnic Institute*

37. Aeroacoustics: Jet and Combustion Noise; Duct Acoustics. 1975 — Henry T. Nagamatsu, Editor, *General Electric Research and Development Center*; Jack V. O'Keefe, Associate Editor, *The Boeing Company*; Ira R. Schwartz, Associate Editor, *NASA Ames Research Center*

38. Aeroacoustics: Fan, STOL, and Boundary Layer Noise; Sonic Boom; Aeroacoustics Instrumentation. 1975 — Henry T. Nagamatsu, Editor, *General Electric Research and Development Center*; Jack V. O'Keefe, Associate Editor, *The Boeing Company*; Ira R. Schwartz, Associate Editor, *NASA Ames Research Center*

39. Heat Transfer with Thermal Control Applications. 1975 — M. Michael Yovanovich, *University of Waterloo*

40. **Aerodynamics of Base Combustion.** 1976

S. N. B. Murthy, Editor
Purdue University
J. R. Osborn, Associate Editor
Purdue University
A. W. Barrows and J. R. Ward,
Associate Editors
Ballistics Research Laboratories

41. **Communication Satellite Developments: Systems.** 1976

Gilbert E. LaVean
Defense Communications Engineering Center
William G. Schmidt
CML Satellite Corporation

42. **Communication Satellite Developments: Technology.** 1976

William G. Schmidt
CML Satellite Corporation
Gilbert E. LaVean
Defense Communications Engineering Center

43. **Aeroacoustics: Jet Noise, Combustion and Core Engine Noise.** 1976

Ira R. Schwartz, Editor
NASA Ames Research Center
Henry T. Nagamatsu,
Associate Editor
General Electric Research and Development Center
Warren C. Strahle,
Associate Editor
Georgia Institute of Technology

44. **Aeroacoustics: Fan Noise and Control; Duct Acoustics; Rotor Noise.** 1976

Ira R. Schwartz, Editor
NASA Ames Research Center
Henry T. Nagamatsu,
Associate Editor
General Electric Research and Development Center
Warren C. Strahle,
Associate Editor
Georgia Institute of Technology

45. **Aeroacoustics: STOL Noise; Airframe and Airfoil Noise.** 1976

Ira R. Schwartz, Editor
NASA Ames Research Center
Henry T. Nagamatsu,
Associate Editor
General Electric Research and Development Center
Warren C. Strahle,
Associate Editor
Georgia Institute of Technology

46. Aeroacoustics: Acoustic Wave Propagation; Aircraft Noise Prediction; Aeroacoustic Instrumentation. 1976

Ira R. Schwartz, Editor
NASA Ames Research Center
Henry T. Nagamatsu, Associate Editor
General Electric Research and Development Center
Warren C. Strahle, Associate Editor
Georgia Institute of Technology

47. Spacecraft Charging by Magnetospheric Plasmas. 1976

Alan Rosen
TRW Inc.

48. Scientific Investigations on the Skylab Satellite. 1976

Marion I. Kent and Ernst Stuhlinger
NASA George C. Marshall Space Flight Center
Shi-Tsan Wu
The University of Alabama

49. Radiative Transfer and Thermal Control. 1976

Allie M. Smith
ARO Inc.

50. Exploration of the Outer Solar System. 1977

Eugene W. Greenstadt
TRW Inc.
Murray Dryer
National Oceanic and Atmospheric Administration
Devrie S. Intriligator
University of Southern California

51. Rarefied Gas Dynamics, Parts I and II (two volumes). 1977

J. Leith Potter
ARO Inc.

52. Materials Sciences in Space with Application to Space Processing. 1977

Leo Steg
General Electric Company

53. **Experimental Diagnostics in Gas Phase Combustion Systems.** 1977

Ben T. Zinn, Editor
Georgia Institute of Technology
Craig T. Bowman,
Associate Editor
Stanford University
Daniel L. Hartley,
Associate Editor
Sandia Laboratories
Edward W. Price,
Associate Editor
Georgia Institute of Technology
James G. Skifstad,
Associate Editor
Purdue University

54. **Satellite Communications: Future Systems.** 1977

David Jarett
TRW Inc.

55. **Satellite Communications: Advanced Technologies.** 1977

David Jarett
TRW Inc.

56. **Thermophysics of Spacecraft and Outer Planet Entry Probes.** 1977

Allie M. Smith
ARO Inc.

57. **Space-Based Manufacturing from Nonterrestrial Materials.** 1977

Gerard K. O'Neill, Editor
Princeton University
Brian O'Leary, Assistant Editor
Princeton University

58. **Turbulent Combustion.** 1978

Lawrence A. Kennedy
State University of New York at Buffalo

59. **Aerodynamic Heating and Thermal Protection Systems.** 1978

Leroy S. Fletcher
University of Virginia

60. **Heat Transfer and Thermal Control Systems.** 1978

Leroy S. Fletcher
University of Virginia

61. **Radiation Energy Conversion in Space.** 1978

Kenneth W. Billman
NASA Ames Research Center

62. **Alternative Hydrocarbon Fuels: Combustion and Chemical Kinetics.** 1978

Craig T. Bowman
Stanford University
Jorgen Birkeland
Department of Energy

63. **Experimental Diagnostics in Combustion of Solids.** 1978

Thomas L. Boggs
Naval Weapons Center
Ben T. Zinn
Georgia Institute of Technology

64. **Outer Planet Entry Heating and Thermal Protection.** 1979

Raymond Viskanta
Purdue University

65. **Thermophysics and Thermal Control.** 1979

Raymond Viskanta
Purdue University

66. **Interior Ballistics of Guns.** 1979

Herman Krier
University of Illinois at Urbana-Champaign
Martin Summerfield
New York University

67. **Remote Sensing of Earth from Space: Role of "Smart Sensors."** 1979

Roger A. Breckenridge
NASA Langley Research Center

68. **Injection and Mixing in Turbulent Flow.** 1980

Joseph A. Schetz
Virginia Polytechnic Institute and State University

69. **Entry Heating and Thermal Protection.** 1980

Walter B. Olstad
NASA Headquarters

70. **Heat Transfer, Thermal Control, and Heat Pipes.** 1980

Walter B. Olstad
NASA Headquarters

71. **Space Systems and Their Interactions with Earth's Space Environment.** 1980

Henry B. Garrett and
Charles P. Pike
Hanscom Air Force Base

72. **Viscous Flow Drag Reduction.** 1980

Gary R. Hough
Vought Advanced Technology Center

73. **Combustion Experiments in a Zero-Gravity Laboratory.** 1981

Thomas H. Cochran
NASA Lewis Research Center

74. **Rarefied Gas Dynamics, Parts I and II (two volumes).** 1981

Sam S. Fisher
University of Virginia at Charlottesville

75.	Gasdynamics of Detonations and Explosions. 1981	J. R. Bowen *University of Wisconsin at Madison* N. Manson *Université de Poitiers* A. K. Oppenheim *University of California at Berkeley* R. I. Soloukhin *Institute of Heat and Mass Transfer, BSSR Academy of Sciences*
76.	Combustion in Reactive Systems. 1981	J. R. Bowen *University of Wisconsin at Madison* N. Manson *Université de Poitiers* A. K. Oppenheim *University of California at Berkeley* R. I. Soloukhin *Institute of Heat and Mass Transfer, BSSR Academy of Sciences*
77.	Aerothermodynamics and Planetary Entry. 1981	A. L. Crosbie *University of Missouri-Rolla*
78.	Heat Transfer and Thermal Control. 1981	A. L. Crosbie *University of Missouri-Rolla*
79.	Electric Propulsion and Its Applications to Space Missions. 1981	Robert C. Finke *NASA Lewis Research Center*
80.	Aero-Optical Phenomena. 1982	Keith G. Gilbert and Leonard J. Otten *Air Force Weapons Laboratory*
81.	Transonic Aerodynamics. 1982	David Nixon *Nielsen Engineering & Research, Inc.*
82.	Thermophysics of Atmospheric Entry. 1982	T. E. Horton *The University of Mississippi*

83. **Spacecraft Radiative Transfer and Temperature Control.** 1982

T. E. Horton
The University of Mississippi

84. **Liquid-Metal Flows and Magnetohydrodynamics.** 1983

H. Branover
Ben-Gurion University of the Negev
P. S. Lykoudis
Purdue University
A. Yakhot
Ben-Gurion University of the Negev

85. **Entry Vehicle Heating and Thermal Protection Systems: Space Shuttle, Solar Starprobe, Jupiter Galileo Probe.** 1983

Paul E. Bauer
McDonnell Douglas Astronautics Company
Howard E. Collicott
The Boeing Company

86. **Spacecraft Thermal Control, Design, and Operation.** 1983

Howard E. Collicott
The Boeing Company
Paul E. Bauer
McDonnell Douglas Astronautics Company

87. **Shock Waves, Explosions, and Detonations.** 1983

J.R. Bowen
University of Washington
N. Manson
Université de Poitiers
A.K. Oppenheim
University of California at Berkeley
R.I. Soloukhin
Institute of Heat and Mass Transfer, BSSR Academy of Sciences

88. **Flames, Lasers, and Reactive Systems.** 1983

J.R. Bowen
University of Washington
N. Manson
Université de Poitiers
A.K. Oppenheim
University of California at Berkeley
R.I. Soloukhin
Institute of Heat and Mass Transfer, BSSR Academy of Sciences

(Other volumes are planned.)

Introduction

Y.B. Zel'dovich[*]
Academy of Sciences, Moscow, USSR

Combustion as a source of heat was a concern even in ancient times, as is attested by the myth of Prometheus, who stole fire from the gods and bestowed it upon mankind. These proceedings, however, are devoted mainly to the dynamic effects of combustion manifested by explosion. High pressure achieved as a result of chemical reaction was apparently realized for the first time ages ago in China, and in the Middle Ages in Europe, by burning powder in closed or semiclosed vessels. A medieval German monk by the name of Berthold Schwartz is credited with the invention of explosive powder--a material whose overall combustion reaction is locally self-sustained under conditions of escalating pressure.

In the 19th century, catastrophic accidents in coal mines made man aware of the effects of detonation in a gaseous medium. In the same period, chemists discovered highly sensitive explosives, such as lead azide, nitrogen iodide, and fulminate of mercury. Soon thereafter followed the development of high explosives, such as dynamite, trotyl, picrin acid, etc., which, on one hand, were sufficiently stable to enable their storage and transportation, and, on the other, could be detonated by the use of explosives of the former group. The year 1981 marks the centenary of the publications of the outstanding contributions of Vielle, LeChatelier, and Mallard. They reported the first results of experimental laboratory studies in which gas detonation phenomena were subjected to a thorough quantitative exploration.

While studying inert gas motion, Riemann realized the inevitable outcome of the flowfield generated by explosions: the formation of a shock wave at its front. In modern

---

Presented at the 8th ICOGER, Minsk, USSR, Aug. 23-25, 1981. Copyright © 1983 by R. I. Soloukhin. Published by the American Institute of Aeronautics and Astronautics with permission.
   [*]Academician, Institute of Chemical Physics.

terms, this means that equations of motion of a nonviscous, non-heat-conducting fluid lead to a catastrophe, a singularity for which these equations are no longer valid. Flow processes in particularly strong shock waves cannot be described adequately by the Navier-Stokes equations because significant disturbances of the Maxwell-Boltzmann molecular and atomic distributions occur in such instances.

It is striking, however, how profound was the effect of simple conservation equations, which were correctly formulated for the first time by Hugoniot, in leading to the development of the detonation theory, formulated by Chapman and Jouguet on the eve of the 20th century. Along with a hypothesis that was quite plausible, albeit not well-grounded upon direct experimental evidence, the Chapman-Jouguet theory provided the basis for calculation of the essential properties of detonation waves, such as density, temperature, pressure, and velocity produced by detonation wave propagation. This was quite sufficient to evaluate the gas motion and the destructive effects caused by detonation.

The initial evolution of the detonation theory, referred to as "hydrodynamic," was followed by a slack period of about 40 years. The discovery of detonation spin in the 1920s and the development of many new explosive materials do not contradict this statement insofar as theoretical knowledge is concerned.

The next major theoretical advance in which I had the good fortune to participate, occurred in the 40s. It resulted from the realization that shock-wave-induced chemical transformation must involve a far greater number of molecular collisions than those required for physical accommodation of molecular structure to the passage of a shock wave. Accounting for the finite rate of chemical transformation enabled us to consider the effects of losses of heat and momentum that occur in the course of the process. In a cylindrical charge, these losses are manifested by the dependence of the detonation velocity on the diameter.

It is thus with a good deal of satisfaction that I noted a number of papers presented at the 8th Colloquium that detail carefully conducted experimental and theoretical studies of these effects.

Proper consideration of heat transfer and gas friction at the wall leads to the establishment of steady-state conditions for which the parameters of state, T, P, ρ and velocity v depend on a combination of physical space ($x - Dt$). Analysis of the equations indicates that the solution can be expressed in terms of not one but two values of D. One of them, as losses decrease, tends to the detonation velocity predicted by the Chapman-Jouguet theory. The second value of D, on the contrary, decreases indefinitely, being essentially unsteady in nature.

This consideration thus furnishes a proper background for a hypothesis governing rate selection, logically quite different from the conventional approach. One may adopt a simple solution by postulating that only steady state is of interest. However, one may include the possibility of other solutions by allowing a more "mathematical" outcome, although not more than that of a singular point in the phase plane of an ordinary differential equation. The standard demonstration of the rate selection hypothesis involves the matching of a steady chemical process with an unsteady rarefaction wave immediately behind it. With finite rate chemical reaction, the process becomes metastable. So far there exists no theory for the determination of an appropriate correction for this phenomenon.

The allowance for finite reaction rates and losses accounts also for the detonation limit and the critical charge diameter. As far back as 1928, Vendlandt considered the process of the initiation of reaction in the shock wave propagating at detonation velocity. In 1939, Khariton formulated the most important principle (named after him) applicable to condensed explosives, namely: The condition for the existence of detonation implies that the time of chemical reaction must be less than the time of charge scattering. In the case of condensed explosives, it is the the development of high-pressure scattering (rather than mechanical wall effects and heat transfer) that is primarily responsible for losses. The scattering time is determined by the ratio of the charge dimension to the sound speed in the products. The Khariton principle accounted for the fact that a charge of 20,000 tons of ammonia nitrate and ammonia sulfate mixture detonated in Onnay in 1922, causing a tremendous amount of damage, whereas small-scale laboratory-tested samples of this mixture behaved as an inert substance.

Despite the appreciable achievements in the 1940s, some difficulties soon became apparent. Earlier, spinning detonation was considered to be a particular property of a limiting process. Troshin showed that at conditions significantly different from the limit conditions, spin structure becomes finer but does not vanish completely. Shchelkin expounded a general hypothesis on the instability of detonation waves at high activation energy. A number of studies followed, dealing with the following two aspects: first, the consideration of one-dimensional steady behavior confirming Shchelkin's ideas and, second, the investigation of a particular reaction mechanism occurring in a spinning detonation, affecting the structure of the flowfield and the pressure and temperature profiles. In this connection, studies by Soloukhin and his associates in Novosibirsk are particularly noteworthy. Thus the one-dimensional pattern of detonation wave turned out to be incorrect. Remarkably, in spite of that, the essential conclusions concerning the state after the chemical reaction remained valid. After all, laws of conservation applied to the whole of the charge cross section could not have changed. This does not mean, however, that the selection rule still holds as well. In spinning detonation, the temperature immediately behind the shock front is not sufficiently high to ignite the gas. Ignition may take place only at certain hot spots on the front or in a second shock wave propagating into the gas compressed behind the front.

The subsequent reaction in the remaining portions of the gas is initiated either by some additional transverse waves or by a turbulent wake behind the hot spot. In my opinion, this problem merits further study. However, for the selection rule to be applied, it is sufficient to know that the chemical reaction takes place somewhere in a gas compressed by the first shock wave.

At present, new and more precise and detailed methods for the study of detonation waves are in the course of intensive evolution, and undoubtedly they will be advanced further in the near future. Calculation techniques for two- and three-dimensional chemical reaction pattern in the detonation wave are in an advanced stage of development (note, in particular, the studies of Fujiwara). High sensitivity of chemical reaction to temperature may cause formation of fine structures. Artificial viscosity included in numerical calculations, as well as other smoothing techniques, must be thoroughly analyzed to make sure that they do not eliminate actual fine details of the wave structure. Do secondary structures on small bright waves of spinning detonation exist? What part does turbulence play? Today these questions are still open, providing promising prospects for our future studies. Of great interest and practical significance are the investigations of detonation in two-phase systems (powder suspensions of Nigmatullin et al. at Moscow State University; fuel on the walls with oxygen diffusion limited chemical reaction of Troshin at the Institute of Chemical Physics). The paper on one-dimensional detonation instability by Zeidel and myself demonstrates the possibility of a superstrong detonation wave propagating over the chemical process of a normal wave. Since the superstrong wave rapidly consumes the reacting substance, the whole structure is essentially unstable. Final or temporary attenuation may follow, depending on the reaction rate that can be sustained at low pressures and temperatures.

Instability of steady-state one-dimensional detonation waves is the most important factor of a realistic theory on detonation limits. In the theory of spinning detonation, the limit depends on the condition at which the hot spot ignition vanishes rather than on general kinetics.

Whereas the detonation pressure and velocity are stable relative to disturbances, the behavior at the limits is a quite sensitive function of the disturbance. The problem of detonation origin and transition from deflagration to detonation has been investigated for over a century. A most noteworthy achievement was due to Shchelkin. While the decisive role of the hydrodynamic factors is obvious, Shchelkin stressed the turbulence factor. As a consequence, the term "Shchelkin spiral" became firmly established in the literature.

Soon after the publication of Shchelkin's paper, I raised the question of the role played by the velocity profile, i.e., on the difference between axial and mean velocities in pipe flow and its effect on the generation of a multidimensional wave structure. Eventually it became clear that in reality both effects, that is, gas turbulization and its order motion, are essential.

In this connection, I wish to single out several significant papers presented at the 8th Colloquium, but not included in the Proceedings. One by Mishuev et al., dealing with the effects of flame shape on the acceleration of the combustion front, formed the most impressive presentation at the poster sessions. The other was the numerical modeling analysis of Ghoniem, Chorin, and Oppenheim presented at the discussion on turbulent combustion, a very elegant computational study of flame propagation in a turbulent flowfield. The method of vortex dynamics applied for this analysis is most promising. Finally, with particular reference to molecular kinetics, the study of Manelis and Dremin on specific processes in strong shock waves is particulary noteworthy. Disturbances in local Maxwell-Boltzmann molecular distribution evidently play an important role in initiating chemical reactions.

To sum up, I recognize that detonation studies have reached a mature age in which definite achievements have been attained, but that there are still some problems to solve. Complete understanding of detonation is necessary for various practical reasons, such as safety rules, technology of explosives, combustion in internal combustion engines, and detonation lasers. In my opinion, it would be desirable to include such practical aspects among the topics to be considered at future colloquia.

Science is like Antqueus, the mythological giant who derived his immense strength from contact with the Earth. It is born from the union of practical tasks and demands for knowledge inherent in man; it derives new strength from its contact with "Earth" - the practial requirements of mankind.

Chapter I. Shock Wave Interactions

The Study of Shock-Induced Signals and Coherent Effects in Solids by Molecular Dynamics

A.M. Karo,* F.E. Walker,† and W.G. Cunningham‡
Lawrence Livermore National Laboratory, Livermore, Calif.
and
J.R. Hardy§
University of Nebraska, Lincoln, Neb.

Abstract

Molecular dynamics calculations are presented that address the extent of microscopic detail that can be deduced from macroscopic gage measurements of shock propagation in condensed systems. We have simulated large asymmetrically shock-loaded lattices, varying the initial temperature of both the loading plates and the lattice. Specifically, we have studied triple-shock loading of a thin lattice; double-shock loading of the same lattice; and triple-shock loading of a thick lattice. In all cases we found strong "memory effects" in that the spall pattern always mirrors the loading history. This is a direct consequence of our two basic results: 1) the energy in finite width shocks is only weakly coupled to motions transverse to their direction of propagation; and 2) coupling between shock motion and random thermal motion is relatively weak, even in lattices that are near their melting temperatures.

Introduction

A series of molecular dynamics calculations have been carried out that address the question of the extent to which

---

Presented at the 8th ICOGER, Minsk, USSR, Aug. 23-26, 1981. Copyright © American Institute of Aeronautics and Astronautics, Inc., 1982. All rights reserved.
*Senior Scientist, Department of Chemistry and Materials Science.
†Deputy Program Director, Non-nuclear Ordinance Program.
‡Computer Scientist, Computation Department.
§Professor of Physics, Department of Physics.

microscopic detail can, in fact, be deduced from macroscopic measurements of shock propagation in condensed systems. The measurements we are considering are those that would typically be taken by Manganin or electromagnetic particle-velocity gages whose signals are then related to shock passage through the materials being investigated. In our computer studies, we have simulated large asymmetrically shock-loaded lattices, varying both the initial temperature and the strength of shock loading.

We have found from examining the results of a wide-ranging series of such calculations that the overall pattern of events is the consequence of a sequence of microscopic (i.e., atomic or molecular) processes occurring in picoseconds over dimensions of angstroms. It would appear necessary, then, that experimental techniques are required simultaneously approaching such temporal and spatial resolutions in order to deduce shock characteristics at the atomic and molecular level. Thus the attempt to measure microscopic shock rise times by methods such as those referred to above is perhaps doomed to failure until gages are developed capable of measuring displacements within the materials on the order of angstroms and coupled with clocks capable of measuring time to tenths of picoseconds. The effort currently expended to measure macroscopic rise times would appear to give only a measure of surface finish, microcrystalline structure, or void size, even with subpicosecond instrumentation. It can also be readily seen that with shock or detonation waves propagating in solids at velocities ranging from 5 to 9 mm/µs, i.e., from 5 to 9 Å in 10^{-13} s, and with impurities or structural irregularities no larger than 10 µ, an apparent shock rise time of from 1 to 2 ns (i.e., the transit time over this 10 µ distance) would be the shortest time measureable. Measurements made at shorter time intervals would only be probing incoherent effects associated with the intrinsic random defect structure of the material. If sample preparation could be improved to the extent that the defects or microcrystalline features in the material under study could be kept to no more than 0.1 µ, the measured or apparent rise time, even with subpicosecond instrumentation, would still appear to be from about 0.01 to 0.01 ns, i.e., about 100 times the periods associated with phenomena occurring on the atomic and molecular scale. Our calculations show that microscopic rise times associated with the shock front itself are occurring in a few measured units of 10^{-13} s at all points of contact between irregular surfaces. Thus the overall rise time in a real system problem is determined merely by the irregularities of the impact points at the microlevel.

The coherence associated with shocks that have been launched into a large host lattice by smaller plates that impact different regions of the surface at different times is readily observed in molecular dynamics. Lateral transfer of shock energy is minimal, even when lattices are initially thermally highly excited. Thus, in our calculations, striking examples have been seen of the localization of shock motion within the lattice and of the degree to which such motion remains uncoupled to the lattice phonons even when the system is thermally highly excited. We have also noted in single- and multiple-plate calculations a very interesting demonstration of a phenomenon that can referred to as a "memory effect," in that the subsequent history of the shocked system "remembers" the details of the initial loading. That is, after shock transit through the lattice and the emergence of well-separated spall from the far wall, there is a relationship displayed by the ejected material with reference to the details of the initial loading.

This is consistent with all our previous calculations on perfect lattices (Karo and Hardy 1981) where shock coherence and stability are evident in the propagating disturbance. A major part of the energy is carried off by microscopic spall as the shock reaches a surface, while the rest of the system proportionally picks up very little energy.

Method

In a molecular dynamics simulation of the behavior of an assembly, or array, of atoms the basic technique is the solution by computer of the Newton's law equations of motion for all of the atoms (Alder and Wainwright 1959). A recent excellent review of the applications of this technique with particular emphasis on shock studies has been given by MacDonald and Tsai (1978). The forces between the atoms are given by the appropriate spatial derivatives of the potential function for the whole system. Ideally this function would be the true many-body potential function derived from a first-principles quantum chemistry calculation: the particle motions would be trajectories on this 3N dimensional hypersurface, where N is the number of atoms considered.

In practice, if one wishes to simulate a large system, as is the case in the present studies, drastic simplification is required to make the problem tractable without requiring excessive computer time.

The simplest approximation is to represent the potential by a sum of short-range pairwise interactions, and it is this that we have employed in the present studies. We have also restricted ourselves to studies on two-dimensional systems which represent further extensions of our earlier work (Karo and Hardy 1981).

Since it is certain qualitative features of shock loading that we wish to demonstrate, these restrictions on potentials and dimensionality are unimportant. Indeed, as we shall see later, there is good reason to believe that our pair potentials provide a surprisingly good approximation to that part of the true hypersurface probed by shock loading. Moreover, by restricting ourselves to two dimensions we are able to simulate much larger crystals and thus answer questions which would require prohibitively large amounts of computer time to address for three-dimensional systems. [In a previous study (Karo and Hardy 1981), we alluded to this approach as part of our general strategy.]

Given these general principles, the computational problem reduces to the solution of the 4N coupled differential equations for the particles' velocities and positions in two dimensions: the initial values of these quantities at time t = 0 constitute the initial conditions of the problem.

Specific Studies

Our main objective in the present work is to examine the remarkable coherence of shock loading and its consequences when the lattice studied is subject to localized asymmetric loading. We also wish to demonstrate that the qualitative features observed are preserved both when the shock transit distance is tripled and/or when the lattice is hot.

In order to achieve the required loading of our lattice, three plates, each with a velocity of 1.4 natural units to the right (one interplanar spacing per unit of t), are initially offset to the left of our lattice by three different distances. They thus strike the lattice at a sequence of times, and each launches a spatially localized shock which transits the lattice and spalls material from the right-hand side. Two lattices were studied: one contained 65 rows and 10 columns; the other 65 rows and 30 columns. Both lattices and the plates are bonded by first- and second-neighbor Morse potentials which are identical for all such bonds in the systems (i.e., both lattices and plates) under study. The atomic masses are all identical, and their value, and the bond potential, is that appropriate to the "model" lattice of Karo and Hardy (1981). In order to simulate the effects of temperature, when required, the atoms are given initial random velocities at the outset of

SHOCK EFFECTS IN SOLIDS 13

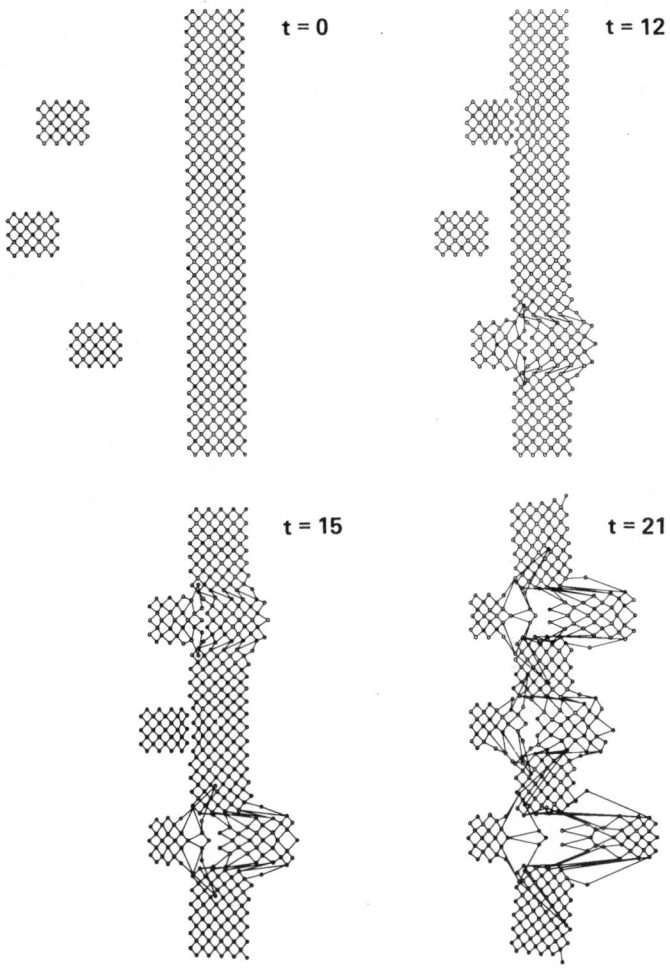

Fig. 1 Configurations of initially quiescent 10 x 65 lattice at a sequence of times t as the shocks launched by a triplet of plates, initially moving towards the lattice at t=0 with a velocity of 1.4 units, transit the lattice. At t=12, the first spall is complete; at t=15, the second spall is complete; and at t=21, the third spall fragment is clearly separated.

the problem. In the case of the plates these are simply added vectorially to the initial uniform translation.

In Figs. 1-3 we show sequences of configurations from the histories of three different simulations for the thin (65 x 10) lattice.

In Fig. 1 both lattice and plates are cold (no initial random motion); in Fig. 2 the plates are "warm" (average atomic kinetic energy ∿5% of the bond energy: this corresponds to approximately 300-400 K) and the lattice is hot (average atomic kinetic energy 18% of the bond energy); in Fig. 3 the plates are again warm, but the lattice is very hot (average atomic kinetic energy ∿72% of the bond energy).

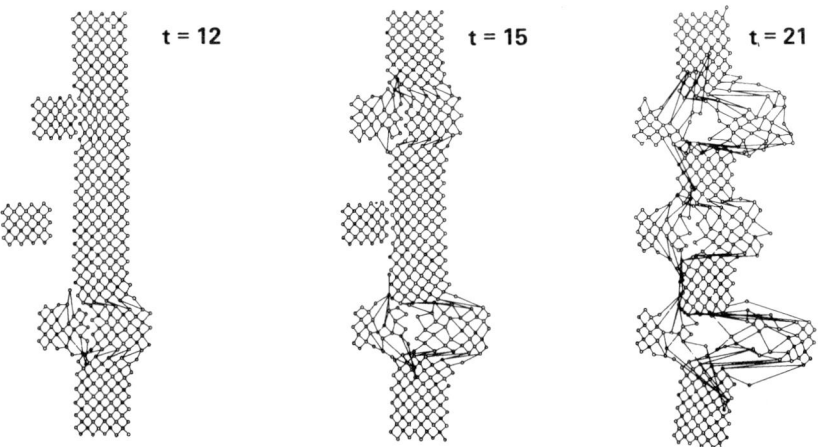

Fig. 2 Configurations corresponding to those in Fig. 1 except for an initial thermal motion per bond ∿5% of the dissociation energy for the plates and ∿18% of the dissociation energy for the lattice. The spall sequence is strikingly similar to that for the quiescent system.

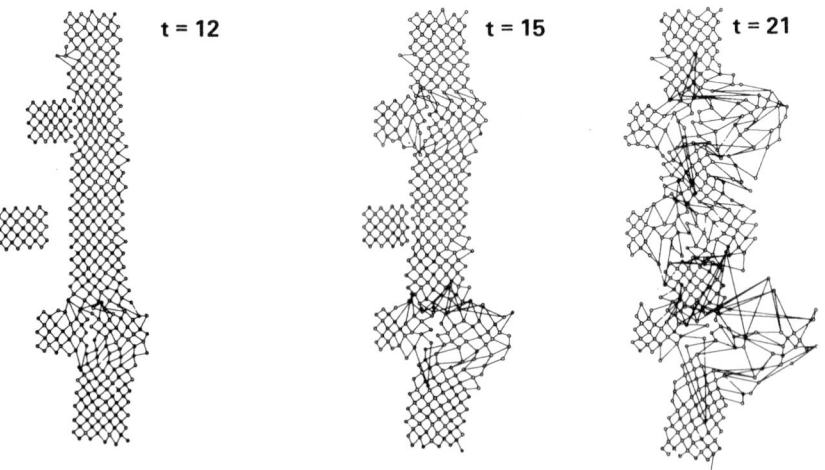

Fig. 3 Configurations corresponding to those in Fig. 1 except for an initial thermal motion per bond ∿5% of the dissociation energy for the plates and ∿72% of the dissociation energy for the lattice.

In all three sequences the time t is shown on each configuration: the initial configuration for all three is that shown for t=0 in Fig. 1. In these and subsequent figures the atoms are shown as dots connected by lines to their first neighbors in the initial configuration. As the simulations proceed these lines are retained, irrespective of separation. Thus, particularly at later times, widely separated pairs of atoms remain connected by these lines even though there is no significant physical bonding remaining between them.

From Fig. 1 we see that the successive shocks launched in the initially quiescent thin lattice proceed through essentially independently. Each produces its own spalled fragment at the right-hand edge, which carries off a major fraction of the energy imparted to the lattice by the

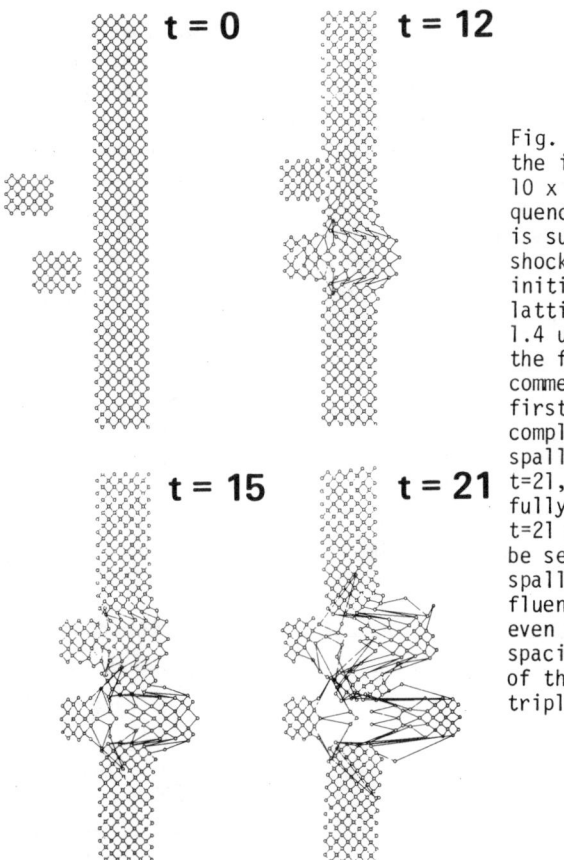

Fig. 4 Configurations of the initially quiescent 10 x 65 lattice at a sequence if times t when it is subjected to double-shock loading by two plates initially moving towards the lattice with velocities if 1.4 units at t=0. At t=12, the first spall has clearly commenced. At t=15, the first spall is essentially complete and the second spall is commencing. At t=21, the second spall is fully developed. From the t=21 configuration it can be seen that the second spall is very little influenced by the first shock even though the interplate spacing is only two-thirds of that used for the triple impact.

appropriate plate impact. The bulk of the lattice remains relatively quiescent, while the plate itself retains some energy as kinetic energy of recoil. The lack of transmission of energy transverse to the shock directions is most clearly evident from the remarkably symmetric nature of each spall pattern. This is best seen for the t=21 configuration. Here even the last (central) spall is developing essentially as if it were taking place in isolation. From this it is clearly evident that the shock energy is highly focused. Also, the little transverse activity that does develop appears to propagate with sonic rather than supersonic or shock speeds.

Figures 2 and 3 show the same finite time sequences for the hot and very hot lattices. Again the clear development of successive isolated spalls is clearly apparent. There are obvious losses of symmetry, but from our earlier work (Karo and Hardy 1981) it also is clearly apparent that these are due mainly to the initial thermal motion. This persistence of spall sequence and the associated "memory effect," whereby the spall pattern "remembers" or reflects the details of the initial loading, is truly remarkable. This is particularly so for the "very hot" lattice, which is, in fact, so thermallly excited that it is in a quasiliquid state. In fact, if left to evolve without shock loading it spontaneously disintegrates. The presence of the "memory effect" in a system initially subject to such chaotic motion appears to us only to be explicable if the coupling between shock motion and thermal lattice motion is inherently a low-order process.

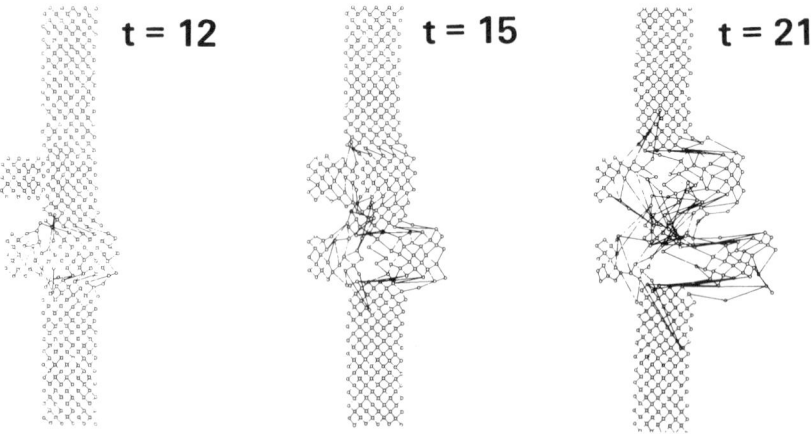

Fig. 5 Configurations corresponding to those in Fig. 4 except for an initial thermal motion per bond ∿18% of the dissociation energy for the lattice and ∿5% for the plates.

In order to test this belief further we carried out a series of simulations involving double- rather than triple-plate impact. However, the initial configuration was such that the vertical separation of the two plates was reduced by 30% in order that the transverse disturbance produced by the first impact would have ample time to reach and interact with the shock produced by the second impact. The histories of three such simulations are shown in Figs. 4-6. Apart from the difference in loading, these correspond exactly to the triple-impact histories shown in Figs 1-3, respectively. These figures show clearly that the same general conclusions we reached for the triple-shocked system remain valid. For the initially quiescent system, the transverse coupling between the shocks is again weak: some loss of symmetry of the second shock is apparent, but its basic integrity is unimpaired. In the cases of the "hot" systems the shock integrities are maintained to the same degree as they are for the corresponding triple-shocked lattices. This is entirely consistent with the results for the "cold" system: if transverse coupling is weak for this quiescent system, one does not expect it to be enhanced by the presence of thermal motion.

In Figs. 7 and 8 we show the history of a triple impact on the 65 x 30 lattice; apart from the large size of the lattice everything else is as it was for the first simulation for the 65 x 10 lattice. Thus, in the history of

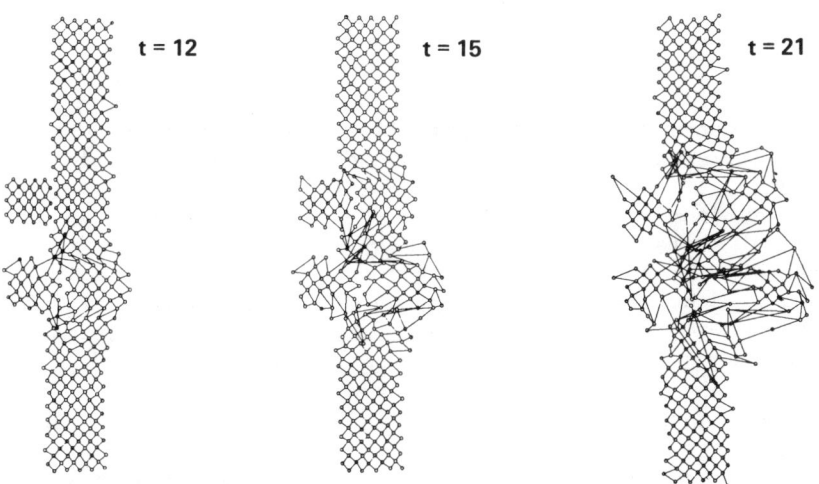

Fig. 6 Configurations corresponding to those in Fig. 4 except for an initial thermal motion per bond ∿72% of the dissociation energy for the lattice and ∿5% for the plates.

t = 0

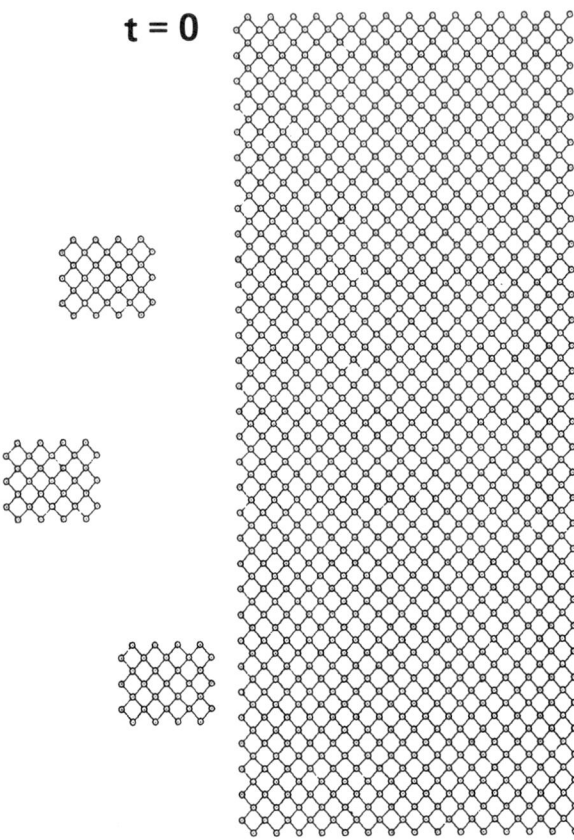

Fig. 7 Configuration of the 30 x 65 lattice system and associated triplet of loading plates.

this initially quiescent system, we see, at t=15, single spall is commencing due to the first (lowest) plate impact; at t=27, clear double spall, due to upper and lower plate impacts, is clearly developed, and the third spall is commencing from the center; at t = 52, it is clear that we have a final state with clear triple spall and memory effect. For this thicker lattice, more transverse energy transfer is indicated by the clear loss of symmetry. However, it should be borne in mind that any loss of symmetry is automatically magnified as the simulation proceeds. Moreover, the horizontal crack that is apparent at t=53 is obviously a highly coherent effect: at t=27 one can see it initiating at the upper horizontal edge of the central shocked region. It is mechanical cracking and is certainly not produced by shock energy degraded into random thermal motion. Moreover, if the upper and lower edges of the plate had been constrained, this crack might well not have developed.

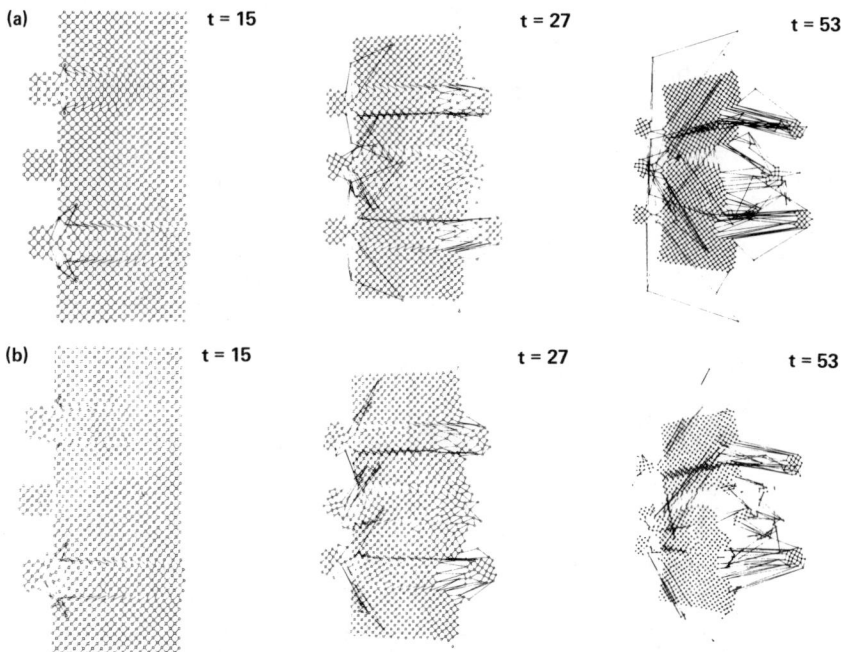

Fig. 8 Configurations of the triple-loaded 30 x 65 system at a sequence of times t for a nonthermal and a thermal situation: a) initially quiescent, and b) thermal motion per bond ~5% of the dissociation energy. For both systems the first spall is about to commence at t=15, and at t=27, double spall is clearly apparent. At t=53, we see the final state of both systems.

In Fig. 8b the same time squence of configurations is shown for a simulation identical in every way except that both plates and lattice were given the same initial "temperatures" as the plates in the 65 x 10 simulation. Thus both are at, or a little above, room temperature. The history is virtually identical to that for the initially quiescent system. This indicates even more strongly than the earlier simulations that the coupling between shock and random thermal motions is very weak; it even appears that the transverse leakage of energy that does occur from shocks of finite transverse width is largely uncoupled to such thermal motion, at least at modest temperatures.

Finally, in Fig. 9 we show typical Hugoniot plots of shock vs particle velocities computed from our molecular dynamics simulations (Karo and Hardy 1981) compared with experimental results. It can be seen that the two molecular dynamics (M.D.) Hugoniots bracket those for a wide range of experimental materials. On this basis we can argue that the

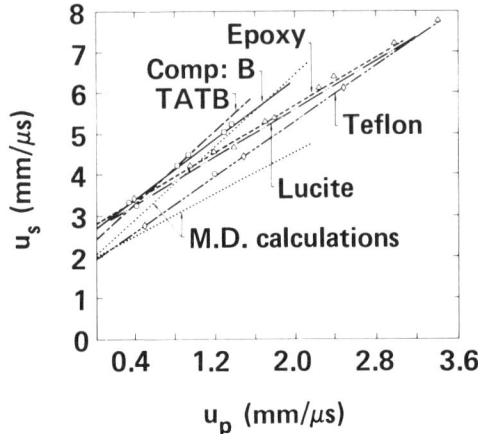

Fig. 9 Theoretical Hugoniots for two different lattice models compared with experimental data. The position of our systems is indicated by the arrow pointing to the upper molecular dynamic Hugoniot, corresponding to our lattice.

pair potential approximation, used in the present studies, provides an excellent first approximation to that part of the true potential surface probed by shock loading. Moreover, the position of the present studies on the appropriate theoretical Hugoniot indicates that they are at the lower limits of those used in experimental studies. It thus follows that all the effects that we have described are not artifacts produced by unrealistically violent loading in our simulations and should be present in real single crystals under typical shock loading. Indeed, since our simulations lie at the lower end of the Hugoniot, we would expect enhancement of the predicted effects in single crystals subjected to the generally more violent loading typical of experiments. We infer this from earlier simulations (Karo and Hardy 1981) which have confirmed the intuitive belief that more violent loading enhances the present type of behavior.

Conclusions

The principal result of the present studies is the demonstration of the manner in which the high degree of localization and coherence of shock energy at the atomic level affects our model systems multiply shocked at several locations and at several times. We find that there is remarkably little interference between different shocks. This indicates that shock energy is localized, not merely in the direction of shock propagation, but also transverse to this direction. As a consequence of this novel result, indicated in earlier studies (Karo and Hardy 1981) and confirmed by the present work, there exists a "memory effect" whereby the final spall pattern from a multiply shocked system "remembers" the details of the initial shock loading. Moreover, this memory effect persists, even when the shocked system is extremely hot.

As a consequence of these results, one can infer that any macroscopic determination of rise time measures not the true microscopic rise time, which we have shown to be $\sim 10^{-12}$-10^{-13} s, but some average rise time of the overall loading which is determined by the surface irregularities of the loading system and the shocked system. We also infer that this loading history will be "remembered" by the shock as it propagates into the loaded system. Although the "memory" is not perfect, and can be degraded by irregularities in the system, what will be retained in the front as it propagates are the average spatial and temporal widths of the initial loading. It is these quantities that will be measured macroscopically and are erroneously regarded as "shock widths" and rise times." The intrinsic values of the quantities are shown by our studies to be $\sim$1-10 Å and $\sim$0.1-1 ps, and these can only be probed by instrumentation having simultaneously spatial and temporal resolutions within or below these bounds.

Acknowledgment

Work performed under the auspices of the U.S. Department of Energy by the Lawrence Livermore National Laboratory under Contract W-7405-ENG-48.

References

Alder, B. J. and Wainwright, T. E. (1959) Studies in molecular dynamics. I. Ground method. J. Chem. Phys. 31, 459.

Karo, A. M. and Hardy, J. R. (1981) Proceedings of the NATO Advanced Study Institute on Fast Reactions in Energetic Systems (edited by C. Capellos and R. F. Walker), p. 611-643. D. Reidel Publishing Co.

MacDonald, R. A. and Tsai, D. H. (1978) Molecular dynamical calculations of energy transport in crystalline solids. Phys. Rep. 46, 1.

Oblique Shock Waves in Two-Phase Flow

Koichi Hayashi*
Princeton University, Princeton, N.J.
and
Melvyn C. Branch†
University of Colorado, Boulder, Colo.

Abstract

Oblique shock waves in two-phase flow were studied, and the relation between gas-particle flow properties ahead of an oblique shock wave and gas-particle flow properties behind the shock wave were derived by a solution of the steady-state conservation equations of mass, momentum, and energy for the mixture phase. Some of the important assumptions made are that the particles are uniform in size, uniformly distributed, and in thermal equilibrium with the gas ahead of the oblique shock wave. Large particles (>10 μm) do not change direction immediately after passing through the shock wave. Particle drag coefficients in two-phase flow were reviewed, and the data used in this study were obtained from aeroballistic range measurements. Results of the oblique shock wave calculations are presented as the relation between the incident angle β of the two-phase flow and the deflection angle θ. The solution considered the effects of the shock wave Mach number, particle velocity lag, particle feeding rate, and particle size. A comparison of the calculations to experimental data was used as a guide to selection of the drag coefficient and gave good agreement between theory and experiment for the available data. It was found that the shock wave structure

---

Presented at the 8th ICOGER, Minsk, USSR, Aug. 23-26, 1981. Copyright © American Institute of Aeronautics and Astronautics, Inc., 1982. All rights reserved.
*Associate Research Staff, Fuels Combustion Research Laboratory.
†Associate Professor, Mechanical Engineering Department.

was influenced by shock wave Mach number, particle feeding rate, and particle velocity lag.

Introduction

Recent studies (Morgenthaler 1962; Pergament and Thorpe 1975; Alkhimov et al. 1977; Hayashi and Branch 1980; Chang 1981) have focused on prediction of the characteristics of nozzles and exhaust plumes of rocket engines in order to evaluate nozzle flow and plume visibility, radiation signatures, and impingement effects. Oblique shock waves occur in these flows and accompany detonation and blast waves in heterogeneous media (Fig. 1). The nature of oblique shock waves in two-phase flow has not been extensively characterized experimentally or theoretically.

Alkhimov et al. (1977, 1978) investigated shock wave structures in nozzle flows with different particle sizes. It was observed that the steady shock wave position in large particle two-phase flow was closer to the nozzle throat than that in small particle flow. From their velocity measurement through a shock front using a Laser Doppler Velocimetry, the shock front thickness was about 1 mm at M=2.8.

Shadowgraphs of several kinds of oblique shock waves (Fig. 2) were obtained by Saltanov and Simanovskii (1978). The curvature of the oblique shock wave was clearer in two-phase flow than that in single-phase flow since the particle flow reduces the gas flow to produce the strong oblique conditions in the shock wave. They pointed out that condensation of droplets was observed in the shock wave relaxation zone.

The theoretical results by Probstein and Fassio (1970) showed the particle trajectory deviations through a shock wave over a wedge were qualitatively independent of void fraction. When the void fraction in a gas-solid mixture was less than $0.92 \sim 0.95$, particle-particle interactions must be taken into account. Ignatov et al. (1978) pointed out that the particle velocity immediately behind a shock wave differed little from that in the incident flow.

Several important results concerned with the shock wave in two-phase flow in the recent review work (Hayashi 1980) are that small particles (≤ 1 μm) follow the gas flow, and large particles (>10 μm) do not change their velocity immediately after passing through the shock wave. The shock wave relaxation zone in two-phase flow is longer than that in single-phase flow. In addition, the gas velocity has been found to be slower than the discrete phase velocity behind the shock wave, and the gas-phase flow has been found

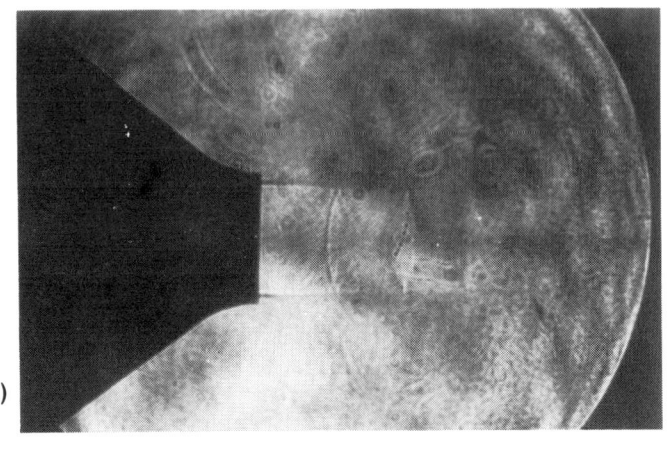

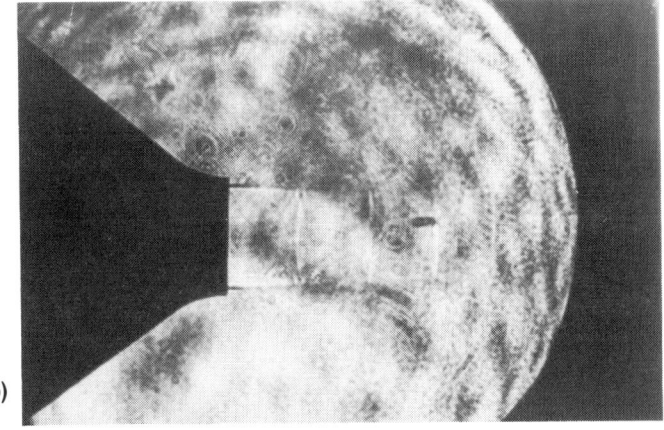

Fig. 1 Shock wave structures in a) two-phase and b) single-phase jet flows.

to decelerate the particle flow. In two-phase flow there are two types of sound velocities--the frozen sound velocity and the equilibrium sound velocity--and two types of shock waves--the supersonic two-phase shock wave and the fully diffuse two-phase shock wave (Kleigel 1963; Rudinger 1969; Gregor and Rumpf 1975).

Oblique Shock Wave Equations in Two-Phase Flow

Particle Drag Coefficient

In this study the most appropriate choice of particle drag coefficient was investigated by considering important

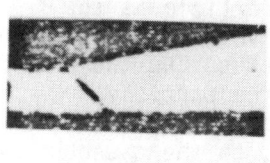

Fig. 2 Shadowgraphs of oblique condensed droplets shock waves (after Saltanov and Simanovskii 1978).

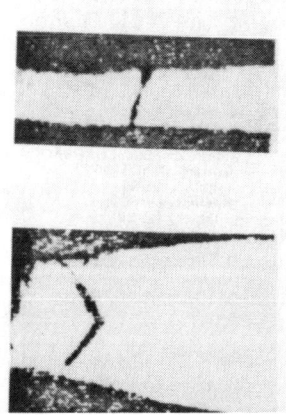

factors that influence the particle drag coefficient--particle velocity lag, relative Reynolds number, relative Mach number, and particle shape and size.

The standard average drag coefficient obtained by Lapple and Shepherd (1940), which is the empirical drag coefficient in steady laminar flow, and the modified drag coefficient, which is obtained by Gilbert et al. (1955) and Ingebo (1956), theoretically and empirically considering the velocity lag between the two phases, have been used over the full range of Mach numbers in two-phase flow problems. Carlson and Hoglund (1964) studied an empirical expression for the drag coefficient of a spherical particle as a function of the relative Mach number as well as the relative Reynolds number. Later, particle drag coefficients were measured in an aeroballistic range at various Reynolds numbers and Mach numbers by many authors. The preliminary calculation of oblique shock wave relations using the standard drag coefficient showed disagreement of computed results with experimental data at low Reynolds numbers (Stokes region).

Particle velocity lag is the difference between the gas velocity and the particle velocity and can be either positive or negative. The relative Reynolds number and relative Mach number are functions of the absolute value of a particle velocity lag. The drag force with a positive velocity lag does not have the same effect on a particle as the equivalent negative velocity lag. Zero velocity lag corresponds ideally to a stationary state between the two phases and results in no drag force acting on the particle.

Relative Reynolds number is the ratio of the relative inertia force, which is always positive, to the gaseous viscous force. The relation between drag coefficient and relative Reynolds number is not linear over the full range of relative Reynolds numbers (Fig. 3). Relative Mach number

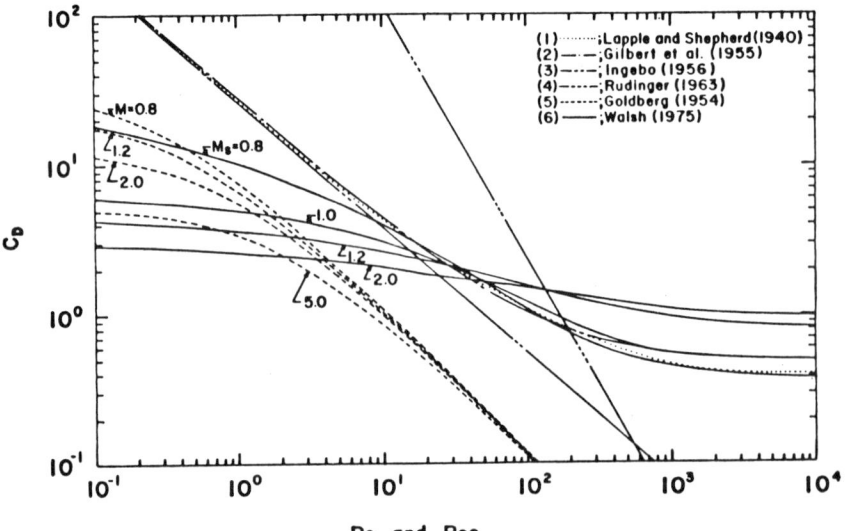

Fig. 3 Drag coefficients (R_e is the Reynolds number and R_{es} the relative Reynolds number).

is the ratio of the relative velocity to the frozen speed of sound in two-phase flow and is an important measure of compressibility in high-speed two-phase flows.

Particle shape is always assumed to be spherical and the particle surface smooth in the present study. Particles used in the aeroballistic range measurement and other measurements are uniformly spherical. It is difficult to account for particle surface roughness in the particle drag coefficient. Particle size is already considered in the expression for the relative Reynolds number. The effects of wake and turbulence on the drag force due to the particle size are not considered.

The standard average drag coefficient, which is the empirical drag coefficient in steady laminar flow, and the modified drag coefficient, which considers the velocity lag between the two phases, have been used until recently over the full range of Mach numbers in two-phase flow problems. Walsh (1975) used the following modified drag coefficient equation fitting the available drag coefficient data by Bailey and Hiatt (1971).

$$C_D = C_{D,C} + (C_{D,FM} - C_{D,C}) \exp(-A \text{ Re}_S^N) \qquad (1)$$

where $C_{D,C}$ and $C_{D,FM}$ are the continuum and free molecular values of the drag coefficient, respectively, and A and N are constant. This expression, the values of $C_{D,C}$, $C_{D,FM}$, A and N obtained by Walsh at relative Reynolds numbers of 10^{-1}

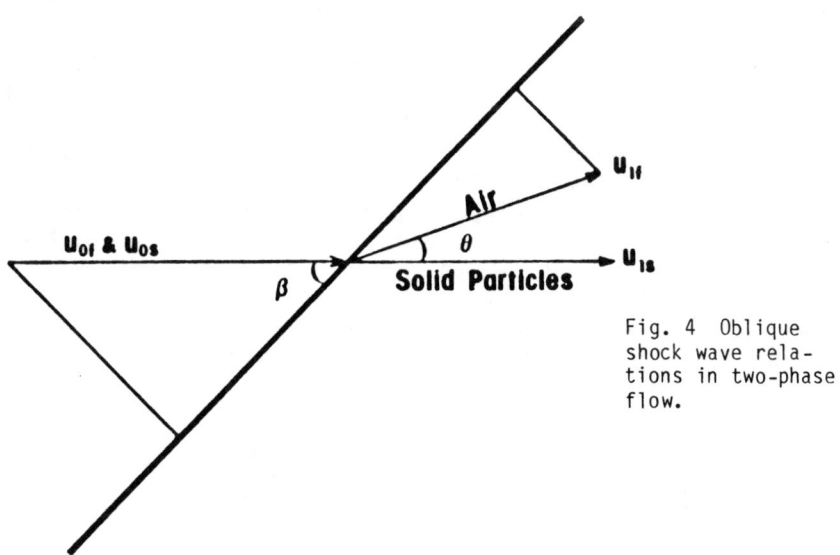

Fig. 4 Oblique shock wave relations in two-phase flow.

to 10^4, and relative Mach numbers 0 to 2.0 were used in the present study. These values have been found by other investigators to give the most extensive agreement to available data (Fig. 3).

Oblique Shock Wave Equations in Two-Phase Flow

Thermodynamic parameters in two-phase flow are dependent on the gas void fraction, i.e., the volume fraction of gas in two-phase flow, and are described elsewhere (Hayashi 1980). When the gas-particle flow properties incident to an oblique shock wave are given, gas-particle flow properties behind the shock front are obtained from the oblique shock wave relations. The resultant shock polar relation between the incident angle β and the deflection angle θ is a function of the equilibrium Mach number, incident particle velocity lag, particle feeding rate, and particle size. Oblique shock wave relations in two-phase flow described in Fig. 4 can be written as

$$u_{1Lf} = u_{1f} \cos(\beta - \theta), \quad u_{1Nf} = u_{1f} \sin(\beta - \theta)$$
$$u_{1Ls} = u_{1s} \cos, \quad u_{1Ns} = u_{1s} \sin \beta \qquad (2)$$

behind the oblique shock wave front and

$$u_{0Lf} = u_{0f} \cos \beta, \quad u_{0Nf} = u_{0f} \sin \beta$$
$$u_{0Ls} = u_{0s} \cos \beta, \quad u_{0Ns} = u_{0s} \sin \beta \qquad (3)$$

in front of the oblique shock wave front, where β is the incident angle, θ the deflection angle, the subscripts 0 and 1 denote, respectively, a condition ahead of the shock wave front and that behind the shock wave front, the subscript L the parallel direction to the shock wave front, and the subscript N the normal direction to the shock wave front.

Several assumptions are made in order to formulate the governing equations in two-phase media and are discussed elsewhere (Hoglund 1962; Krier and Mozaffarian 1978). The gas is assumed to be ideal and perfect; specific heats of gas and particles and the viscosity of gas are constant through the shock wave front; particles are uniform in size, spherical, incompressible, of uniform temperature, and uniformly distributed in the flowfield. In addition, particles are assumed to be in thermal equilibrium with the gas incident to the oblique shock wave; the wakes of particles do not affect the shock wave structure; and the gas void fraction ε does not change through the wave front. The last assumptions are: the particles do not change direction immediately after passing through the shock wave; the velocity of gas parallel to the oblique shock wave front does not change its magnitude through the shock wave front (oblique shock wave conditions); and external forces (e.g., gravity force) are neglected.

The general mass, momentum, and energy conservation equations of shock waves are

$$(\varepsilon \rho_f u_f) = 0 \quad [(1-\varepsilon)_s u_s] = 0 \tag{4}$$

$$[\varepsilon \rho_f u_f^2 + (1-\varepsilon)\rho_s u_s^2 + \varepsilon P_f + (1-\varepsilon)\tau_s] = 0 \tag{5}$$

and

$$[\varepsilon \rho_f u_f E_f + (1-\varepsilon) \rho_s u_s E_s + (1-\varepsilon) u_s \tau_s] = 0 \tag{6}$$

where E is the total energy per unit mass and τ_s is the particle drag force in the form

$$\tau_s = (\pi/8) C_D \rho_f (u_f - u_s) |u_f - u_s| \tag{7}$$

From the assumptions and oblique shock wave relations, mass conservation equations of gas and solid particles become

$$\varepsilon_1 \rho_{1f} u_{1Nf} = \varepsilon_0 \rho_{0f} u_{0Nf} \tag{8}$$

and

$$(1-\varepsilon) \rho_{1s} u_{1Ns} = (1-\varepsilon_0) \rho_{0s} u_{0Ns} \tag{9}$$

Normal and parallel components of mixture momentum equations are

$$\varepsilon_1 \rho_{1f} u^2_{1Nf} + (1-\varepsilon_1) \rho_{1s} u^2_{1Ns} + \varepsilon_1 P_{1f} + (1-\varepsilon_1) \tau_{1Ns}$$
$$= \varepsilon_0 \rho_{0f} u^2_{0Nf} + (1-\varepsilon_0) \rho_{0s} u^2_{0Ns} + \varepsilon_0 P_{0f} + (1-\varepsilon_0) \tau_{0Ns} \tag{10}$$

and

$$\varepsilon_1 \rho_{1f} u_{1Lf} + (1-\varepsilon_1) \rho_{1s} u_{1Ns} u_{1Ls} + (1-\varepsilon_1) \tau_{1Ls}$$
$$= \varepsilon_0 \rho_{0f} u_{0Nf} u_{0Lf} + (1-\varepsilon_0) \rho_{0s} u_{0Ns} u_{0Ls} + (1-\varepsilon_0) \tau_{1Ls} \tag{11}$$

where τ_{iNs} and τ_{iLs} ($i=0$ and 1) are components of the particle drag force between mixture and particles and are described by assuming that the velocity of gas represents the velocity of mixture since the solid particle loading is small as follows

$$\tau_{iNs} = (\pi/8)\, C_{Di}\, \rho_{if}\, (u_{iNf} - u_{iNs})\, |u_{iNf} - u_{iNs}| \quad (12)$$

and

$$\tau_{iLs} = (\pi/8)\, C_{Di}\, \rho_{if}\, (u_{iNf} - u_{iNs})\, |u_{iLf} - u_{iLs}| \quad (13)$$

Drag coefficients in mixture phase C_{Di} are given with constants obtained from data by Bailey and Hiatt (1971). The two-phase mixture energy equation is

$$\varepsilon_1\, \rho_{1f}\, u_{1Nf}\, [1/2\, (u^2_{1Lf} + u^2_{1Lf}) + C_{vf}\, T_{1f} + P_{1f}/\rho_{1f}]$$
$$+ (1-\varepsilon_1)\, \rho_{1s}\, u_{1Ns}\, [1/2\, (u^2_{1Ns} + u^2_{1Ls}) + C_{vs}\, T_{1s}$$
$$+ P_{1f}/\rho_{1f}\, (u_{1Nf}/u_{1Ns})] + (1-\varepsilon_1)\, u_{1Ns}\, [(\pi/8)\, C_{D1}\, \rho_{1f}\, u^2_{1Nf}$$
$$\times\, (1 - u_{1Ns}/u_{1Nf})\, |1 - u_{1Ns}/u_{1Nf}|]$$
$$= \varepsilon_0\, \rho_{0f}\, u_{0Nf}\, [1/2\, (u^2_{0Nf} + u^2_{0Lf}) + C_{vf}\, T_{0f} - P_{0f}/\rho_{0f}]$$
$$+ (1-\varepsilon_0)\, \rho_{0s}\, u_{0Ns}\, [1/2\, (u^2_{0Ns} + u^2_{0Ls}) + C_{vs}\, T_{0s}$$
$$+ P_{0f}/\rho_{0f}\, (u_{0Nf}/u_{0Ns})] + (1-\varepsilon_0)\, u_{0Ns}\, [(\pi/8)\, C_{D0}\, \rho_{0f}\, u^2_{0Nf}$$
$$\times\, (1 - u_{0Ns}/u_{0Nf})\, |1 - u_{0Ns}/u_{0Nf}|] \quad (14)$$

The gas void fraction ε does not change through the oblique shock wave front, and the oblique shock wave condition is preserved in two-phase flow. Hence,

$$\varepsilon_1 = \varepsilon_0 \quad (15)$$

and

$$u_{1Lf} = u_{0Lf} \quad (16)$$

The ratio of gas density behind the shock wave front to density ahead of the shock wave front becomes

$$\frac{\rho_{1f}}{\rho_{0f}} = \left\{ 1 + \frac{\tan(\beta-\phi)}{\tan\beta} \pm \left[1 + 2\frac{\tan(\beta-\phi)}{\tan\beta} \right.\right.$$
$$\left. + \left(\frac{\tan(\beta-\phi)}{\tan\beta}\right)^2 - 8\frac{\tan(\beta-\phi)}{\tan\beta}\left(\frac{u_{0s}}{u_{0f}}\right)\right.$$
$$\left.\left. + 4\frac{\tan(\beta-\phi)}{\tan\beta}\left(\frac{u_{0s}}{u_{0f}}\right)^2 \right]^{1/2} \right\} \frac{\tan(\beta-\phi)}{\tan\beta} \quad (17)$$

Three unknowns ρ_1, P_1, and θ behind the oblique shock wave front are calculated numerically from these equations with known initial conditions ahead of the shock wave front (see Hayashi 1980).

Results and Discussion

The important feature of a two-phase oblique shock wave is that the solid phase has lags in flow properties behind the shock wave front although it does not have any lag in flow properties ahead of the shock wave front. This implies that these flow properties are dependent upon the relative Reynolds number and the relative Mach number. Initial conditions used in the present study are summarized in Table 1, and the results are discussed in the following three sections.

Comparison of Numerical Results with Experimental Results

Morgenthaler (1962) analyzed and measured the steady oblique shock wave over a wedge in two-phase flow. He found that the oblique shock wave angle changed when two-phase flows passed through the shock wave. Table 2 shows the comparison of the present result with both his experimental and theoretical results.

The present computational results agree well with his experimental results. The velocity lag ahead of the oblique shock wave front was not measured in his experiment. Since alumina particles were small (<1 μm), no incident particle velocity lag was assumed in the present calculation.

Table 1 Initial conditions used in the present study

Equilibrium Mach number	M_M	1.2, 2.0, 10.0
Particle velocity lag	$U_0 = u_{0s}/u_{0f}$	0.9, 0.95, 1.0
Particle feeding rate	wt.%	1, 5, 10, 30
Particle size	d_s, μm	1, 27, 70
Particle type		fly ash, alumina

Table 2 Comparison of computational results with experimental results

Feeding rate mass wt.%	ε	β,deg	θ,deg[a]	θ,deg[b]	θ,deg[c]	θ,deg[d]
27.0	0.999880	47.250	7.6604	5.7500	7.3476	10.2891
27.0	0.999880	41.533	2.6778	3.9833	5.4879	5.5241
28.8	0.9998682	54.133	12.3475	12.5167	14.9893	14.5601
28.8	0.9998682	55.500	13.0461	12.7000	15.1924	15.2109
28.9	0.9998676	44.300	5.3694	5.4333	7.4035	7.9680
28.9	0.9998676	42.783	4.6004	4.5333	6.6470	6.6605
29.0	0.9998669	48.250	8.5710	5.8833	8.3214	11.0098
29.0	0.9998669	46.867	7.5098	4.7500	7.1231	10.0042

[a] Present results. [b] Experimental results by Morgenthaler (1962).
[c] Computational results by Morgenthaler. [d] Single-phase (air) flow case (ε=0).

Morgenthaler did not consider the effect of void fraction in his analysis so that his computational results did not agree with his experimental results as well as the present results. The comparison between this computational result and the experimental result verifies that the present model and the choice of the drag coefficient formula are appropriate.

Shock Polar Relations between the Incident Angle β and the Deflection Angle θ

The shock polar relations describing gas-particle two-phase properties behind the oblique shock wave consider the equilibrium shock wave Mach number, particle loading in weight percent, particle velocity lag, particle size, and particle type. Incident properties are given: The gas and particle temperatures T_{0f} and T_{0s} are 300 K; the gas pressure P_{0f} is 1 atm; the solid sphere particle densities $\rho_{0s\ fly\ ash}$ and $\rho_{0s\ alumina}$ are 0.737 and 3.965 g/cm^3; and the solid particle heat capacities $C_{s_2 fly\ ash}$ and $C_{s\ alumina}$ are 8.0×10^6 and 8.96×10^6 cm^2/s^2 K.

Solid particles pull the gas toward the solid particle flow, as Saltanov and Simanovskii (1978) pointed out in their shock wave observation, since the momentum of solid particles does not change much through the shock wave front. When the particle concentration increases, the deflection angle of gas becomes smaller. Figure 5 shows the effect of the particle feeding rate on the shock polar relation. The deflection angles of the gas in two-phase flow are smaller than those in single-phase flow and become smaller when the particle feeding rate increases.

When the shock wave is weaker, it is influenced more by solid particles. This phenomenon was observed in the

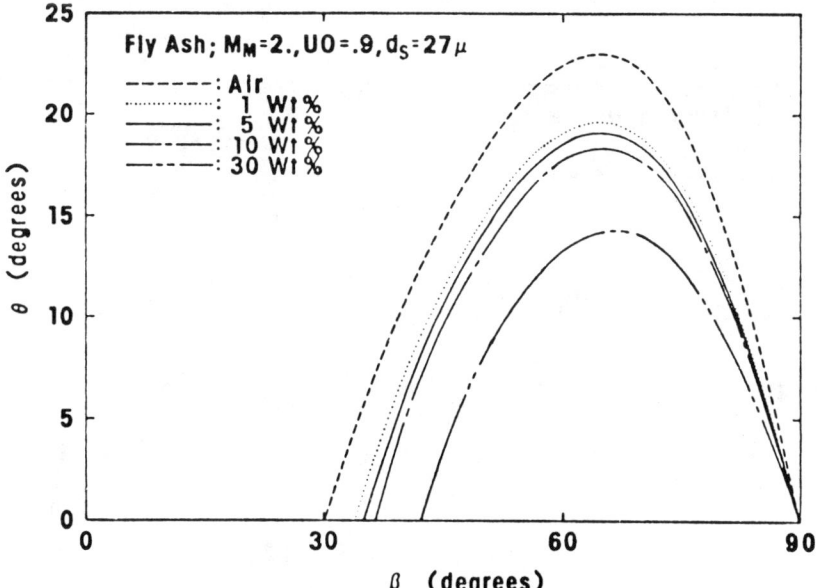

Fig. 5 Shock polar relations to the particle feeding rate by weight in two-phase flow (M_M is the equilibrium Mach number, U_0 the relative velocity, u_{0s}/u_{0f}, d_s the solid particle diameter).

Schlieren photographs of shock waves (Fig. 1) and is predicted in the computational results on oblique shock wave in two-phase flow (Fig. 6). The difference in deflection angles of the gas between two-phase and single-phase flow becomes larger as the equilibrium Mach number decreases.

One more important factor that influences the shock wave structure is the relative particle velocity. In many cases, at the nozzle exit, the flow is accelerated, and there is a velocity lag between solid particles and gas (Hetsroni and Sokolov 1971). Figure 7 describes the effect of relative particle velocity on the shock polar relation. The deflection angle of the gas becomes smaller when the particle velocity lag $1-U_0$ increases. When the particle velocity lag is large, the drag force reduces the momentum of the gas and hence the deflection angle of the gas. Although the incident velocities of the gas and the particles are identical ($U_0 = 1.0$), the deflection angle of the gas in two-phase flow is smaller than that in single-phase flow since particles reduce the gas flow.

The effects of particle size and particle type (fly ash and aluminum oxide) on the shock wave are found to be small, but not negligible. The major difference between the fly

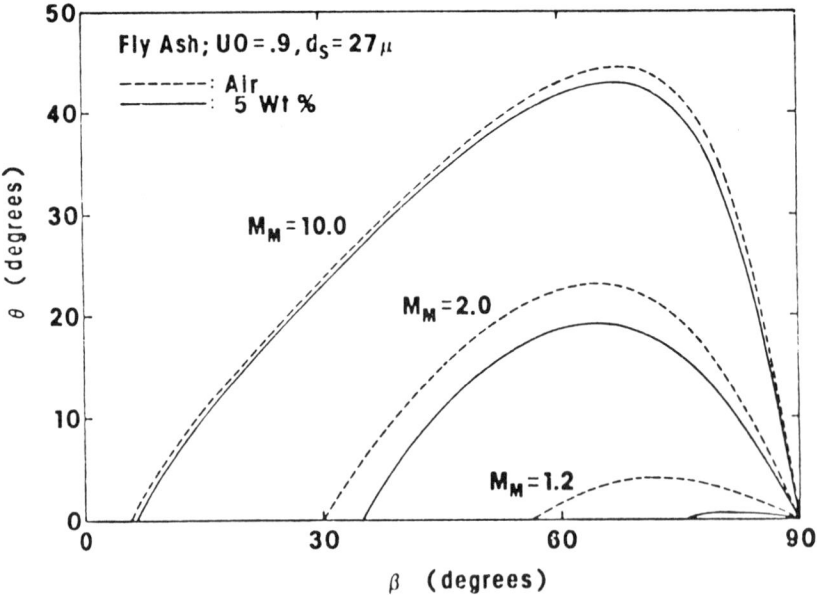

Fig. 6 Shock polar relations to the equilibrium Mach number in two-phase flow.

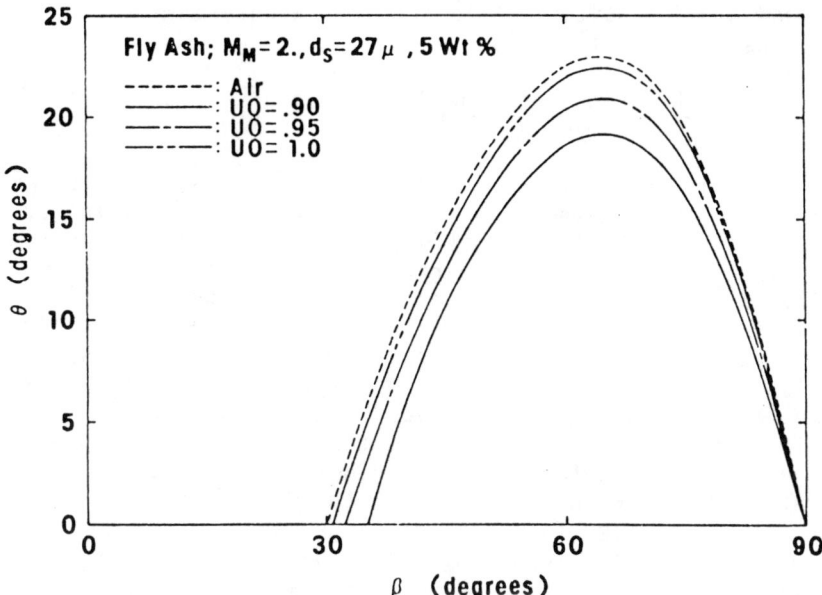

Fig. 7 Shock polar relations to the particle relative velocity in two-phase flow.

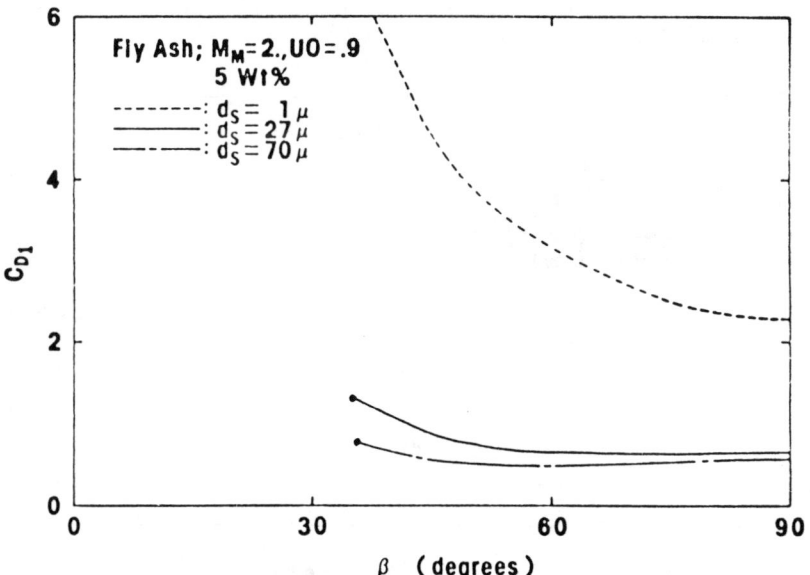

Fig. 8 Drag coefficients behind the shock wave for a range of incident angles (· is the limit of compression wave).

ash and alumina particles was density, with alumina 5.4 times as dense as fly ash.

Shock Polar Relations of Flow Properties to the Incident Angle β

The flow properties behind the oblique shock wave front were also investigated. As mentioned earlier, the velocity of the solid particles is different from that of the gas in the direction of the solid particle flow since the gas flow deflects behind the oblique shock wave. The drag coefficient is a function of the relative Reynolds number and the relative Mach number and is affected by solid particles as well. Figure 8 shows the drag coefficient behind the shock wave front for a range of incident angles. The drag coefficient of a 1 µm particle is much larger than that of large particles at the relative velocity U_0 of 0.9. If the relative Reynolds number is fixed, the drag force of a 1 µm particle is smaller than that of large particles since the velocity lag of small particles is smaller than that of large particles. The drag coefficient of large particles does not change significantly for a large range of incident angles since the relative Reynolds number of large particles becomes larger and their drag coefficients become constant (Fig. 3). The end points in Figs. 8-11 imply the limit of a compression wave.

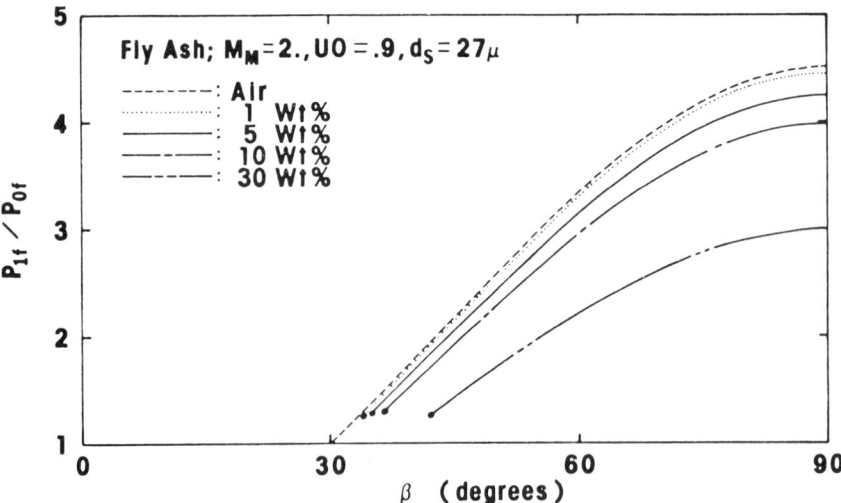

Fig. 9 Ratio of the gas pressure behind the oblique shock wave to that ahead of the shock wave for a range of incident angles.

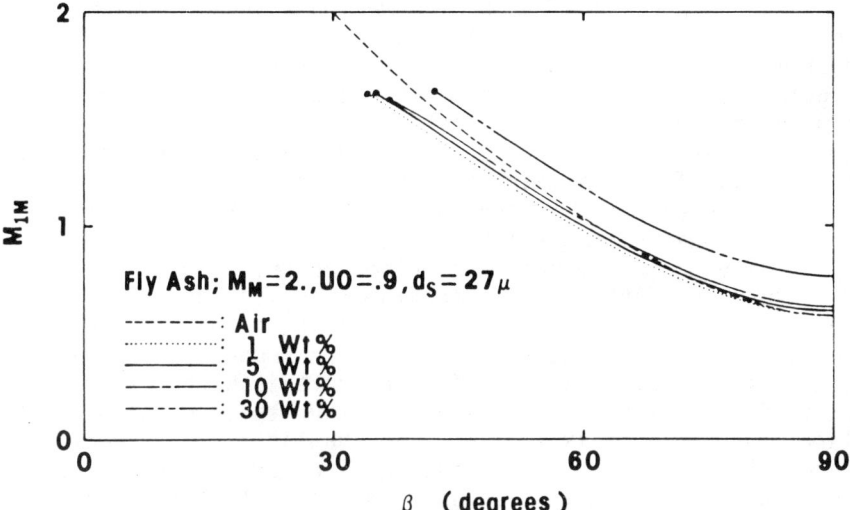

Fig. 10 Equilibrium Mach number behind the shock front for different incident particle feeding rates for a range of incident angles.

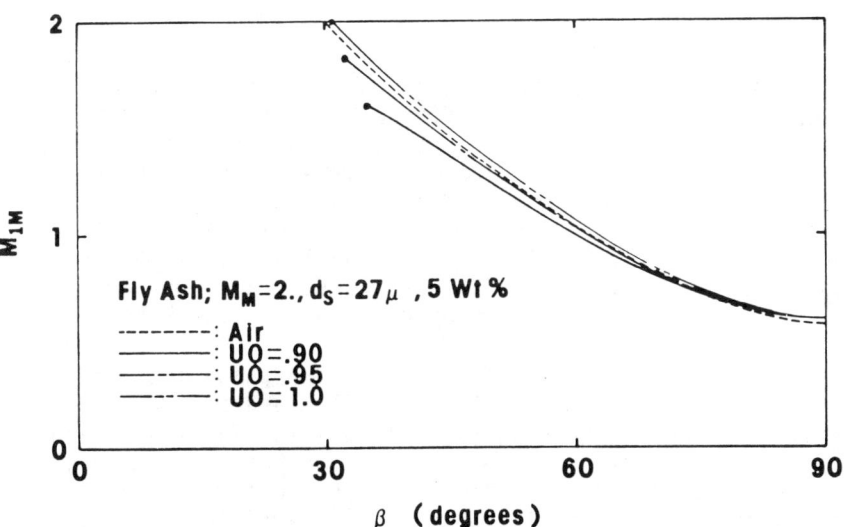

Fig. 11 Equilibrium Mach number behind the shock front for different incident relative velocity for a range of incident angles.

The suspension of particles weakens the shock wave so that the gas in two-phase flow is less compressed in the oblique shock wave front than that in single-phase flow. Figure 9 shows that the ratio between the gas pressure behind the shock front and that ahead of the shock front decreases at any incident angle as the particle feeding rate increases. The same characteristics were seen in the ratio between the gas density behind the oblique shock wave front and that ahead of the shock front.

The equilibrium Mach number behind the shock front for different incident particle feeding rates is shown in Fig. 10 and that for different relative incident velocity is shown in Fig. 11. The equilibrium Mach number behind the shock wave front at a large feeding rate is higher than at a small feeding rate since the equilibrium speed of sound at a large feeding rate is smaller than that at a small feeding rate. Figure 11 shows that the equilibrium Mach number with small particle velocity lag is higher than that with large particle velocity lag. The particle velocity lag causes the drag force and, hence, weakens the shock wave. The equilibrium Mach number at a 90-deg incident angle is dependent upon the particle feeding rate, but is independent of the relative particle velocity.

Conclusions

The two-phase oblique shock wave was investigated theoretically. The drag coefficient was obtained from aeroballistic range measurements as a function of relative Reynolds number and relative Mach number. The calculations made using this drag coefficient showed good agreement between theory and experiment for the available data of Morgenthaler (1962). The computational study of oblique shock waves in two-phase flow showed several important effects of two-phase flow properties on the shock polar relations.

The deflection angles of gas in two-phase flow are smaller than those in single-phase flow and become smaller when the particle feeding rate increases. The suspension of particles affects weak shock waves more than strong shock waves. The deflection angle of the gas becomes smaller when the particle velocity lag increases. Particle size and particle type have little effect on the oblique shock wave relations in two-phase flow.

The drag coefficient of large particles is smaller than that of small particles and does not change significantly over a large range of incident angles. The pressure ratio between that ahead of the shock wave and that behind the shock wave in two-phase flow is smaller than the ratio in single-phase flow and decreases at any incident angle as the particle feeding rate increases. An increase in the suspension of particles strengthens the oblique shock wave, but the large particle velocity lag weakens the oblique shock wave in two-phase flow.

This study provides some fundamental characteristic features of the oblique shock wave in two-phase flow and suggests an appropriate choice of drag coefficient for the problem. The equations developed in this study may be used further for the problems of triple shock waves, detonation, and blast waves in multiphase media.

Acknowledgments

This work was supported by the Air Force Office of Scientific Research through Grant AFOSR 77-3107.

References

Alkhimov, A. P., Papyrin, A. N., Predein, A. L., and Soloukhin, R. I. (1977) Experimental investigation of the effect of velocity lag of particles in a supersonic gas stream. Zh. Prikl. Mekh. Tekh. Fiz. 4, 80-88.

Alkhimov, A. P., Boiko, V. M., Papyrin, A. N., and Soloukhin, R. I. (1978) Diagnostics of supersonic two-phase streams from scattered laser radiation. Zh. Prikl. Mekh. Tekh. Fiz. 2, 36-46.

Bailey, A. B. and Hiatt, J. (1971) Free-flight measurements of sphere drag at subsonic, transonic, supersonic and hypersonic speeds for continuum transition and near-free-molecular flow conditions. AEDC-TR-70-291 (AD-721-208).

Carlson, D. J. and Hoglund, R. F. (1964) Particle drag and heat transfer in rocket nozzles. AIAA J. 2, 1980-1984.

Chang, I. S. (1981) Three-dimensional two-phase supersonic nozzle flow. AIAA 14th Fluid and Plasma Dynamic Conference, Palo Alto, Calif.

Gilbert, M., Davis, L., and Altman, D. (1955) Velocity lag of particles in linearly accelerated combustion gases. Jet Propul. 25, 2530.

Gregor, W. and Rumpf, H. (1975) Velocity of sound in two-phase media. Int. J. Multi. Flow 1, 753-769.

Hayashi, K. and Branch, M. C. (1980) Concentration, velocity and particle size measurements in gas-solid two-phase jets. J. Energy 4, 193-198.

Hayashi, K. (1980) Measurement and calculation of properties of gas-solid two-phase jets. Ph.D. Thesis, University of Colorado, Boulder, Colo.

Hetsroni, G. and Sokolov, M. (1971) Distribution of mass, velocity and intensity of turbulence in a two-phase turbulent jet. J. Appl. Mech. 38, 315-327.

Hoglund, R. F. (1962) Recent advances in gas-particle nozzle flows. ARS J. 32, 662-671.

Ignatov, S. H., Mironchuk, N. S., and Khramov, N. E. (1978) Flow around a sphere by a two-phase. Mekhan. Zhidk. Gaza 6, 171-176.

Ingebo, R. D. (1956) Drag coefficients for droplets and solid spheres in clouds accelerating in airstreams. NACA TN-3762.

Kleigel, J. R. (1963) Gas particle nozzle flows. Ninth Symposium (Int.) on Combustion, pp. 881-826. The Combustion Institute, Ithaca, N. Y.

Krier, H. and Mozaffarian, A. (1978) Two-phase reactive particle flow through normal shock waves. Int. J. Multi. Flow 4, 65-79.

Lapple, C. E. and Shepherd, C. B. (1940) Calculation of particle trajectories. Ind. Eng. Chem. 32, 605-617.

Morgenthaler, J. H. (1962) Analysis of two-phase flow in supersonic exhausts. Detonation and Two-Phase Flow: AIAA Progress in Astronautics and Rockerty edited by S. S. Penner and F. A. Williams, Vol. 6, pp. 145-171, AIAA, New York.

Pergament, H. S. and Thorpe, R. D. (1975) A computer code for fully-coupled rocket nozzle flows (FULNOZ). Aerochem TP-322 (AFOSR-TR-75-1563).

Probstein, R. F. and Fassio, F. (1970) Dusty hypersonic flows. AIAA J. 8, 772-779.

Rudinger, G. (1969) Relaxation in gas-particle flow. Non-Equilibrium Flows (edited by P. P. Wegener), pp. 119-161. Marcel Dekker, New York.

Saltanov, G. A. and Simanovskii, G. P. (1978) Two-dimensional mixed flows of supersaturated and two-phase medium with nonequilibrium phase transitions. Mekhan. Zhidk. Gaza 4, 87-93.

Walsh, M. J. (1975) Drag coefficient equations for small particles in high speed flows. AIAA J. 13, 1526-1528.

Equilibrium Shock Wave Properties in Dusty and Clean Air

Allen L. Kuhl*
R&D Associates, Marina del Rey, Calif.

Abstract

This study considers shock propagation in a two-phase medium of gas and solid dust particles. First, it is shown that small particles reach equilibrium with the surrounding flow in relatively short distances (e.g., a few millimeters for 10-μ particles). When flowfield changes are small over this same dimension, then such cases can be modeled accurately as equilibrium dusty gas flows. Next, the theoretical shock jump conditions, shock polars, and isentropic flow relations are derived for dusty gas flows, assuming velocity and temperature equilibrium between the dust and the gas. These relations were then used to evaluate flow conditions in a mixture of real air and SiO_2 particles by comparing dusty and clean flows at equal shock overpressures. It was found that the dynamic pressure, stagnation pressure, and normal shock reflection factor increase significantly as the dust loading and preshock air temperature are increased. It was concluded that the dust mass and thermal heat capacity should be included in the flow analysis if the dust loading is greater than about one-tenth the ambient air density.

Nomenclature

a   = sound speed = $\sqrt{\gamma p/\rho}$
a' = effective dusty gas sound speed = $a/\sqrt{1+\kappa}$

Presented at the 8th ICOGER, Minsk, USSR, Aug. 23-26, 1981. Copyright © American Institute of Aeronautics and Astronautics, Inc. 1983. All rights reserved.
 *Senior Research Scientist, Continuum and Fluid Dynamics Department.

A	= sound speed ratio across a shock = a_y/a_x
C	= specific heat capacity of dust particle
C_p	= specific heat at constant pressure for the gas
C_d	= drag coefficient for the particle
d_p	= dust particle diameter
e	= internal energy of the gas
e_p	= internal energy of the dust particles = CT_p
h	= enthalpy of the gas = γe
k_i	= $[\gamma_i/(\gamma_i-1)][1 + \kappa C/C_p]$ for $i = x,y$
ℓ_v	= velocity equilibration length = $4\rho_s d_p / 3\rho C_d$
ℓ_T	= thermal equilibration length = $(3/2)\text{Pr}\,\ell_v$
M	= Mach number
M_x	= $V_x/a_x = (W-u_x)/a_x$
M_y	= $V_y/a_y = (W-u_y)/a_y$
p	= pressure
P	= pressure ratio across a shock = p_y/p_x
Pr	= Prandtl number of the gas = $\mu C_p/k$
q	= dynamic pressure = $\tfrac{1}{2}\rho u^2$
r	= dust particle radius
R_p	= individual gas constant
Re	= Reynolds number = $\rho(u-u_p)d_p/\mu$
t	= time
T	= temperature
u	= gas velocity
u_p	= dust particle velocity
U	= change in velocity across a shock
	= $(V_x-V_y)a_x = (u_y-u_x)/a_x$
V_x	= upstream velocity in shock fixed coordinates
	= $W-u_x$
V_y	= downstream velocity in shock fixed coordinates
	= $W-u_y$
W	= shock wave velocity
x_p	= particle position
α	= dust density/air density at ambient temperatures
	= ρ_{dust}/ρ_a
β	= $(1-1/2k_y)(1+\kappa)\gamma_x U^2 - (1-k_x/k_y)$
β'	= $(1+\kappa)\gamma_x M_x^2 (1-1/k_y) - 1$
γ	= effective gamma for the gas = $1 + p/\rho e$
κ	= dust loading parameter = ρ_{dust}/ρ_{air}
μ	= gas viscosity
ν	= specific volume ratio across a shock = ρ_x/ρ_y
ρ	= gas density
ρ_s	= density of a solid particle

Subscripts

0	= stagnation conditions
1, x	= conditions upstream of a shock
1d	= state 1 with heated preshock conditions
2, y	= conditions downstream of a shock
3	= conditions behind a stationary shock in supersonic flow
1,2,3,4,5	= states associated with the interaction of two shock waves (see insert in Fig. 3).
a	= ambient STP conditions

Introduction

Shock propagation in a mixed medium of solid dust particles and gas is a physical phenomenon often accompanying real explosions. For example, blast propagation over a soil surface can induce a dusty compressible boundary layer near ground. Dust explosions in mines and industrial facilities also contain dusty shocked flows. Knowledge of shock properties in dusty gases is very useful in assessing the mechanical damage effects of such explosions. In particular, it is important to determine what dust loadings are required to significantly influence the flow.

A discussion of dusty gas flows under various nonequilibrium assumptions can be found in the review article by Marble (1970), while various authors [Carrier (1955) originally, and Von Schubert (1969) more recently] have calculated the shock structure in such media.

In this paper we investigate the macroscopic properties of one-dimensional compressible flow in hot dusty gases. The analysis starts by estimating the characteristic dimensions required to achieve equilibrium between the gas and the dust. Next the shock jump conditions, shock polars, and isentropic flow relations are derived for dusty gases under the assumption of velocity and temperature equilibrium. The theoretical relations are applied to a mixture of SiO_2 dust and real air. Various flow characteristics are then compared parametrically for clean and dusty air to determine the influence of dust loading and ambient temperature effects. Important findings are summarized in the conclusion section.

Analysis

Equilibrium Condition Investigation

Dusty gas flowfields are governed by separate conservation equations of mass, momentum, and energy for the gas and the dust; these are coupled through the drag and heat-transfer terms. In general, at a given point the dust and air can have different velocities and temperatures; that is, a velocity and temperature slip condition can exist. Surface drag accelerates the particle and decelerates the gas, thus reducing the slip; surface heat transfer acts likewise. Let us calculate the distance required for equilibration.

Consider a spherical particle with radius r_p, velocity u_p, and density ρ_s, imbedded in a gas with velocity u and density ρ. The momentum equation for the particle gives

$$\frac{4}{3} \pi r_p^3 \rho_s \frac{du_p}{dt} = \frac{1}{2} \rho (u-u_p)^2 C_d \pi r_p^2 \qquad (1)$$

Rearranging and utilizing the definition of the particle velocity $u_p \equiv dx_p/dt$, we find the following relation:

$$\frac{3}{2} \frac{\rho C_d}{\rho_s r_p} dx_p = \frac{u_p/u}{(1-u_p/u)^2} d(\frac{u_p}{u}) \qquad (2)$$

The drag coefficient C_d is a function of the Reynolds number, $Re = \rho(u-u_p)d/\mu$; for $Re < 10^5$ it increases as the relative velocity decreases. To be conservative, one can use the minimum drag coefficient (the one associated with the maximum Reynolds number, $Re = \rho u d/\mu$). Assuming also that the gas velocity and density are constant, we can integrate Eq. (2) to give

$$\frac{x_p}{\ell_v} = \frac{u_p/u}{1-u_p/u} + \log_e (1-u_p/u) \qquad (3)$$

where ℓ_v is the characteristic velocity equilibration length scale

$$\ell_v = \frac{8}{3} \frac{\rho_s r_p}{\rho C_d}$$

or nondimensionalizing with the particle diameter, we find

$$\frac{\ell_v}{d_p} = \frac{4}{3}\frac{\rho_s}{\rho C_d} \quad (4)$$

According to Eq. (3), the velocity equilibration length is the distance required for the particle to reach 68% of the gas velocity. It depends linearly on the solid particle/gas density ratio and inversely on the drag coefficient.

As shown by Marble (1970), the thermal equilibration length ℓ_T is related to the velocity length according to

$$\ell_T = (3/2) \Pr \ell_v \quad (5)$$

where Pr denotes the gas Prandtl number ($\Pr = \mu C_p/k$). For most gases, the Prandtl number is about 0.7, so that the thermal and velocity equilibration lengths are equal.

Figure 1 gives the velocity equilibration lengths (where $u_p = 0.68u$) for spherical particles as a function of shock strength and particle diameter. These results were calculated from Eq. (4) by using flow conditions behind

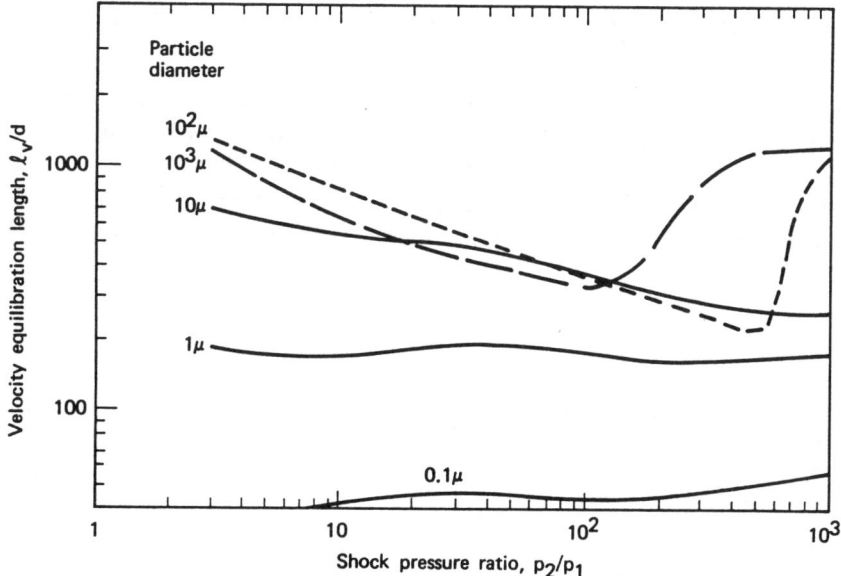

Fig. 1 Velocity equilibration length for spherical particles as a function of particle size and shock strength (note, $u_p = 0.68$ u when $x = \ell_v$).

various strength shocks and spherical drag coefficients (Schlicting 1968) which were evaluated at the maximum Reynolds number, Re = $\rho u d/\mu$. 1-μ particles equilibrate in about 200 diam. (0.2 mm); 10-μ particles equilibrate in from 300 to 700 diam. (from 3 to 7 mm), while 100-μ particles require from 200 to 1500 diam (from 2 to 15 cm) for equilibration. Clearly, if flowfield gradients are small over a distance of ℓ_v and ℓ_T, then such flowfields can be analyzed by means of the equilibrium equations for dusty gasdynamics.

The shape of the curves in Fig. 1 is determined by the particle drag law dependence on the Reynolds number. More precise values of the velocity equilibration length could be evaluated from Eq. (2) written in the following form:

$$\frac{x_p}{d_p} = \frac{4}{3} \frac{\rho_s}{\rho} \int \frac{u_p/u \; d(u_p/u)}{(1-u_p/u)^2 C_d(Re,M)} \quad (6)$$

where the drag coefficient C_d is a function of the local Reynolds number and Mach number [see Hayashi and Branch (1981)] for an equation for C_d (Re,M)

Shock Jump Conditions for a Dusty Gas

For the remainder of the analysis we shall assume that the dust particles are in velocity and temperature equilibrium with the gas. Let us consider the conservation of mass, momentum, and energy across a shock discontinuity. For a dust loading of $\kappa = \rho_{dust}/\rho_{air}$, the jump conditions in shock fixed coordinates acquire the following form for a dusty gas:

$$(1+\kappa) \rho_y V_y = (1+\kappa) \rho_x V_x \quad (7)$$

$$p_y + (1+\kappa) \rho_y V_y^2 = p_x + (1+\kappa) \rho_x V_x^2 \quad (8)$$

$$h_y + \kappa e_{p,y} + \tfrac{1}{2}(1+\kappa) V_y^2 = h_x + \kappa e_{p,x} + \tfrac{1}{2}(1+\kappa) V_x^2 \quad (9)$$

where ρ, p, and h denote the gas density, pressure, and enthalpy, respectively; V equals the relative velocity (V = W-u, where W is the shock wave velocity and u is the gas particle velocity); e_p represents the internal energy of the dust; and x and y denote conditions upstream and downstream of the shock.

The above system of equations may be closed by specifying the equations of state for the gas and the dust. In this analysis, a variable γ equation of state was used for the air:

$$\gamma \equiv 1 + p/\rho e \qquad (10)$$

This effective γ was a tabulated function of density and internal energy as calculated by Gilmore (1955) for equilibrium air. The effective gamma function $\gamma \equiv \gamma(\rho,e)$ is displayed in Fig. 2. The enthalpic equation of state for air then becomes

$$h = \gamma e = \frac{\gamma}{\gamma-1} \frac{p}{\rho} \qquad (11)$$

An ideal caloric equation of state was used for the dust particles:

$$e_p = CT_p \qquad (12)$$

Since by assumption the dust temperature equals the air temperature, the thermal equation of state for the gas ($p/\rho = RT$) can be used to relate the dust internal energy to the gas pressure and density:

$$e_p = \frac{C}{R} \frac{p}{\rho} = \frac{\gamma}{\gamma-1} \frac{C}{C_p} \frac{p}{\rho} \qquad (13)$$

For a mixture of SiO_2 and air, the specific heat ratio is $C/C_p = 1.833$ at ambient temperature.

To manipulate the shock jump relations, it is convenient to use nondimensional variables. Following Oppenheim (1970), we define the thermodynamic variables of pressure, density, and sound speed as

$$P = p_y/p_x$$

$$\nu = \rho_x/\rho_y$$

$$A = a_y/a_x$$

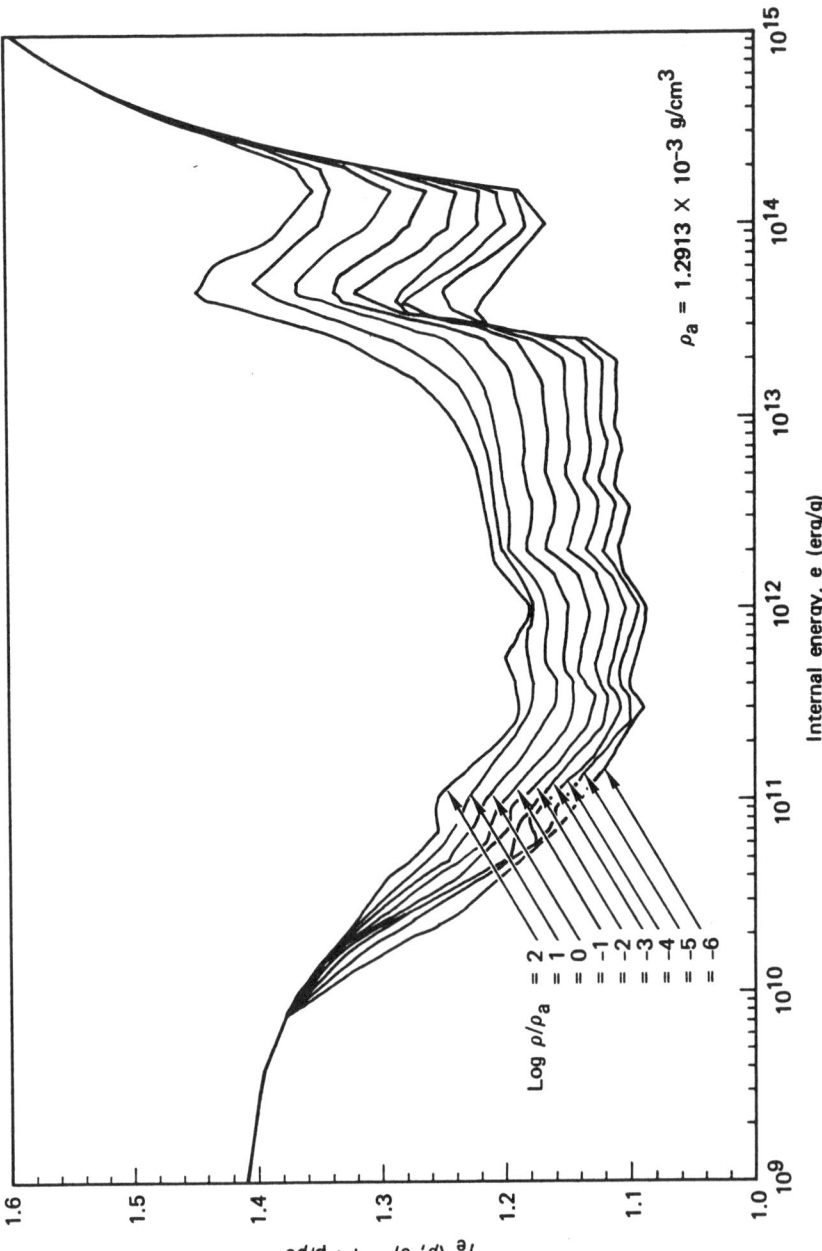

Fig. 2 Effective γ as a function of internal energy and density for equilibrium air.

and the kinematic variables of flow Mach numbers and change in flow velocity across the shock as

$$M_x = V_x/a_x = (W-u_x)/a_x$$

$$M_y = V_y/a_y = (W-u_y)/a_y$$

$$U = (V_x-V_y)/a_x = (u_y-u_x)/a_x$$

Utilizing the above variables, the shock jump conditions become

$$U = M_x (1-\nu) \quad \text{(mass)} \tag{14}$$

$$(1+\kappa) \gamma_x M_x^2 = (P-1)/(1-\nu) \quad \text{(Rayleigh line)} \tag{15}$$

$$k_y P\nu - k_x = \tfrac{1}{2} (P-1)(1+\nu) \quad \text{(energy)} \tag{16}$$

where

$$k_i = \frac{\gamma_i}{\gamma_i-1} \left(1 + \frac{\kappa C}{C_p}\right) \quad i = x,y \tag{17}$$

The effect of dust mass shows up in the Rayleigh line, Eq. (15); this can be suppressed by defining a dust sound speed $a_x'^2 = a_x^2/(1+\kappa)$. The effect of dust also occurs in the energy equation in the k_i terms. This can only be removed if the dust has no thermal capacity (i.e., $C = 0$).

Shock Polars for a Dusty Gas

The nonsteady shock polars are obtained from Eqs. (14-16) by solving for U or P:

$$U = f_1(P) = \pm \sqrt{\frac{(P-1)^2(k_y-1) + (P-1)\cdot(k_y-k_x)}{\gamma_x(1+\kappa)\,[(P-1)(k_y-\tfrac{1}{2}) + k_y]}} \tag{18}$$

or

$$P = f_2(U) = f_1^{-1}$$

$$= 1 + \frac{\beta}{2(1-1/k_y)} \cdot \left[1 + \sqrt{1 + 4(1-1/k_y)(1+\kappa)\gamma_x U^2/\beta^2}\right] \quad (19)$$

where

$$\beta = (1 - 1/2k_y)(1+\kappa)\gamma_x U^2 - (1-k_x/k_y)$$

Given P and U, the remaining flow parameters may be determined from the following:

$$\nu = 1 - (1+\kappa)\gamma_x U^2/(P-1) \quad (20)$$

$$M_x = U/(1-\nu) \quad (21)$$

$$A = \sqrt{P\nu\gamma_y/\gamma_x} \quad (22)$$

The above relations have been used to generate clean air shock polars displayed in Fig. 3, assuming the effective γ equation of state of Gilmore (1955) from Fig. 2. These polars can be used to determine the states 3 and 4 resulting from the collision of shock waves with incident pressures p_2 and p_5 up to 1 kbar in ambient air, state 1. Similar shock polars can be found in monographs on gasdynamics (Oppenheim 1970 and Owczarek 1964) for the case of an ideal gas with a constant specific heat ratio. Dusty gas shock polars could be evaluated and plotted similar to the clean air polars of Fig. 3; however, it is more convenient to evaluate characteristic parameters of the polars (such as normal shock reflection factors), which are described parametrically in a following section.

For steady flows it is useful to solve Eqs. (14-16) for pressure in terms of the flow Mach number; this yields the following relation:

$$P = 1 + \frac{\beta'}{2-1/k_y}$$

$$\times \left[1 + \sqrt{1 + 4(1-1/2k_y) \cdot (1+\kappa)\gamma_x M_x^2 \cdot (1-k_x/k_y)/\beta'^2}\right] \quad (23)$$

where $\beta' = (1+\kappa)\gamma_x M_x^2 (1-1/k_y) - 1$.

Isentropic Flow Relations for Dusty Gases

Also of interest is the isentropic stagnation pressure. This is derived from the steady flow energy equation.

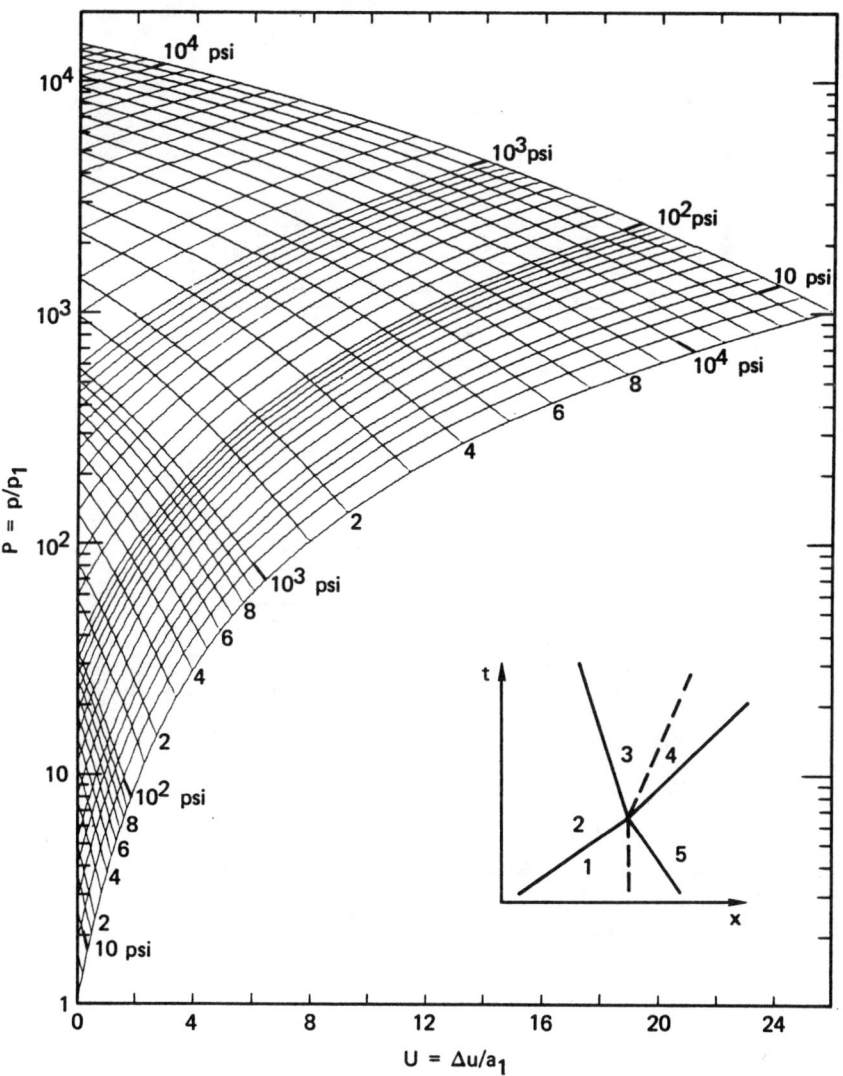

Fig. 3a Pressure-velocity shock polars for real air.

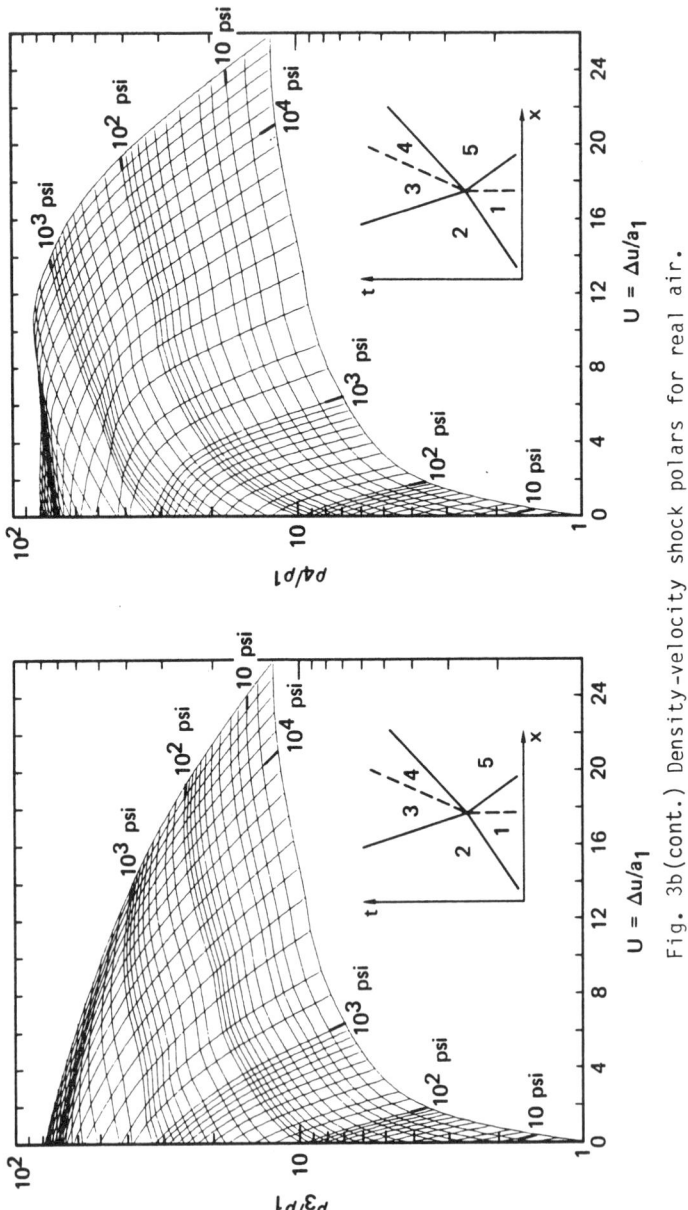

Fig. 3b(cont.) Density-velocity shock polars for real air.

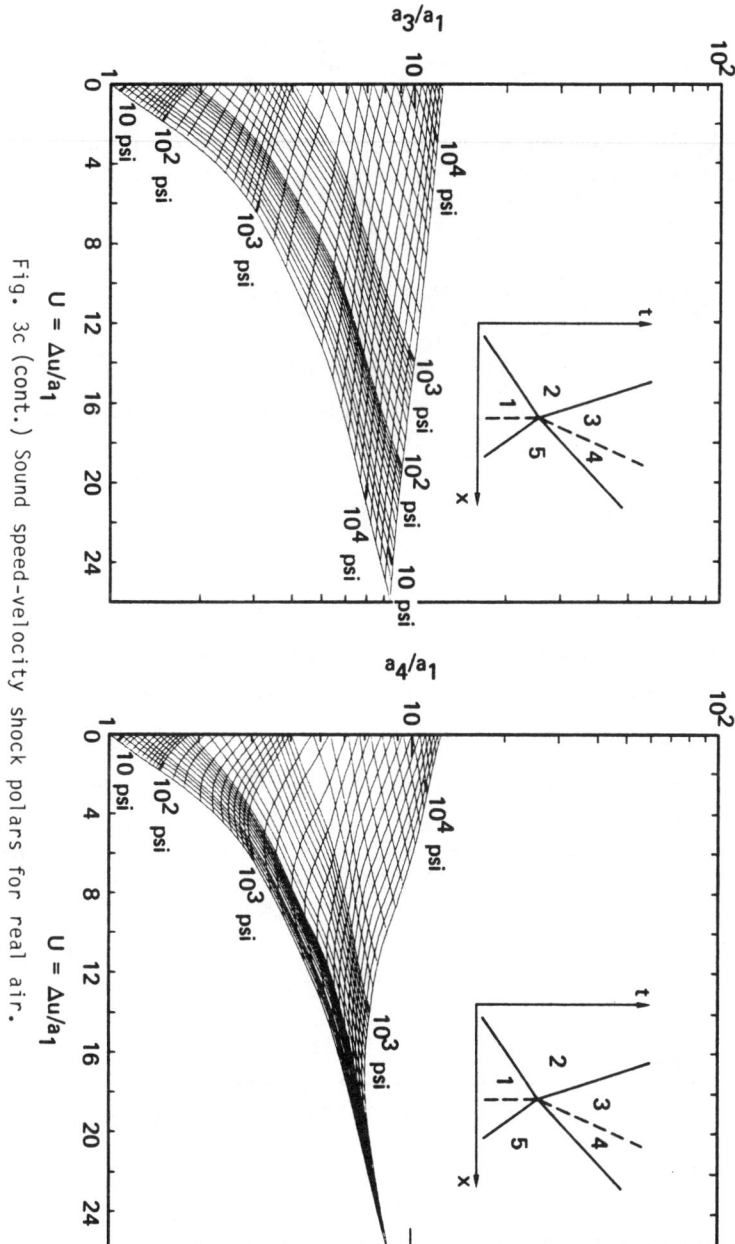

Fig. 3c (cont.) Sound speed-velocity shock polars for real air.

Taking into account the thermal heat capacity of the dust, it acquires the following form for hot dusty gas flows:

$$h + \kappa e_p + \tfrac{1}{2}(1+\kappa) V^2 = h_o + \kappa e_{p,o} \qquad (24)$$

Again assuming velocity and temperature equilibrium between the dust and the gas, and applying this to conditions behind the shock, one finds

$$k_y + \tfrac{1}{2}(1+\kappa)\gamma_y M_y^2 = k_y \frac{p_{oy}}{p_y}\left(\frac{\rho_{oy}}{\rho_y}\right)^{-1} = k_y \left(\frac{p_{oy}}{p_y}\right)^{(\gamma_{\dot y}-1)/\gamma_{\dot y}} \qquad (25)$$

where the latter equality assumes adiabatic, isentropic stagnation.[†] Thus one obtains the isentropic stagnation pressure relation for a dusty gas:

$$\frac{p_{oy}}{p_x} = P\left[1 + \frac{1}{2k_y}(1+\kappa)\gamma_y M_y^2\right]^{\gamma_y/(\gamma_y-1)} \qquad (26)$$

The effect of dust density can again be suppressed into the dusty sound speed [$a_y'^2 = a_y^2/(1+\kappa)$], while the thermal capacity contributions are contained in k_y. Of course, for higher shock strengths, the flow is supersonic behind the shock; in dusty gases this means $(1+\kappa)M_y^2 > 1$. In such cases, fluid particles pass through another shock during stagnation, which causes a loss in stagnation pressure. If this is a normal shock, the following equation applies:

$$\frac{P_{03}}{P_1} = P_{32} \cdot P_{21}\left[1 + \frac{1}{2k_3}(1+\kappa)\gamma_3\left(\frac{u_3}{a_3}\right)^2\right]^{\gamma_3/(\gamma_3-1)} \qquad (27)$$

where subscripts 1 and 2 denote flow conditions ahead of and behind the shock, while subscript 3 denotes conditions behind a stationary shock which has state 2 as an upstream boundary condition.

Also of interest is the dynamic pressure behind the original shock, which acquires the following form for a

---

[†] Note that the γ in the exponent is really an adiabatic γ. For the present calculations we have used values corresponding to Gilmore's effective air γ; this is an approximation.

dusty gas

$$q_2/p_1 = \tfrac{1}{2}(1+\kappa)\gamma U^2/\nu \qquad (28)$$

Density increases due to dust appear in the κ term in the above equation.

Parametric Results

A series of calculations were performed for shock waves propagating in clean air and in dusty air. In the clean air case, the ambient conditions were taken as:

$$p_1 = p_a \equiv 1.01325 \text{ bars}$$
$$\rho_1 = \rho_a \equiv 1.224 \times 10^{-3} \text{ g/cm}^3 \qquad (29)$$
$$T_1 = T_a \equiv 293 \text{ K}$$
$$u_1 = \kappa = C = 0$$

In the dusty air cases, the medium was characterized by the preshock gas temperature T_{1d} and the dust loading factor $\alpha \equiv \rho_{dust}/\rho_a$, which is evaluated at ambient temperature $T_a = 293$ K. The ambient conditions ahead of the wave then become

$$p_1 = p_a$$
$$\rho_1 = \rho_a T_a/T_{1d} \quad \text{(gas density)}$$
$$\rho_{1d} = \kappa \rho_1 \quad \text{(dust density)}$$
$$\rho_{1,total} = (1+\kappa)\rho_1 \quad \text{(total density)} \qquad (30)$$
$$T_1 = T_{1d}$$
$$u_1 = 0$$
$$\kappa = \alpha T_{1d}/T_a$$

Note that the parameter α denotes the dust loading at ambient temperature; the actual preshock dust loading κ is

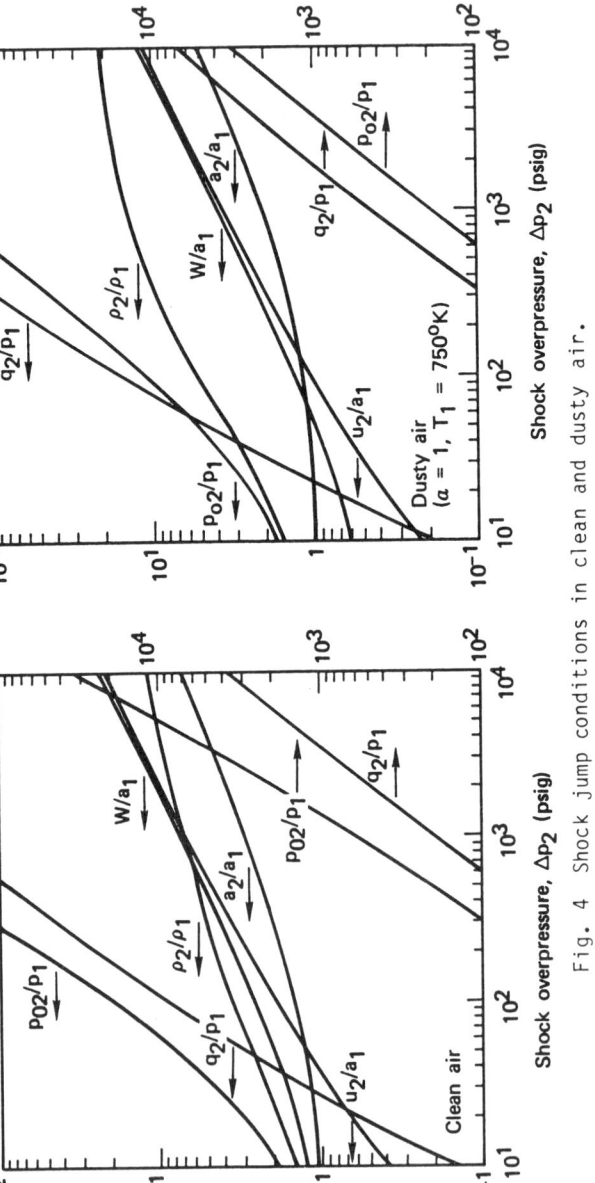

Fig. 4 Shock jump conditions in clean and dusty air.

greater than α for elevated preshock temperatures since $\kappa = \alpha T_{1d}/T_a$. The dust particles were assumed to be SiO_2, with a specific heat $C = 1.833\ C_p$.

The flow conditions behind shock waves propagating in clean ambient air and in hot dusty air ($\alpha = 1.0$ and $T_1 = 750$ K) have been evaluated as a function of shock overpressure. Results are presented in Fig. 4. Comparing flow conditions at equal shock overpressure, the presence of dust causes the following changes:

$$\frac{u_2}{a_1} \quad \frac{W}{a_1} \quad \frac{a_2}{a_1} \quad \frac{p_{02}}{p_1} \quad \text{(decrease)}$$

$$\frac{\rho_2}{\rho_1} \quad \frac{q_2}{p_1} \quad \text{(increase)} \tag{31}$$

The primary effect of dust is to decrease the fluid velocity behind the shock. This deceleration slows the propagation of the front. In fact, low-pressure shock waves in dusty gases propagate slower than the clean air sound speed (i.e., subsonic relative to a_a). As the shock strength approaches the acoustic regime, the shock Mach number approches 1 when referred to the appropriate dusty air ambient sound speed. Flow deceleration due to dust along with the dust mass tends to increase the density behind the shock, and consequently the mixture sound speed is lowered.

Normal shock reflections in clean and hot dusty air were calculated by the shock polar technique as follows. For a given shock pressure ratio P_{21}‡ the gas velocity was determined from the incident P-U shock polar

$$U_{21} = f_1(P_{21}) \tag{32}$$

and then A_{21} was determined from Eqs. (20) and (22). The

---

‡Two-digit subscripts denote states upstream and downstream of the shock, e.g., $P_{yx} = p_y/p_x$, and similarly for the other variables.

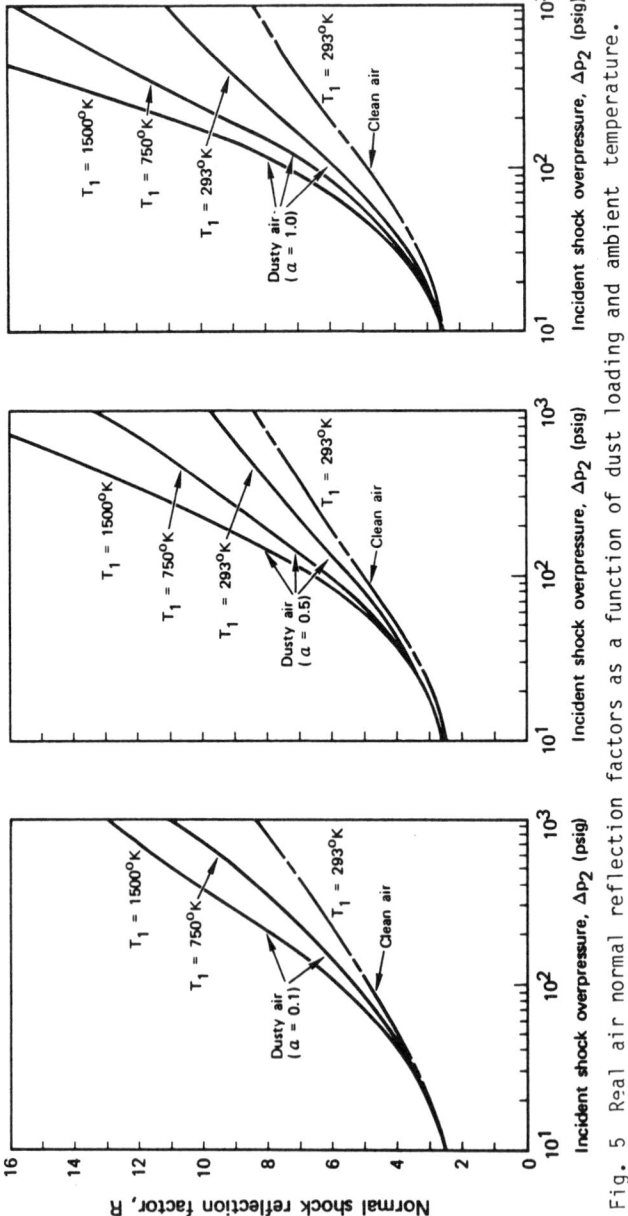

Fig. 5 Real air normal reflection factors as a function of dust loading and ambient temperature.

SHOCK WAVE PROPERTIES IN DUSTY AND CLEAN AIR

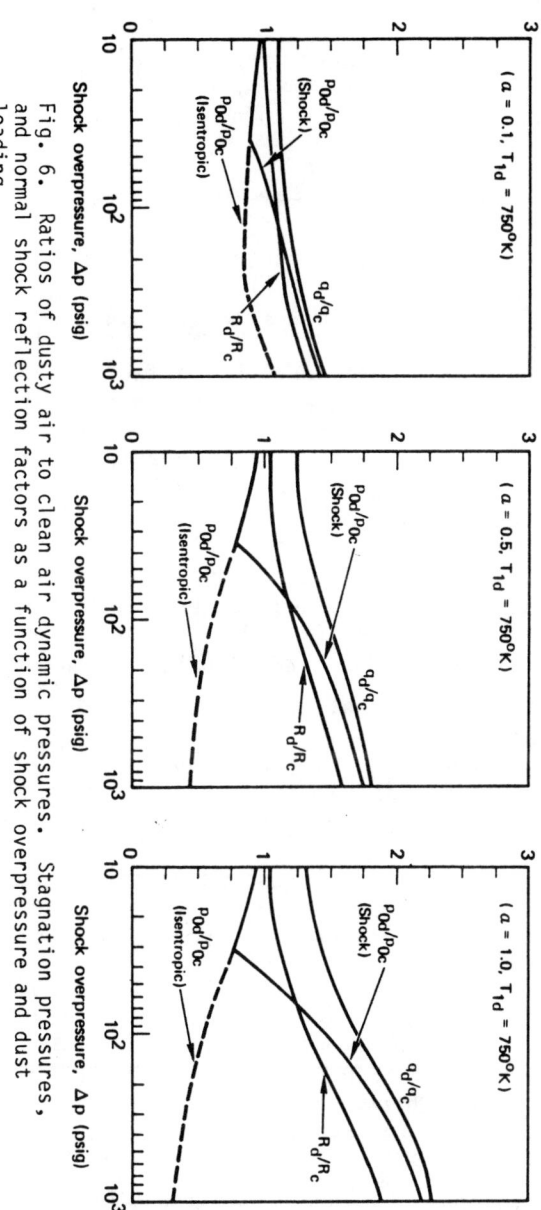

Fig. 6. Ratios of dusty air to clean air dynamic pressures, Stagnation pressures, and normal shock reflection factors as a function of shock overpressure and dust loading.

reflected shock is defined as that shock which decelerates the flow to zero velocity, $u_3 = 0$. Thus state 3 is related to state 2 by the following:

$$U_{32} = (u_3-u_2)/a_2 = -U_{21}/A_{21} \tag{33}$$

The P-U shock polar then determines the reflected shock pressure

$$P_{32} = f_2(U_{32}) \tag{34}$$

The normal shock reflection factor was calculated from

$$R = (p_3-p_1)/(p_2-p_1) = (P_{32}P_{21}-1)/(P_{21}-1) \tag{35}$$

Real air normal shock reflection factors are presented parametretically in Fig. 5 as a function of dust loading (α = 0.1, 0.5, 1.0) and ambient temperature. Examination of this figure shows that dust and ambient temperature effects can significantly increase the reflected shock pressures at high dust loadings.

The dynamic flow parameters are compared parametrically in Fig. 6. The ratios of dusty to clean air reflection factors and dynamic and stagnation pressures are compared for various dust loadings (α = 0.1, 0.5, 1.0) and one temperature ratio (T_1/T_{1d} = 293/750 K). We observe that at equal shock overpressures, dynamic pressures are increased by the pressence of dust in proportion to dust loading and ambient temperature T_{1d}. Although dust reduces gas velocities, the increase in total density of the shocked state more than compensates for this, causing a net increase in dynamic pressure.

Isentropic stagnation pressure decreases for dusty shocks relative to clean air values because the flow Mach number $M_2 = u_2/a_2$ decreases with increasing dust density and initial temperature. Lower isentropic stagnation pressures indicate more entropy change across dusty shocks; i.e., the dust introduces more irreversibilities into the flow. If the flow is supersonic and decelerated nonisentropically (e.g., through a stationary shock), the stagnation pressure increases - a result of the fact that dust increases the stationary shock pressures ratio P_{32}.

The above analysis investigated the combined effects of dust mass and enhanced preshock temperatures on equilibrium shock properties. To examine the effects of preshock temperature by itself, the dust mass and thermal capacity were

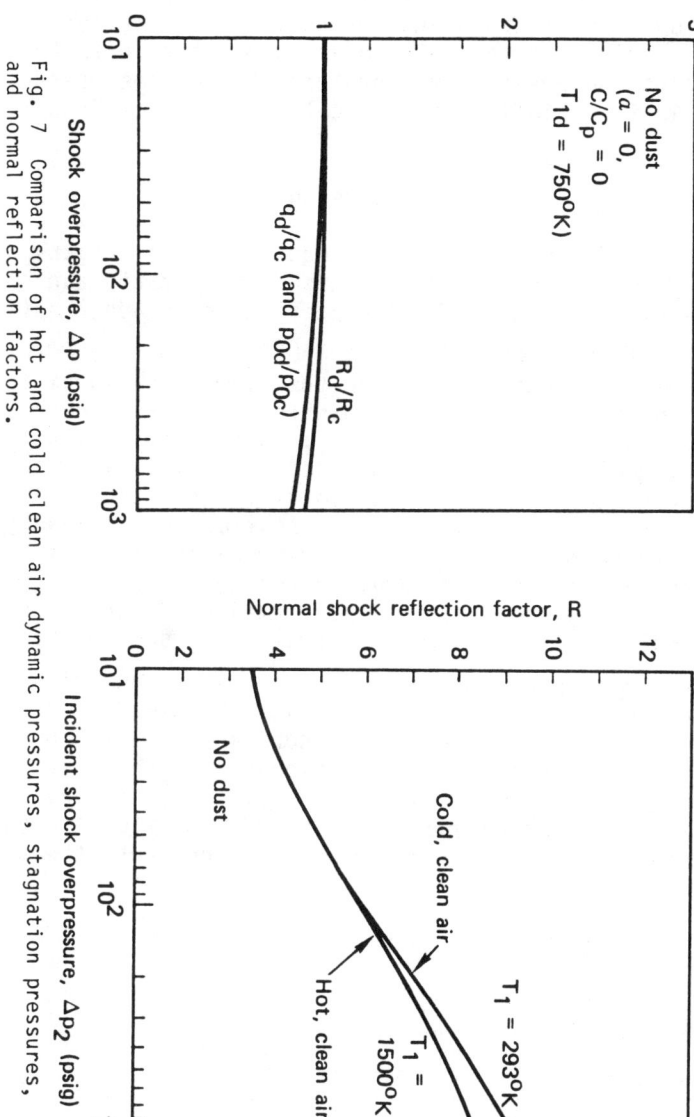

Fig. 7 Comparison of hot and cold clean air dynamic pressures, stagnation pressures, and normal reflection factors.

set to zero. Figure 7 gives a comparison of shocked flow properties for hot and cold clean air. This figure shows that large increases in the values of the preshock air temperature cause relatively small decreases in the values of the dynamic pressure and the normal shock reflection factor. This is a real air equation-of-state effect; increasing the preshock temperature changes the preshock γ, which, in turn, influences the calculated shock jump conditions.

Conclusions

It was found that velocity equilibration lengths depend linearly on the solid particle/gas density ratio and inversely on the particle drag coefficient. Spherical dust particles (from 1 to 10^3 µ) require from 100 to 1500 diam to reach velocity equilibrium with the flow, with specific values being dependent on the particle size and shock strength considered. According to this analysis, the velocity equilibration for 10-µ particles is less than about 7 mm. For many applications, flowfield gradients are small over length scales of ℓ_v; hence such flows can be analyzed most conveniently as equilibrium flows with the dust velocity and temperature equal to the gas values.

Equilibrium analysis was performed to determine shock flow conditions in a mixture of small SiO_2 dust particles and air. By performing comparisons of dusty and clean flows at equal shock overpressures, it was determined that the dynamic pressure, stagnation pressure and normal shock reflection factor increase significantly as the dust loading and preshock air temperature increase. The dust mass and thermal heat capacity should be included in flow analyses if the dust loading is greater than about one-tenth the ambient air density.

References

Carrier, G. F. (1955) Shock waves in a dusty gas. <u>J. Fluid Mech.</u> 4, 376-382.

Gilmore, F. R. (1955) Equilibrium composition and thermodynamic properties of air to 24,000 K. RM-1543, Rand Corporation, Santa Monica, Calif.

Hayashi, K. and Branch, M. C. (1983) Oblique shock waves in two-phase flow. <u>Shock Waves, Explosions, and Detonations: AIAA Progress in Astronautics and Aeronuattcs</u> (edited by Bowen, Manson, Oppemheim, and Soloukhin), Vol.87, pp.22-40. AIAA, New York.

Marble, F. (1970) Dynamics of dusty gases. *Annual Review of Fluid Mechanics* (edited by M. Van Dyke), Vol. 2, pp. 397-446. Annual Reviews Inc., Palo Alto, Calif.

Oppenheim, A. K. (1970) *Introduction to Gasdynamics of Explosions* (course No. 48 of the International Center for Mechanical Sciences held at Udine, Italy). Springer-Verlag, Berlin.

Owczarek, J. A. (1964) *Fundamentals of Gas Dynamics*, International Textbook Company, Scranton, Pa.

Schlicting, H. (1968) *Boundary Layer Theory* (6th ed.), p. 17. McGraw Hill, New York.

Von Schubert, B. S. (1969) Existence and uniqueness of normal shock waves in gas-particle mixtures. *J. Fluid Mech.* 28, 633-655.

Shock Waves in Water Induced by Focused Laser Radiation

O.G. Martynenko,* N.N. Stolovich,† G.I. Rudin,† and S.A. Levchenko‡
Academy of Sciences, Minsk, USSR

Abstract

The mechanisms of excitation of acoustic disturbances occurring at the interaction of a powerful laser radiation with absorbing liquids are discussed. Laser breakdown is the most effective mechanism underlying the excitation of strong shock waves. Wide-band (above 50 MHz) thin-film piezoelectric transducers were used to monitor pressure pulses which arise in underwater laser breakdown (explosion). Breakdown is initiated by monopulse ruby laser radiation (a duration level of from 0.5 to (4-5) x 10^{-8} s, and energy of from 0.1 to 1 J). The total time resolution of the transducer, including delays of the oscillographic and amplifying devices, is 30-40 ns. It is shown that in underwater laser explosions the first pressure pulse can be distinctly identified and is followed by several compression waves. A marked scattering in the values of peak pressures at the shock wave front as compared with the predicted pressures is noted. Hypotheses are discussed which, may furnish an explanation for quantitative deviations from experiment to experiment. An analogy between the hydraulic phenomena observed in underwater explosions of chemicals, electric discharge, and optical breakdown (explosion) is suggested.

Presented at the 8th ICOGER, Minsk, USSR, Aug. 23-26, 1981. Copyright © American Institute of Aeronautics and Astronautics, Inc., 1982. All rights reserved.

*Professor, Luikov Heat and Mass Transfer Institute.
†Candidate Science (Thermal Engineering), Luikov Heat and Mass Transfer Institute.
‡Junior Researcher, Luikov Heat and Mass Transfer Institute.

Introduction

Exposure of absorbing media, such as water, to laser radiation is an efficient means of exciting acoustic disturbances (Askariyan et al. 1963; Bunkin and Tribelsky 1980; Lyamshev and Sedov 1981). The following mechanisms contribute, depending on the specific conditions of the experiment, to the acoustic excitation of a liquid: thermal, evaporational, electrostrictional, and laser breakdown of substance.

The thermal mechanism present at any laser radiation intensity results from heating of the substance explosed to laser radiation and generates a pressure rise in the light energy absorption zone. Subsequently, these pressure changes propagate in the medium in the form of an acoustic wave (Bunkin and Tribelsky 1980; Lyamshev and Sedov 1981). The evaporational mechanism is of importance when the temperature of the surface heated by a laser beam focused on the surface exceeds the boiling temperature of the liquid. At this point surface evaporation, during which vapor escapes from the surface and a reactive response pulse, acting on the target, is formed, begins to play a considerable role in heat balance. There is a substantial difference between the thermal and evaporational mechanisms. While in the former the dimensions of the region of acoustic disturbances are determined by the diffusion velocity of the temperature field, in the latter this region is a relatively narrow-band the surface where the temperature is extremely high (i.e., in the very zone of radiation incidence).

The electrostrictional mechanism is especially important in nonabsorbing liquids for which thermal and evaporative processes are absent and hence no acoustic excitations are induced by these mechanisms. Laser breakdown of a substance is the most efficient mechanism of excitation of high-power shock waves. Breakdown has a pronounced threshold character in that it occurs at a specific radiation intensity for each substance. In a liquid it is accompanied by various processes, such as cavitation, cavity luminescence, and high-power shock wave perturbations.

Breakdown of a weakly absorbing liquid has been shown to be initiated by various fine inhomogeneities (dust, soot and other particles with a size of 1-10 μm) which have large optical radiation absorption coefficients described in the references.§ As a result, the temperature of such an inhomogeneity may be of the order of 10^4K; and the dust particle material and the adjoining liquid layer are converted to a plasma state. Since the plasma cloud temperature is sufficiently high, thermal radiation to the surrounding medium occurs, and the absorption of this energy leads to the heating of the medium. As the laser radiation energy is introduced during short periods of time (on the order of 10^{-8} s), and a very high pressure (10-100 kbars and more) region is formed in the plasma cloud, a powerful shock wave is generated, propagates throughout the liquid, and causes motion of the liquids.

Experimental Apparatus

An investigation of shock waves produced during optical breakdown in water was conducted on an apparatus which is schematically shown in Fig. 1. A pulse ruby laser (Soviet commercial holographic apparatus UIG-1m-I), produced under the modulated Q-factor conditions, monopulse generation at λ = 0.6493 μm with a duration level of semiheight) 40 ns ± 5% and with energy of 0.2-1 J, controlled by a solid calorimeter IKT-1m with a maximum error of measurement of 10%. Modulation of the Q factor of the pulse laser cavity was achieved by a passive gate cryptocyanine alcohol solution (the initial transmission coefficient of the modulator was 20%).

A short focal lens (5 cm in air) was used to focus the laser radiation in water which is contained in a plexiglass cuvette with built-in plane-parallel glass plate mirrors and with dimensions of 15 x 15 x 15 cm. Pressures were monitored with a pulse pressure gage with a very high time resolution and an upper boundary frequency of at least 50 cycles/s.π The total time resolution, including time delay of the oscillograph and amplifiers used in the experiments, was approximately 20-40 ns. The total error in recording of peak values (in the shock wave front) of pressure and their changes in time behind the shock wave front did not exceed 10%. The plane surface of a sensititve element of the pressure gage was located above the focal point and was normal to the laser beam axis.

---

§See Askariyan et al. (1963); Bell and Landt (1967); Bunkin and Tribelsky (1980); Buzukov and Teslenko (1970); Egerev (1980); Ioffe et al. (1970); Lyamshev and Sedov (1981); Teslenko (1977).

πThe pressure gage consisting of a thin-film piezoelectric transducer was developed and kindly supplied for our experiment by L. M. Dorozhkin.

SHOCK WAVES IN WATER INDUCED BY LASER RADIATION

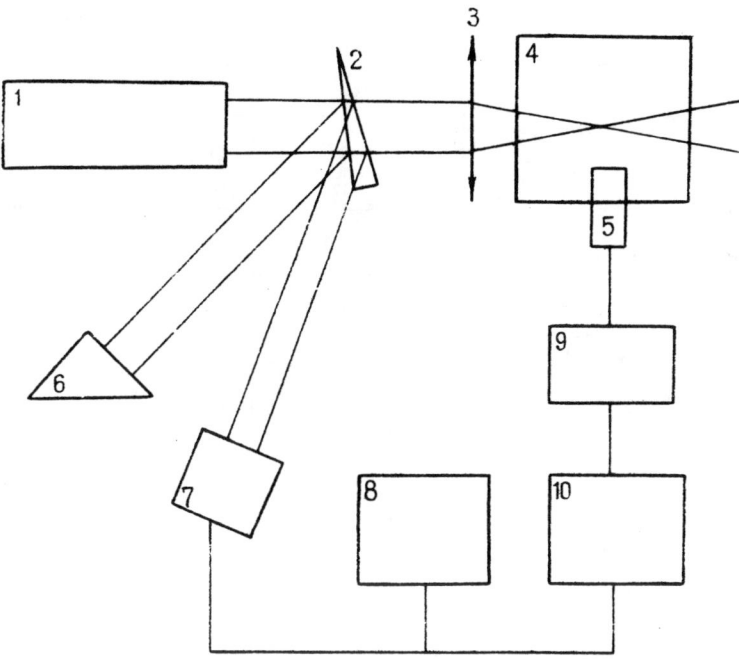

Fig. 1 Schematic of the experimental apparatus:
1) laser, 2) optical wedge, 3) lens, 4) cuvette,
5) pressure gage, 6) solid calorimeter IKT-IM,
7) photoreceiver FEK-15, 8) oscillograph C8-12,
9) broadband amplifier with the output signal fed
to the electron-beam tube of the oscillograph
OK-17(10).

Discussion of Results

The experimental results are presented in Fig. 2. The considerable scatter of the experimental points with respect to the corresponding predictions probably indicated that for each experiment a definate number of impurities (inhomogeneities) is present along the axis of the breakdown-initiating laser radiation. Upon interaction with radiation, these inhomogeneities become the source of the shock waves. The experimental data of Bell and Landt (1967) appear to support this hypothesis.

The presence of several sources of shock waves, randomly situated in the vicinity of the focal spot (the focusing lens caustic), must lead to the following consequences. As a result of the superposition of these shock waves at the observation point, pressure variations inside and behind the shock wave front, apart from the quantitative deviations

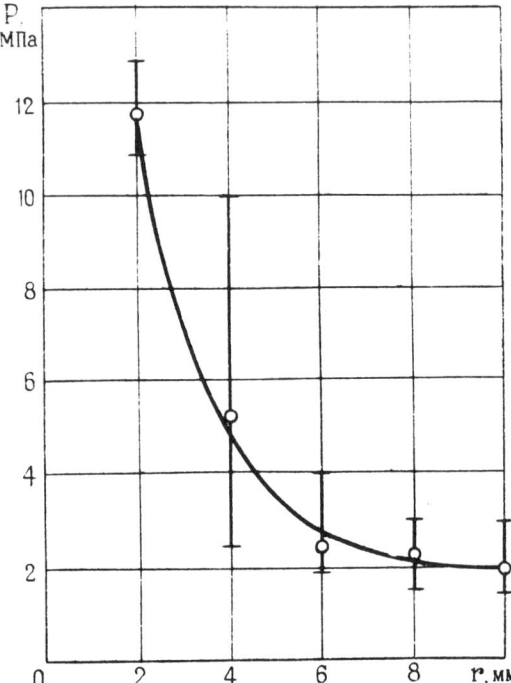

Fig. 2 Dependence of the peak pressure P_m in the shock wave front on the distance from the observation point r at radiation energy Q = 0.1 J.

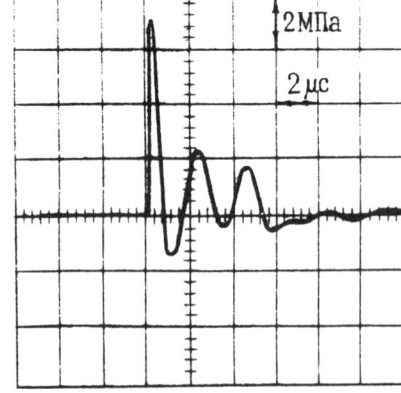

Fig. 3 Pressure oscillogram recorded with a gage in water at distance r = 4 mm at Q = 0.1 J.

in the course of the experiments, must be of a sufficiently complex form, as is confirmed by oscillographic investigations (see Fig. 3).

The pressure oscillogram behind the shock wave during an underwater laser explosion, presented here as an example, show that, just as in the case considered by Teslenko

(1977), the distinctly detectable first pressure pulse is followed by a few more compression waves. This can be adequately explained by a nonsimultaneous appearance of disturbing centers in the focal spot zone.

This hypothesis is confirmed by experiments with distilled and filtered water. With purified water the scatter of the shock wave pressures and the intensity of secondary compression waves are noticeably reduced. The data of other authors also support this observation.§

Moreover, under the conditions of this work, the nonlinear effects of radiation interaction with the liquid, together with the spatial radiation inhomogeneity and insufficient liquid purity, which cause the instability effects, must lead to a breakdown strictly at the focal point (Buzukov and Teslenko 1970; Teslenko 1977). This in turn causes the ambiguity of specific threshold powers and the corresponding radiation field strengths, and provides a possible reason for a considerable scatter of pressures in the site of the shock wave front in the course of the experiments.

Within the framework of the present study no consideration was given to vapor-gas cavity dynamics of the underwater laser explosion. As in the case of chemical explosive explosions and electric discharge, the medium is subjected to several pulsations accompanied by generation of sufficiently powerful compression waves which attack the cavity. While in the underwater explosions of explosives and electric pulse high-current discharge the densities of energy and power are approximately equal [10^3-10^4 J/cm^3 and 10^8-10^9 W/cm^3 being the energy and power upper estimates, respectively (Letokhov and Ustinov 1980)], the densities of energy and power of underwater laser explosion may reach the values of 10^{10}-10^{12} J/cm^3 and 10^{20}-10^{22} W/cm^3, equal to or even exceeding those of nuclear explosions even though the total energy is small (Letokhov and Ustinov 1980). Therefore the explanation of certain differences in hydrodynamics of the above-mentioned underwater explosions should probably be sought in the differences of velocities, peculiarities, and achieved densities of the energy of explosions, as was previously mentioned by Bunkin and Tribelsky (1980), Lyamshev and Sedov (1981), Egerev (1980), and Letokhov and Ustinov (1980).

---

§See Askariyan et al. (1963); Bell and Landt (1967); Bunkin and Tribelsky (1980); Buzukov and Teslenko (1970); Egerev (1980); Ioffe et al. (1970); Lyamshev and Sedov (1981); Teslenko (1977).

References

Askariyan, G. A., Prokhorov, A. A., Chanturiya, G. F. and Shipulo, G. P. (1963) Beam of an optical quantum generator in liquid. Zh. Eksp. Teor. Fiz. 44, 2177-2182.

Bell, C. E. and Landt, J. A. (1967) Laser-induced high-pressure shock waves in water. Appl. Phys. Lett. 10, 46-68.

Bunkin, F. V. and Tribelsky, M. E. (1980) Nonresonant interaction of powerful optical radiation with liquid. Usp. Fiz. Nauk. 130, 193-239.

Buzukov, A. A. and Teslenko, V. S. (1970) Pressure at the shock wave front in the near zone of laser spark breakthrough in water. Zh. Prikl. Mekh. Tekh. Fiz. 11, 123-124.

Egerev, S. V. (1980) Optical excitation of ultrasonic pulses in a liquid. Acoustic Institute of the USSR, Academy of Sciences, Moscow, USSR.

Ioffe, A. I., Melnikov, N. A., Naugolnykh, K. A., and Upadyshev, V. A. (1970) Shock wave on optical breakdown in water. Zh. Prikl. Mekh. Tekh. Fiz. 11, 125-127.

Letokhov, V. S. and Ustinov, N. D. (1980) High-power lasers and their applications. Izd. Sov. Radio, Moscow, USSR.

Lyamshev, L. M. and Sedov, L. V. (1981) Optical sound generation in liquid. Thermal mechanism. Akust. Zh. 27, 5-29.

Teslenko, V. S. (1977) Study of light-acoustic and light-dynamic parameters of laser breakdown in liquids. Kvantovaya Elektron. 4, 1732-1737.

Ignition of Small Particles Behind Shock Waves

V.M. Boiko,* A.V. Fedorov,† V.M. Fomin,‡ A.N. Papyrin,§ and R.I. Soloukhin^π
Academy of Sciences, Novosibirsk, USSR

Abstract

The results of a theoretical and experimental study of motion and ignition of solid particles behind shock waves are reported. Rapid pulse methods of laser diagnostics are described. The drag coefficients of particles are found; and the volumetric concentration effect on particle acceleration in a supersonic gas flow is determined. A mathematical numerical model is proposed to describe particle ignition behind a shock wave, and a comparison is made between theoretical and experimental data for the magnesium particle ignition delay times.

Nomenclature

a = sound velocity
C_D = drag coefficient
C_{of} = volume concentration of oxidant in a gas
C = specific heat of substance
C_p = heat capacity
d = diameter
D = shock wave velocity
d_m = mean diameter
E = activation energy
K_o = pre-exponential factor
l = distance between particles
L = latent heat of vaporization

Paper presented at the 8th ICOGER, Minsk, USSR, Aug. 23-26, 1981. Copyright © American Institute of Aeronautics and Astronautics, Inc., 1982. All rights reserved.
  *Research Assistant, Institute of Theoretical and Applied Mechanics, Novosibirsk.
  †Senior Scientist, Institute of Theoretical and Applied Mechanics, Novosibirsk.
  ‡Deputy Director, Institute of Theoretical and Applied Mechanics, Novosibirsk.
  §Laboratory Head, Institute of Theoretical and Applied Mechanics, Novosibirsk.
  ^πDirector, Heat and Mass Transfer Institute, Minsk.

Le = Lewis number
M = shock Mach number
M* = relative Mach number
Nu = Nusselt number
P = Pressure
P_* = characteristic pressure
q = reaction heat effect per unit consumed oxygen mass
Re* = relative Reynolds number
R = universal gas constant
t = time
T = temperature
v = velocity
x = coordinate
α = thermal diffusivity
δ = oxide film thickness
ε = radiating surface emissivity
ϕ = volume concentration of particles
λ = thermal conductivity
ρ = density
σ = Stefan-Boltzmann constant
τ_i = ignition delay (induction) time
τ_b = burning time

Subscripts

s = particle parameters
1 = initial conditions
2 = shock conditions

Introduction

In the study of heterogeneous combustion and detonation in gas-droplet and gas-solid particle systems, practical problems associated with explosion prevention in industrial dust-air mixtures are of great interest. The theoretical analysis of the problem involves essential difficulties of diverse complex nonequilibrium physical and chemical processes responsible for the development of fast combustion and detonation in gas suspensions. In constructing numerical mathematical models of detonation processes in gas-particle systems, many basic details associated with acceleration, disintegration, heating, ignition, and burning of particles behind the shock wave will be considered, and the characteristic times of these processes will be established. Up to now, a considerable amount of research has dealt with different aspects of the problem. However, only detonation in gas-droplet mixtures and pertinent problems of detonation and ignition of liquid particles have been investigated thoroughly (see, for example, Gelfand

1977). The laws of ignition of solid particles behind shock waves are more obscure, and the scant amount of studies in this field (Fox et al. 1976; Strauss 1968; Ryzhik 1980) do not allow a reliable comparison of the results to be made, particularly on the ignition time delays.

The present contribution reports some experimental results on motion and ignition of solid particles behind shock waves obtained with the use of rapid pulse laser diagnostic techniques. The drag coefficients of particles are obtained; and the effects of the particle number concentrations on their motion in a supersonic gas flow are demonstrated. A numerical model of particle ignition is suggested, and the experimental and theoretical data on the ignition time delays are compared.

Experimental Diagnostic Techniques

The experimental setup is shown in Fig. 1. Its main elements include a shock tube with gas pumping and injection systems, a laser diagnostic complex to observe particles in a gas flow, and a synchronization and control systems.

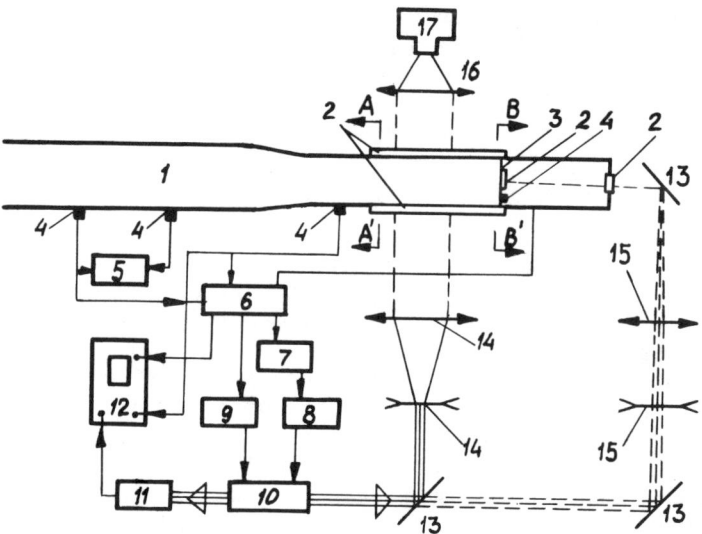

Fig. 1 Experimental setup: 1) driven shock chamber; 2) windows; 3) shock reflecting wall; 4) piezoelectric pressure gages; 5) chronometer; 6) clock pulse generator; 7) master oscillator; 8) high-voltage pulse generator; 9) flash lamp unit; 10) laser; 11) photoelectric calorimeter; 12) oscilloscope; 13) mirrors; 14) telescopic system; 15) system generating "plane" light beam; 16) lens; 17) streak camera.

The shock tube consists of driver and driven sections, respectively 1.5 and 4.5 m long, separated by a pneumatically monitored valve. The driven section is a 7.8-cm-diam tube smoothly connected with a square 56 x 56 mm^2 section.

Solid particles have been introduced into the gas flow in two ways. In the first version, they are delivered instantaneously with a striker on the bottom wall which tossed a required number of particles to the prescribed height. In the second version, a vertical flow of free-falling particles of uniform concentration was formed with the aid of a vibrating grid. The specially shaped diaphragm being set on the top tube wall, a cloud of particles can be obtained with a given geometry and concentration. The test section was supplied with glass windows (2 x 200 mm^2) for optical measurments; and piezoelectric pressure gages and an electronic chronometer were designed for shock wave velocity measurements and for the diagnostic designs synchronization.

Helium at from 30 to 100 atm was used as a driver; and the oxidants (air, oxygen) were utilized at the initial pressure p_1 =0.05 + 1 atm. The initial shock Mach number M = 2.5 + 5 was obtained in the experiments.

The dynamic behavior of particles behind shock waves has been studied with the use of laser visualization (shadow or schlieren techniques) as well as with the scattered light, the so-called laser "knife" method. A ruby laser, operating in a stroboscope mode by the use of a periodically modulated Kerr cell, served as the light source. The stroboscopic laser system was capable of providing a series of 3.5 x 10^{-8} s pulses (up to 50). The time intervals between pulses were regulated within 3-500 µs as accurately as 0.2 µs. A cloud of moving solid particles was recorded by schlieren photography, permitting particle image representation together with registration of gas density perturbations. A multi-frame record was obtained with the use of a high-speed camera. The exposure time and frame frequency were governed by laser radiation parameters. For obtaining particle trajectories, the laser "knife" method is more convenient, using multi-exposure registration (Boiko and Papyrin 1981) with a series of exposures fixed on a frame. Intrinsic radiation of igniting and burning particles was observed with the aid of a conventional streak camera. As the duster operated in the 10^{-1} + 10^{-3}-s time period necessary for a cloud of particles of a certain configuration to be formed, an electric signal was set to the pneumatic valve control system, and the laser light pulses were generated at the required moment by controlled signals of piezoelectric gages supplied to the inlet of a six-channel time delay generator via the conventional amplifiers.

Particle Motion Behind the Shock Wave

Consider the experimental results concerning some aspects of small particle acceleration behind the shock wave. The drag coefficient of a particle motion, C_D, is one of the important parameters, but the available data on C_D and $C_D = f(Re^*, M^*)$ (see, for example, Yanenko et al. 1980) are inconsistent, showing discrepancies in scanty shock tube

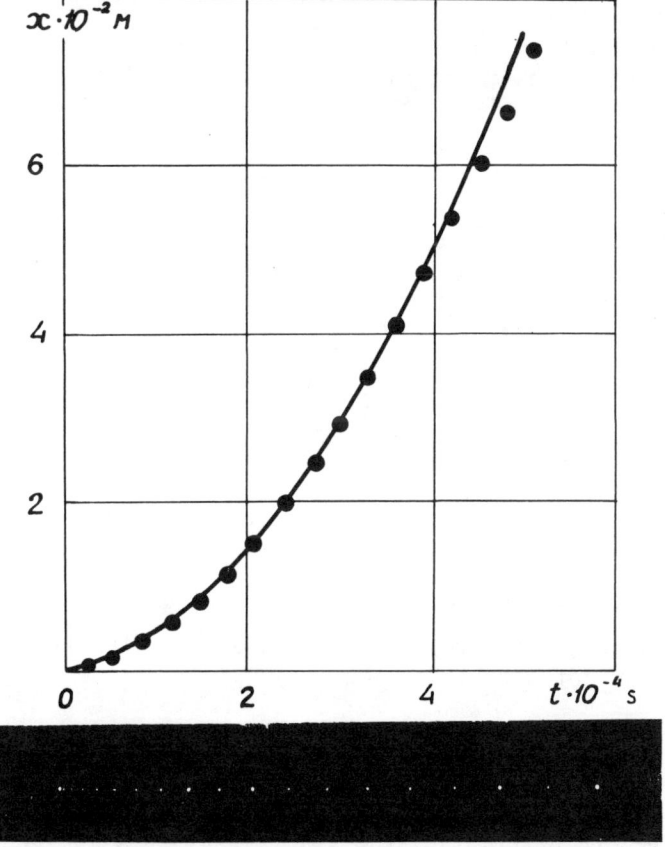

Fig. 2 Relationships between pressure ratios in reflected (Θ_2) and incident (Θ_1) shock waves propagating through continuous and two-phase media. 1) Acoustic waves in a continuous liquid; 2) nonisothermal gas; 3) isothermal gas; 4) two-phase isothermal medium with an insoluble and uncondensable gas; 5), 6) two-phase fluid with condensable or soluble gas; $\beta_5 > \beta_6$.

experiments (Crowe 1967; Selberg and Nickolls 1968; Bailey and Hiatt 1972; Rudinger 1965). Also, an important problem concerning particle concentration effect on their acceleration in a supersonic gas flow is still unclear. These factors have promoted new studies using the laser diagnostic techniques with high space and time resolutions.

The equation of a single spherical particle motion in a viscous fluid flow is of the form (Saw 1971)

$$\frac{1}{6}\pi d_s^3 \rho_s \frac{dv_s}{dt} = F_D + F_P + F_M + F + F_T \quad (1)$$

A simple analysis shows that the aerodynamic drag

$$F_D = (1/8)\pi d_s^2 \rho_R C_D |v_2 - v_s|^2 \quad (2)$$

makes the dominant contribution to solid particle acceleration. Thus, according to a simple estimate (Yanenko et al. 1980) the effect of

$$F_P = \frac{1}{6}\pi d_s^3 \frac{\partial P}{\partial x} \quad (3)$$

due to the pressure gradient in the shock front can be neglected as compared to F_D. This results from the particle flight time over the shock front area being much less than the relaxation time due to F_D. The force

$$F_M = \frac{\pi d_s^3}{12} \rho_2 \frac{dv_s}{dt} \quad (4)$$

including the joined mass and the Basse force

$$F_B = -\frac{3}{2} d_s^2 \sqrt{\pi \rho_2 \mu_2} \int_0^t (t-\tau)^{-1/2} \frac{dv_s}{d\tau} d\tau \quad (5)$$

are insignificant since $\rho_s \gg \rho_2$. Allowing for the smallness of the gravity force for the particles with $d_s = 1 \div 100$ μm as compared to F_D, Eq. (1) can be written as

$$\frac{1}{6}\pi d_s^3 \rho_s \frac{dv_s}{dt} = \frac{1}{8} \pi d_s^2 \rho_2 C_D |v_2 - v_s|^2 \quad (6)$$

Hence the characteristic velocity relaxation time for the particle affected by the aerodynamic drag force is

$$\tau_D \cong \rho_s d_s / [C_D \rho_2 (v_2 - v_s)] \quad (7)$$

Double integration of Eq. (6) yields

$$x = v_2 t - \left[\frac{3\rho_2}{4d_s \rho_s} C_D \log 1 + \frac{3\rho_2}{4d_s \rho_s} C_D v_2 t\right]^{-1} \quad (8)$$

at $t = 0$, $x = 0$, $v_s = 0$.

Figure 2b shows a representative multiexposure picture obtained by the laser "knife" technique and illustrating single particle acceleration (M = 2.6; ρ_s = 8.6 g/cm^3; d_s = 180 μm) behind the shock wave. The comparison of the experimental and calculated data, $x = f(t)$, for $C_D = 1$ is given in Fig. 2a and shows their good agreement. The values of C_D = 0.6 - 1.0 for Re* = (1.3-1.9) x 10^4 and M* = 0.8 + 1.2 are obtained owing to the interpretation of a great number of such records. These data agree well with the results for C_D measured for the supersonic two-phase nozzle (Yanenko et al. 1980) and shock tube (Crowe 1967) flows. This supports the validity of the relation $C_D = f(Re^*, M^*)$ (Yaneko et al. 1980) suggested by Henderson (1976) for calculating particle motion in a supersonic gas flow.

The above considerations hold true for the low volumetric disperse phase concentrations, ϕ, when particle interactions and their effects on the gas state could be neglected. The laws of particle motion become rather complicated with ϕ increasing to some value of ϕ^*. In this case, as shown in Yanenko et al. (1980) the constraint effects become essential, causing changes in the gas state and two-phase particle flow parameters. These effects may be attributed to changes in the wave structure arising in a flow around a particle cloud.

In the shock tube experiments, at M > 2, a supersonic flow around a particle is observed at the initial stages of their acceleration. Therefore interacting shock waves are formed around each particle. Under certain conditions, these waves may overlap each other and create a collective shock front. The value of ϕ at which this effect is expected may be estimated from the following considerations. Examine the particles in one plane normal to the incident gas flow. Owing to the work of Blagosklonov et al. (1979), to assume that the collective front shock is formed when transonic shock layers of individual particles are overlapped. Then, the condition for its appearance is $1/d_s < 1^*/d_s$ where 1^* is the characteristic size of the transonic region depending on the relative flow Mach number of a

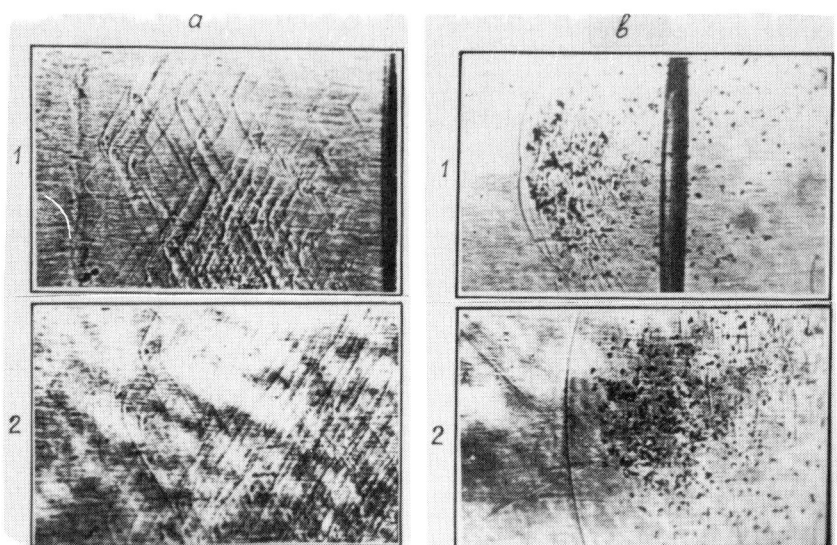

Fig. 3 Schileren records illustrating change in the wave structure at different volumetric concentrations of cloud particles: M = 2.7; p_1 = 1 atm; time intervals between frames 1 and 2 are Δt = 60 µs.
a) Bronze particles, d_{sm} = 80 µm, ρ_s = 8.6 g/cm³; $\phi 1_3$ = 0.3%.
b) Magnesium particles, d_{sm} = 300 µm; ρ_s = 1.74 g/cm³; ϕ_1 = 2%.

moving particle. In shock experiments, the maximum M* values are attained at an initial stage of acceleration and constitute 1.1 + 1.3 for typical conditions in the described experiments. Then, according to Blagosklonov et al. (1979), the value of l^*/d_s is estimated to be about 5. Hence the collective shock wave is to expected at $l/d_s < 5$, the volumetric particle concentrations uniquely related to $1/d_s (1/d_s \sim \phi^{-1/3})$ being about 1%.

Multiframe schlieren pictures obtained by studying the dynamics of a particle cloud with different ϕ (see Fig. 3) support these considerations. With initial concentrations, $\phi_1 \cong 0.3\%$ (Fig. 3a, frames 1 and 2), only local shocks around individual particles are seen. At $\phi_1 \cong 2\%$ (Fig. 3b, frames 1 and 2), the wave configuration changes. A collective front shock appears, and local bow shock angles increase. It is clear that under this condition, the gas state parameters in the gas suspension cloud are changed together with particle flow characteristics.

Special experiments were performed by registering the particle velocity change effect at $\phi \geq 1\%$ in order to study the cloud dynamics, with the dispersed phase concentrations ranging between 0.1 to 10%. The particle cloud was formed as a vertical column of free-falling particles and constituted a transverse cross-section 2 x 10 mm^2 rectangle with its long side normal to the incident flow. Figure 4 presents a set of multiframe pictures obtained for two particle concentrations at $\phi_1 < 1\%$ and $\phi_1 > 1\%$.

Because of the difficulties involved in experimental observations of individual particle motion at large concentrations, another technique was chosen to measure the shift of the cloud front relative to the incident flow. As is seen in Fig. 4, at the beginning of motion the longitudinal "spreading" of the cloud proceeds owing to polydispersed composition of particles, with the smaller particles accelerating and the large ones remaining at the front boundary. It may be observed in all of the frames in Fig. 4, as the velocity scattering with d_s changing in a certain range is less for larger particles. In view of these observations, the velocity of the cloud front is assumed to be close to the larger particle velocity.

Figure 5 presents experimental data for acceleration of single magnesium particles of d_{so} = 300 μm, ρ_s = 1.74 g/cm^3, at $\phi_1 \ll 1$. The figure also plots a history of the polydispersed particle motion at d_s = 5 + 300 μm, with the largest sizes, d_{so}, for the cloud front. It is seen that an appreciable discrepancy in the acceleration of "single" and collective" particles begins at $\phi > 1\%$ and ϕ_1 = 6% for t = 150 μs, $\Delta x/x \cong 30\%$.

Thus the experimental data show that with the particle cloud accelerating behind the shock wave, the description of the single particle motion by Eq. (1) is valid at $\phi < \phi^*$ 1%. Also, at $\phi > 1\%$ ($1/d_s$ < 5 s) the flow "constraint" effects become important, causing a decrease in velocity as compared to "single" particle motion. A detailed analysis of these effects must allow for $\phi^* = f(M^*)$ varying with particle acceleration (Blagosklonov et al. 1979).

Ignition of Particles

As far as detonation in the gas-solid particle mixture is concerned, the ignition of particles measuring between 1 and 100 μm with small induction times of $\tau_i \sim 10^{-3}$-10^{-5} s (Fox et al. 1976; Strauss 1968; Ryzhik et al. 1980) is shown

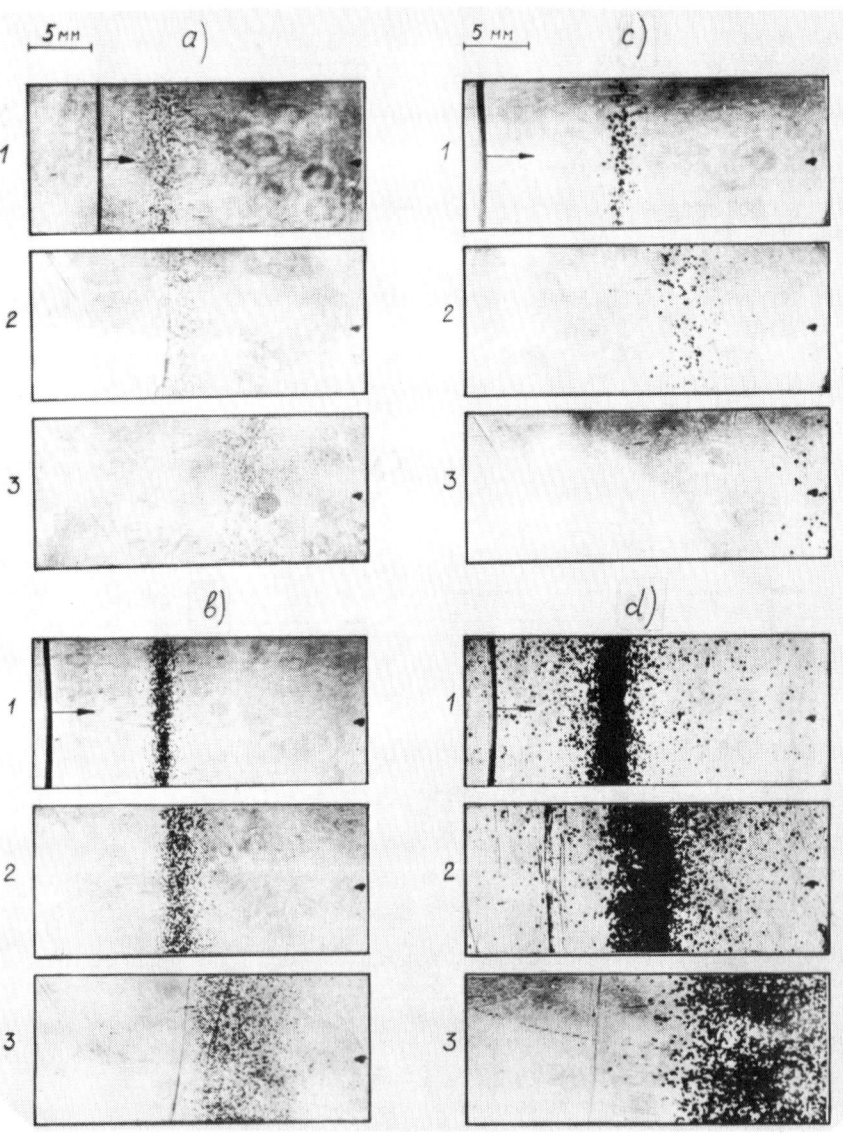

Fig. 4 Schileren records of the particle cloud motion at different initial concentrations: M = 2.8; p_1 = 1 atm; time intervals between frames 1, 2, and 3 are 100 μs. a), b) Bronze particles, d_s = 40-100 μm; ρ = 8.6 g/cm^3; a)$\phi_1 \sim 0.1\%$, b) $\phi_1 \sim 3\%$. c), d) plexiglass particles, d_s = 100-500 m; ρ = 1.2 g/cm^3; c)$\phi_1 \sim 0.3\%$, d)$\phi 1 \sim 10\%$.

to be most interesting. For the τ_i estimation, most promising are the shock tube studies, which provide highly accurate measurements owing to the "instantaneous" injection of particles into a high-temperature gas flow. Consider the problem of particle ignition in a high-temperature gas flow behind the incident shock wave. The particles behind the shock propagating in a suspension of particles in the rest gas are accelerated and heated because of gas-to-particle heat transfer and oxidation. Assume that all particles are spherical and of the same size, their concentration being low ("single" particle regime and disintegration absent). Also, setting as valid other conventional assumptions for independent gas and particle continua (Yanenko et al. 1980), and relying on the conservation law for each of the phases, write the equations

$$\frac{d\rho_c}{dt} + \rho_c \frac{\partial v}{\partial x} = 0 \qquad (9)$$

$$\frac{\pi}{6} d_s^3 \rho_s \frac{dv_s}{dt} = \sum_i F_i \qquad (10)$$

$$\frac{\pi}{6} d_s^3 \rho_s C_s \frac{dT_s}{dt} = Q_c + Q_{ch} + Q_r + Q_v + Q_{int} \qquad (11)$$

The heat flux in the heat balance equation includes the convective heat transfer to gas,

$$Q_c = \pi d_s^2 \lambda \, Nu(T - T_s) \qquad (12)$$

the radiation losses,

$$Q_r = \pi d_s^2 \varepsilon \sigma (T_s^4 - T^4) \qquad (13)$$

the chemical reaction contribution Q_{ch} dependent on the metal oxidation rates and oxide film properties (Merzhanov 1975); heat losses for particle vaporization,

$$Q_v = \frac{\lambda L}{gC_p} \frac{\pi d_s^2}{Le} \frac{P_*}{P} \exp\left(-\frac{L}{RT_s}\right) \qquad (14)$$

and heat losses for heating and melting of a particle,

$$Q_{int} = 2\pi d_s \lambda_s (T_s - T) \exp\left(-\frac{12\alpha_s}{d_s^2} t\right) \qquad (15)$$

For the magnesium particles, the expression for heat influx due to chemical reaction has the form (Gurevich and Stepanov 1968)

$$Q_{ch} = \pi d_s^2 q \rho_s C_{ok} K_o \exp(-E/RT_s) \qquad (16)$$

In estimating the induction time, three ignition mechanisms have been taken into account; namely, 1) the disruption of thermal equlibrium (Merzhanov 1975; Blashenko et al. 1974); 2) an achievement of critical ignition temperature (Ezhovsky et al. 1979); and 3) the specific mechanism for including melting and disintegration of particles when moving in a high-velocity gas flow (see Fox et al. 1976).

The ignition problem is solved numerically using equation system (1) with the following initial conditions:

$$t = 0 \qquad v_s = D \qquad T_s = T_1 \qquad \delta = \delta_o$$

Thus the problem of estimating the velocity, temperature, and thickness of the oxide film of a particle moving in a high-temperature gas flow behind a shock wave has been reduced to the Cauchy problem, implying estimates of the functions $(v(t), T(t),$ and $\delta(t)C'(0,\tau_i))$ to agree with Eqs. (9-11) and initial conditions in $[0,\tau_i]$. The A-stable numerical Gear method has been used for solution.

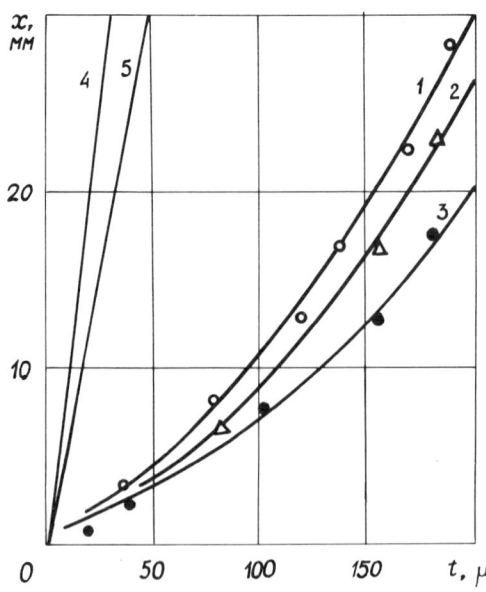

Fig. 5 x-t diagram of particle cloud front motion vs concentration ϕ: M = 2.7; P_1 = 1 atm; d_s = 40-200 μm; ρ_s = 1.2 g/cm^3; 1) $\phi_1 \cong$ 1%; 2) $\phi_1 \cong$ 3%; 4) x-t diagram for the shock front, 5) x-t diagram for the gas flow; 0 - values of x = f(t) for monodispersed particles; d = 200 μm at $\phi_1 \ll$ 1 ("single" particle regime).

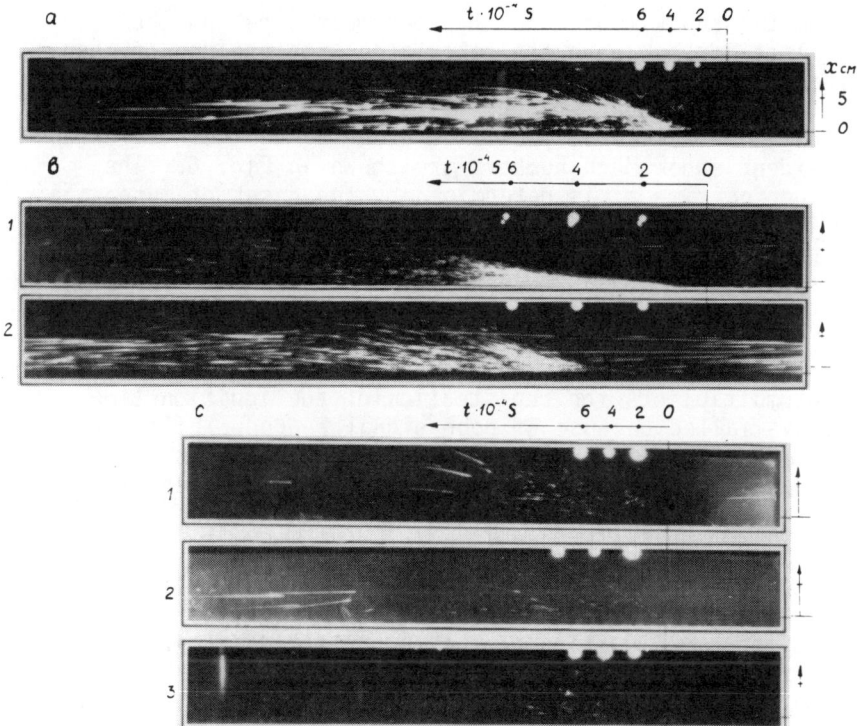

Fig. 6 Representative photographic sweeps of magnesium particle ignition and burning behind the reflected shock. a) d_s = 40 + 100 μm, M = 4.7, p_1 = 0.62 atm; b) M = 4.4, P_1 = 0.1 atm; frame 1, d_s = 1 + 40 μm; frame 2, d_s = 40 + 100 μm. c) d_s = 90 μm, frame 1, M = 4.4, P_1 = 0.1 atm; frame 2, M = 3.9, P_1 = 0.2 atm; frame 3, M = 3.4, P_1 = 0.3 atm.

The numerical modeling shows a great dependence of the ignition delay on the velocity nonequilibrium effects in the gas and particle motion, in addition to other factors. Therefore it seems reasonable to eliminate this effect at the first stage of studies in order to verify the thermal particle ignition theory in the shock tube experiments and to choose correctly the values of the kinetic rate constants q, K_o, and E. To this end, experimental observations have been made behind the reflected shock waves, where gas and particles are almost at rest. The experiments were performed with magnesium particles for which ignition under "static" conditions has been studied most thoroughly (Pokhil et al. 1972; Gurevich and Stepanov 1968; Ezhovsky et al. 1979).

A cloud of the magnesium particles containing in total from 10 to 100 particles was injected at a distance of 10 mm

from the shock tube end. Gas parameters behind the
reflected shock wave are calculated with the use of the
measured initial shock wave velocity data.

Typical streak records illustrating the ignition and
burning of magnesium particles dpending on their sizes and
incident shock Mach numbers are shown in Fig. 6. The
induction time τ_i is determined by the onset of intensive
emission relative to the time of the shock reflection. To
estimate τ_i and calibrate the seep duration, a set of three
laser pulses with a 200 μs interval between them is supplied
to the streak camera. 200 μs later, as the signal appeared
at the tube end pressure gage (Fig. 1) they are easily seen
at the top of each frame as three spots. The attempt to use
photomultipliers for registration of the ignition time
delays failed because of poor signal reproducibility.

The streak records presented in Figs. 6a and 6b are
obtained for fractions of polydispersed particles with
different d_{s_m} and a certain size deviation Δd_s. Glowing
traces which correspond to the burning of individual
particles are easily seen; the glowing duration and
intensity are increased with time. This may be attributed
to successive ignition of the larger sized particles, as the
burning time, τ_b, of the magnesium particles is in proportion to their squared diameter (Derevyaga et al. 1978).

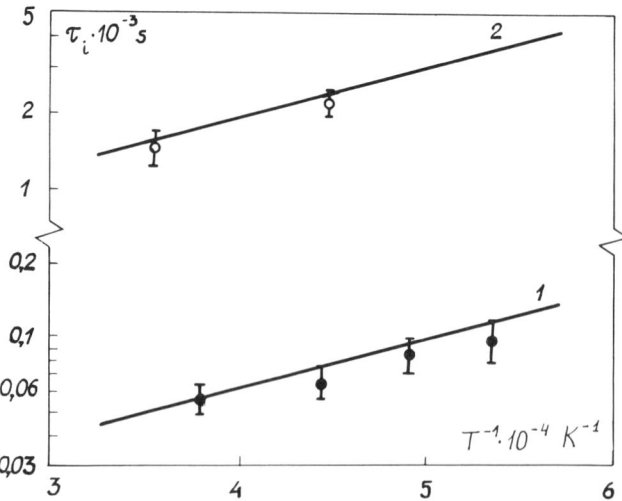

Fig. 7 Magnesium particle ignition induction time vs temperature
behind the reflected shock: 1), 2) theoretical $\tau_i = f(1/T)$ for d_s =
15 and 90 μm, respectively; ⊕, experimental data for particle d_s =
1 + 40 μm; ⊖, for particles d_s = 90 μm (mean values and scatter of
τ_i are given).

These pictures give data on $\tau_i = f(\tau_b)$ and allow $\tau_b \sim d_s^2$ to be plotted, taking into account that $\tau_i = f(d_s)$. However, it is difficult to eliminate the influence of the factors associated with successive excitation of ignition by the particles already ignited.

In this connection, experiments were performed with fractions of monodispersed particles (scattering in sizes being no more than 5%). Typical records for particles with d_s = 90 μm at different incident shock Mach numbers are presented in Fig. 6c. The total number of the injected particles in the field of observation amounts to 10. It is easily seen that τ_i increases with decreasing M, so that no ignition is observed at M = 3.4 (Fig. 6c, frame 3).

Note that a weak flash of shorter length tracks is registered at the beginning of the streak record (Fig. 6c). This is presumably attributed to smaller particles ($d_s \approx 1$ μm) being retained on initial particles (d_s = 90 μm) on walls by electrostatic forces and flying off as the shock wave passed by.

The results of interpreting a large number (about 50) of such records are plotted in Fig. 7 as $\tau_i = f(1/T)$ for d_s = 15 and 90 μm. For d_s = 15 μm, τ_i is estimated by the ignition delay in the polydispersed fraction d_s = 1 + 40 μm at d_s = 15 μm because of certain difficulties in separating

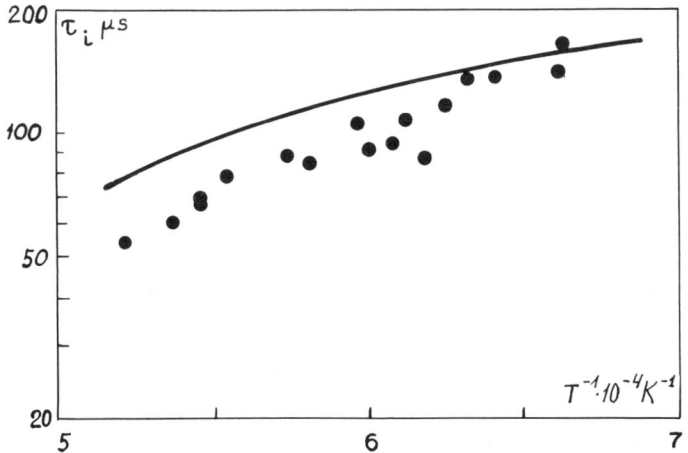

Fig. 8 Comparison of theoretical and experimental (Fox et al. 1976) data for $\tau_i = f(1/T)$ behind the incident shock wave, magnesium particles d_s = 17 μm, P_1 = 0.135 atm.

these particles. The same figure illustrates the calculated dependences of these particles. The values of kinetic constants were taken from Gurevich and Stepanov (1968) to be $q = 9.18$ cal/g, $E = 4.5 \times 10^4$ cal/mole, and $K = 0.86 \times 10^{11}$ cm/s. Theoretical and experimental τ_i induction times are seen to agree satisfactorily. The increase in τ_{ie} as compared to τ_{ip} for particles of d = 15 μm with falling temperature may be attributed to the presence of smaller particles with lower critical temperature in this fraction.

Figure 8 shows the comparison of theoretical and experimental data for τ_i (Fox et al. 1976) obtained in experiments on the ignition of magnesium particles with d = 17 μm behind the incident shock wave. The dependence, C_D = f(Re*, M*) is determined by the Henderson (1976) formula.

It should be stressed in conclusion that the experiments performed support the validity of the developed physical and numerical model for describing particle motion and ignition behind a shock wave.

References

Bailey, A. B. and Hiatt, J. (1972) Sphere drag coefficient for a broad range of Mach and Reynolds numbers. AIAA J. 10, 1436-1440.

Blagosklonov, V. I., Kuznetsov, V. M., Minailos, A. N., Stasenko, A. L., and Chekhovsky, V. F. (1979) On interaction of hypersonic non-one-phase flows. PMTF 5, 59-67.

Bloshenko, V. N., Merzhanov, A. G., and Khaikin, B. I. (1974) Interdependence between reaction kinetics and laws of metal particle ignition in a gas. Theory and Technology of Metallurgical Processes, pp. 22-30. Novosibirsk, Nauka, Moscow.

Boiko, V. M. and Papyrin, A. N. (1981) Fast laser diagnostics of heteorgeneous flows. Modern Experimental Methods of Heat and Mass Transfer Study. Itmo An Bssr, Minsk, USSR.

Crowe, C. T. (1967) Drag coefficient of particles in a rocket nozzle. AIAA J. 5, 257-258.

Derevyaga, M. E., Stesik, L. M., and Fedorin, E. A. (1978) Magnesium combustion regimes. Fiz. Goreniya Vzryva 14, 3-10.

Ezhovsky, G. K., Ozerov, E. S., and Roshchenya, Y. V. (1979) Critical conditions of magnesium and circonium powder gas suspension ignition. Fiz. Goreniya Vzryva 15, 97-102.

Fox, T. W., TeVelde, J. A., Nockolls, J. A. (1976) Shock wave ignition of metal powders. Proc. 1976 Heat Transfer and Fluid Mech. Inst., Univeristy of California at Davis, Davis, Calif., 241-256.

Gelfand, B. E. (1977) Modern state-of-art and detonation problems in droplet-gas system. Chemical Physics of Combustion and Explosion. Detonation. pp. 23-29. Chernogolovka, USSR Academy of Sciences, Moscow.

Gurevich, M. A. and Stepanov, A. M. (1968) Ultimate conditions for metal particle ignition. Fiz. Goreniya Vzryva 4, 189-195.

Henderson, C. B. (1976) Drag coefficient of spheres in continuum and rarefied flows. AIAA J. 14, 701-708.

Merzhanov, A. G. (1975) Thermal theory of metal particles ignition. AIAA J. 13, 209-214.

Pokhil, P. F., Belyaev, A. F., et al. (1972) Combustion of powdered materials in active media, p. 294. Nauka, Moscow.

Rudinger, G. (1965) Some effects of finite particle volume on the dynamics of gas-particle mixtures. AIAA J. 3, 3-10.

Reflection of Shock Waves at Rigid Walls in Two-Phase Media

A.A. Borisov,* B.E. Gel'fand,† R.I. Nigmatulin,* K.A. Rakhmatulin,*
and E.I. Timofeev‡
Academy of Sciences, Moscow, USSR

Abstract

Reflection of shock waves at rigid walls in gas-liquid bubble media is considered. Small amounts of gas bubbles introduced into a liquid lead to the pressure rise behind the reflected wave in a liquid without bubbles. Under certain conditions the pressure behind reflected waves in a two-phase medium is higher than that behind the reflected waves in a gas. The pressure increase is particularly high when shock waves propagate in a liquid containing bubbles filled with a soluble or condensable gas. Experimental facts are satisfactorily explained by the hypothesis that gas inclusions may completely collapse behind the wave front.

Introduction

The magnitude of the pressure behind reflected waves is used as a guideline to estimate the destructive action of shock waves on various structures, in estimating the intensity of explosion generated waves in industrial practice, and to recommend protection measures for safe operation of explosion-hazard installations. Shock waves suffer reflection at various obstructions, valves, flow bends, interfaces, etc. For continuous media (gases and liquids), the pressure behind a reflected wave is estimated by the well-known formulas. For a medium of weak compressibility, the relationship between the pressure rise at the front of an incident wave, $\Delta p_1 = p_1 - p_0$, and the pressure rise

---

Paper presented at the 8th ICOGER, Minsk, USSR, Aug. 23-26, 1981. Copyright © American Institute of Aeronautics and Astronautics Inc., 1982. All rights reserved.

*Head of Laboratory, Institute of Chemcial Physics.
†Senior Researcher, Institute of Chemical Physics.
‡Institute of Chemical Physics.

at the front of a wave, reflected at a rigid wall, $\Delta p_2 = p_2 - p_1$, is quite simple: $p_2 - p_1 = p_1 - p_0$ or $p_2 - p_0 = 2(p_1 - p_0)$. Here, p_0, p_1, and p_2 are the pressures in an undisturbed medium, behind the incident waves, and behind the reflected waves, respectively. Pressure ratios at the fronts of incident and reflected waves are connected by the relationship $\Theta_2 + \Theta_1^{-1} = 2$. Here $\Theta_2 = p_2/p_1$ and $\Theta_1 = p_1/p_0$. In media of high compressibility (gases) the pressure behind reflected waves is much larger. In an ideal gas, with an arbitrary adiabatic exponent γ, the ratios Θ_2 and Θ_1 are related by the formula (Mori et al. 1975)

$$\Theta_2^{1/\gamma}(\Theta_1^{1/\gamma} - 1)(\Theta_2^{1/\gamma} - 1)^{-1} = \Theta_1(\Theta_2 - 1)(\Theta_1 - 1)^{-1}$$

The maximum of Θ_2 is attained at $\gamma = 1$, with $\Theta_1 = \Theta_2$; while for $\gamma \neq 1$, $\Theta_1 > \Theta_2$. In an ideal gas, as $\Theta_1 \to \infty$, $\Theta_2 \to (3\gamma-1)(\gamma-1)^{-1}$; hence Θ_2 is finite for $\gamma > 1$. For instance, at $\gamma = 1.4$ one has $\Theta_2 = 8$. A relationship of the first type between Θ_1 and Θ_2 is typical not only for gases. The pressure increase behind reflected waves in two-phase media, say, liquids containing bubbles or foams, for some values of Θ_1, may be as high as Θ_2 in a gas. The measured parameters of reflected waves in gas-liquid media with a small volume content of gas bubbles ($\beta < 20\%$) (Mori et al. 1975; Gel'fand et al. 1978) and in liquid and solid foams (Kudinov et al. 1977; Gel'fand et al. 1975; Voskoboinikov et al. 1977) along with the calculations (Parkin et al. 1961) and pioneer experimental data (Campbell and Pitcher 1958) have confirmed this possibility. But however high may be the pressure rise behind a reflected wave in continuous liquids containing compressible inclusions, in all the cases observed in the previously cited literature, $\Theta_1 > \Theta_2$; and $\Theta_1 = \Theta_2$ is considered as a limiting case. At the same time, the equations of conservation do not impose such a restriction, and their general interpretation follows as a consequence from the hypothesis assumed for the behavior of gas bubbles.

The main limitation on the magnitude of the pressure behind a reflected wave arises from the assumption that the mass content of a gas in liquid is constant. This constancy implies that the mass of the gas cannot flow through the bubble walls or foam cells into the liquid. But one cannot exclude the transport of the gas mass in gas bubbles when the contents of bubbles condense or dissolve owing to a higher pressure behind the compression wave front. In many practical cases, suspensions of soluble gas or condensable vapor bubbles in liquids are used as working media. Ammonia and carbon dioxide dissolve readily in water. Vapor bubbles are inevitably found in cryogenic liquids which are used as working media in laboratory and industrial installations.

Until recently there has been much doubt whether the content of bubbles in liquids can rapidly condense or dissolve under a sudden rise of pressure in a time ($\sim 10^{-4}$ s) typical of compression in shock waves (Parkin et al. 1961). The available data on the dynamics of condensation and dissolution processes are inadequate to answer this question. When shock waves travel in gas-liquid mixtures, a fragmentation of bubbles takes place which promotes fast cooling and collapse of gas bubbles because of the increased interphase surface. This process is rather intense even in weak waves when $\Delta p_1 = 0.2$ MPa and $P_0 = 0.1$ MPa. In stronger waves, gas bubbles of the size 10^{-3} m that break up into 10^{-4} m fragments lead to a 10^2 increase in interphase surface. At present, experiments are the only way to answer the question of whether or not a gas can condense or dissolve in shock waves propagating through a two-phase medium. Shock tube experiments have revealed certain facts which suggest the gas bubbles may condense or dissolve behind the shock wave front in a number of two-phase systems.

Reflection of Shock Waves at Rigid Walls in Liquid Containing Bubbles Filled with a Readily Soluble Gas

In shock tube experiments carried out according to the scheme proposed by Gel'fand et al. 1978, the gas pressure behind a reflected wave is found to be much higher in systems with bubbles of an easily soluble gas than in those with bubbles of poorly soluble gases. The system CO_2-H_2O is the simplest two-phase medium containing bubbles filled with an easily soluble gas. The solubility of carbon dioxide in water is known to be from 150 to 200 times that of nitrogen. For example, the pressure behind a reflected wave in water with carbon dioxide bubbles for $\beta = 4.5\%$, $\Theta_1 = 12.1$, and $p_0 = 0.1$ MPa is $p_2 = 14.5$ MPa, that is, $\Theta_2 = 12$. In water containing nitrogen bubbles for $\Theta_1 = 13$, $p_0 = 0.1$ MPa, and $\beta = 3.5\%$ one finds that $\Theta_2 = 6.15$, since $p_2 \approx 8$ MPa. Thus the use of a readily soluble gas instead of a poorly soluble gas in the bubbles leads to a significant increase of pressure behind the reflected waves. The values of Θ_2 measured in water-containing carbon-dioxide bubbles and in water-containing nitrogen bubbles for β = 4-3.5% are shown in Fig. 1. Line 1 ($\Theta_1 = \Theta_2$) separates the plane (Θ_1, Θ_2) into two parts. In all the experiments conducted with liquids containing poorly soluble gas bubbles it has been found that $\Theta_2 < \Theta_1$ (line 2). Some of the experimental points (line 1) for the system with readily soluble gases fall into the region $\Theta_2 > \Theta_1$; and all the experimental data in this case appear to be higher than those for systems with poorly soluble gases. This result cannot be explained in the framework of the model with a constant mass content of a gas in a liquid.

Reflection of Shock Waves at a Rigid Wall in Liquids Containing Bubbles Filled with Vapor

It is quite natural to expect, even in water systems with such readily soluble gases as carbon dioxide or ammonia, that bubbles do not vanish completely. It is more probable to expect a complete collapse for vapor bubbles. The parameters of reflected waves have been measured in

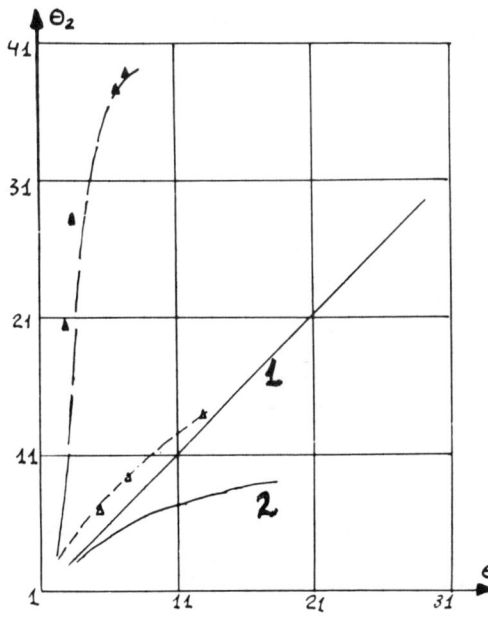

Fig. 1 Pressure ratio in a reflected wave Θ_2 as a function of pressure ratio in an incident wave Θ_1. 1) Campbell and Pitcher (1958); 2) gas; △ - CO_2 in water; ▲ - boiling liquid nitrogen.

boiling water containing vapor bubbles. The results of these measurements are shown in Fig. 1, by two dots (line 4). In the experiments, for $\Theta_1 \cong 1.2$ and $p_o = 0.1$ MPa, it is found that $\Theta_2 \cong 30$, since $p_2/p_o \cong 36$ and $p_2 \approx 4.3$ MPa. Comparing this value of Θ_2 with those for the systems with soluble gas bubbles, one finds a very high increase of Θ_2 in the case of a condensable gas. The data in the figure show that Θ_2 may be as high as ~ 50. It is noteworthy that these high values of Θ_2 have been obtained for weak shock waves at $1.2 < \Theta_1 < \Theta_2$. An abnormally high increase in the pressure behind reflected waves has also been observed in experiments on shock wave propagation in boiling liquid nitrogen (Borisov et al. 1976). For $1.2 < \Theta_1 < 4$ in boiling liquid nitrogen it has been found that $\Theta_2 \approx 40$ for $p_o = 0.1$ MPa.

Table 1 Asymptotic pressure ratios for reflected waves in two phase media with complete void collapse

β	0.05	0.1	0.2	0.3	0.4	0.5	0.6	0.7	0.8	0.9
Θ_2^*	2.05	2.11	2.25	2.42	2.66	3.0	3.5	4.33	6.0	11.0

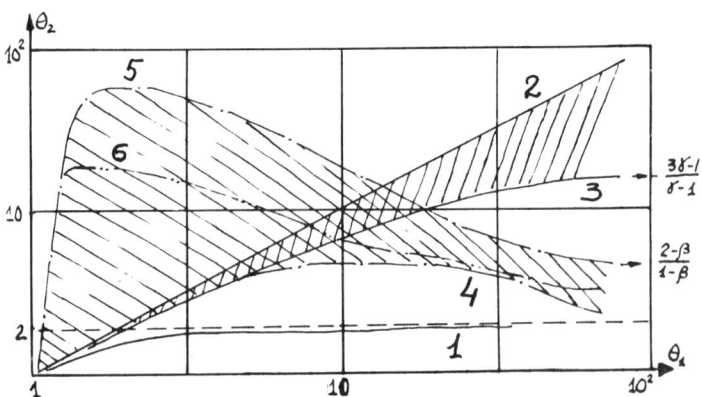

Fig. 2 Relationships between ratios in reflected (Θ_2) and incident (Θ_1) shock waves propagating through continuous and two-phase media. 1) Acoustic waves in a continuous liquid; 2) nonisothermal gas; 3) isothermal gas; 4) two-phase isothermal medium with an insoluble amd uncinsable ot soluble gas; $\beta_5 > \beta_6$.

Discussion and Conclusions

The reason for a perceptible increase in the pressure behind reflected waves, observed in experiments, is not trivial. The hypothesis that the pressure increases owing to intensive vaporization is untenable since the mass content of gas in liquids is very small ($f \ll 1$ for two-phase media of bubble structure, and $f \leq 0.1$ for two-phase media of foam structure). The heat stored in the compressed gas is too small to intensify the evaporation of the liquid substantially. For instance, for $\Theta_1 \approx 100$ and $p_0 = 0.1$ MPa, this heat is sufficient to raise the temperature of the liquid by several degrees only. The heat of condensation of the gas component does not affect the process. The most reasonable explanation of these effects stems from the assumption of the collapse of gas bubbles behind the propagating wave front. In this model, the pressure behind a reflected wave is calculated by the scheme proposed by Gel'fand et al. (1978) and Parkin et al. (1961), with an additional assumption that the liquid behind a shock wave becomes practically continuous. Then the reflected shock wave velocity for $\Theta_2 < 10^3$ is equal to the sound velocity in liquid, without bubbles, and the density is equal to that of a liquid. To allow for the effect of liquid compressibility, it is useful to express the relationship between the velocity and pressure drop in the incident wave in the form

$$D^2 = C_f^2 (p_1-p_0) \left((1-\beta)Bn \quad 1-(1-\beta)[1 + (p_1-p_0)B^{-1}n^{-1}]\right)^{-1} \quad (1)$$

Here $B = 304.5$ MPa and $n = 7.15$ are the constants in the equation of state of a liquid. For water, $\rho_f = 10^3$ kg/m^3 and $C_f = 1500$ m/s; while for liquid nitrogen, $C_f = 960$ m/s and $\rho_f = 850$ kg/m^3. The pressure behind the reflected wave is

$$p_2 = p_1 + \rho_f (C_f + u_1) u_1$$

Since $C_f \gg u_1$, one obtains

$$\Theta_2 = 1 + C_f[(1 - \beta)D]^{-1}(1-\Theta_1^{-1})$$

For stronger waves, when $\Theta_1 \to \infty$ and $D \to C_f$, one has

$$\Theta_2^* = (2-\beta)(1-\beta)^{-1}$$

where Θ_2^* is the asymptotic value.

This relation is listed in Table 1.

The Θ_2^* values show that for $\beta > 0$ the pressure behind the reflected wave in a porous medium is always higher than that in an incompressible medium. The increase of Θ_2^* with increasing porosity results in, for $\beta \approx 0.8$, the value of Θ_2^* approaching that of a gas (in a gas for $\gamma = 1.4$, $\Theta_2 = 8$; for $\gamma = 1.3$, $\Theta_2 = 9.66$; and for $\gamma = 1.2$, $\Theta_2 = 13$). The expression $\Theta_2 = f(\Theta_1)$ for a medium with a complete collapse of voids shows that at some particular value of the incident wave strength $\Theta_1 = \Theta_1^*$, Θ_2 is maximum. Applying the expressions $\Theta_1 = (D/C)^2$ and $C^2 = \rho_0[\beta(1-\beta)\rho_f]$, one can show that $\Theta_1^* = 1.5$. This means that a maximum increase of the pressure behind a reflected wave in two-phase systems with a condensable gas occurs in weak waves when $\Theta_1 = 1.5$. This conclusion agrees very well with the experiments carried out with boiling water or boiling liquid nitrogen, where maximum Θ_2 is observed for $1.2 < \Theta_1 < 3$.

The analysis performed permits the construction of a general pattern of relationship between the reflected and incident waves parameters, as shown in Fig. 2. The relationship between the pressure ratios at the fronts of incident Θ_1 and reflected Θ_2 shock waves in an incompressible medium is presented in Fig. 2 by line 1. The functional relationship between Θ_1 and Θ_2 on line 1 reads $\Theta_1 + \Theta_2^{-1} = 2$. Line 1 gives a lower limit of possible Θ_2 values in two-phase media. The values of Θ_2 for gaseous systems lie in the area between line 2, where $\Theta_1 = \Theta_2$, and line 3, where $\Theta_1 > \Theta_2$. The asymptotic Θ_2 value for a gas with $\gamma > 1$ is $\Theta_2 = (3\gamma-1)(\gamma-1)^{-1}$ when $\Theta_1 \to \infty$. The relationship $\Theta_2 = f(\Theta_1)$ for an ideal gas has no maximum. For two-phase gas-liquid systems, the possible values lie in the region between lines 4 and 5. Line 4 corresponds to Θ_2 values for noncollapsing inclusions, while line 5 corresponds to those for the complete collapse of gas voids behind the front of an incident wave. The relationships of the types shown on lines 4 and 5 have maxima at particular values of Θ_1. The maximum and asymptotic values of Θ_2 in two-phase media decrease with the decreasing gas volume content in a liquid. The relationships $\Theta_2 = f(\Theta_1)$ obtained for the systems with soluble or condensable gas inclusions show that the reflection of extremely weak shock waves at a high rigid wall may be a nonacoustic nature. Moreover, they indirectly testify to the fact that a gas may dissolve and condense in liquid in low-intensity compression waves.

References

Borisov, A. A., Gel'fand, B. E., Gubaidullin, A. A., et al. (1976) Amplification of shock waves in a liquid containing vapor bubbles. Proc. Siberian Thermophys. Seminar, Novosibirsk, USSR.

Campbell, L. and Pitcher, A. (1958) Shock waves in a liquid containing gas bubbles. Proc. R. Soc. London Ser. A., 243, 534.

Gaydon A. G. and Hurle I. R. (1963) The shock tube in high temperature chemical physics, Chapman and Hall Ltd., London, England.

Gel'fand, B. E., Gubin, S. A., Kogarko, S. M., and Popov, O. E. (1975) Investigation of propagation and reflection of pressure waves in porous media. Zh. Prikl. Mekh. Tekh. Fiz. 16, 74-77.

Gel'fand, B. E., Gubin, S. A., and Timofeev, E. I. (1978) Reflection of plane shock waves at a rigid wall in gas bubbles-liquid systems. Izv. Akad. Nauk SSSR Mekh. Zhidk. Gaza 13, 174.

Kudinov, V. M., Palamarchuk, B. I., Gel'fand, B. E. and Gubin, S. A. (1977) Shock waves in gas-liquid foams. Sov. Appl. Mech. 13, 279.

Mori, Y., Hijikata, K., and Komine, A. (1975) Propagation of pressure wave in two-phase flow. Int. J. Multiphase Flow 2, 139-152.

Parkin, B. R., Gilmore, F. R., and Brode, H. L. (1961) Shock waves in bubbled water. Memorandum R.M. - 2795-RR (Abridged).

Voskoboinikov, I. M., Gogulia, M. F., Voskoboinikova, N. F., and Gel'fand, B. E. (1977) A possible scheme for description of shock compression of porous specimens. Dokl. Akad. Nauk SSSR 236, 75.

Relaxation Phenomena in a Foamy Structure

V.M. Kudinov,* B.I. Palamarchuk,† V.A. Vakhnenko,‡ A.V. Cherkashin,§
S.G. Lebed,π and A.T. Malakhov‡
Academy of Sciences, Kiev, USSR

Abstract

The gas-liquid foams, unlike the single-phase media of a gas and liquid type, display a number of relaxation properties that determine the structure and parameters of wave disturbances propagating in such a medium. The present work analyzes the effect of thermal and kinematic relaxation between phases on the shock wave parameters and wave reflection from the rigid wall. The problem of a peak stage of explosion in such a medium and the dependence of the generated shock wave parameters on explosion energy and medium density are considered. The estimated results are compared with the experimental data obtained with piezoelectric and electric pressure gages in foams with liquid mass concentration from 2 to 50 kg/m^3.

The theoretical analysis presented herein provides a physical explanation for the experimentally observed phenomena of the shock dispersion, pressure growth at the incident and reflected wave front with increasing duration of the wave disturbance, and the increase of the shock wave damping factor with distance in foam explosion.

Presented at the 8th ICOGER, Minsk, USSR, Aug. 23-26, 1981. Copyright © 1983 by V. M. Kudinov, B. I. Palamarchuk, V. A. Vakhnenko, S. G. Lebed, A. T. Malakhov, A. V. Cherkashin. Published by the American Institute of Aeronautics and Astronautics, Inc. with permission.
 *Corresponding Member.
 †Deputy Chief of Explosion, Welding Institute and Cutting Dept.. E. O. Paton Welding Institute.
 ‡Associate, E. O. Paton Welding Institute.
 §Senior Engineering, E. O. Paton Welding Institute.
 πChief of Explosives Bureau, E. O. Paton Welding Institute.

Nonmenclature

a_e = equilibrium sound velocity in a two-phase mixture
a_o = sound velocity of a gas filling the foam cells
C_p = specific heat of a gas phase at a constant pressure
D = shock wave velocity
d = density of a condensed phase
E = specific internal energy of a mixture
H = specific enthalpy of a two-phase mixture
k = pressure damping coefficient
P = pressure
q = mass of explosive charge
r = radius
T = gas temperature
t = time
t_o = characteristic time
u = gas phase velocity
v = condensed phase velocity
Γ = ratio of specific heats of a mixture
Γ_f = effective index of shock adiabat for gas-liquid foam
γ = index of adiabat of a gas
δ = ratio of specific heat of a condensed phase and specific heat of a gas phase at a constant pressure
ε = volume fraction of a condensed phase in a mixture
η = ratio of mass concentrations of condensed and gas phases
ρ = density of mixture
ρ_g = gas density
σ = mass concentration of a condensed phase
τ = temperature of a condensed phase

Introduction

Recent studies** of dynamic processes in gas-liquid foams have revealed that the phenomena of shock and detonation wave propagation in these systems differ significantly in the following way from the phenomena observed when these waves propagate in a homogeneous media of either gas or liquid:

(1) The structure of a shock wave in foam depends on velocity of their propagation. At $D < a_o$ the pressure

---
**See for example Borisov et al. (1978), Krasinski et al. (1978), Kudinov et al. (1976), Kudinov et al. (1977a, 1977b), Palamarchuk (1979), Palamarchuk et al. (1979), and Saint-Cloud et al. (1976).

profile of the wave has a two-front configuration.

(2) At a fixed shock wave velocity the amplitude of pressure depends upon a time of wave disturbance.

(3) The structure of shock wave front depends on the conditions of its excitation; shock waves of equal amplitude generated in the shock tube and by a charge explosion of explosives are significantly different.

(4) Damping of explosive waves occurs more rapidly in foams than in gas, liquid, or bubble media.

(5) Even though the detonation wave velocity is constant in detonable foams at from 2 to 15 kg/m^3, the pressure behind its front decreases as liquid concentration increases.

(6) Heterogeneous detonation waves in foam do not have clearly defined "chemical peaks;" the concentration limits of foam detonation considerably exceeding those in condensed gaseous media.

(7) The pressure generated by the reflection of a foam detonation wave from the rigid wall tends to the pressure in the incident detonation wave as liquid concentration increases.

Mathematical models for such media are substantially complicated by a variety of processes of interphase interaction. At the same time, Palamarchuk et al. (1979) have shown that the phenomena could not be explained within the scope of classical thermodynamics and that any mathematical model must account for the relaxation of detonation waves in foam. Analysis of the experimental data on detonation wave propagation in foams has led to the conclusion that the characteristic times for deformation and breakage of foam cells are considerably shorter than those for thermal and kinematic relaxation processes between gaseous and condensed phases. Palamarchuk (1979) has constructed a qualitative model which simulates the dynamic behavior of foam and explains the observed relaxation phenomena in the propagation of shock and detonation waves in foams.

For investigation of stationary and quasistationary waves in foams if an effective index of foam adiabat is defined:

$$\Gamma_f = \gamma \; \frac{1 + \eta\delta(\tau/T)}{1 + \gamma\eta\delta(\tau/T)}$$

then, mathematical the state equation for foam takes a form:

$$E = P(1-\varepsilon)(\Gamma_f - 1)\rho$$

Solution of the one-dimensional conservation equations across a discontinuity in the foam leads to a family of Hugoniot abiabats for a relaxing two-phase medium (Palamarchuk et al. 1980):

$$p^* = \frac{\Gamma+1}{\Gamma-1} - \frac{1-\eta_0}{\rho^*} + \frac{2\gamma}{\gamma-1} \times \frac{\eta_0-\eta}{1+\eta_0}(1-\varepsilon_0)Q^* - \varepsilon_0 \frac{2\Gamma}{\Gamma-1} \Big/$$

$$\frac{\Gamma_f+1}{\Gamma_f-1} \frac{1+\eta_0}{\rho^*} - 1 - \varepsilon_0 \frac{2\Gamma_f}{\Gamma_f-1} \frac{\eta}{1+\eta} \frac{1+\eta_0}{\eta_0}$$

$$\Gamma = \gamma \frac{1+\eta d}{1+\gamma \eta d}, \quad Q^* = \frac{Q}{C_p T_0}$$

$$\rho^* = \frac{\sigma+(1-\varepsilon)\rho_g}{(1-\varepsilon_0)\rho_{go}}, \quad p^* = \frac{P_1}{P_0}$$

The analysis of Rankine-Hugoniot relations for shock and detonation waves for foams showed that variation of volume fraction of a condensed phase did not result in a considerable change of the velocity of wave propagation and pressure in it. If the volume fraction is neglected, mass velocity and a density of the mixture are changed considerably.

The present work, a continuation of the previous investigations, is concerned with the effect of kinematic and thermal relaxation and also the volume fraction upon the parameters of shock wave reflections in foam and a strong stage of explosion in foam. A model of the kinetics of relaxation processes in foams is proposed.

Experimental Equipment

Measurements of shock wave parameters in foam were performed in the following installations.

1) A horizontal shock tube of 67 mm i.d. 5.9 m long with a 2.6-m-long driver section separated by a diaphragm: In the end plate and in the walls of the shock tube, pressure transducers, whose characteristics will be given below, were located. The tube was filled with water foam, the liquid concentration in foam varying from 2 to 30 kg/m^3.

2) A 1600-mm diam cylindrical vessel: Spherical shock waves were generated by the explosion of bulk density

cyclotrimethylene trinitramine (RDX) charges, weighing from 0.5 to 2.8 kg.

3) 529-mm-diam steel cylinders: These were used to study shock waves in the nearest explosion zone.

Recording of the parameters of spherical shock waves was accomplished with the help of the piezoceramic transducer of the "knife" type, representing a flat steel rod, on which piezoceramic 15-mm elements were attached. In the shock tube shock waves with a gradual pressure rise were recorded by mass-produced piezoceramic transducers whose resonance frequency is 30 kHz. Reproducibility of the impulse shape of pressure by piezoceramic transducers is not possible because of the influence of blast furnace processes on the charge generated under ceramic compression (Khokhlov et al. 1978). Electret pressure gages were developed to measure the loading impulses on the rigid wall during the reflection of strong waves. The sensitive element of this gage was an electret rigid vinyl-plastic. These electrets are produced from rigid PVC by the application of an intensive electric field to the polymer, which is maintained at temperatures in access of the vitrification temperature for several hours, then removed from the electric fields and cooled. In a series of independent experiments, it was established that the piezomodule of rigid PVC electrets with one-sided axial loading depends on the polarization condition which remains unchanged during load increment from 10^{-7} to 10^{-3} s. Output characteristics of electret gages were linear at pressures up to 200 MPa. The piezomodule of electret PVC approaches piezomodules of the natural piezoelectrets: quartz and tourmaline. In this work electret gages with values of resonance frequencies 0.4 and 1.7 MHz were used. The gages were calibrated in the shock tube, and also with the explosions of spherical charges of explosives both in water and air. Correlation of the gage response was performed by the source repeater and matching coaxial cable. Recording of signals was made by the memory oscilloscope C 8 -2.

Reflection of Plane Shock Waves

Borisov et al. (1978) and Palamarchuk (1979) have shown that shock waves in foam had developed relaxation zones. To study shock wave reflection from a rigid wall, the characteristic dimension of the test section should be several times the length of the relaxation zone. The dimensions of relaxation zone depend on an intensity of wave and on a concentration of liquid and range from 0.5 to 1.0 m. The test section length restricts the minimum possible lengths of the driver section, since a sufficiently long period of a stationary zone behind the wave front is needed to avoid interaction between the rarefaction wave from the driver and the developed relaxation zones. In this investigation, experimental data reported by Borisov et al. (1978) was used to select test and driver section lengths of 5.9 and 2.6 m, respectively.

The relationship between the propagation velocity of weak quasiacoustic pressure waves and the liquid concentration of the foam is shown in Fig. 1. Here, the calculated equilibrium sound velocity in foam (Rudinger 1965) is also shown by the solid line

$$a_e = \sqrt{\Gamma P_0/\rho_0(1-\varepsilon_0)}$$

As is seen, a good correlation is observed between the calculated and experimental data.

As in the previous investigations (Borisov et al. 1978), two types of shock waves were observed: a disperison wave with a two-front configuration at $D < a_0$ and a one-front one at $D > a_0$.

The calculated and experimental pressure ratios at the shock wave front for foams with different initial densities are plotted against the Mach value, $M = D/a_e$ in Fig. 2.

The data indicate that, within the measurement error, the observed velocity of incident shock waves agrees with predictions based on an equilibrium model whose parameters can be calculated through the model for dynamic behavior of foam (Palamarchuk 1979; Palamarchuk et al. 1979).

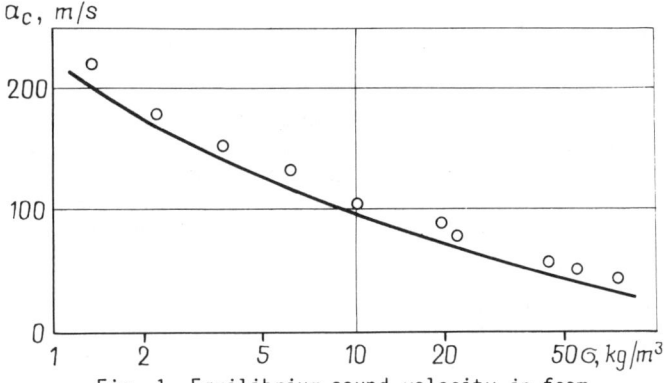

Fig. 1 Equilibrium sound velocity in foam.

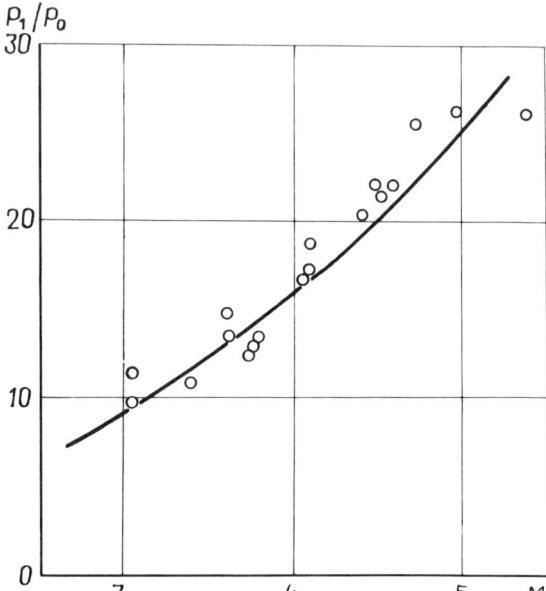

Fig. 2 Pressure behind incident shock wave in foam.

Figure 3 shows the evolution of the reflected shock wave for foam of $\sigma = 27$ kg/m^3 and the record of the incident shock wave. Locations of the pressure gages are also shown in this figure.

As seen in Fig. 3, qualitative changes occur in the pressure profiles of the reflected wave subsequent to reflection. The maximum pressure achieved in the wave, is 29 MPa, increasing during 2.5 ms. Unlike the incident shock wave, the reflected wave has a triangular pressure profile. The precursor wave before the sharp use in pressure is the incident wave; this wave decreases in amplitude as the wave approaches the end wall.

As a distance of 65 mm from the end, the precursor wave becomes shorter in time. Maximum pressure in the wave amounts to 23 MPa. Further on, the wave front becomes abruptly steeper and, at a 0.5 m distance from the end, the wave has a steep shock front, the decrease of pressure beyond the front becoming more flat. The wave amplitude drops, too. The reflected wave velocity D_r on the base of 0-65 mm is 95 m/s, decreases to 55 m/s on the base of 65-135 mm, and abruptly rises to 140 m/s between gages 3 and 4 and 470 m/s between gages 4 and 5.

Experimental and predictions of pressure data on reflection of shock waves from the rigid wall for foams with

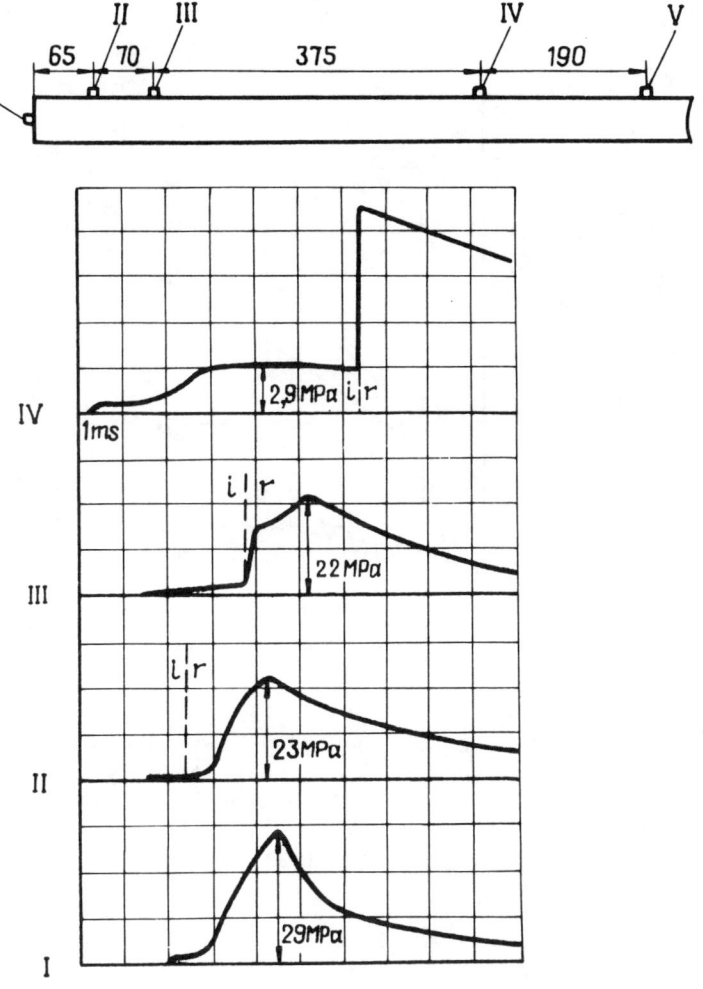

Fig. 3 Evolution of a reflected shock wave in foam.

different initial density and air are shown in Fig. 4. As this figure shows, the coefficient of reflection (P_2/P_1) of shock waves in foam exceeds the maximum possible for air.

To analyze the stationary of reflected shock waves in foam, calculations were made of the effect of the volume fraction of liquid on the values of the reflection coefficients, on the velocity of the reflected wave, and on its other conditions. In this case it was assumed that 1) gas was considered as an ideal one; 2) pressure in the mixture was determined by the gas only; 3) flow was one-dimensional

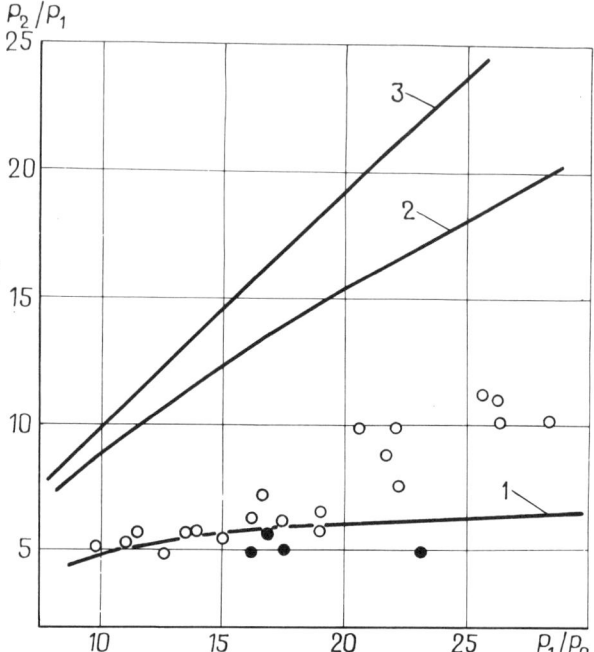

Fig. 4 Pressure behind reflected shock waves in foam and air. Curves 1,2,3: prediction for air and foam with σ = 2.5 and 27 kg/m^3, respectively; o: this investigation; o: observations due to Borisov et al. (1978).

and the effects of the wall boundary layer are neglected; 4) a condensed phase with a constant heat capacity is supposed to be incompressible; and 5) no mass exchange exists between the phases.

The calculations indicated that the pressure in the reflected wave did not, in fact, depend on the volume fraction of the condensed phase. However, if this fraction is ignored, the predicted velocity of the reflected wave decreases considerably (see Fig. 5). A qualitative difference in the relationship between the velocity of the reflected wave and that of the incident one then results. For sufficiently strong shock waves, when the volume fraction is ignored, the velocity of the reflected wave drops as the intensity of the incident wave increases and, if the volume fraction is considered, the velocity rises.

In Fig. 4, the calculated curves of the reflection coefficients (P_2/P_1), are plotted for foams at σ = 2.5 and 27 kg/m^3. Closed circles indicate the results of Borisov et al. (1978) obtained on shock tube with a 1.5-m-long test section. As is seen, the calculated curves lie considerably higher than the experimental data. At the same time, the increase of test and driver section lengths led to the

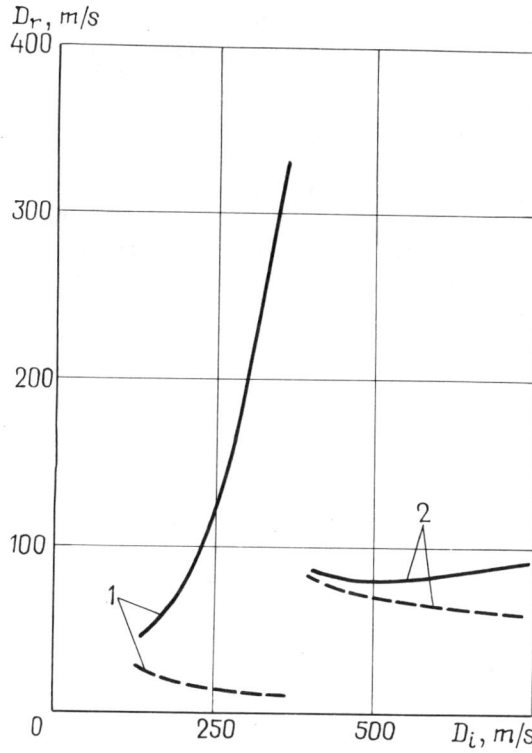

Fig. 5 Prediction of reflected shock wave velocity in foam. (——): taking into consideration the volume fraction of liquid; (---): without taking into consideration the volume fraction of liquid. For curve 1, $\sigma = 27$ kg/m^3; for curve 2, $\sigma = 2.5$ kg/m^3.

increase of both the absolute pressure value and the reflection coefficients as compared to the data obtained earlier.

The difference is not directly associated with the volume fraction of the liquid. Instead, the source of this difference should be found in the nature of the reflection of the waves, which have developed relaxation zones.

To date, it is remarkable to date that in all experiments stationary reflected waves have not been attained. The profile variation of a reflected shock wave in foam suggests that at a distance of 0.3-0.5 m from reflection the wave propagates not in foam, but in a pushing gas which is attributed to an abrupt wave front characteristic for the shock waves in gases.

The formation of the reflected wave occurs in a narrow zone near the test section end plate, and is complicated by the superimposed disturbances, which are generated at the contact surface and which should be, in view of the acoustic impedances of foam and gas, rarefaction waves. The decrease of pressure beyond the front of the reflected wave suggests the existence of rarefaction waves. Analysis shows these interactions can be minimized if the test section is 30 m long or more.

To evaluate the effect of kinematic and thermal nonequilibrium between the two phases on the parameters of the reflected wave, the jump equations, the laws of conservation of mass, momentum, and energy, the equation of state for each of the phases, and the equation for a mixture density was solved for two-velocity and two-temperature media. The suppositions about the mixture properties are similar to those indicated for the equilibrium waves. The shock wave conditions were assumed to be stationary in time. The system of equations in a coordinate system fixed on the shock front is

$$\frac{\rho_o}{1+\eta_o} u_o = \frac{\rho}{1+\eta} u$$

$$\frac{\rho_o \eta_o}{1+\eta_o} v_o = \frac{\rho \eta}{1+\eta} v$$

$$P + \frac{\rho}{1+\eta}(u^2 + \eta v^2) = P_o + \frac{\rho_o}{1+\eta_o}(u_o^2 + \eta_o v_o^2)$$

$$\frac{u_o^2}{2} + C_p T_o + \eta_o \frac{v_o}{u_o}(\frac{v_o^2}{2} + \delta C_p \tau_o + \frac{P_o}{d})$$

$$= \frac{u^2}{2} + C_p T + \eta_o \frac{v_o}{u_o}(\frac{v^2}{2} + \delta C_p \tau + \frac{P}{d})$$

$$\frac{1}{\rho} = \frac{P_o T}{PT_o} \frac{1}{\rho_{go}} + \frac{\eta}{d}$$

The system of equations of stationary equilibrium discontinuities is obtained from this system at equal temperatures and velocities of the phases on both sides of the discontinuity.

Numerical solutions indicate that small deviations from kinematic equilibrium in the incident wave cause significant reductions of the reflection coefficient. As an example, if the mass velocity of the condensed phase is 10% less than the gas velocity, in laboratory coordinates for a wave of 2 MPa intensity for air foam at $\sigma = 27$ kg/m^3, the reflection coefficient is 60% smaller than that of equilibrium foam at the same conditions. Differences in the temperature of the phases in the incident wave also lead to a decrease of the reflection coefficient. In the extreme case of the absence of initial heating of liquid in the incident and reflected shock waves, its value corresponds to the reflection coefficient for gas filling foam cells.

A further study of the complete reflection of shock waves in foams should be conducted at higher shock wave intensity, when the waves have narrower relaxation zones.

Blast Waves in Foam

For prediction of the parameters of shock waves generated by the ignition of explosives in foam the nature

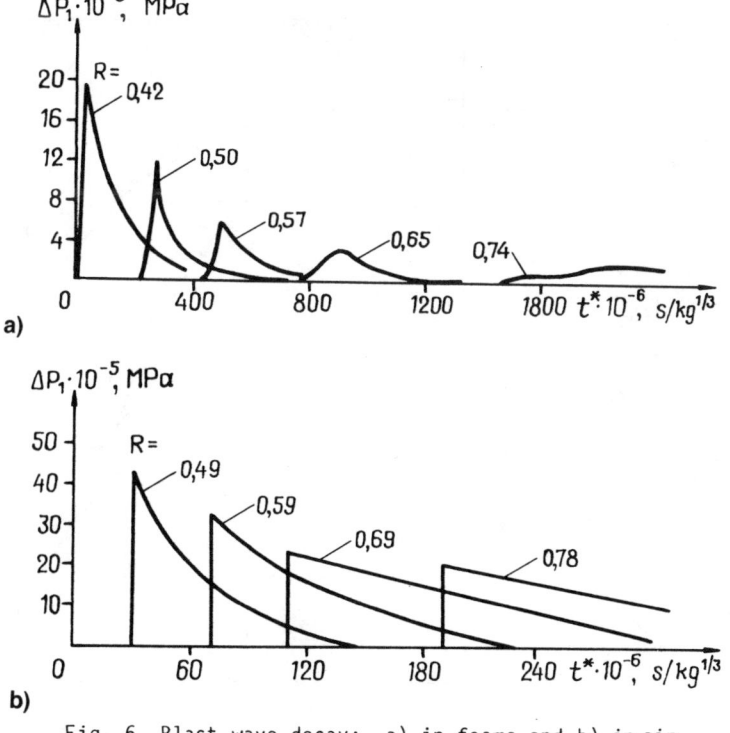

Fig. 6 Blast wave decay: a) in foams and b) in air.

of the relaxation interaction between gas and liquid must be understood. At the explosion in foam, the degree of completion of relaxation processes depends both upon thermophysical properties of phases and the energy of the explosion determining the period of wave disturbance.

To find the nature of the relaxation interaction between the phases, the conditions of shock waves excited by the explosion of 0.5-28-kg mass RDX charges were experimentally investigated.

Figure 6 shows the evolution of a shock wave in foam. The abscissa is the relative time; t^* is $s/kg^{1/3}$ from the moment of the wave entering the relative distance $R = 0.42$ $m/kg^{1/3}$.

For comparison the similar diagram has been plotted for air as reported by Adushkin (1963). The blast wave quickly damped more rapidly in foam than in air; the pressure profile decays from a triangular wave to a two-front dispersed wave similar to that observed in the shock tube.

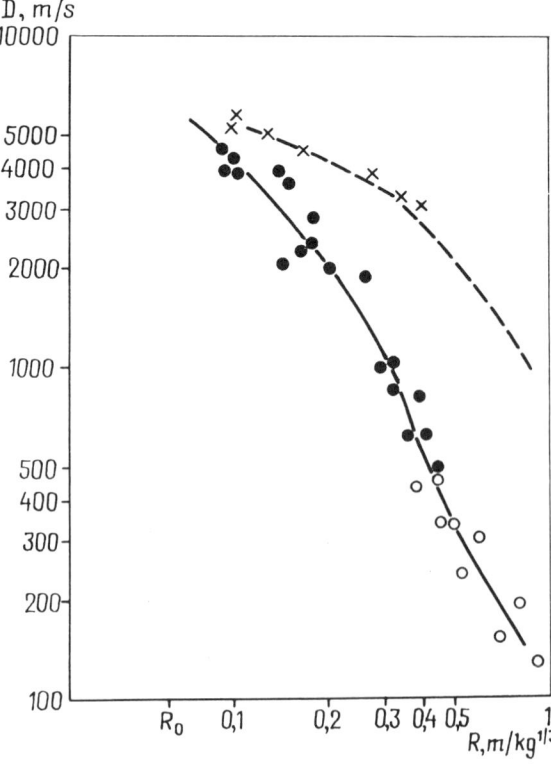

Fig. 7 Spherical shock front propagation velocity. (---): air; (———): foam; •: q = 0.5 kg, o: q = 0.5-2.8 kg.

Near the point of explosion an abrupt rise of pressure in foam was observed. In the explosive wave with a pressure drop of 2 MPa for R = 0.42, the time of pressure increase is a maximum, i.e. of the order to 20 μs. The relaxation zone of a stationary shock wave of the same amplitude generated on the shock tube has a period of about 2-3 ms.

Figures 7 and 8 show the relationships between the velocity and the pressure drop for spherical shock waves in foam and air.

Analysis of the experimental data indicates that the relationship between the maximum pressure ratio at the of a shock wave is

$$P_1/P_0 = M^2$$

where $M = D/a_e$, and a_e is the equilibrium sound velocity in foam. In the zone nearest to the explosion site the pressure ratio at the wave front is close to the ratio for kinematic equilibrium between the phases (Palamarchuk et al. 1979).

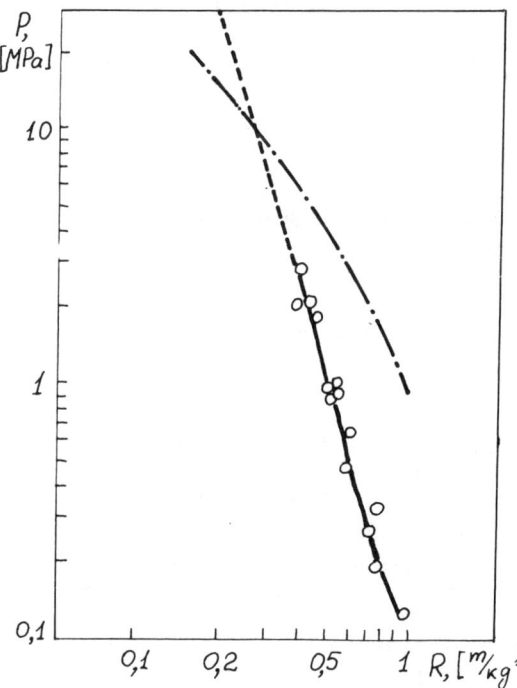

Fig. 8 Pressure drop as a function of a radius. (-·-·-): air, (---): estimated values for foam.

Direct measurements of pressure were not made for R < 4. Extrapolation of the data for R > 4 indicates that in the nearest zone the pressure at the front in foam exceeds that at the wave front in air. The data obtained agree with the predictions of the initial parameters of a shock wave at the interface explosion products of RDX-foam. From the analysis of shock abiabat for foam at $\sigma = 15$ kg/m^3 and isentrope of the expansion of the RDX explosion products for D = 6000 m/s and P = 500 MPa, the conditions of the wave weakly depend upon a degree of the completion of heat relaxation between the phases.

The abrupt reduction of the wave parameters in foam observed, as compared to a gas, is associated with the processes of heat exchange between liquid and gas (Palamarchuk et al. 1979). In this case the increase of the coefficients of damping of shock wave parameters, depending on a distance, leads to the conclusion that the course of the heat relaxation occurs more slowly than in the instance for which the two phases are kinematic equilibrium.

For a further theoretical analysis of the explosive waves in foam it should be noted that for the conditions of this experiment, the thermophysical properties of water, particularly the heat capacity, do not change significantly and evaporation of liquid between the phases is insignificant.

A Model of Foam Blast Waves

Analysis of the experimental data suggests the following assumptions in theoretical studies of relaxation processes of a strong blast wave:

1) A gas-liquid foam consists of a gas phase with liquid particles distributed in it.

2) No mass transfer takes place between the gas and the liquid particles.

3) The velocities of gas and condensed phase are equal.

In addition, it is assumed that the density and a specific heat of liquid are constant, the volume fraction of liquid phase and liquid particles partial pressure are negligibly small, and the gas is described by the equation of state for a perfect gas with constant specific heats. With these assumptions, the wave processes beyond the shock wave front are described by hydrodynamic equations (Baum et al. 1975; Sedov 1972) and the equations of a total balance of complete energy (Sedov 1972; Kastenboim et al. 1974):

$$\frac{\partial \rho}{\partial \tau} + u \frac{\partial \rho}{\partial r} + \rho(\frac{\partial u}{\partial r} + \frac{2}{r} u) = 0$$

$$\rho(\frac{\partial u}{\partial t} + u \frac{\partial u}{\partial r}) + \frac{\partial P}{\partial r} = 0 \quad (1)$$

$$\frac{\partial}{\partial t} \rho E' + \frac{\partial}{\partial r} \rho u E' + P(\frac{\partial u}{\partial r} + \frac{2}{r} u) = 0$$

$$4\pi \int_0^{r_\phi} (\rho E' + \tfrac{1}{2}\rho u^2) r^2 \, dr = q_0$$

Here, the state equation for foam can be reduced to

$$E' = P/\rho(\Gamma'_f - 1) \quad (2)$$

where Γ'_f is the effective index of shock abiabat of foam

$$\Gamma'_f = \gamma \frac{1 + n\delta(\tau - T_0)/T}{1 + \gamma n\delta(\tau - T_0)/T} \quad (3)$$

On the front of the strong shock wave at "freezing" ($\tau = T_0$) of relaxation processes $\Gamma'_f = \gamma$, while at thermodynamic equilibrium ($\tau = T$) between the phases $\Gamma'_f = \Gamma$. Here, for concentrations of a condensed phase corresponding to foams, Γ is close unity.

Model kinetics of Γ'_f time variation were used as

$$\Gamma'_f = \Gamma + (\gamma - \Gamma) \exp(-\theta'/t_0) \quad (4)$$

The exponential relationship was based on the assumption that energy transfer to the liquid phase occurs through heat conduction which follows the exponential law.

The residence time θ' of a microvolume in the wave is described by the differential equation

$$\frac{\partial \theta'}{\partial t} + u \frac{\partial \theta'}{\partial r} = 1 \quad (5)$$

At the shock front, owing to "freezing" of the relaxation processes

$$\theta'(r_\phi) = 0 \qquad (6)$$

From the Rankine-Hugoniot conditions the relations for the leading front at the strong explosion limit are as follows

$$\rho = \frac{\gamma+1}{\gamma-1}\rho_0 \qquad u = \frac{2}{\gamma+1}D \qquad P = \frac{2}{\gamma+1}\rho_0 D^2 \qquad (7)$$

where D is the velocity of shock front. At the center for all t

$$u = 0 \qquad (8)$$

As at the initial moment the medium does not display relaxation properties, the system of the differential equations has a similarity solution (Sedov 1972; Kastenboim et al. 1974).

The system of Eqs. (3-5) is transformed through variables $\xi = r/r_\phi$ and χ where χ is determined by the equation $d\chi/dt' = 1-Z$ and $\chi = 0$ at $t' = 0$.

$$\Phi = \frac{\rho}{\rho_0} \qquad V = \frac{U}{D} \qquad P = \frac{P}{\rho_0 D^2} \qquad \theta = \frac{\theta'}{t_0} \qquad t' = \frac{t}{t_0} \qquad (9)$$

to the form

$$\chi(1-Z)\frac{\partial \Phi}{\partial \chi} + (V-\xi)\frac{\partial \Phi}{\partial \xi} + \Phi(\frac{\partial V}{\partial \xi} + \frac{2}{\xi}V) = 0$$

$$\Phi[\chi(1-Z)\frac{\partial V}{\partial \chi} + (V-\xi)\frac{\partial V}{\partial \xi} + ZV] + \frac{\partial P}{\partial \xi} = 0$$

$$\chi(1-Z)\frac{\partial P}{\partial \chi} + (V-\xi)\frac{\partial P}{\partial \xi} + 2ZP + \Gamma'P(\frac{\partial V}{\partial \xi} + \frac{2}{\xi}V) + \frac{\Gamma'-\Gamma}{\Gamma'-1}\chi = 0$$

$$\psi = \int_0^1 (\frac{P}{\Gamma'-1} + \frac{V^2}{2})\xi^2 d\xi \qquad (10)$$

$$\chi(1-Z)\frac{d\psi}{d\chi} + \psi(2Z + 3) = 0$$

$$\chi(1-Z)\frac{\partial\theta}{\partial\chi} + (V - \xi)\frac{\partial\theta}{\partial\chi} = \chi$$

$$\Gamma_f' = \Gamma + (\gamma-\Gamma)\exp(-\theta)$$

where $Z = (r_\Phi/D^2)(dD/dT)$ and, at $t'=0$, $Z = -1.5$. The boundary conditions then take the form $V = 0$ at $\xi = 0$:

$$\Phi = \frac{\gamma+1}{\gamma-1} \qquad V = \frac{2}{\gamma+1} \qquad \theta = 0 \text{ at } \xi = 1 \qquad (11)$$

At the initial moment $\chi = 0$, system (10) has a similar solution (Sedov 1972; Kastenboim et al. 1974).

The system of nonlinear differential equations (10) is hyperbolic (Godunov 1979; Rozhdestvenskii and Yanenko 1978) for $\chi > 0$, $0 < \xi < 1$. The characteristic roots of the determinant whose coefficients are the partial derivatives with respect to ξ are real. In spite of the fact that there are multiple roots, there still exists the transformation with the determinant differing from zero, which reduces the system of the differential equations to a canonic form. In this case, the necessary condition of the uniqueness of the solution requires

$$\Gamma_f' > 1 \qquad \frac{\partial \Gamma_f'}{\partial t} \leq 0 \qquad (12)$$

To find the solutions of the quasilinear partial differential equations, numerical methods must be used.

At the symmetry center and at the initial moment, the system of differential equations (10) has some difficult limit points. At the center, a saddle point is observed and causes difficulty in calculations of this region. At the initial moment the order of the equation system increases and complicates the calculations.

Near the center, beginning from some ξ, pressure P was kept constant, along the coordinate, while the change of velocity and density follows the asymptotic formulas

$$V \sim \xi \qquad \Phi \sim \xi^s$$

where s > 0. The error caused by this approximation is not significant, since a contribution of the central region to the energy integral is small.

The method of solution used to overcome the singularity associated with the disappearance of terms involving partial derivatives with respect to x is as follows: At the initial moment, when the relaxation processes have little effect on the flowfield as compared to its initial value, all changes in the dependent variables can be considered as small disturbances. The unknown variables can be written as

$$\Phi = \Phi_1(1 + \delta_\Phi \chi) \qquad V = V_1(1 + \delta_V \chi) \qquad P = P_1(1 + \delta_p \chi)$$

$$\Gamma'_f = \Gamma + (\gamma - \Gamma) \exp(-\delta_\Gamma \chi) \qquad \theta = \delta_\Gamma \chi \qquad (13)$$

$$\psi = \psi_1(1 + \delta_\psi \chi) \qquad Z = -1.5 + \delta_Z \chi$$

where variables δ_Φ, δ_V, δ_p, δ_Γ depend on the coordinate only, while δ_ψ and δ_Z are constants, the initial values of the variables of the flow being designated by index I.

If the expressions (13) are substituted into the equation system, the resulting equations are expanded into a series with respect to a small parameter, $\delta \cdot \chi$, and if only first-order terms are retained; a linear system of differential equations results.

This system was solved by a finite-difference scheme which approximates differential equations within the second order of accuracy.

When the condition $\delta_\chi \ll 1$ is not satisfied, the system (10) must be solved subject to boundary conditions (11). Implicit difference schemes used for flow computer calculations are approximated by the first-order system of Eqs. (10).

It is necessary to note that Z both in the method of calculation at the initial time moment and at successive times is determined by the method of successive approximations. The chosen value Z_0 is substituted for the equation, and density, velocity, and pressure are calculated. Substitution of these values permits the calculation of energy integral ψ and, consequently, a new value of Z_1. This iteration scheme quickly converges. For times $t \sim t_0$,

Z is linear in x, with

$$\delta_Z = -0.199 \tag{14}$$

The parameter Z characterizes a degree of wave damping, since

$$Z \frac{r_\phi}{D^2} \frac{dD}{dt} = \frac{1}{2} \frac{d \ln P}{d \ln r_\phi} \tag{15}$$

So, at propagation of a strong shock wave in a nonrelaxing medium we have Z = -1.5, which characterizes the rate of damping because of the geometric divergence of the wave. In a relaxing medium which obeys the kinetic model equations (2, 4), the shock wave damps faster and at the initial moment the degree of damping is expressed as

$$\frac{d \ln P}{d \ln r_\phi} = 2Z = -(3 + 5\delta_Z \frac{t}{t_0}) = -(3 + 0.995 \frac{t}{t_0}) \tag{16}$$

With this relationship a characteristic time t_0 may be determined from the slope of the curve in Fig. 8. For the charges of explosives with mass of 0.5-2.8 kg, t_0 is approximately 150 µs.

It is interesting to compare the change in a pressure drop in a passing wave for a relaxing medium, foam, and for a nonrelaxing one, for example, air. The coefficient of damping is defined as the pressure as a relation of the pressure in a nonrelaxing medium P_n to that in a relaxing medium P_f at the same relative distance from the point explosion:

$$k = P_n/P_f \tag{17}$$

The pressure damping coefficient is shown in Fig. 9. For distance $R \approx 0.4$ m/kg$^{1/3}$, the explosion in air can be considered point explosion (Adushkin 1973). Analysis of the damping coefficient for a point explosion yields

$$k = \exp(-2\delta_Z t/t_0) \tag{18}$$

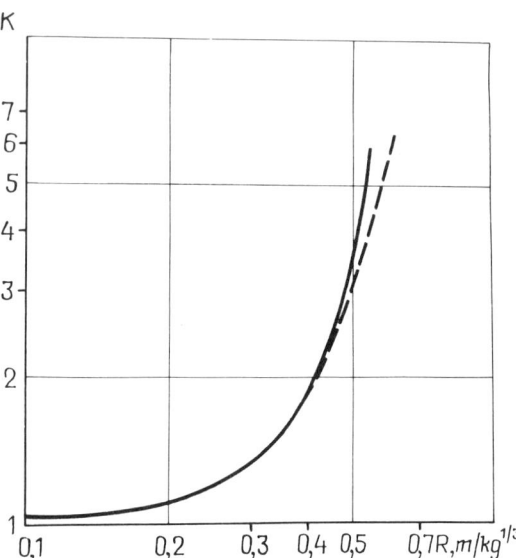

Fig. 9 Pressure damping coefficient for a blast wave in foam. (---): experimental values; (——): estimated values.

A better agreement between the pressure damping coefficients calculated from the experimental results, Fig. 18, and estimated from the ratio (18) is observed for t_o = 180 μs (Fig. 9).

Since the thermal relaxation times obtained by two independent methods agree, it appears that an adequate description of the relaxation processes in foam is possible within the framework of the proposed theory.

Conclusions

The results of the experimental and theoretical investigations of the relaxation phenomena which accompany the propagation of shock waves in foam indicate that within the scope of relaxation gasdynamics it is possible to explain the phenomena observed. Further investigations of the kinetics of relaxation processes in foams will allow construction of strict mathematical models.

References

Adushkin, V. V. (1963) About formation of shock wave and spreading of explosion products in the air. Prikl. Mekh. Tekh. Fiz. 5, 107-114.

Baum, F. A., Orlenko, L. P., Stanjukovich, K. P., Chel'shev, V. P., and Shehter, B. I. (1975) Explosion Physics, p. 704, Nauka, Moscow, USSR.

Borisov, A. A., Gel'fand, B. E., Kudinov, V. M., Palamarchuk, B. I., Timofeyev, E. I., Stepanov, V. V., and Khomik, S. V. (1978) Shock waves in water foams. Acta Astron. 5, 1027-1033.

Godunov, S. K. (1979) Equations of Mathematical Physics, p. 391. Nauka, Moscow, USSR.

Kastenboim, K. S., Roslyakov, G. S., and Chudov, L. A. (1974) Spot Explosion, p. 254. Nauka. Moscow, USSR.

Khokhlov, N. F., Mineev, V. N., Ivanov, A. G., and Luchinin, V. I. (1978) Dynamic piezomodule of TsTS-19 ceramics. Fiz. Gor. Vzr. 14, 146-149.

Krasinski, J. S., Khosla, A., and Ramesh, V. (1978) Disperions of shock waves in liquid foams of high dryness fraction. Arh. Mekh. Stos. 30, 461-475.

Kudinov, V. M., Palamarchuk, B. I., Gel'fand, B. E., and Gubin, S. A. (1976) Parameters of shock waves at explosion of charges of explosives in foam. Dokl. Akad. Nauk SSR 228, 555-557.

Kudinov, V. M., Palamarchuk, B. I., Lebed, S. G., Borisov, A. A., and Gel'fand, B. E. (1977a) Structure of detonation waves in two-phase media. Detonation, pp. 107-111. Chernogolovka, Institute of Chemical Physics, Ac. of Scs. of USSR.

Kudinov, V. M., Palamarchuk, B. I., Lebed, S. G., Borisov, A. A., and Gel'fand, B. E. (1977b) Pecularities of detonation wave propagation in water-mechanical foam formed by a combustible gas mixture, Dokl. Akad. Nauk SSR, 234, 1977.

Palamarchuk, B. I. (1979) Investigation of shock wave flows in two-phase media of a foamy structure. Thesis, p. 14. Institute of Hydrodynamics, Academy of Science of UkrSSR, Kiev.

Palamarchuk, B. I., Kudinov, V. M., Vakhnenko, V. A., and Lebed, S. G. (1980) Effect of condensed phase volume fraction on the detonation parameters in dispersed media. Detonation, pp. 92-96. Chernogolovka, Institute of Chemical Physics, Ac. of Scs. of USSR.

Palamarchuk, B. I., Vakhnenko, V. A., Cherkashin, A. V., and Lebed, S. G. (1979) Effect of relaxation processes on shock wave damping in water foams. Proceedings of IV International Symposium on Explosive Working of Metals, pp. 398-408, CSSR.

Rozhdestvenskii, B. L. and Yanenko, N. N. (1978) Systems of Quasilinear Equations, p. 687, Nauka, Moscow, USSR.

Rudinger, G.(1965) Some effects of finite particle volume on the dynamics of gas-particle mixtures, AIAA J. 3, 1217-1222.

Saint-Cloud, J. P., Guerraud, C., Moreau, M., and Manson, N. (1976) Experiences sur la propagation des detonations dans un milieu biphasique. Astron. Acta 3, 781-794.

Sedov, L. I. (1972) Mechanics of Similarity of Dimensions in Mechanical Engineering, p. 440 Nauka, Moscow, USSR.

Chapter II. Blast Waves

Self-Similar Blast Waves Incorporating Deflagrations of Variable Speed

R.H. Guirguis,* M.M. Kamel,† and A.K. Oppenheim‡
University of California, Berkeley, Calif.

Abstract

Solutions to a class of self-similar blast waves incorporating variable speed deflagrations are presented. Their scope covers a full regime bounded on one side by an adiabatic strong explosion and on the other by a deflagration propagating with an infinite acceleration. Within these limits, results for a representative set of accelerations are displayed, extending over the full range of propagation speeds from zero to those corresponding to the Chapman-Jouguet deflagrations. As a consequence of self-similarity, the results are applicable essentially to the case when the ambient atmosphere into which the wave front propagates is at negligibly low pressure and temperature. The formulation is presented in the Eulerian form, while that in the Lagrangian form is invoked only to derive the induction time-temperature dependence that would conform to the condition of self-similarity. The restrictions imposed by gasdynamic conditions on the deflagration speed are determined on the basis of the properties of the singularity of governing equations. The formulation includes the case of blast waves driven by deflagrations of constant speed whose fronts can propagate into atmospheres of finite pressure and temperature.

Presented at the 8th ICOGER, Minsk, USSR, Aug. 23-26, 1981. Copyright © 1983 by A.K. Oppenheim. Published by the American Institute of Aeronautics and Astronautics, Inc. with permission.
*Assistant Research Engineer.
†Professor of Mechanical Engineering, Cairo University.
‡Professor of Mechanical Engineering.

Nomenclature

a = sonic speed = $\sqrt{\gamma p/\rho}$
a,b,c = parameters defined in Eq. (7)
C = logarithmic time ratio = $\mu \ln(t) t_s$
D = function defined in Eq. (4)
E = energy deposited in blast wave per unit area for j=0, per unit polar and axial length for j=1, and per unit steric angle for j=2.
f = nondimensional particle velocity = u/W_s
F = reduced coordinate = f/x
g = nondimensional pressure = $p/\rho_a W_s^2$
h = nondimensional density = ρ/P_a
j = geometric index: 0, 1, or 2 for plane, cylindrical, or spherical symmetry
J = energy integral, defined in Eq. (23)
m = induction time density exponent
M = Mach number
p = pressure
P = function defined in Eq. (6)
q = exothermic energy per unit mass
Q = function defined in Eq. (5)
r = radius
R = gas constant
S = propagation speed of deflagration = $W_d - U_u$
t = time
T = temperature
u = particle velocity
W = wave speed
X = similarity variable = r/r_s
y = front coordinate = $(a_a/W_s)^2$
Z = reduced coordinate = g/hX^2
α = induction time-temperature exponent
β = energy distribution index = $(d\ln q/d\ln r)_a$
γ = specific heat ratio
θ = nondimensional temperature = T/T_n
λ = decay coefficient = $d\ln y/d\ln r_s$
μ = velocity index = $d\ln r_s/d\ln t$
ν = exponent in Eq. (10) = $\lambda/(j+1)$
ρ = density
τ = induction time dependence on temperature
ϕ = equivalence ratio
ω = nondimensional exothermic energy = q/W_d^2

Subscripts

a = ambient state

b	=	state immediately behind deflagration front
CJ	=	Chapman-Jouguet condition
i,k	=	reference states
n	=	state immediately behind shock front
p	=	state immediately ahead piston face
u	=	state immediately ahead deflagration front

Locations

d	=	deflagration
1	=	reference point
p	=	piston
s	=	shock front
0	=	reference point on the shock trajectory

Superscripts

CJ	=	cases associated with λ_{CJ}
0	=	steady flow conditions

Singularities

A	=	$D=0$, $Q=0$, $P=0$
C	=	$F=1$, $Z=0$
D	=	$Q/Z=0$, $Z=\infty$
E	=	$F=1$, $Z=\infty$

Introduction

Studies of the gasdynamic properties of unsteady flames have been pursued for a century, as manifested by the classical paper of Mallard and Le Chatelier (1881), while the interest of gasdynamic features accompanying flame propagation was steadily growing.

The notion of an unsteady double-discontinuity system was presented by Oppenheim at the Seventh International Congress of Applied Mechanics in 1948, and its analysis was presented at the Fourth (International) Symposium on Combustion (Oppenheim 1952) as a model of the most prominent gasdynamic process associated with the transition from deflagration to detonation.

In their monograph, Shchelkin and Troshin (1965) included a chapter on "nonstationary double discontinuities." Korobeinikov (1969) used this concept in his analysis of a point explosion in a denotating gas. Using dimensional analysis he derived the restrictions imposed by self-similarity on the chemical kinetic rate expressions used for the induction process. Bishimov,

Korobeinikov, and Levin (1970) obtained a rule for the growth of the induction zone in the early stages of explosion, when chemical energy released is sufficiently small in comparison to the energy of explosion.

Oppenheim and Kamel (1973), using a high-speed Schlieren system, recorded the decoupled shock-flame system in a cinematographic sequence of Schlieren records. Ignition was achieved by a short duration (20-s) laser light pulse focused on a point, producing thus a sharp, perfectly spherical, double discontinuity. Kuhl et al. (1973) treated the problem of a steady flame driving a blast wave in a combustible medium. Oppenheim et al. (1978), using a finite-difference scheme, analyzed the nonsteady flowfield formed in an inert atmosphere surrounding an exploding cloud. The problem of self-similar explosions with energy deposition at the front was treated by Barenblatt et al. (1980).

In the study reported here, a systematic approach to the problem of self-similar blast waves incorporating nonsteady flames is developed. In order to emphasize their gasdynamic characteristics, the flames are treated as deflagrations. Their scope is restricted by the Chapman-Jouguet condition at which the deflagration propagates at local sonic speed relative to the flow behind it.

The solutions presented cover a regime, bounded on one side by an adiabatic strong explosion and, on the other, by deflagration propagating at an infinite acceleration. Within these limits, results for a representative set of accelerations are displayed, extending over the full range of propagation speeds from zero to those corresponding to the Chapman-Jouguet deflagration. In order to satisfy the self-similarity requirement, blast waves associated with nonsteady flames have to propagate into an atmosphere of zero pressure and temperature. However, this restriction does not apply to blast waves of constant velocity which are included within the scope of the analysis.

The formulation is presented in the Eulerian form, while that in the Lagrangian form is invoked only to derive the induction time-temperature dependence that would conform to the condition of self-similarity. On the other hand, the restriction imposed by gasdynamics on the propagation speed of the deflagration is obtained by studying the singularities of the governing equations.

Formulation

Governing Equations

Equations governing the self-similar unsteady one-dimensional flowfield of a perfect gas can be expressed in terms of the following dimensionless variables (see Oppenheim et al. 1971, 1972):

$$X \equiv r/r_s$$

where r is the space coordinate, while the subscript s denotes the shock front position;

$$f \equiv u/W_s$$

where u is the flow velocity, while $W_s = dr_s/dt$, and t is the time;

$$h \equiv \rho/\rho_a$$

where ρ is the density, while subscript a denotes ambient state; and

$$g \equiv p/\rho_a W_s^2$$

where p is the pressure.

Then, by introducing the two reduced coordinates

$$F \equiv f/X \quad Z \equiv g/hX^2 \tag{1}$$

the conservation equations are transformed into the following autonomous set:

$$\frac{dF}{d\ln X} = -\frac{Q(F,Z)}{D(F,Z)} \tag{2}$$

and

$$\frac{dZ}{d\ln x} = -\frac{Z}{1-F}\frac{P(F,Z)}{D(F,Z)} \tag{3}$$

where

$$D(F,Z) \equiv \gamma Z - (1-F)^2 \tag{4}$$

$$Q(F,Z) = \gamma(j+1)(F-b)Z - (a-F)(1-F)F \tag{5}$$

and

$$P(F,Z) \equiv [(j+1)(\gamma-1)+2](C-F)D(F,Z) + (\gamma-1)Q(F,Z) \quad (6)$$

where the parameters a, b, and c, controlling the singularities of the above system of O.D.E.s, are specified as follows:

$$a \equiv (\lambda+2)/2$$

$$b \equiv \lambda/(j+1)\gamma \quad (7)$$

$$c \equiv (\gamma+2)/[(j+1)(\gamma-1)+2]$$

while $j = 0$, 1, or 2 for the plane, line, or point symmetrical flowfields, respectively. In the above,

$$\lambda \equiv \frac{d\ln y}{d\ln r_s} = -2\,[(dW_s/dt)r_s]/W_s^2 \quad (8)$$

is the so-called decay coefficient, while

$$y \equiv (a_a/W_s)^2 = \gamma Z_a \quad (9)$$

where a is the local sonic speed. For a self-similar wave, $\lambda y = 0$.

One should note that γ is the local specific heat ratio. Thus, referring to Fig. 1, in the induction zone, $\gamma = \gamma_u$, the specific heat ratio of the reactants, while in the zone behind the flame, $\gamma = \gamma_b$, the specific heat ratio of the products.

Within the frame of the above formulation, the energy equation can be integrated in closed form to yeild

$$\frac{g}{g_i}\left(\frac{h}{h_i}\right)^{-\gamma} = [\frac{h}{h_i}\left(\frac{x}{x_i}\right)^{j+1}\frac{1-F}{1-F_i}]^{-\nu} \quad (10)$$

where $\nu \equiv \lambda/(j+1)$ while subscript i denotes a reference point situated so that the integration path between it and the point in question does not cross any discontinuity.

Boundary Conditions

Flame Front. In order to emphasize its gasdynamic characteristics, the flame is treated as a deflagration.

BLAST WAVES INCORPORATING DEFLAGRATIONS

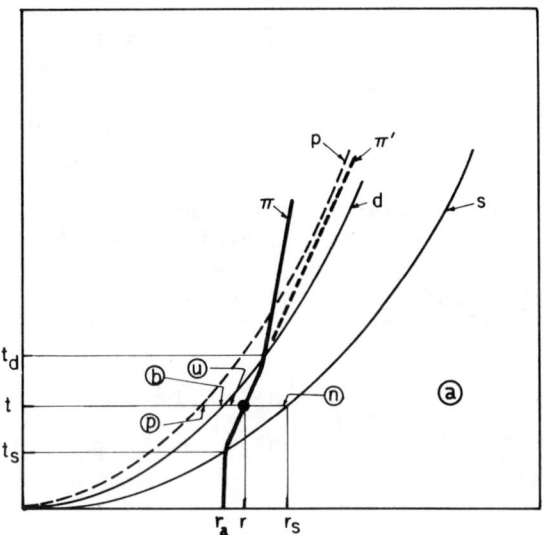

Fig. 1 Shock wave S, deflagration d, virtual position p, particle path π, trajectories in the time-space domain. π' denotes the particle path in the absence of a deflagration front.

Denoting its position by a subscript d, self-similarity requires that

$$X_d \equiv r_d/r_s = \text{const}$$

One can write then

$$W_d \equiv \frac{dr_d}{dt} = \frac{d(x_d r_s)}{dt} = X_d \frac{dr_s}{dt} = X_d W_s \quad (11)$$

and

$$\frac{dW_d}{dt} = X_d \frac{dW_s}{dt}$$

whence, from Eq. (8),

$$\lambda_d \equiv -2r_d \frac{dW_d}{dt} / W_d^2 = \lambda$$

Jump conditions across the deflagration front have to satisfy the conservation of mass, momentum, and energy expressed respectively as

$$\rho_b(W_d - U_b) = \rho_u(W_d - U_u)$$

$$p_b + \rho_b(W_d - U_b)^2 = p_u + \rho_u(W_d - U_u)^2$$

$$\frac{\gamma_b}{\gamma_b-1}\frac{p_b}{\rho_b} + \tfrac{1}{2}(W_d - U_b)^2 = \frac{\gamma_u}{\gamma_u-1}\frac{p_u}{\rho_u} + \tfrac{1}{2}(W_d - U_u)^2 + q$$

where subscripts u and b denote, respectively, the states immediately ahead and behind the deflagration front, as illustrated in Fig. 1, and q is the exothermic energy per unit mass of reactants consumed. In terms of nondimensional variables, continuity equation can be written as

$$h_b(1 - F_b) = h_u(1 - F_u) \qquad (12)$$

which, with the momentum equations, yields the Rayleigh line equation

$$Z_b/(1-F_b) + (1-F_b) = Z_u/(1-F_u) + (1-F_u) \qquad (13)$$

while the energy equation is expressed as follows:

$$\frac{\gamma_b}{\gamma_b-1} Z_b + \tfrac{1}{2}(1-F_b)^2 = \frac{\gamma_u}{\gamma_u-1} Z_u + \tfrac{1}{2}(1-F_u)^2 + \omega \qquad (14)$$

where $\omega \equiv q/W_s^2$, a constant for a given self-similar wave.

Our solutions are obtained by evaluating ω from Eq. (14), in contrast to the practical problem where one wishes to determine (F_b, Z_b) for a given ω, once (F_u, Z_u) is specified, by the intersection of the locii expressed by Eqs. (13) and (14).

Shock Front. Similar equations to those describing the jump across the deflagration can be derived when subscript u is replaced by a, while b is replaced by n, denoting the state immediately behind the shock front, and $q = 0$. Moreover, $\gamma_a = \gamma_n = \gamma_u$; $F_a = 0$ according to Eq. (1); and by virtue of Eq. (9), $Z_a = y/\gamma_u$. Solving under such circumstances the energy and the Rayleigh line equations for F_n and Z_n in terms of y, one obtains

$$F_n = \frac{2}{\gamma_u+1}(1 - y)$$

$$Z_n = \frac{2}{(\gamma_u+1)^2}(2 - \frac{\gamma_u-1}{\gamma_u}y)(\frac{\gamma_u-1}{2} + y)$$

(15)

By eliminating y from the above set one obtains the locus of the states immediately behind the shock front, the Rankine-Hugoniot curve,

$$Z = (\frac{\gamma_u - 1}{2\gamma_u} F + \frac{1}{\gamma_u})(1 - F) \qquad (16)$$

Equations (1) and (15) combined with the continuity equation then yield

$$h_n = \frac{\gamma_u + 1}{\gamma_u - 1 + 2y}$$

$$g_n = \frac{2}{\gamma_u + 1}(1 - y) + \frac{y}{\gamma_u} \qquad (17)$$

When y = 0, point n is located on the strong shock limit. Equations (15) and (17) then reduce to the classical limit

$$F_n = g_n - 2/(\gamma_u + 1)$$

$$h_n = (\gamma_u + 1)/(\gamma_u - 1) \qquad (18)$$

$$Z_n = 2(\gamma_u - 1)/(\gamma_u + 1)^2$$

Wave Mach Number

As a consequence of the self-similarity requirement, the locii of constant Mach numbers coincide with lines of constant X, so that in view of Eq. (11),

$$M \equiv (W-u)/a = (XW_s - U)/a$$

which, with the use of Eq. (1), yields

$$M = (1 - F)/\sqrt{\gamma Z} \qquad (19)$$

It follows then from Eq. (4) that $D(F,Z)=0$ is the locus of sonic states. As a consequence of Eq. (19), if S denotes the speed of propagataion of the deflagration front,

$$M_u \equiv \frac{S_u}{a_u} \equiv \frac{W_d - U_u}{a_u} = \frac{1 - F_u}{\sqrt{\gamma_u Z_u}} \quad (20)$$

while

$$M_b \equiv \frac{W_d - U_b}{a_b} = \frac{1 - F_b}{\sqrt{\gamma_b Z_b}} \quad (21)$$

Mass Integral

The global mass conservation can be expressed as follows:

$$\int_0^{r_d} \rho r^j \, dr + \int_{r_d}^{r_s} \rho r^j \, dr = \rho_a r_s^{j+1}/(j+1)$$

or in a dimensionless form,

$$\int_0^{x_d} h x^j \, dx + \int_{x_d}^1 h x^j \, dx = 1/(j+1) \quad (22)$$

Energy Integral

The total amount of energy contained within the blast wave per unit area, or per unit length and polar angle, or per unit solid angle for $j = 0, 1,$ or 2, respectively, can be expressed as follows:

$$E \equiv \int_0^{r_d} \left(\frac{P}{\gamma_b - 1} + \rho \frac{u^2}{2} \right) r^j \, dr + \int_{r_d}^{r_s} \left(\frac{P}{\gamma_u - 1} + \rho \frac{u^2}{2} \right) r^j \, dr$$

$$= P_a r_s^{j+1} W_s^2 J \quad (23)$$

where

$$J \equiv \int_0^{X_d} \left(\frac{g}{\gamma_b-1} + \frac{hf^2}{2}\right) x^j dx + \int_{X_d}^1 \left(\frac{g}{\gamma_u-1} + \frac{hf^2}{2}\right) x^j dx$$

is referred to as the energy integral, a constant for a given self-similar blast wave. Taking into account the ambient integral energy engulfed by the propagating shock front, it follows from the principle of global energy conservation that

$$\frac{dE}{dt} = \rho_u S r_d^j q + W_s r_s^j \frac{p_a}{\gamma_u - 1}$$

$$= \rho_a r_s^j w_s^3 [h_u(1-F_u)X_d^{j+3} + y/\gamma_u(\gamma_u-1)] \quad (24)$$

Differentiating Eq. (23) with respect to time yields

$$\frac{dE}{dt} = \rho_a r_s^j W_s^3 (j+1 - \lambda)J \quad (25)$$

Comparing Eqs. (24) and (25), one obtains

$$(j+1 - \lambda) J = h_u(1-F_u) X_d^{j+3} \omega + y/\gamma_u(\gamma_u-1) \quad (26)$$

It should be noted that for the case of a strong adiabatic point explosion ($y=0$, $\lambda=j+1$), Eq. (26) yields $\omega=0$, as it should.

General Properties

Referring to Fig. 1, if $t_s = t_s(r,b)$ denotes the time at which one particle currently at radius r crossed the shock front, a particle path is identified by t_s = const. As a consequence, one has

$$\left(\frac{\partial r}{\partial t}\right)_{t_s} = u(r,t)$$

Noting that $X \equiv r/r_s(t)$,

$$\left(\frac{\partial r}{\partial t}\right)_{t_s} = r_s\left(\frac{\partial X}{\partial t}\right)_{t_s} + X\frac{dr_s(t)}{dt} = r_s\left(\frac{\partial X}{\partial t}\right)_{t_s} + XW_s(t)$$

and since $f(x,t) \equiv u(r,t)/W_s(t)$, the above relations yield

$$\left(\frac{\partial X}{\partial t}\right)_{t_s} = [f(x,t) - x]\frac{W_s(t)}{rx(t)} \qquad (27)$$

The velocity modulus

$$\mu \equiv \frac{d\ln r_s}{d\ln t} = \frac{W_s t}{r_s} \qquad (28)$$

is a constant for a given self-similar wave; it is related then to λ as follows (Oppenheim et al. 1972):

$$\mu = 2/(\lambda+2)$$

From Eqs. (27) and (28) one can deduce

$$\int_{t_s}^{t} \mu \frac{dt}{t} = \int_{x=1}^{X} \frac{dX}{f(X,t) - X}$$

which for a self-similar wave yields the time t at which a particle originally crossing the shock at t_s reaches X, in the form

$$t/t_s = \exp[C(X)/\mu] \qquad (29)$$

where

$$C(X) \equiv \int_{1}^{X} \frac{dX}{f(X) - X} \qquad (30)$$

is a function of x only, f being independent of t for self-similar waves. Integrating Eq. (28), the shock trajectory is expressed as

$$r_s/r_{so} = (t/t_o)^\mu \qquad (31)$$

whence

$$W_s/W_{so} = (t/t_o)^{\mu-1} \qquad (32)$$

where (r_{so}, t_o) is a reference point on the shock trajectory and $W_{so} \equiv \mu r_{so}/t_o$.

Referring to Fig. 1 again, a particle that has crossed the shock wave at time t_s releases its stored energy at time t_d, once it reaches the deflagration front. From Eq. (29),

$$t_d/t_s = \exp[C(X_d)/\mu] = \text{const}$$

Whence, by virtue of Eq. (32),

$$W_s(t_d)/W_s(t_s) = \text{const} \qquad (33)$$

Equation (14) demonstrates that $q/W_d^2 = \text{const}$ for a self-similar wave. Noting that $W_d(t) = X_d W_s(t)$,

$$q(t) = \text{const } W_s^2(t)$$

where $q(t)$ is the exothermic energy released at time t, or more accurately, at $t_d = t$. In view of Eq. (22), the above relation may be rewritten as

$$q(t_d) = \text{const } W_s^2(t_s) \qquad (34)$$

Eliminating t/t_o between Eqs. (31) and (32),

$$W_s/W_{so} = r_s/r_{so}^{(\mu-1)/\mu}$$

which yields, with Eq. (34), since q is invariant along the particle path,

$$q_a = \text{const } r_a^\beta \qquad (35)$$

where $\beta \equiv 2(\mu-1)/\mu$, while subscript a stresses the fact that Eq. (36) describes the distribution of stored energy in the undisturbed medium, which determines the acceleration of the double discontinuity system. For instance, in a stratified charge, q is proportional to the equivalence ratio ϕ; hence, from Eq. (35), $\beta = d\ln \phi/d\ln r$, yielding

$$\mu = \frac{2}{2-\beta} = \left(1 - \frac{d\ln\sqrt{\phi}}{d\ln r}\right)^{-1} \qquad (36)$$

Equation (36) demonstrates that a decelerating deflagration ($\mu<1$) propagates in a mixture which is rich towards the center, while an accelerating deflagration ($\mu>1$) is associated with a mixture leaner towards the center. Moreover, only a constant speed double discontinuity will be formed in a uniform charge.

It is important to note, however, that the above conclusions apply only to the case $y=0$, approached in practice when $p_n \gg p_a$, although the predicted trend is in agreement with experimental observations. Moreover, since $M_u \equiv S_u/a_u = $ const, self-similarity requires that the deflagration front propagates with a speed, S_u, proportional to $\sqrt{T_u}$, where T denotes the temperature.

A double discontinuity configuration may also result owing to the release of exothermic energy upon the elapse of an induction period after crossing the shock wave (Korobeinikov 1969; Lundstrom and Oppenheim 1969; Bishimov et al. 1970). While induction time under steady flow conditions is determined from an Arrhenius-expression, self-similarity requires a power temperature law, deduced as follows. If Δt denotes the induction time, let

$$\Delta t^0 = \rho^{0^{-m}} \tau^0(T^0)$$

where τ is a function of a form yet to be determined, superscript 0 denotes steady flow conditions, and m is an arbitrary constant (m>0). When the flow conditions change with time, a quasisteady assumption yields the induction time $\Delta t = t_d - t_s$ as the solution of the integral equation

$$\int_{t_s}^{t_d} \frac{dt}{\rho^{-m}(t)\,\tau^0[T(t)]} = 1 \qquad (37)$$

analogous to $\Delta t^0/\rho^{0^{-m}} \tau^0(T^0) = 1$, where $T(t)$ and $\rho(t)$ are the variations along the particle path. Defining

$$\theta(x,t) \equiv T(t,t_s)/T_n(t)$$

which is a function of alone for a self-similar wave, one has to invert Eq. (29) to solve the integral in Eq. (37). Instead we transform t to x in Eq. (37) which is rewritten for this purpose as follows:

$$\int_1^{X_d} \left(\frac{\partial t}{\partial X}\right)_{t_s} \frac{\rho_a^m [h(X)]^m \, dX}{\tau^0 [T_n(t)\theta(X)]} = 1$$

where X_d is regarded as the locus of the end states of the induction process. The above relation yields, when combined with Eqs. (27) and (28),

$$\int_1^{X_d} \rho_a^m \frac{t}{\mu} \frac{[h(X)]^m \, dX}{[f(X) - X]\tau^0 [T_n(t) \theta(X)]} = 1$$

For X_d = const, the function $\tau^0(T)$ has to be of such form as to admit separation of variables, i.e., $\tau^0(T_n\theta) = \tau^0(T_n)\tau^0(\theta)$, and also as to reduce $\tau^0[T_n(t)]/t$ to a constant value. Thus

$$\int_1^{X_d} \frac{[h(x)]^m dx}{[f(x) - x] \tau^0[\theta(x)]} = \frac{\mu \tau^0 [T_n(t)]}{\rho_a^m t} = \text{const}$$

yields a constant value for X_d, since both f and θ are functions of X only.

Since for a perfect gas, $p_n/\rho_n = RT_n$, Eq. (17) specifies in the case of y=0,

$$T_n = [2(\gamma_u-1)/(\gamma_u+1)^2] W_s^2$$

The above relation, combined with Eq. (32), yields $t = \text{const } T_n^\alpha$, where $\alpha \equiv (\mu-1)^{-1}/2$, hence

$$\tau^0(T) = (\text{const}/\mu) T^\alpha \qquad (38)$$

admitting separation of variables. Since induction time is known to decrease with temperature, $\alpha < 0$. Consequently, $\mu < 1$, in accordance with the fact that decoupled detonations may occur only in decaying wave systems.

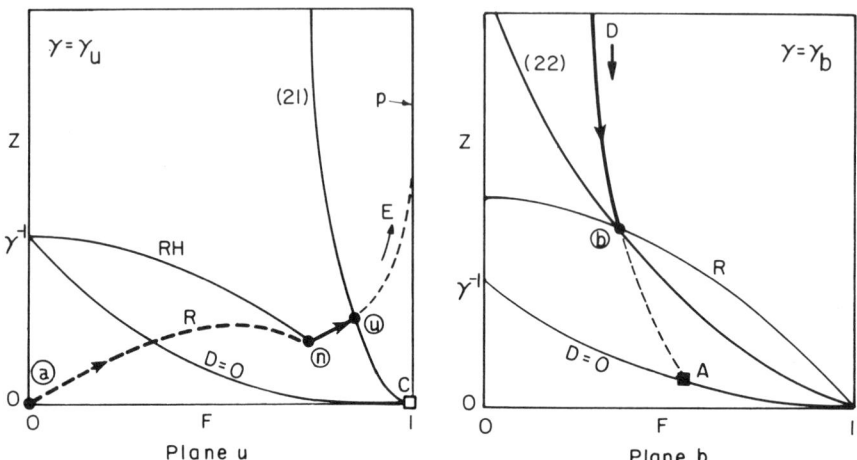

Fig. 2 Sketch of solution on the phase plane; y = 0. ■ - singularity A; □ - singularity C; D denotes singularity D; E denotes singularity E; p is the piston face; R is the Rayleigh line; RH is the Rankine-Hugoniot curve. Equations (20) and (21) are the locii of constant Mach numbers M_u and M_b, respectively. D = 0 is the sonic line.

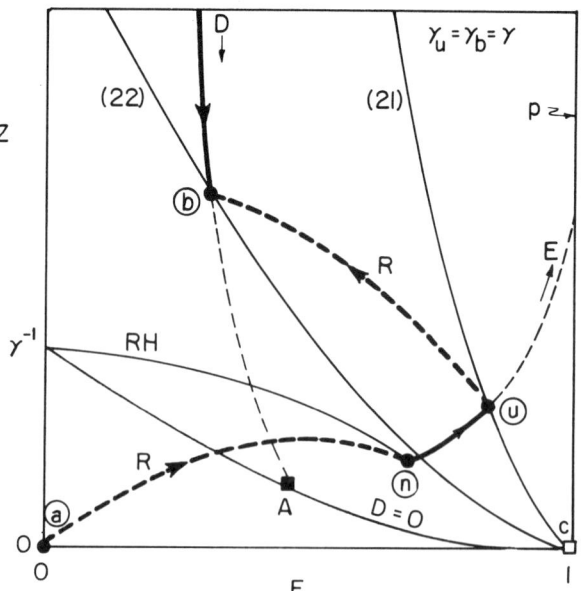

Fig. 3 Sketch of solution on the phase plane; $\gamma_u = \gamma_b = \gamma$, y = 0. Nomenclature used is the same as in Fig. 2.

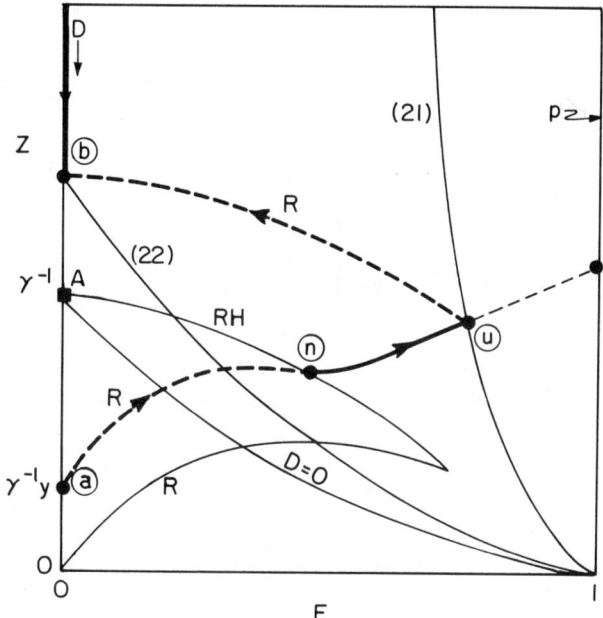

Fig. 4 Sketch of solution on the phase plane; $\gamma_u = \gamma_b = \gamma$, $y = 0$. Nomenclature used is the same as in Fig. 2.

Solutions

Integral Curves

Equations (2) and (3) yield a single first-order nonlinear O.D.E. governing the solution in the phase of the reduced coordinates, F and Z, namely,

$$\frac{dZ}{dF} = \frac{Z}{1-F} \frac{P(F,Z)}{Q(F,Z)} \qquad (39)$$

As a consequence of different values of γ ahead and behind the deflagration, Eq. (39) refers to two phase planes. Its integration is carried out as follows.

Referring to Fig. 2, the integral curve starts on plane u ($\gamma = \gamma_u$) at point n,, the strong limit ($y = 0$) of the Rankine-Hugoniot (RH) curve. It proceeds until the intersection with the locus of constant Mach number M_u, expressed by Eq. (20), thus fixed point u. The similarity variable X can be found by the quadrature of Eqs. (2) or (3), with $\gamma = \gamma_u$, starting from point n, where $x = 1$, until

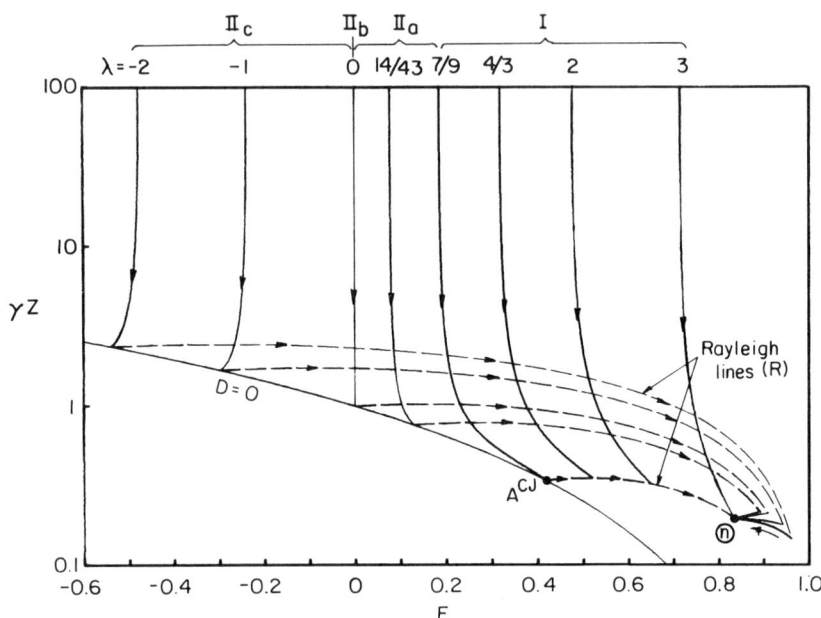

Fig. 5 Solution on the phase plane; $j = 2$, $\gamma_u = \gamma_b = \gamma = 1.4$. (——) - integral curve; (---) - jump across deflagration. A_{CJ} denotes singularity A when $\lambda = \lambda_{CJ} = 7/9$. $D = 0$ is the sonic line.

point u, where X_d is determined. The density h is evaluated using

$$\frac{h}{h_i} = \left[\frac{Z}{Z_i} \frac{1-F}{1-F_i} \left(\frac{x}{x_1}\right)^{(j+1)\nu+2} \right]^{1/(\gamma k - 1 - \nu)} \quad (40)$$

where $i = n$, $l = s$, and $k = u$, obtained from Eq. (10), and the definition of Z, rewritten as

$$\frac{g}{g_i} = \frac{Z}{Z_i} \frac{h}{h_i} \left(\frac{X}{X_i}\right)^2 \quad (41)$$

used also to evaluate the pressure g, since g_n is given by Eq. (18).

Because of the absence of source terms in the energy equation, the structure of induction zone is identical to that created ahead of a piston corresponding to the same value of the decay parameter λ. In fact, if integration proceeds beyond point u, it will eventually reach the line F = 1, representing the state at piston face, denoted by

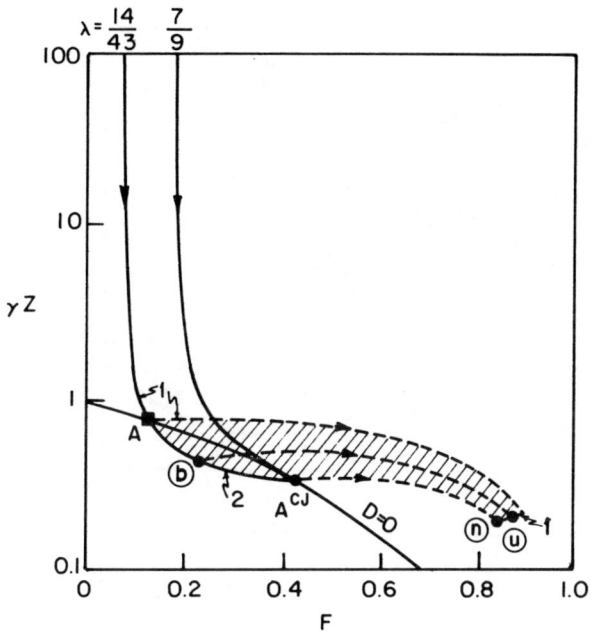

Fig. 6 Solution on the phase plane for $\lambda = 14/43$ as a typical case of group II; $j = 2$, $\gamma_u = \gamma_b = \gamma = 1.4$. For integral curve 1, $M_b = 1$, whereas for curve 2, $X_d = 1$ and $M_b > 1$. Nomenclature used is the same as in Fig. 5; ■ - singularity A. Hatched region is empty of solutions.

subscript p. From the continuity at the piston face $U_p = W_p$ and since $W_p = X_p W_s$, in view of Eq. (11), $F_p = 1$. Substituting $f(X_p) = X_p$ in Eq. (30) yields $C(X_p) = \infty$, whence, according to Eq. (29), a particle reaches the piston face at infinite time (along path π' shown by broken line in Fig. 1). The point of intersection can coincide either with singularity $E(Z=\infty)$ or singularity $C(Z=0)$, (Oppenheim et al. 1972). Once the point (F_u, Z_u) is determined, Eq. (13) specifies the locus of point b, on plane b$(\gamma = \gamma_b)$ to which the integration is transferred at this stage.

The integral curve starts again at singularity $D[Z_D = \infty, F_D = \lambda/(j+1)\gamma_b]$, representing the center of symmetry, and proceeds until it intersects the Rayleigh line (R), Eq. (13), fixing point b. The value of h_b is then specified by Eq. (12), h_u/h_n being given by Eq. (40) and h_n by Eq. (19), while is found using Eq. (14) and the Mach number M_u is obtained from Eq. (21).

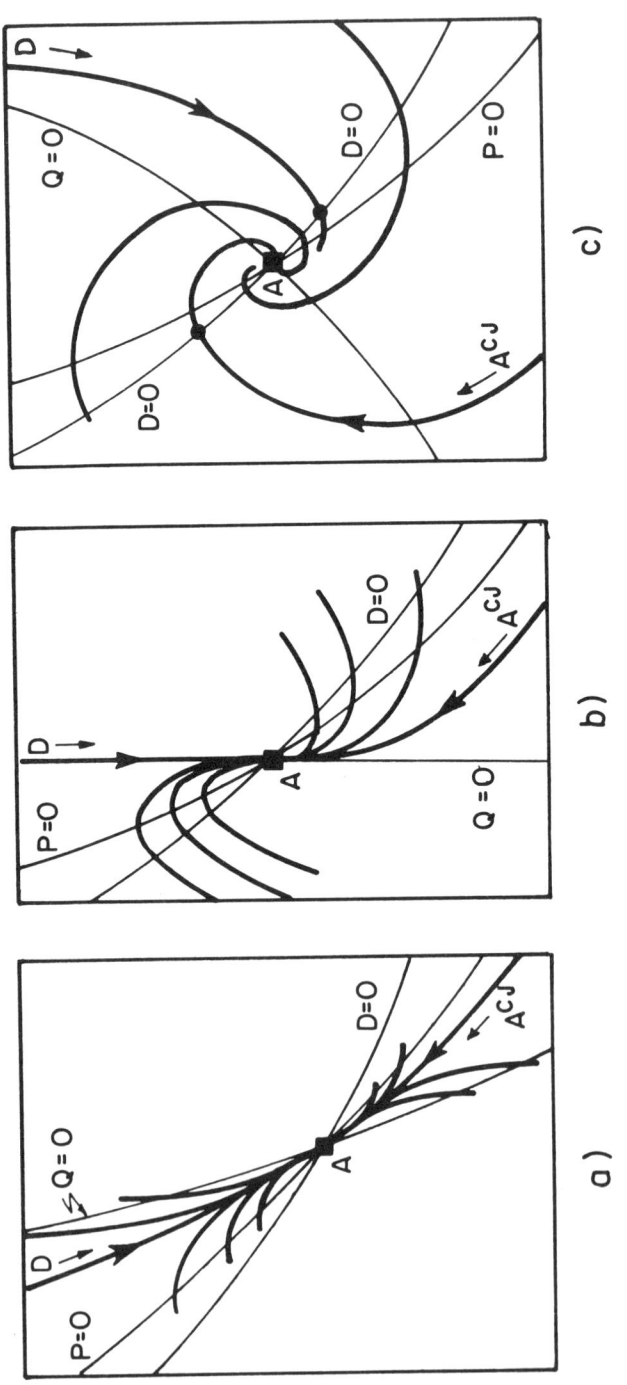

Fig. 7 Singularity A ■ specified by shape of Q = 0 and P = 0 locii: a) λ > 0, A is a nodal point singularity; b) λ = 0, A is a degenerate nodal point singularity; c) λ < 0, A is a spiral point singularity. D = 0 is the sonic line. ACJ is the singularity A when λ = λ$_{CJ}$ = 7/g; D is the singularity D.

The values of X are found by the quadrature of Eq. (2) or Eq. (2), with $\gamma = \gamma_b$, starting from point b where $X = X_d$. The density h is obtained from Eq. (40), when i = b, l = d, and k = b, while the pressure g is evaluated using Eq. (41), with i = n.

If one proceeds with the integration beyond the point b, the integral curve eventually attains singularity A, at the intersection with the sonic line, D = 0.

The accuracy of the integration is checked using Eq. (22), and Eq. (26) with y = 0. The results reported in this study are for the case when $\gamma_u = \gamma_b = \gamma$. As a consequence, the two planes of Fig. 2 coincide, as shown in Fig. 3. One can have then a Rayleigh line joining points u and b.

The above solution applies to the case of y = 0, irrespectively of the value of λ. Another class of self-similar solutions is obtained for $\lambda = 0$, while y ≠ 0, representing the constant speed double discontinuity. A similar strategy can be followed then to derive the solution. As demonstrated on Fig. 4, the integral curve from the center coincides with the axis F = 0. As a consequence, $u = U_b = 0$, $h = h_b$ and $g = g_b$ behind the

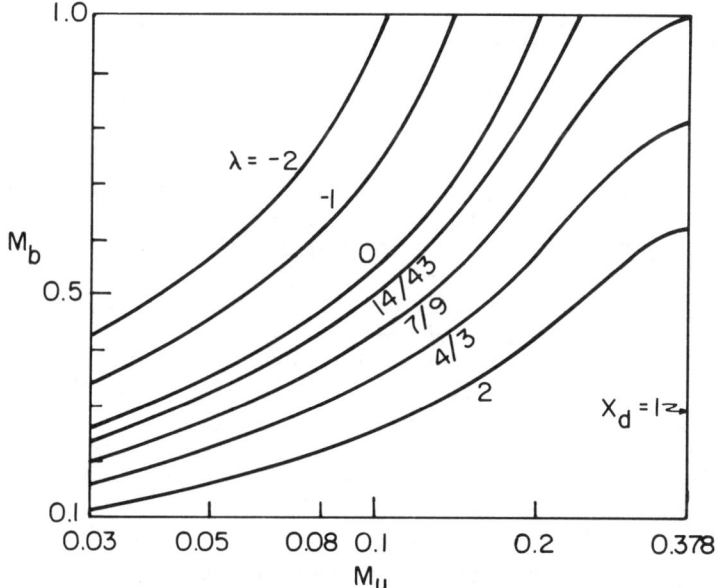

Fig. 8 Relationship between Mach numbers on the two sides of the deflagration; j = 2, $\gamma_u = \gamma_b = \gamma = 1.4$.

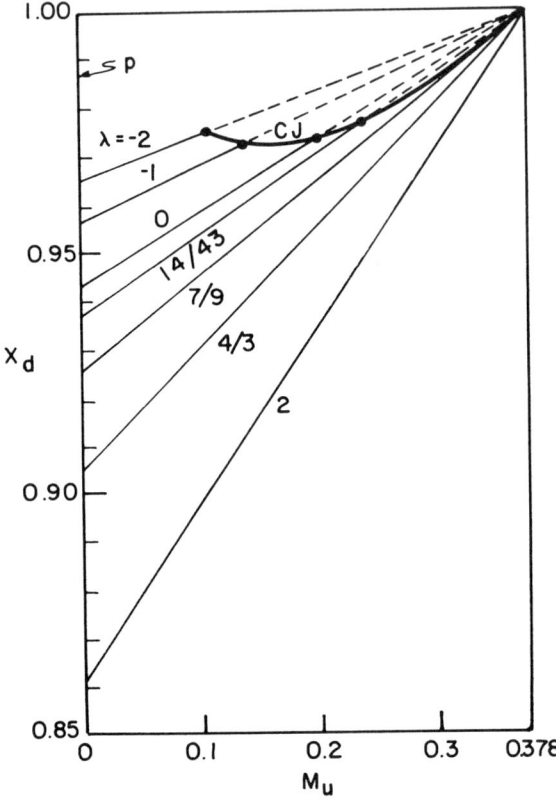

Fig. 9 Deflagration location X_d as a function of its propagation Mach number, M_u; $j = 2$, $\gamma_u = \gamma_b = \gamma = 1.4$. (——) - possible solutions; (---) - hypothetical solutions corresponding to strong deflagrations. p is the piston; CJ is the locus of Chapman-Jouguet deflagrations; $M_b = 1$.

deflagration. One does not need therefore to integrate Eqs. (39), (2), (3), or use Eqs. (40) and (41); and the solution can be obtained without considering plane b.

Limits of Solution ($\gamma_u = \gamma_b = \gamma$)

Weak deflagrations are limited by the Chapman-Jouguet (CJ) state. As a consequence, $0 \leq M_u \leq M_{CJ}$, while $0 \leq M_b \leq 1$. According to Eq. (20), $M_u = 0$ corresponds to $F_u = 1$, so that point u is on a piston face. Since at the Chapman-Jouguet state, $M_b = 1$, point b is located on the line $D(F,Z) = 0$. However, M_{CJ}, yielding $M_b = 1$, is not known a priori. The integration procedure to be used is therefore switched around. Integral curve starts at singularity D on plane b until it reaches the sonic line, thus fixing point b. Integration is restarted then on plane u from point n towards the piston face until it is intercepted by the

BLAST WAVES INCORPORATING DEFLAGRATIONS

Rayleigh line, thus fixing point u. The limiting Mach number M_{CJ} is then determined from Eq. (20).

The CJ limit cannot be reached, however, if singularity A is located below the Rayleigh line through state n. This restriction divides the set of solutions into two groups, as shown in Fig. 5.

Group I: $\lambda_{CJ} \leq \lambda \leq j+1$

Group I extends between $\lambda = j+1$, corresponding to adiabatic point explosion ($\lambda > j+1$ corresponds to blast waves with energy withdrawal, excluded since deflagrations are heating discontinuities), and λ_{CJ}, when singularity A is located on the Rayleigh line, through state a; i.e., when

$$F_A^{CJ} = 1/(\gamma+1)$$

and

$$Z_A^{CJ} = \gamma/(\gamma+1)^2$$

obtained by solving Eqs. (4) and (13), when u = a, simultaneously. Superscript CJ is used to denote the values associated with $\lambda = \lambda_{CJ}$. Since singularity A is located at the intersection of $D(F,Z)=0$ and $Q(F,Z)=0$, one obtains upon solving Eqs. (4) and (5) simultaneously (see Oppenheim et al. 1972),

$$F_A = \frac{1}{2}(\frac{2-\gamma}{2\gamma j}\lambda + 1) - \sqrt{\frac{1}{4}(\frac{2-\gamma}{2\gamma j}\lambda + 1)^2 - \frac{\lambda}{\gamma j}}$$

Equating F_A to F_A^{CJ}, the expression of λ_{CJ} is obtained:

$$\lambda_{CJ} = \frac{2}{3}\frac{\gamma j}{\gamma+1}$$

For $\gamma = 1.4$ and $j = 2$, $\lambda_{CJ} = 7/9$.

Integral curves of this group, starting from the center, cannot reach the sonic line, since they are intercepted by the Rayleigh line through point n. This implies that for the cases of this group associated with the highest possible Mach number M_u, u = n so that $X_d = 1$. By virtue of Eqs. (18) and (19) one has then

$$M_u = M_n = (1-F_n)/\sqrt{\gamma Z_n} = \sqrt{(\gamma-1)/2\gamma} \qquad (42)$$

For $\gamma = 1.4$, $M_n = 0.378$. Moreover, for these limiting cases, point b lies on the Rayleigh line between point n

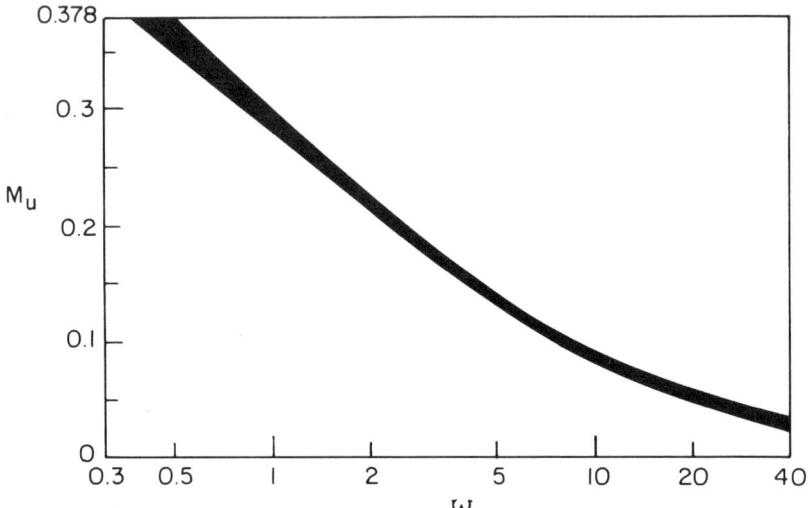

Fig. 10 Relation between propagation Mach number M_u and specific exothermic energy ω; $j = 2$, $\gamma_u = \gamma_b = \gamma = 1.4$. Black strip covers the full scope of λ.

when $\lambda = j+1$, yielding then $M_b = M_n$, and singularity A^{CJ} when $\lambda = \lambda_{CJ}$, resulting in $M_b = 1$ in this case. One should note that although at these limiting cases $X_d = 1$, there is still a jump $a \to n$ across the shock front followed by a drop $u \to b$ across the deflagration front, both occurring at the same location in space, thus maintaining the double discontinuity configuration.

Group II: $-2 \leq \lambda \leq \lambda_{CJ}$

Group II extends between λ_{CJ} and $\lambda = -2$, corresponding to an infinitely accelerating deflagration. Figure 6 shows two limiting integral curves, marked 1 and 2. They are limiting curves in the sense that for both $M_b = 1$, corresponding to CJ deflagrations. As illustrated in Fig. 6, for any integral curve in between, point b is located below the sonic line, $D = 0$, whence, according to Eq. (21), $M_b > 1$. Since strong deflagrations have been ruled out, the hatched region in Fig. 6 is devoid of any possible solution, thus representing a gap in the values of X_d and M_u extending between those values associated with integral curve 1 and $X_d = 1$, $M_u = M_n$.

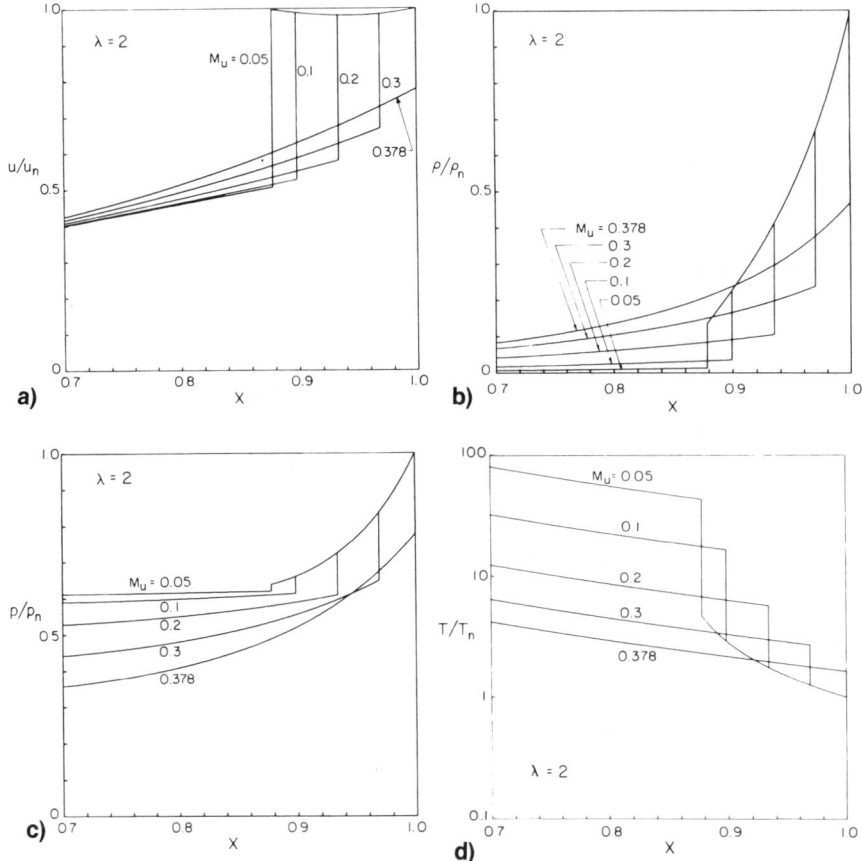

Fig. 11 For $\lambda = 2$, $j = 2$, $\gamma_u = \gamma_b = \gamma = 1.4$, and $M_u = 0.05$, 0.1, 0.2, 0.3, 0.378: a) particle velocity; b) density; c) pressure; d) temperature profiles.

The cases of group II associated with $X_d = 1$ have been investigated in detail by Barenblat et al. (1980). For these, $u = n$ and $b = A^{CJ}$. Moreover, when $0 > \lambda > -2$, the solution embodies an internal weak shock wave in the vicinity of singularity A, as explained below, thus adding a third discontinuity. As a consequence, solutions in group II can be divided into three subgroups according to the nature of singularity A, which depends on the shape of $Q(F,Z)=0$, representing the locus of infinite slopes of integral curves, and $P(F,Z)=0$, representing the locus of their zero slopes [see Eq. (39)]. As demonstrated on Fig. 7, for $0 < \lambda \leq \lambda_{CJ}$, singularity A is a nodal point; at $\lambda = 0$

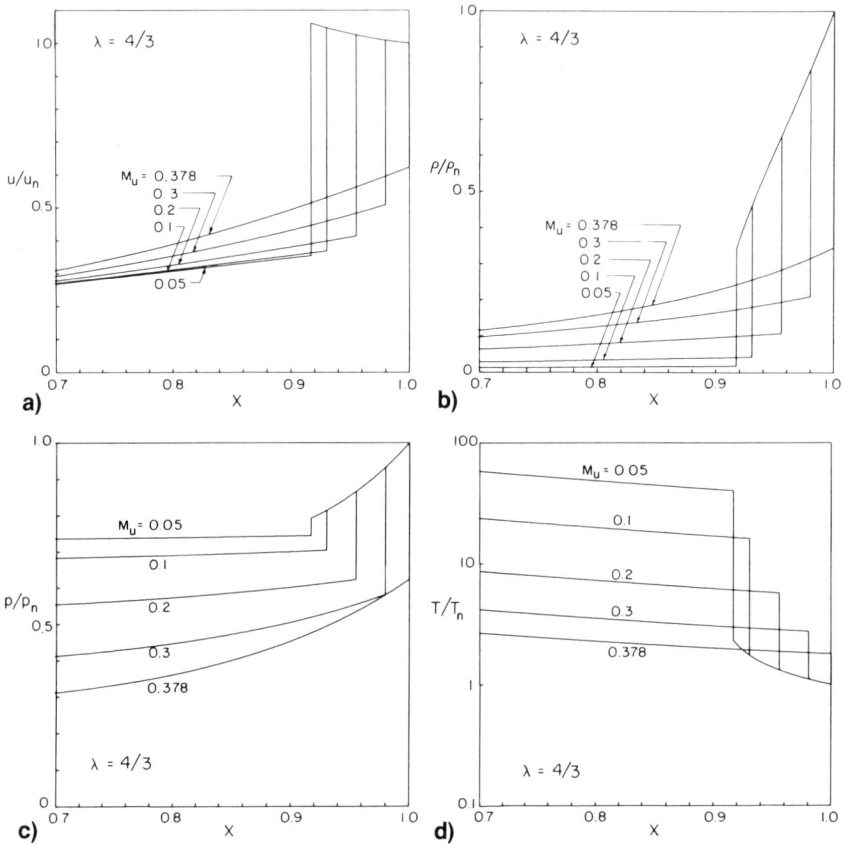

Fig. 12 For $\lambda = 4/3$, $j = 2$, $\gamma_u = \gamma_b = \gamma = 1.4$, and $M_u = 0.05$, 0.1, 0.2, 0.3, 0.378: a) particle velocity; b) density; c) pressure; d) temperature profiles.

it becomes a degenerate nodal point, while for $-2 \leq \lambda < 0$, it is a spiral point singularity. The importance of the subdivision is associated with the place where integral curves intersect the sonic line $D = 0$. If at this point $Q(F,Z) = 0$, then, according to Eq. (2), $d\ln x/dF = 0$; if on the other hand $P(F,Z) = 0$, then, according to Eq. (3), $d\ln x/dZ = 0$ and $Q = 0$ imply $P = 0$. Therefore, unless the integral curve crosses $D = 0$ at singularity $A(Q(F_A, Z_A)=0$, $P(F_A, Z_A)=0)$, $dX=0$, i.e., X exhibits a local maximum or minimum. Under such circumstances, in the neighborhood of the point of intersection, two different points on the integral curve are associated with each value of X. This is impossible unless one admits the presence of a discontinuity between some point on the integral curve and another, located on the other side of the sonic line D = 0, thus circumnavigating it. Both points should have the same value of X and since the zone above D = 0 is subsonic while that below is supersonic such a discontinuity can represent a shock front. In other words, one can always locate two points, one on the integral curve between singularities D and A and the other on that between singularities D and A and the other on that between singularity A and point A^{CJ} connected by a Rayleigh line for which Eq. (14) yields ω = 0. According to Fig. 7, one has thus the following subgroups delineated earlier in Fig. 5.

Group IIa: $0 \leq \lambda \leq \lambda_{CJ}$. The integral curve starting from the center as well as that from A^{CJ}, both cross the sonic line D = 0 at singularity A. Consequently, there is no internal shock for those cases associated with X_d = 1 and M_b = 1.

Group IIb: λ = 0. The integral curve starting from the center as well as the Q = 0 locus coincide with the axis F = 0, which intersect the sonic line D = 0 at singularity A. Since the Q = 0 line, the locus of infinite slopes, is vertical, the integral curve starting from A^{CJ} cannot cross it and therefore must intersect the sonic line at singularity A, making a continuous solution possible when X_d = 1 and M_b = 1.

Group IIc: $-2 \leq \lambda \leq 0$. Both integral curves do not intersect the sonic line D = 0 at singularity A, imposing thereby a requirement for an internal shock discontinuity in those cases associated with X_d = 1 and M_b = 1.

Results

The phase plane of the solutions for j = 2 and γ = 1.4 is presented in Fig. 5. Included there are only the integral curves corresponding to the highest Mach numbers M_u. The arrows indicate the direction of increasing X. Starting from the center at singularity D located at Z = ∞, each integral curve embodies first the deflagration discontinuity represented by a Rayleigh line (R), and then proceeds towards the strong shock point n. The sonic line, D = 0, delineates the cases for which the Chapman-Jouguet (CJ) deflagration is attainable, namely, those associated with $-2 < \lambda < 7/g$. Point A^{CJ} denotes the position of singularity A when $\lambda = \lambda_{CJ}$ (λ_{CJ} = 7/g for j = 2 and γ = 1.4).

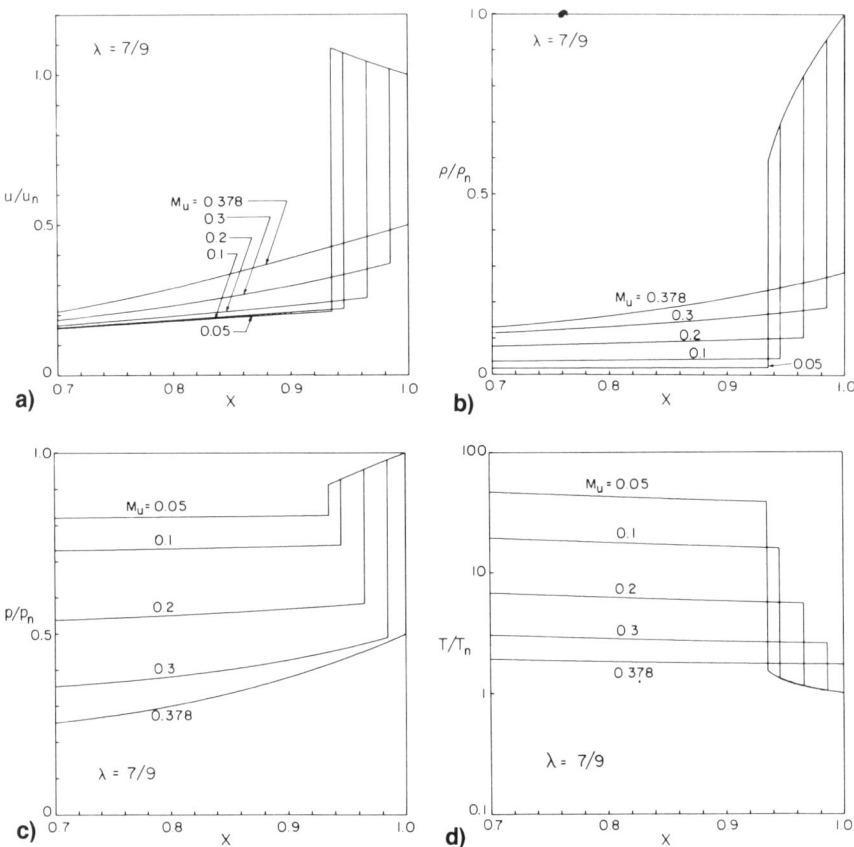

Fig. 13 For $\lambda = \lambda_{CJ} = 7/9$, $j = 2$, $\gamma_u = \gamma_b = \gamma = 1.4$, and M_u = 0.05, 0.1, 0.2, 0.3, 0.378: a) particle velocity; b) density; c) pressure; d) temperature profiles.

The relationship between the Mach numbers on the two sides of the deflagration discontinuity is depicted in Fig. 8. According to Eq. (42), M_u = 0.378 (for $\gamma = 1.4$) is the locus of cases when the flame is coupled to the shock front, i.e., $X_d = 1$, while $M_b = 1$ is the locus of cases attaining CJ condition. The point (M_u = 0.378, M_b = 1) is common to all solutions associated with $-2 \leq \lambda \leq 7/9$. As explained ealier, there is a gap empty of solutions extending between M_u at the upper end point of these curves and M_u = 0.378.

The flame location X_d, as a function of propagation Mach number, M_u, is presented in Fig. 9, displaying the amazingly straight line relationship between these two parameters. Noting that M_u = 0 represents a piston face, denoted by a subscript p, and $M_n = 0.378$, one has

$$X - X_p \cong (M/M_n)(1 - X_p) \qquad (43)$$

The accuracy of the above relation has been tested when $\lambda = j+1$ (3 for $j = 2$), for which $X_p = 0$, providing, therefore, testing values of X as far as $X = 0$. Equation (43) proved to be accurate near $X = 1$, losing accuracy as the point is moved away from it. For $\lambda = 2$, $X_p \cong 0.86$. For similar values of λ, the piston is closer to the front, indicating that almost all of the domain of interest lies within the zone of good accuracy.

Recalling Eq. (2), one can write, near $X = 1$,

$$\frac{dF}{dX} \cong - \frac{Q(F_n,Z_n)}{D(F_n,Z_n)} \equiv C_n$$

Yielding, upon integration starting from point n,

$$(1-F) \cong (1-F_n) - C_n(X-1) \qquad (44)$$

Equation (44) may be used to approximate X_p, since $F = 1$ at the piston face. Thus

$$C_n \cong (1-F_n)/(X_p-1)$$

and Eq. (44) becomes, after substituting for C_n from above,

$$(1-F)/(1-F_n) \cong (X-X_p)/(1-X_p) \qquad (45)$$

With reference to Eq. (19), the Mach number can be approximated as follows:

$$M \cong (1-F)/\sqrt{\gamma Z_n} \qquad (46)$$

This is due to the fact that the square root is a weak function and because the integral curves on the phase plane are flat in the vicinity of point n. Equations (42) and (46) substituted in Eq. (45) yield Eq. (43) above, expressing the straight line relationship between X_d and M_u.

It should be noted here that Eq. (44) is still valid near $F = 1$, although the integral curves on the phase plane are tangent to the $F = 1$ line there, because of the term $(1 - F)$ in Eq. (3). As a consequence of this term, when $F = 1$, $d\ln X/dZ = 0$, i.e., $X = $ const. The value of X_p is fixed

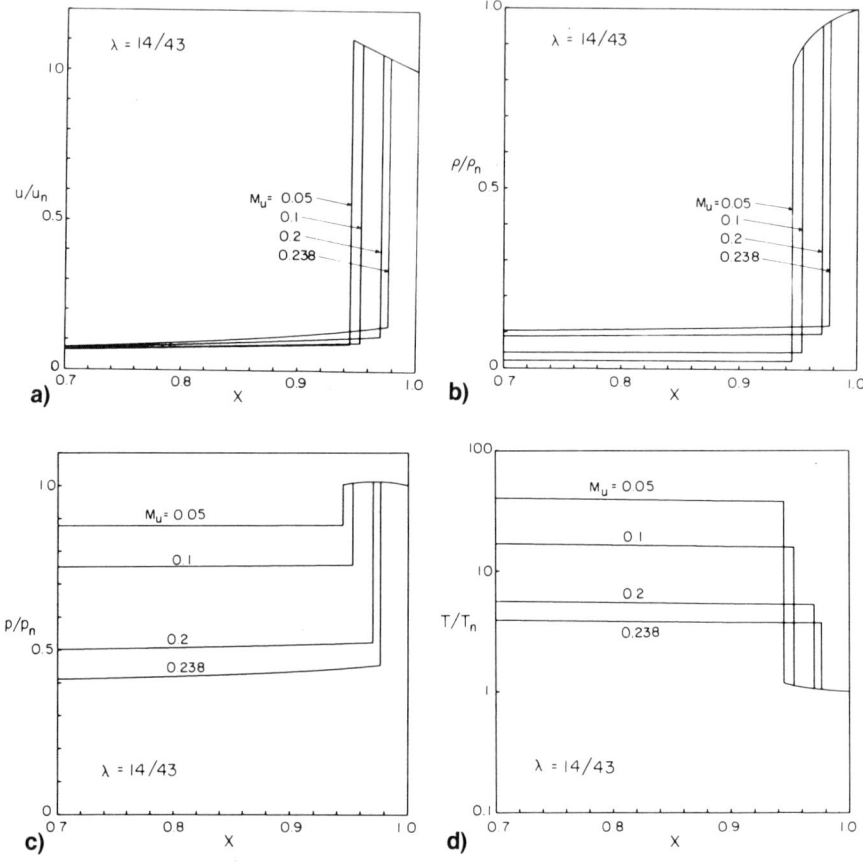

Fig. 14 For $\lambda = 14/43$, $j = 2$, $\gamma_u = \gamma_b = \gamma = 1.4$, and $M_u = 0.05, 0.1, 0.2, 0.238$: a) particle velocity; b) density; c) pressure; d) temperature profiles.

therefore, with sufficient accuracy, as soon as the integral curve comes close enough to F = 1.

The locus CJ of $M_b = 1$ is demonstrated on Fig. 9. It is tangent to the line of $\lambda = 7/9$ at $M_u = 0.378$. The broken lines above $M_b = 1$ are hypothetical solutions possible only for stong deflagrations. Again, the point ($M_u = 0.378$, $X_d = 1$) is common to all solutions, and the gap empty of solutions mentioned earlier is exhibited as the region above the CJ locus.

Figure 10 presents the relationship between the propagation Mach number of the deflagration and its specific exothermic energy. The black strip covers the full scope of

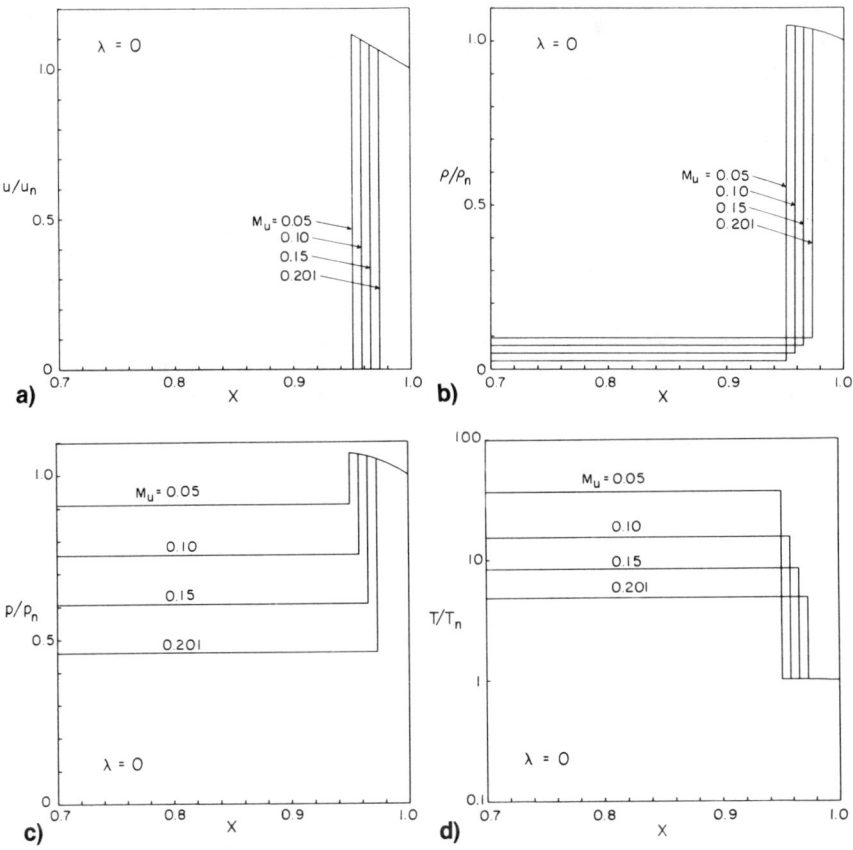

Fig. 15 For $\lambda = 0$, $j = 2$, $\gamma_u = \gamma_b = \gamma = 1.4$, and $M_u = 0.05$, 0.1, 0.15, 0.201: a) particle velocity; b) density; c) pressure; d) temperature profiles.

λ. In principle, for given values of M_u and ω one should be able to determine λ by locating the proper curve passing through the point of intersection fo the two corrdinates. However, since all the solutions are confined to a narrow strip, can shift from one value to another with a very small perturbation in either M_u or ω, thus manifesting the instability of deflagration waves due to gasdynamic effects.

The particle velocity, density, pressure, and temperature profiles for $j = 2$ and $\gamma = 1.4$ are depicted in Figs. 11-17, for the decay coefficients $\lambda = 2$, 4/3, 7/g, 14/43, 0, -1, -2 corresponding to velocity moduli $\mu = 0.5$, 0.6, 0.72, 0.86, 1, 2, ∞, respectively. The deflagration is associated by a drop in particle velocity, density, and

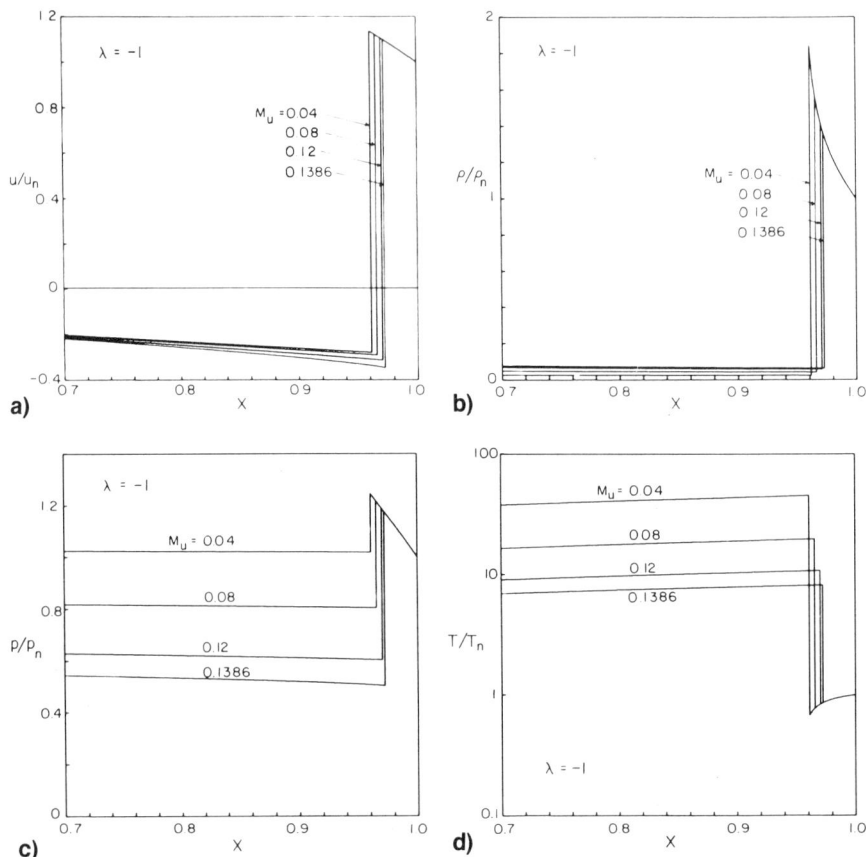

Fig. 16 For $\lambda = -1$, $j = 2$, $\gamma_u = \gamma_b = \gamma = 1.4$, and $M_u = 0.04$, 0.08, 0.12, 0.1386: a) particle velocity; b) density; c) pressure; d) temperature profiles.

pressure and a rise in termperature. One should note that the rarefaction, due to the blast wave expansion in a spherical geometry, is compensated by the pressure waves emanating from the deflagration. As a consequence, for values of $\lambda = 14/43$, 0, -1, -2, pressure increases behind the shock, reaching a maximum immediately ahead of the deflagration front, a phenomenon which at $\lambda \leq 0$ is accompanied by density increase.

Summary

The problem of blast waves incorporating deflagrations of variable speed was formulated under conditions soft

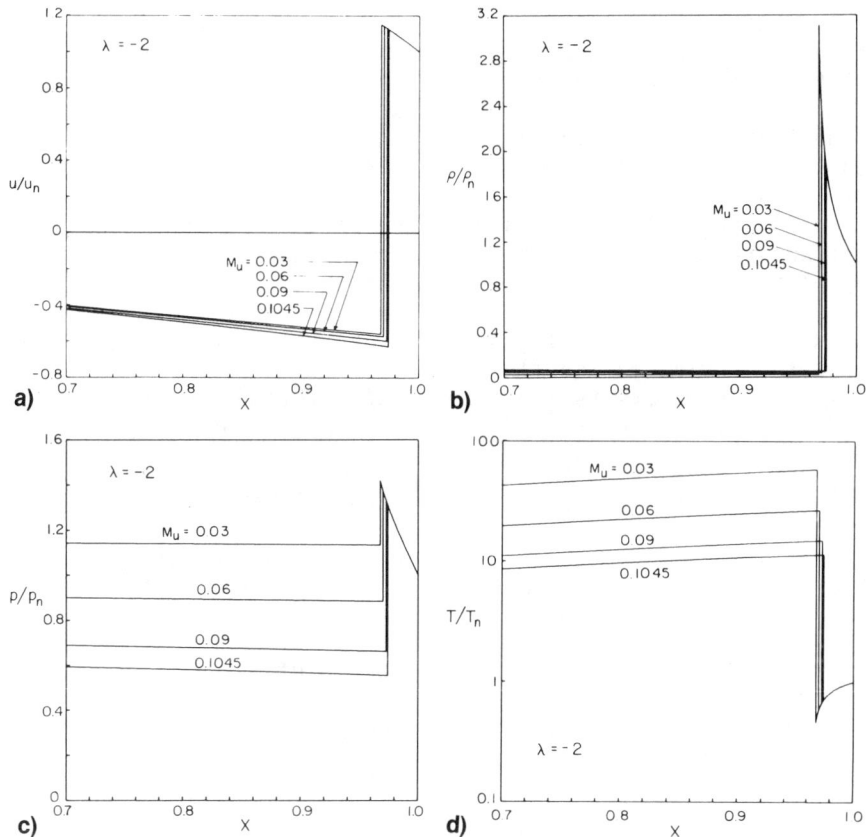

Fig. 17 For $\lambda = -2$, $j = 2$, $\gamma_u = \gamma_b = \gamma = 1.4$, and $M_u = 0.03$, 0.06, 0.09, 0.1045: a) particle velocity; b) density; c) pressure; d) temperature profiles.

self-similarirty, while the reactants and products were considered to behave as perfect gases. Although the strategy proposed to solve the problem was general, in the set of solutions obtained here the specific heat of reactants was taken to be equal to that of products.

The formulation was presented in the Eulerian form. That in the Lagrangian form was invoked only to derive the following conclusions:

1) The distribution of stored energy in the undisturbed medium determines the acceleration of the deflagration-shock wave system.

2) If the exothermic energy is released upon elapse of a certain induction time after crossing the shock wave,

self-similarity requires a power temperature law for the temperature dependence of the induction time, rather than that specified by the classical Arrhenius expression.

By studying the singularities of the governing equations, the restrictions implied by gasdynamic conditions on the propagation speed of the deflagration were found to be as follows:

1) For a certain range of their decelerations, the deflagrations cannot reach the Chapman-Jouguet (CJ) velocity.

2) An accelerating detonation wave (deflagration coincident with a leading shock front) must always contain a weak shock wave in the flowfield.

3) There exists a gap empty of solutions between the CJ deflagration and the CJ detonation except at $\lambda = \lambda_{CJ} = 2/3\ \gamma j/(\gamma+1)$, where they both coincide. This gap can be filled only by the physically unacceptable strong deflagrations.

Finally, it is of interest to note, our results reveal the existence of a simple relation between the location of the deflagration and its Mach number.

Appendix: Numerical Solution

Integration of Eq. (39) is carried out on plane (Z^{-1} - F) if $Z>1$ or (Z-F) if $Z<-1$. F is considered an independent variable if $|dZ/dF|<1$ (or $|dZ^{-1}/dF|<1$); otherwise it is replaced by Z (or Z^{-1}). In the actual computations the bound on the derivative is relaxed to 1.1 to insure a smooth transition from F to Z (or Z-1) or vice-versa. The fourth-order Runge-Kutta method is used.

The space coordinate X is found by simultaneous quadrature of either Eqs. (2) or (3), depending upon whether F or Z is the independent variable.

Integral curves from point n to point u are first obtained. At saddle point singularity D [$F=\lambda/(j+1)\alpha$, Z-1=0], representing the center of wave, the slope of the integral curve on the phase plane is evaluated by applying l'Hospital rule. One obtains, then,

$$\left(\frac{dF}{dZ^{-1}}\right)_D = \frac{(a-b)(1-b)^2 b}{(j+1)(1-b) + [(j+1)(\gamma-1) + 2](c-b)}$$

In the vicinity of singularity D, up to some arbitrary point O, a first-order asymptotic expansion is used. The integral curve becomes

$$F = F_D + \left(\frac{dF}{dZ^{-1}}\right)_D Z^{-1} + \sigma(Z^{-2})$$

Numerical solution then proceeds from point 0. X_0 being unknown a priori, the differential $d\ln X$ in Eqs. (2) and (3) is replaced by $d\ln(X/X_0)$. The value X_0 is adjusted so that X at point b equals X_d. The density h and the pressure g are then determined using Eqs. (40) and (41), sequentially.

The mass and energy integrals, expressed in Eqs. (22) and (23), respectively, are evaluated by the trapezoidal rule, starting from point 0. Their value there is obtained by first-order asymptotic expansions of integrals in terms of X, yielding

$$\int_0^{X_0} hx^j dx = h_0 x_0^{j+1} \frac{(\gamma-\nu)}{(j+1)}$$

and

$$\int_0^{X_0} \left(\frac{g}{\gamma-1} + \frac{hf^2}{2}\right) x^j dx = x_0^{j+1} \left(\frac{g_0}{(\gamma-1)(j+1)} + \frac{h_0}{2} F_0^2 x_0^2 \frac{(\gamma-\nu)}{(j+3)\gamma-2}\right)$$

It should be noted that in those cases associated with Chapman-Jouguet deflagrations, point b coincides with singularity A, if $\lambda > 0$. Being located at the intersection of Q = 0 and P = 0 locii, dZ/dF [Eq. (39)], changes rapidly in its vicinity. The step of integration is reduced whenever the integral curve tends to cross either Q = 0 or P=0, so that it reaches singularity A smoothly. On the other hand, when $\lambda < 0$, the integral curve from the center first crosses the sonic line D = 0, thus fixing point b, and the step is reduced when the integral curve approaches the D = 0 line.

Acknowledgments

This work was supported by the Office of Energy Research, Office of Basic Energy Sciences, Division of Basic Engineering Research of the Department of Energy under Contract W-7405-ENG-48, the National Science Foundation under Grant ENG 787-12372, and the NASA Lewis Research Center under Grant NAG 3-131.

References

Barenblatt, G. I., Guirguis, R. H., Kamel, M. M., Kuhl, A. L., Oppenheim, A. K., and Zel'dovich, Y. B. (1980) Self-similar explosion waves of variable energy at the front. J. Fluid Mech. 99, 841-858.

Bishimov, E., Korobeinikov, V. P., and Levin, V. A. (1970) Strong explosion in combustible gaseous mixture. Acta Astron. 15, 267-273.

Korobeinikov, V. P. (1969) The problem of point explosion in a detonating gas. Acta Astron. 14, 411-419.

Kuhl, A. L., Kamel, M. M., and Oppenheim, A. K. (1973) Pressure waves generated by steady flames. Fourteenth Symposium (International) on Combustion, pp. 1201-1215. The Combustion Institute, Pittsburgh, Pa.

Lundstrom, E. A. and Oppenheim, A. K. (1969) On the influence of non-steadiness on the thickness of the detonation wave. Proc. R. Soc. London A310, 463-478.

Mallard, E. and Le Chatelier, H. (1881) Sur la vitessse de propagation de l'inflammation dans les melanges explosifs. C. R. Acad. Sci. Paris 93, 145-148.

Oppenheim, A. K. (1952) A contribution to the theory of the development and stability of detonation in gases. J. Appl. Mech. 19, 63-71.

Oppenheim, A. K., Lundstrom, E. A., Kuhl, A. L., and Kamel, M. M. (1971) A systematic exposition of the conservation equations for blast waves. J. Appl. Mech. 38, 783-794.

Oppenheim, A. K., Kuhl, A. L., Lundstron, E. A., and Kamel, M. M. (1972) A parametric study of self-similar blast waves. J. Fluid Mech. 52, 657-682.

Oppenheim, A. K. and Kamel, M. M. (1973) Photographic laboratory studies of explosions. Aerotec. Missili Spazio 2, 122-134.

Oppenheim, A. K., Kurylo, J., Cohen,, L. M., and Kamel, M. M. (1978) Blast waves generated by exploding clouds. Proceedings of the Eleventh International Symposium on Shock Tubes and Waves, pp. 67-200. University of Washington Press, Seattle, Wash.

Shchelkin, K. I. and Troshin, Y. K. (1965) Gasdynamics of Combustion. Mono. Book Corporation, Baltimore, Md. (translated from Russian).

Analysis of Reactive Blast Waves Propagating Through Gaseous Mixtures with Spatially Distributed Chemical Energy

S. Ohyagi* and A. Ohsawa†
Saitama University, Saitama, Japan

Abstract

Accidental spills of fuel gases and liquids into the open air form combustible vapor clouds which can explode. The fuel concentration in such vapor clouds usually varies with the distance from the fuel source. The present analysis makes such vapor cloud explosions one-dimensional point symmetric reactive blast waves (detonation waves) in which the energy of combustion is released instantaneously at their fronts. The distribution of the fuel concentration is taken into account by variation of the detonation energy as a function of distance from the center of explosion. As a model distribution, $Q = Q_0 \exp(-\beta x)$ is used where Q and x denote the chemical energy and the distance from the explosion center and Q_0 and β are constant. The quasi-similar technique is used to solve approximately the non-self-similar blast wave equations, but even then it cannot be solved straightforwardly without an iterative technique when the variation of Q is considered. The analysis shows that the overdriven detonation waves decay to the local Chapman-Jouguet detonation wave solution when β is small, and that they decay to the sound wave as a nonreactive blast wave for large β.

Introduction

When a finite amount of energy is released instantaneously at a point in a combustible gas, a blast wave

---

Paper presented at the 8th ICOGER, Minsk, USSR, Aug. 23-26, 1981. Copyright © American Institute of Aeronautics and Astronautics, Inc., 1982. All rights reserved.

*Lecturer, Department of Mechanical Engineering, Faculty of Engineering.

†Student of Mechanical Engineering, Department of Mechanical Engineering, Faculty of Engineering.

accompanied with a combustion wave is formed. If the
initiation energy exceeds a certain amount (which is called
a critical energy), the blast wave consists of a single wave
front in which combustion is completed. This case has been
termed a detonation wave model. Korobeinikov (1969);
Bishimov, Korobeinikov, and Leven (1970); and Korobeinikov
(1972) treated analytically and numerically the phenomenon
for the limiting cases: the detonation model and two-front
model, which modeled the blast wave as coupled shock and
combustion wave discontinuities; and the chemo-kinetic
model, which incorporated a model of finite chemical
reaction. Lee (1972) termed the wave with a single front a
reactive blast wave and obtained solutions through application of the various solution methods which had been
developed for the blast wave.

In recent years, research on the nonideal explosions as
extensions of classical studies has been prompted because of
accidents resulting from spills of fuel gases and liquids.
In an actual situation such as occurs during an accidental
spill of fuel gases into open air, the initial conditions
are no longer uniform and may be very complex. In particular, the concentration of the fuel gases may vary with the
distance from the fuel source. This nonuniformity will
affect the propagation of the wave considerably and is
essential in real explosions. This fact was pointed out by
Geiger (1979) in his discussion of explosion hazards. In
the present paper, the propagation of reactive blast waves
through a medium of a spatially nonuniform concentration is
analyzed. For simplicity, the fuel gas concentration
distribution is assumed to be symmetrical with respect to
the center of the explosion. With this assumption, blast
wave propagation is a spatially one-dimensional problem. In
addition, the change of the concentration is expressed by
that of the energy released at the wave front. The chemical
heat release Q is determined by the chemical compositions
and the thermodynamic states of the reactants and products.
In this analysis, Q is assumed to depend only on the initial
chemical composition, which varies with the distance from
the explosion center. While the transient diffusion problem
may be used to determine the concentration profile, the
distribution $Q(x)$, with x the distance from the fuel source,
is given a priori. For simplicity, changes in concentration
due to transient diffusion are neglected, since the
diffusional velocities are very much smaller then the shock
wave velocity.

The blast wave equations are solved by the quasi-similar technique developed by Oshima (1960). This technique gives a solution which is reasonably accurate for strong and moderate shock strengths. The effect of combustion heat release appears only in the boundary conditions at the wave front and varies as the wave propagates so that the boundary conditions include the unknown function which controls the propagation of the wave. To circumvent this difficulty, an approximate and iterative method is developed to solve the basic equations.

Formulation of the Problem

At a time $t = 0$ and a point $x = 0$, a finite amount of energy E_0 is liberated by some energy sources to form a nonsteady, symmetrically expanding shock wave, i.e., a blast wave as is shown in Fig. 1. If the medium is an inert gas, the blast wave decays to the sound wave. In the exothermic reactive medium, if the energy E_0 exceeds a critical level, the wave will decay to the Chapman-Jouguet (CJ) wave, whose propagation velocity is determined by the chemical energy of the medium.

In the present problem, the concentration of the fuel gas changes with the distance from the center of explosion. As is mentioned in the previous section, we give the distribution of chemical energy instead of that of the concentration. Here, for simplicity, the distribution is assumed to be symmetric with respect to the center so that the symmetrical nature of the propagation is assumed. The chemical energy release at the wave front is given as a function of x:

$$Q = Q(x) \tag{1}$$

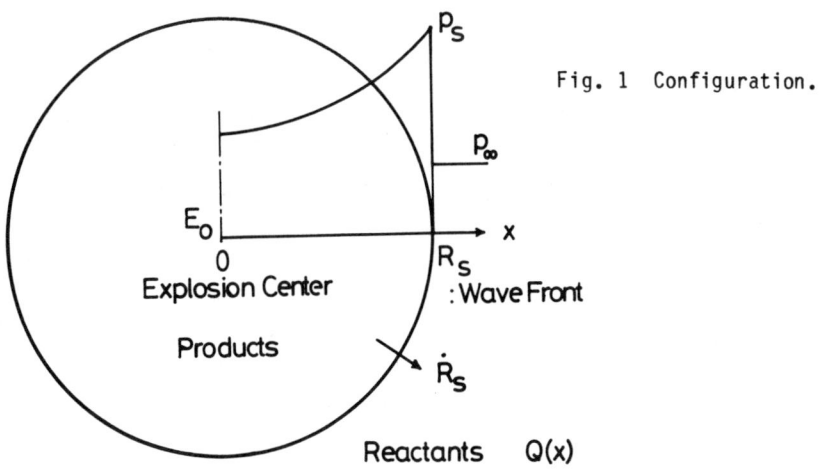

Fig. 1 Configuration.

Real explosive gas mixtures exhibit detonability limits beyond which a detonation cannot be established. Here, the chemical reaction is always assumed to be complete at the wave front, and the complex physico-chemical phenomena occurring in the reaction zone are neglected. If the detonability limits are taken into account, the value of the limiting concentration or the limiting chemical energy should be given a priori.

In addition to these fundamental assumptions, we assume that the molecular weights of the fuel gas and the oxidizer gas are equal to each other so that the density and the pressure of the medium before the front are uniform even if the concentrations are not uniform.

As in the usual blast wave problem, the flow is assumed to be symmetrical, inviscid and nonheat-conducting. The gas mixture is assumed to be thermally and calorically perfect with constant specific heats. The governing equations are

$$\frac{\partial \rho}{\partial t} + \frac{1}{x^j} \frac{\partial}{\partial x} (x^j \rho v) = 0 \tag{2}$$

$$\frac{\partial}{\partial t}(\rho v) + \frac{1}{x^j} \frac{\partial}{\partial x} [x^j (\rho v^2 + p)] = \frac{j}{x} p \tag{3}$$

$$\frac{\partial}{\partial t}(\rho I) + \frac{1}{x^j} \frac{\partial}{\partial x} [x^j \rho v (I + \frac{p}{\rho})] = 0 \tag{4}$$

where ρ, v, p, and I denote respectively the density, velocity, pressure and specific total internal energy defined as

$$I = \frac{1}{\gamma - 1} \frac{p}{\rho} + \tfrac{1}{2} v^2 \tag{5}$$

and γ and j denote the specific heat ratio and geometric factor, which is equal to 0, 1, and 2 for the planar, cylindrical, and spherical wave, respectively.

The boundary conditions at the wave front, $x = R_S(t)$, are given by the modified Rankine-Hugoniot relations as follows:

$$\rho (\dot{R}_S - v) = \rho_\infty \dot{R}_S \tag{6}$$

$$p + \rho(\dot{R}_S - v)^2 = p_\infty + \rho_\infty \dot{R}_S^2 \tag{7}$$

$$\frac{\gamma}{\gamma-1} \frac{p}{\rho} + \tfrac{1}{2}(\dot{R}_S - v)^2 = \frac{\gamma_\infty}{\gamma_\infty - 1} \frac{p_\infty}{\rho_\infty} + Q(\dot{R}_S) + \tfrac{1}{2} \dot{R}_S^2 \tag{8}$$

where $\dot{R}_S$ is the velocity of propagation of the wave, and the subscript ∞ denotes the properties before the front which are assumed to be constant. The boundary condition at the center, $x = 0$, is given by the symmetricity of the flow as follows: at $x = 0$,
$$v = 0 \qquad (9)$$

The integral form of the energy conservation in the region between the wave front and the center is written as

$$E_0 = \int_0^{R_S} \rho I k_j x^j dx - \int_0^{R_S} \rho_\infty [I_\infty + Q(x)] k_j x^j dx \qquad (10)$$

where k_j is equal to 1, 2π, and 4π for $j = 0$, 1, and 2, respectively. This equation can be used as a condition instead of the zero particle velocity condition at the center, i.e., Eq. (9). In principle, the solutions which are obtained with either criterion should be identical. But the result may depend on the method of solution adopted. For the quasi-similar method used in this analysis, Lee (1972) argued that the zero velocity criterion leads to errors which are a little larger than those found for the constant energy criterion, i.e., Eq. (10). In this analysis, Eq. (9) is adopted because this criterion is simple and physically acceptable. If errors exist in the quasi-similar profiles of the dependent variables, they might be accumulated in the integration of Eq. (10).

Transformation

Similarity considerations suggested that the conservation equations for the blast wave model be transformed as follows. The independent variables are

$$\xi = x/R_S \qquad \eta = c_\infty^2/\dot{R}_S^2 = 1/M_S^2 \qquad (11)$$

where c_∞ and M_S denote the sound speed of the undisturbed medium and the Mach number of the wave propagation.

The dependent variables are

$$\psi = \rho_\infty/\rho \qquad f = \dot{f}/\dot{R}_S \qquad h = p/(\rho_\infty \dot{R}_S^2) \qquad (12)$$

With these variables substituted into Eqs. (2-4) and with the assumption that ∞ and c_∞ are the constants used, the following nondimensional basic equations can be obtained:

$$(f - \xi) \frac{\partial \psi}{\partial \xi} - \psi \frac{\partial f}{\partial \xi} - j\psi \frac{f}{\xi} + \lambda \eta \frac{\partial \psi}{\partial \eta} = 0 \tag{13}$$

$$(f - \xi) \frac{\partial f}{\partial \xi} + \psi \frac{\partial h}{\partial \xi} - \tfrac{1}{2}\lambda f + \lambda \eta \frac{\partial f}{\partial \eta} = 0 \tag{14}$$

$$(f - \xi) \frac{\partial h}{\partial \xi} + \gamma h \frac{\partial f}{\partial \xi} - \lambda h + j\gamma h \frac{f}{\xi} + \lambda \eta \frac{\partial h}{\partial \eta} = 0 \tag{15}$$

where λ is the decay coefficient which is defined as

$$\lambda = \frac{d(\log \eta)}{d(\log R_S)} = -2 \frac{d(\log M_S)}{d(\log R_S)} \tag{16}$$

This parameter is a function of η and describes the trajectory of the wave. When λ is positive, the wave decelerates; and when it is negative, the wave accelerates. The problem is reduced to find a functional form of $\lambda(\eta)$. The boundary conditions Eqs. (6-8) are transformed and solved for each property just behind the wave front of strong detonation:

$$f(1,\eta) = \frac{\gamma_\infty + \gamma \eta + \gamma_\infty s}{\gamma_\infty (\gamma + 1)} \tag{17}$$

$$\psi(1,\eta) = \frac{\gamma \gamma_\infty + \gamma \eta - \gamma_\infty s}{\gamma_\infty (\gamma + 1)} \tag{18}$$

$$h(1,\eta) = \frac{\gamma_\infty + \eta + \gamma_\infty s}{\gamma_\infty (\gamma + 1)} \tag{19}$$

where

$$s = \sqrt{(1 - \frac{\gamma}{\gamma_\infty}\eta)^2 + 2(\gamma^2 - 1)\eta \left[\frac{\gamma_\infty - \gamma}{\gamma_\infty(\gamma-1)(\gamma_\infty-1)} - q \right]} \tag{20}$$

with the nondimensional chemical heat release at the wave front defined as

$$q = Q(R_S)/c_\infty^2 \tag{21}$$

The heat release at the wave front, q, depends on the wave position R_S so that it is a function of η.

The boundary condition at the center is written as

$$f(0,\eta) = 0 \tag{22}$$

The problem is to solve Eqs. (13-15) with the boundary conditions Eqs. (17-19) and (22) to find a functional form of $\lambda(\eta)$ for a given function of $q(R_S)$. For this purpose, the relation between R_S and η must be given. Integration of Eq. (16) gives

$$R_S = R_{S,0} \exp\left(\int_{\eta_0}^{\eta} \frac{d\eta}{\lambda\eta}\right) \quad (23)$$

where η_0 and $R_{S,0}$ are arbitrary constants which express the initial condition of the wave Mach number and position, respectively. When the value of η_0 is very small, the self-similarity of the flow is held at this initial stage so that the initial position $R_{S,0}$ is proportional to $E_0^{1/(j+1)}$, where E_0 denotes the initiation energy. With Eq. (23) substituted into Eq. (21), q is expressed in term of η. As is seen from Eqs. (21) and (23), the boundary conditions as well as the basic equations include the unknown function $\lambda(\eta)$ through q.

Quasi-Similar Method of Solution

The quasi-similar method developed by Oshima (1960) is based on the method of separation of the variables. The functions of ξ and η are assumed to be expressed as

$$f(\xi,\eta) = X(\xi) Y(\eta) \quad (24)$$

etc. Owing to this assumption, the partial derivatives with respect to η are approximated by ordinary derivatives for the fixed ξ evaluated at the wave front:

$$\frac{1}{f(\xi,\eta)} \frac{\partial}{\partial \eta} f(\xi,\eta) = \frac{1}{f(1,\eta)} \frac{d}{d\eta} f(1,\eta) \quad (25)$$

etc., whose right-hand sides are calculated from Eqs. (17-19) as

$$\frac{1}{\psi(1,\eta)} \frac{d\psi(1,\eta)}{d\eta} = \frac{\gamma - \gamma_\infty(ds/d\eta)}{\gamma\gamma_\infty + \gamma\eta - \gamma_\infty s} \equiv \Gamma_1(\eta) \quad (26)$$

$$\frac{1}{f(1,\eta)} \frac{df(1,\eta)}{d\eta} = \frac{-\gamma + \gamma_\infty(ds/d\eta)}{\gamma_\infty - \gamma\eta + \gamma_\infty s} \equiv \Gamma_2(\eta) \quad (27)$$

$$\frac{1}{h(1,\eta)} \frac{dh(1,\eta)}{d\eta} = \frac{1 + ds/d\eta}{\gamma_\infty + \eta + \gamma_\infty s} \equiv \Gamma_3(\eta) \qquad (28)$$

where the derivative $ds/d\eta$ is expressed as

$$\frac{ds}{d\eta} = \frac{1}{s} \left[\frac{\gamma}{\gamma_\infty} (\frac{\gamma}{\gamma_\infty} \eta - 1) + \frac{(\gamma+1)(\gamma_\infty-1)}{\gamma_\infty(\gamma_\infty-1)} - (\gamma^2-1)q(1 + \frac{1}{\lambda}\frac{d(\log q)}{d(\log R_S)}) \right] \qquad (29)$$

Here, the function $q(R_S)$ is assumed to be differentiable.
With these relations substituted into Eqs. (13-15), the basic equations are reduced to the ordinary differential equations, which are written as

$$(f - \xi)\psi' - \psi f' - j\psi(f/\xi) + \lambda\eta\psi\Gamma_1 = 0 \qquad (30)$$

$$(f - \xi)f' + \psi h' - \tfrac{1}{2}\lambda f + \lambda\eta f\Gamma_2 = 0 \qquad (31)$$

$$(f - \xi)h' + \gamma hf' - \lambda h + j\gamma h(f/\xi) + \lambda\eta h\Gamma_3 = 0 \qquad (32)$$

where the primes denote the differentiation with respect to ξ. This set of the simultaneous ordinary differential equations is solved numerically by the Runge-Kutta-Gill method. Since $\lambda(\eta)$ is not known initially, the value of λ for each η is assumed at first and integrations are started from the wave front $\xi = 1$, where the boundary conditions are given by Eqs. (17-19) to the center of explosion $\xi = 0$, where the particle velocity is equal to zero. If this condition is not satisfied for the assumed value of λ, the value of λ is corrected a little and the procedure is repeated until Eq. (22) is satisfied. These are the usual steps to find the decay coefficient of the blast wave. Here, the difficulty arises because the integrated term of $\lambda(\eta)$ appears in the boundary conditions. The steps of the solution in this case are as follows. To obtain $\lambda(\eta)$, η is assumed to be equal to η_0. From Eq. (23), $R_S = R_{S,0}$, and q is calculated from Eq. (21) so that the decay coefficient $\lambda(\eta_0)$ is obtained. Next, η is increased by a small amount, $\Delta\eta$, and R_S is calculated by numerical integration of Eq. (23). The same procedure as described previously leads to the solution $\lambda(\eta_0 + \Delta\eta)$. A further solution will be found by repetition of this procedure. The numerical integrations are performed with the trapezoidal rule. The error in this procedure should be reduced if the increment $\Delta\eta$ is small.

Results and Discussions

As an example, a simple, but fundamental, form of the function $Q(x)$ is adopted. In the nondimensional variables, it is

$$q(x) = q_0 \exp[-\beta (x/R_{S,0})] \qquad (33)$$

where q_0 and β are nondimensional parameters, and $R_{S,0}$ is the reference radius defined by Eq. (23) which is proportional to $E_0^{1/(j+1)}$. If β is positive, the fuel concentration becomes smaller with the distance from the center and the chemical energy released at the wave front decreases exponentially as the wave propagates. The values of the parameters used in this calculation are as follows: $j = 2$ (spherical wave); $q_0 = 20$ [if β is equal to zero, the wave will decay to the Chapman-Jouguet wave, with the propagating Mach number M_S equal to 6.202 ($n_{CJ} = 0.026$)]; $\beta = 0.001$, 0.05, 0.1, 0.15, 0.2, and 0.5; $n_0 = 0.001$; and $\gamma = \gamma_\infty = 1.4$. So, in this calculation, the medium has sufficient chemical energy to support a detonation near the explosion center and the energy decreases exponentially with the distance. When

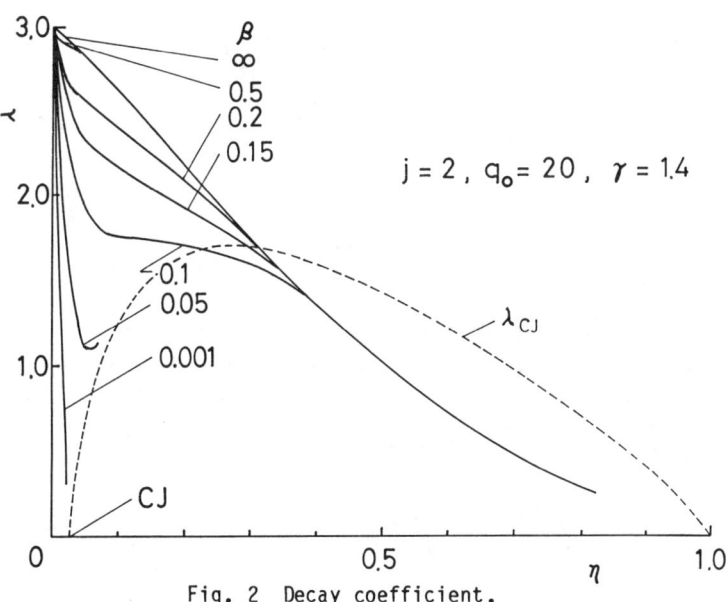

Fig. 2 Decay coefficient.

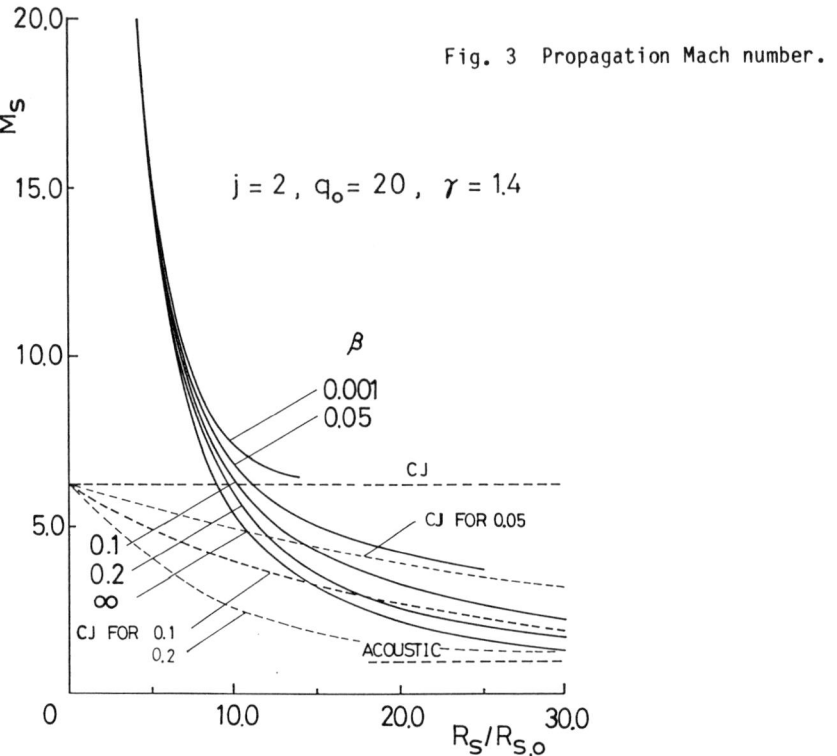

Fig. 3 Propagation Mach number.

β is equal to unity, the characteristic radius of decreasing energy is identical to that of the initation energy.

The step size for the Runge-Kutta-Gill integration for ξ-direction $\Delta\xi$ is chosen to be 0.01. For the η-direction, the step size $\Delta\eta$ is 0.0001.

Figure 2 shows the decay coefficient $\lambda(\eta)$ for each β. When η is equal to zero, the self-similar solution is fulfilled so that λ is equal to j+1 (equal to 3, in this case). As η increases, λ decreases generally, which means that the wave approaches the steady wave. For β as small as 0.001, the heat liberated at the wave front is almost constant, and the wave behaves as the reactive blast wave in the uniform combustible mixture for the regime where the solution can be found. It seems to decay to the CJ wave ($\eta_{CJ} = 0.026$), but the solution cannot be found near the CJ wave because of the singularity at the wave front in the CJ wave. In the CJ wave, the gradients of ψ, f, and h for the η direction becomes infinity, as is indicated by Eqs. (26-29). As β increases, $|d\lambda/d\eta|$ decreases; and a transition from the

reactive blast wave to the nonreactive one occurs with
increasing η. For β = 0.05, λ has a minimal value 1.09 at
the value of η ≃ 0.05. It appears that the wave, decaying
as a reactive blast wave, loses its energy as it propagates
and in turn decays more rapidly as a nonreactive blast wave.
But for this case, the solution cannot be found for η >
0.07 because of the limit of the numerical precision. For β
= 0.1, 0.15, 0.2, and 0.5, the complete solutions show the
transition from reactive blasts to nonreactive ones. For β
= 0.5, the solution is almost identical to that for $q_0 = 0$
(or β → ∞), i.e., the nonreactive blast wave except for the
very small η.

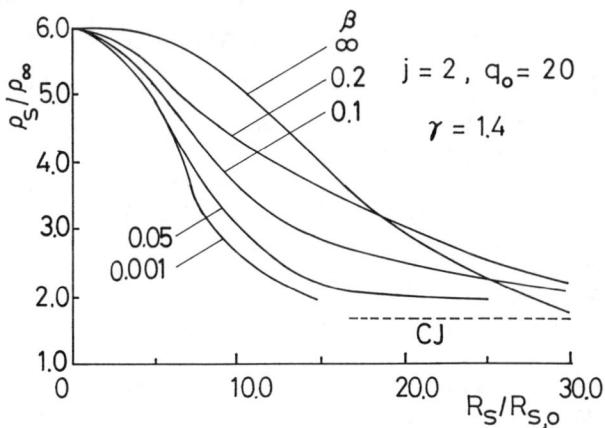

Fig. 4 Peak density.

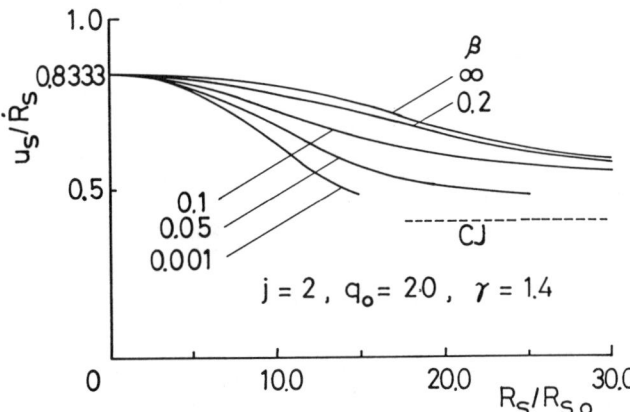

Fig. 5 Peak velocity.

Figure 3 shows that the shock Mach number decreases with an increase in the nondimensional radius. The curves for all β's are calculated with the same initial condition; i.e., for $R_S = R_{S,0}$, $M_S = 100$. For $\beta = 0.001$, the Mach number M_S seems to decay to the Chapman-Jouguet value 6.202 in this calculation. In reality, it should decay to the sound wave as the wave propagates infinitely apart from the center. As β increases, M_S seems to decrease to unity in the far field. When $\eta > 0.35$ (or $M_S < 1.7$), the values of the decay coefficient for $\beta > 0.1$ are almost the same, as is seen from Fig. 2, because, there, the chemical energy q is almost equal to zero. From Fig. 3, it can be seen that for $R_S/R_{S,0} > 25$, M_S is less than 1.7 for $\beta > 0.1$. Then, in the regime $R_S/R_{S,0} > 25$, although q is almost zero for those β's, M_S for larger β is larger than that for smaller β. It can be said that this is the effect of the history which the wave has experienced.

The dashed lines in Fig. 3 indicate the Chapman-Jouguet Mach number obtained by s = 0 [see Eq. (20)]. The CJ Mach number for $\beta = 0.1$ varies with the distance because the chemical energy q is changing, and it can be called the local CJ Mach number. It is evident the Mach numbers for this case, as well as for other β, decay to their own local CJ values.

Figures 4-6 show the variations of the peak values of the nondimensional density, velocity, and pressure with respect to $R_S/R_{S,0}$. They change from the values of the

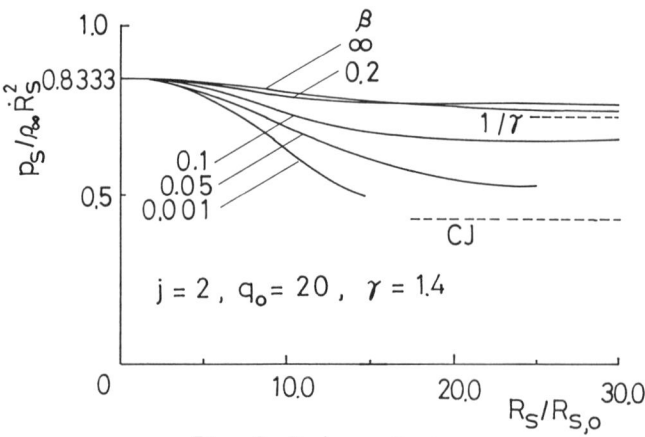

Fig. 6 Peak pressure.

BLAST WAVES WITH DISTRIBUTED CHEMICAL ENERGY 169

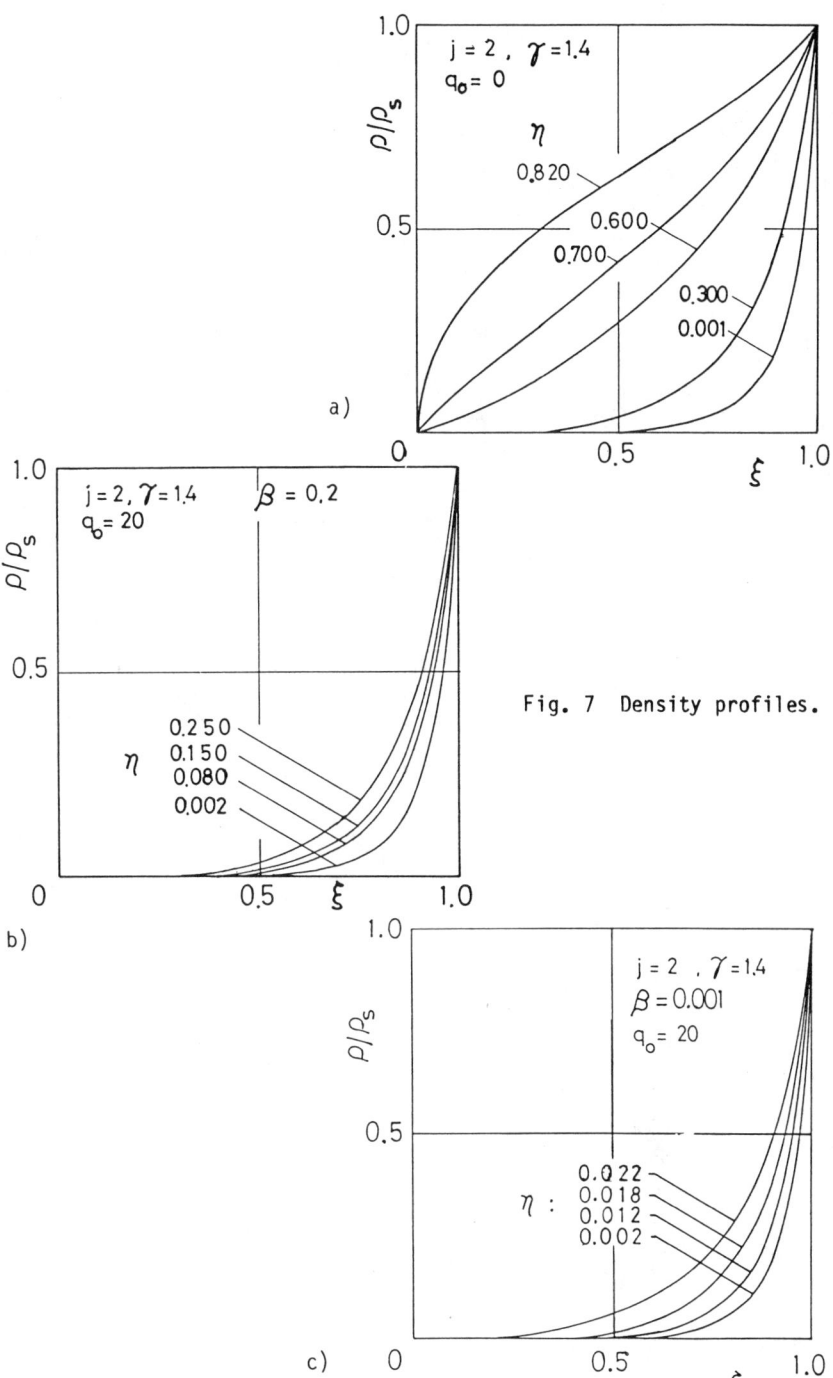

Fig. 7 Density profiles.

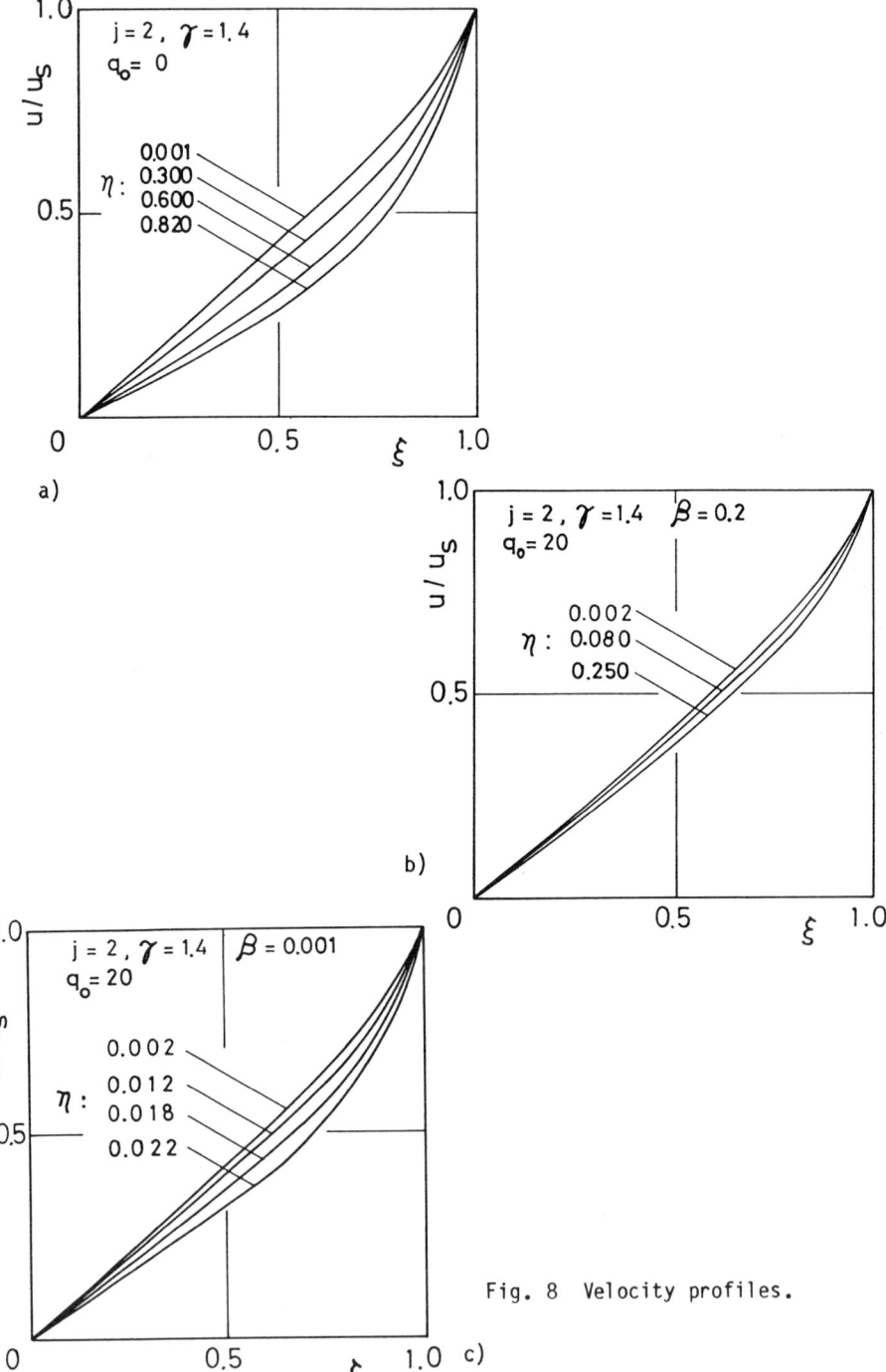

Fig. 8 Velocity profiles.

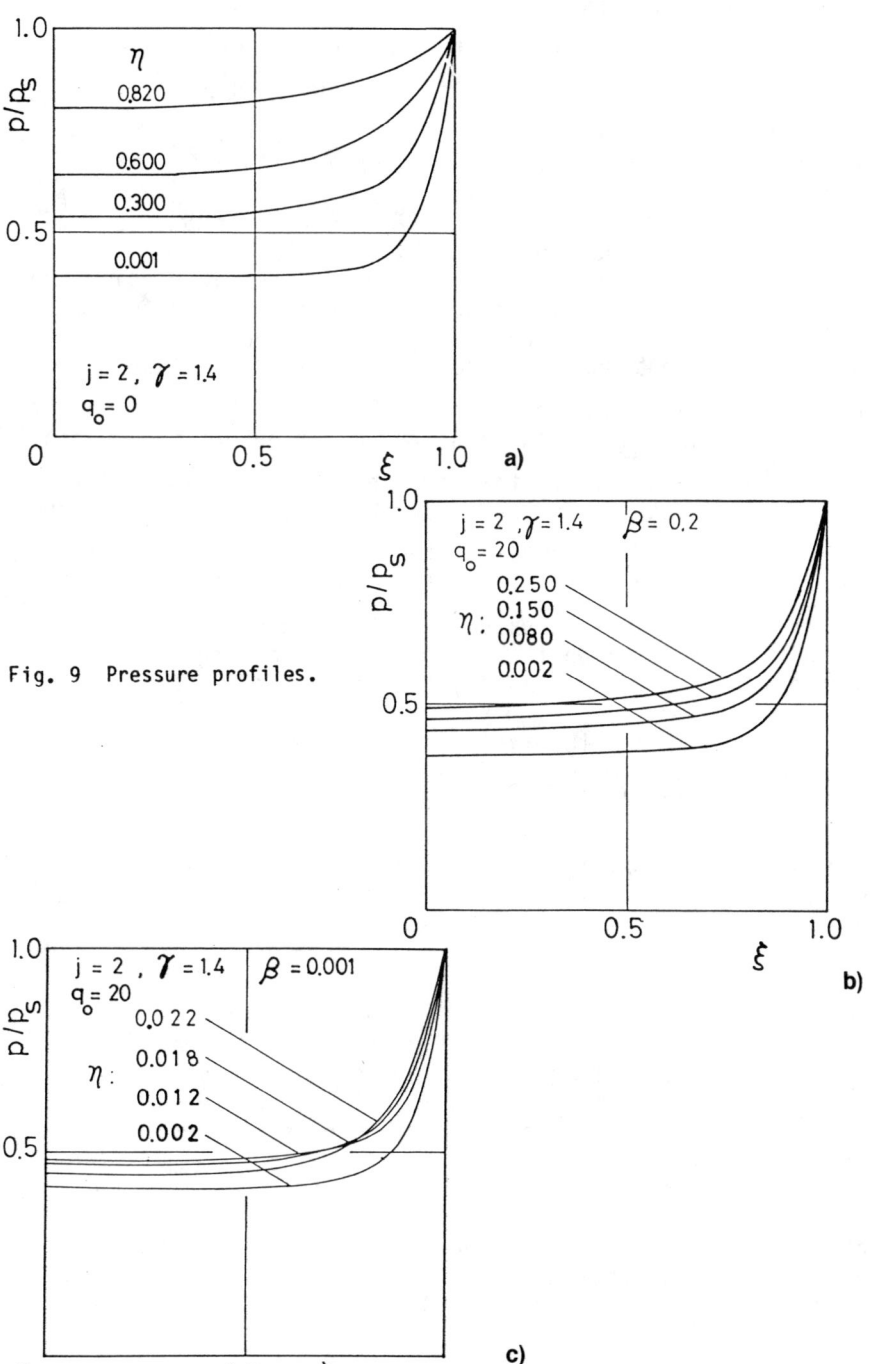

Fig. 9   Pressure profiles.

strong blast wave limits (i.e., $\rho_S/\rho_\infty = 6.0$, $u_S/\dot{R}_S = 0.8333$, and $p_S/\rho_\infty \dot{R}_S^2 = 0.8333$ for $\gamma = 1.4$) to the values of the CJ wave (i.e., $\rho_S/\rho_\infty = 1.683$, $u_S/\dot{R}_S = 0.4508$, and $p_S/\rho_\infty \dot{R}_S^2 = 0.4244$ for $\gamma = 1.4$ and $q = 20$) for the reactive blast wave in the uniform medium while they approach those of the sound wave (i.e., $\rho_S/\rho_\infty = 1.0$, $u_S/\dot{R}_S = 0.0$, and $p_S/\rho_\infty \dot{R}_S^2 = \gamma^{-1}$) for the wave in the nonuniform medium.

From Fig. 4, it can be said that in the early stage the peak density at the wave front decays more rapidly for small β than for large β with respect to the distance from the center (note that the distance $R_S/R_{S,0}$ is not a reduced distance which is measured by the characteristic distance of the explosion. In the early stage, M_S is determined by the initiation energy rather than the chemical energy. For the same M_S, the peak value for small chemical energy is larger than for large q, as is calculated by the Rankine-Hugoniot relation. In the later stage of the propagation, M_S approaches the local CJ value and it is determined by the chemical energy. Then the peak density for large β has a tendency to decay more rapidly than that for large β, as is shown by Fig. 4.

Figure 7 shows the typical density profiles normalized with respect to the peak values at the wave fronts for $\beta \to \infty$ (or $q_0 = 0$) and $\beta = 0.2$ and 0.001. The nondimensional density near the center increases as $\eta \to 1$ for $\beta \to \infty$ and $\beta = 0.2$ (here, the profiles of $\beta = 0.2$ for $\eta > 0.25$ are omitted because they are almost same as those for $\beta \to \infty$). For $\beta = 0.001$, ρ/ρ_S near the center increases very little as $\eta \to \eta_{CJ}$ (when $\eta = 0.022$, $\eta/\eta_{CJ} = 0.846$). In these calculations of the point blast wave, the density near the center remains zero in the regime when the solutions can be found. It corresponds to the fact that the velocity gradient $\partial u/\partial x$ does not vanish. If it is possible to extend the calculation for larger η, the density may increase nearer to the center.

Figure 8 shows the particle velocity profiles. The core in which the velocity vanishes never appears in this calculation. This is an error and a consequence of the quasi-similarity approximation, since it is well known that if $q_0 = 0$ and $\eta > 0.6$, then $f(\xi)$ has a negative region near the center.

Figure 9 shows the pressure profiles for $\beta \to \infty$, $\beta = 0.2$, and $\beta = 0.001$, respectively.

In this model, using the quasi-similar approximation, there must be some errors in the profiles essentially. But the decay coefficient obtained seems to be reasonable in this approximation; and this method permits any type of the heat release function $q(R_s)$; and it is very convenient to find the decay coefficient.

In this example of calculation, where q decreases exponentially with the distance from the center, the nature of wave propagation is determined by the ratio of the characteristic radius of the initiation energy to that of the decreasing initial chemical energy, β; and it is revealed that for β as small as 0.001 and 0.05 (small initiation energy or weakly decreasing initial chemical energy), the wave behaves as a detonation wave, and for β as large as 0.5, it decays as a nonreactive blast wave.

Summary and Conclusions

A model was formulated to analyze the decay of a one-dimensional blast wave in a combustible mixture where the fuel concentration (and therefore the heat release at the detonation front) decayed exponentially with the distance from the center of the explosion. The resulting blast wave equations were solved approximately by application of Oshima's quasi-similar technique. Parameteric calculations were performed to determine how the heat release function influenced the character of the solution. The results show that the detonation wave decays to the local Chapman-Jouguet value, which itself decays with distance owing to variations in the detonation energy of the combustible medium.

With the formulation, it is possible to calculate the feature of the wave propagation for the other forms of the concentration distribution, which may give insight into hazard assessments of the real vapor cloud explosions.

Acknowledgments

The authors with to express their gratitude to Professor Koichi Oshima of the Institute of Space and Astronautical Science, Tokyo, for the very stimulating and helpful discussions.

References

Bishimov, E., Korobeinikov, V. P., and Levin, V. A. (1970) Strong explosion in combustible gaseous mixture. **Astronaut. Acta** 15, 267-273.

Geiger, W. (1979) Explosion hazards of transport gases - research within the reactor safety program of the Federal Republic of Germany. Discussion on Explosion Hazards at 7th ICOGER, Bericht 23/1979, edited by H. G. Wagner, pp. 49-56. Max-Planck-Institut für Strömungsforschung, Gottingen, FRG.

Korobeinikov, V. P. (1969) The problem of point explosion in a detonating gas. Astronaut. Acta 14, 411-419.

Korobeinikov, V. P., Levin, V. A., Markov, V. V., and Chernyi, G. G. (1972) Propagation of blast waves in a combustible gas, Astronaut. Acta 17, 529-537.

Lee, J. H. (1972) Gasdynamics of detonations. Astronaut. Acta 17, 455-466.

Oshima, K. (1960) Blast wave produced by exploding wire. Report No. 358, Aeronautical Research Institute, University of Tokyo, Tokyo, Japan, pp. 137-194.

On the Use of General Equations of State in Similarity Analysis of Flame-Driven Blast Waves

Allen L. Kuhl*
R&D Associates, Marina del Rey, Calif.

Abstract

Considered here are constant velocity, one-dimensional, thin flames propagating in an infinite medium. Flowfields of such waves are self-similar and can be analyzed most conveniently by means of integral curves in an appropriately defined phase plane. A new flame model is proposed where it is demonstrated that the Hugoniot relation can be solved to give the heat release in terms of the pressure and specific volume changes across the flame for fairly general equations of state for the combustion products. The mechanical conditions alone are then sufficient to specify the ratios of pressure and specific volume across the flame in terms of known parameters just ahead of the flame. Hence this flame model allows one to evaluate the heat release and burning velocity of the flame throughout the blast wave and thus identify the flame location. Similarity analysis was applied to flame-driven blast waves in a methane/air mixture (heat addition of 2.755 MJ/kg). The equilibrium enthalpy of the combustion products was fit with a tenth-order polynomial in temperature and used in the new flame model. Solutions were obtained for planar, cylindrical, and spherical waves for a range of burning velocities from 1 to 100 m/s. For the case of a burning velocity of 20 m/s, the shock overpressures were calculated to be 0.113 and 0.039 atm for cylindrical and spherical waves, respectively; while the corresponding overpressures just ahead of the flame were 0.339 and 0.261 atm, respectively. These results are from 30 to 50% higher than previous similarity results which used a constant gamma flame model.

Presented at the 8th ICOGER, Minsk, USSR, Aug. 23-26, 1981. Copyright © American Institute of Aeronautics and Astronautics, Inc., 1983. All rights reserved.

*Senior Research Scientist, Continuum and Fluid Dynamics Department.

Nomenclature

a	=	sound speed
b_i	=	polynomial fit coefficients
C_p	=	specific heat at constant pressure
F	=	phase plane coordinate $\equiv u/xw_2$
h	=	enthalpy
j	=	geometric factor =0,1,2 for plane, line, and point symmetry, respectively.
J_1	=	nondimensional mass integral
M	=	molecular weight
p	=	pressure
P	=	pressure ratio across the flame
q	=	heat released per unit mass by the flame
Q	=	nondimensional heat addition = q/RT_1
r	=	radius
R	=	individual gas constant
S	=	burning velocity of the flame $\equiv w_3 - u_3$
t	=	time
T	=	absolute temperature
u	=	particle velocity
v	=	specific volume
w	=	wavefront velocity
x	=	nondimensional radius; Eulerian space similarity variable $\equiv r/r_2$
y	=	shock Mach number parameter $\equiv (a_1/w_2)^2$
Z	=	phase plane coordinate = $(a/xw_2)^2$
ν	=	specific volume ratio across the flame
ρ	=	density
τ	=	nondimensional time; Eulerian time similarity variable $\equiv t/t_2 = 1/x$

Subscripts

1	=	state ahead of shock front
2	=	state immediately behind shock front
3	=	state immediately ahead of flame
4	=	state behind flame
AD	=	adiabatic flame temperature
p	=	piston

Superscripts

p = products
r = reactants

Introduction

In classical combustion theory, flames are subsonic discontinuities accompanied by chemical energy release. Specific volumes and temperatures increase dramatically across the flame; pressures drop; and there are large changes in flow velocity. Boundary conditions, such as a center of symmetry for a spherical flame or an end wall in a shock tube problem, can constrain the flow. Under such circumstances, continuously propagating flames can generate pressure waves and even shock waves. The burned gases expand during combustion, and owing to boundary constraints, they push the unburned gases ahead of the flame in a "piston-like" action. The strength of the pressure wave depends on the flame velocity (which is controlled by turbulence), the amount of heat release, and the flow geometry.

Flame-driven blast waves are an idealization of such flame-shock systems. They are here defined as spatially one-dimensional, time-dependent, compressible flowfields containing an infinitely thin flame which drives a shock discontinuity into an infinite combustible medium. As shown by Sedov (1959), the flowfields associated with flames which propagate with a constant velocity (which is a fraction of the shock velocity) are self-similar, all gasdynamic parameters are constant along similarity lines $x = r/r_2$, where r_2 denotes the shock front position. Taylor (1946) first performed a similarity analysis of spherical flame-shock systems for a particular gas and derived an acoustic solution which is valid for laminar burning velocities. Kuhl et al. (1973) analyzed constant velocity, flame-driven blast waves as a function of burning velocity. More recently, Guirguis et al. (1981) have extended this similarity analysis to include the case of variable wave velocity.

All previous similarity analyses have been limited to the modeling assumption of a constant value for the specific heat of the reactants and a different constant value for the combustion products. In this paper we propose a generalized flame model which can accommodate a wide variety of equations of state for the combustion products. Analysis is then performed for constant velocity, flame-driven blast waves in a stoichiometric mixture of methane and air. Results are presented parametrically for planar, cylindrical, and spherical waves in terms of burning velocity.

Similarity Analysis

Considered here are constant velocity, flame-driven blast waves depicted in Fig. 1. The ambient medium, state 1, as well as the combustion products behind the flame, state 4, are assumed to be at rest. State 2 denotes conditions immediately behind the shock, while state 3 represents the conditions immediately in front of the flame. The reactants are assumed to behave as a perfect gas with a constant ratio of specific heats, γ.

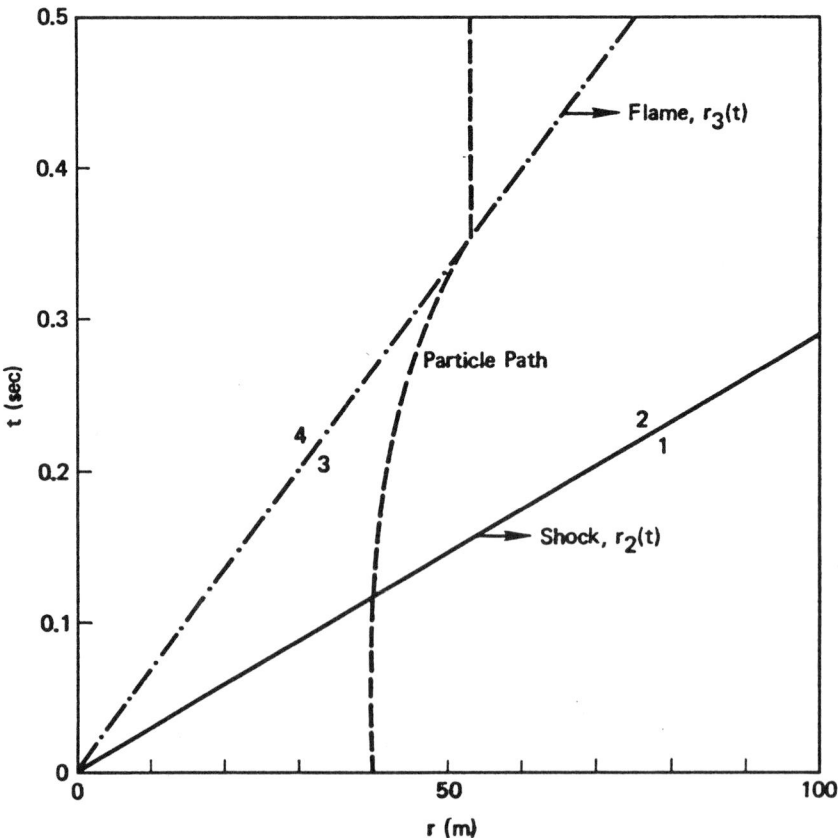

Fig. 1 Time-distance diagram for a spherical self-similar flame-driven blast wave (example given for a burning velocity of 20 m/s).

As we have shown previously (Kuhl et al. 1973), the flowfield of such blast waves can be analyzed most conveniently in terms of the following phase plane coordinates:

$$F \equiv u/xw_2 \quad Z \equiv (a/xw_2)^2 \qquad (1)$$

where a and u denote the gas sound speed and particle velocity, w_2 represents the shock front velocity, and $x = r/r_2$ denotes the spatial coordinate nondimensionalized by the shock front radius, $r_2(t)$. For constant velocity waves, the conservation equations reduce to a single ordinary differential equation in the Z-F phase plane:

$$\frac{dZ}{dF} = \frac{Z}{F} \frac{2D + j(\gamma-1)F(1-F)}{D + jZ} \qquad (2)$$

and a quadrature:

$$\ln x = -\int \frac{D}{F(jZ+D)} dF \qquad (3)$$

where $D = Z-(1-F)^2$, and $j = 0, 1,$ or 2 for plane, line, or point symmetry flow, respectively. Integration of Eqs. (2) and (3) between the shock front locus

$$Z_2 = (1-F_2)[1 + 2F_2/(\gamma-1)] \qquad (4)$$

to the piston boundary

$$F = 1 \quad Z = Z_p \qquad (5)$$

determines the flowfield corresponding to a uniformly expanding piston. Once $Z=Z(F)$ and $x=x(F)$ are known, the blast wave flowfield can be evaluated from the isentropic flow relation:

$$\frac{T}{T_2} = \left(\frac{a}{a_2}\right)^2 = \left(\frac{p}{p_2}\right)^{(\gamma-1)/\gamma} = \left(\frac{\rho}{\rho_2}\right)^{\gamma-1} = x^2 \frac{Z}{Z_2} \qquad (6)$$

where p, ρ, and T denote the thermodynamic pressure, density, and temperature, respectively, while the velocity is determined from

$$u/u_2 = xF/F_2 \qquad (7)$$

which satisfies the definition of F. The above relations give Eulerian space profiles; Eulerian time profiles are determined by replacing x in Eqs. (3), (6), and (7) with $1/\tau$ where $\tau \equiv t/t_2$. Changes across the shock front are specified by the Rankine-Hugoniot relations

$$\frac{p_2}{p_1} = \frac{\gamma-1}{\gamma+1}(\frac{2\gamma}{\gamma-1} y^{-1} -1)$$

$$\frac{\rho_2}{\rho_1} = \frac{\gamma+1}{\gamma-1 + 2y} \quad (8)$$

$$\frac{u_2}{a_1} = \frac{F_2}{\sqrt{y}}$$

where y denotes the shock Mach number parameter

$$y \equiv (a_1/w_2)^2$$
$$= 1 - [(\gamma + 1)/2] F_2 \quad (9)$$

The solution as derived above, specifies the flowfield corresponding to a uniformly expanding piston. As shown by Sedov (1959) and Kuhl et al. (1973), however, the flowfield between a constant velocity flame and the shock is identical to that of a piston-driven wave, where the piston moves with a velocity slightly smaller than the flame velocity. For flame-driven blast waves, one must then determine a relation which specifies the flame location along the integral curve $Z=Z(F)$ in the phase plane.

Generalized Flame Model

The pressure ratio, $P \equiv p_4/p_3$, and the specific volume ratio, $\nu = v_4/v_3$, across the flame satisfy the mechanical jump conditions for discontinuities:

$$\nu = (w_3-u_4)/(w_3-u_3) \quad (10)$$

$$\gamma(w_3-u_3)^2 a_3^2 = (P-1)/(1-\nu) \quad (11)$$

These ratios, as well as the burning velocity of the flame, $S \equiv w_3 - u_3$, can be expressed in terms of the phase plane coordinates:

$$\nu = \nu(F) = (1-F_4)/(1-F_3) \qquad (12)$$

$$P = P(F,Z) = 1 - \gamma_3(F_3-F_4)(1-F_3)/Z_3 \qquad (13)$$

$$S/a_1 = x_3(1-F_3)y^{-1/2} \qquad (14)$$

Next consider the enthalpy of the reactants $h^r(T_3)$ and the products $h^p(T_4)$. The change in enthalpy across the flame is given by the Hugoniot relation, which can be written in the form

$$h^p(T_4) - h^r(T_3) = q_3 + p_3 v_3 (P-1)(1+\nu)/2 \qquad (15)$$

the heat release q_3 at elevated temperature T_3 is related to the heat of combustion q_1 at ambient temperature T_1 by the relation

$$q_3 + h^p(T_3) - h^p(T_1) = q_1 + h^r(T_3) - h^r(T_1) \qquad (16)$$

as can be readily verified by geometric considerations in the temperature-enthalpy diagram, Fig. 2.

For most flame-shock systems, peak pressures are less than 10 bars, hence the reactants can be considered to behave as thermally and calorically perfect gases:

$$pv = RT \qquad (17)$$

$$h^r = C_p T = [\gamma/(\gamma-1)]pv \qquad (18)$$

For many hydrocarbon fuels of interest, the molecular weight changes due to combustion are negligible; therefore the products gases can also be considered to behave as a thermally perfect gas with the same gas constant as the reactants. Let us leave caloric the equation of state for the products unspecified, so that the enthalpy is a general function of temperature: $h^p = h^p(T)$. Then Eqs. (15) and

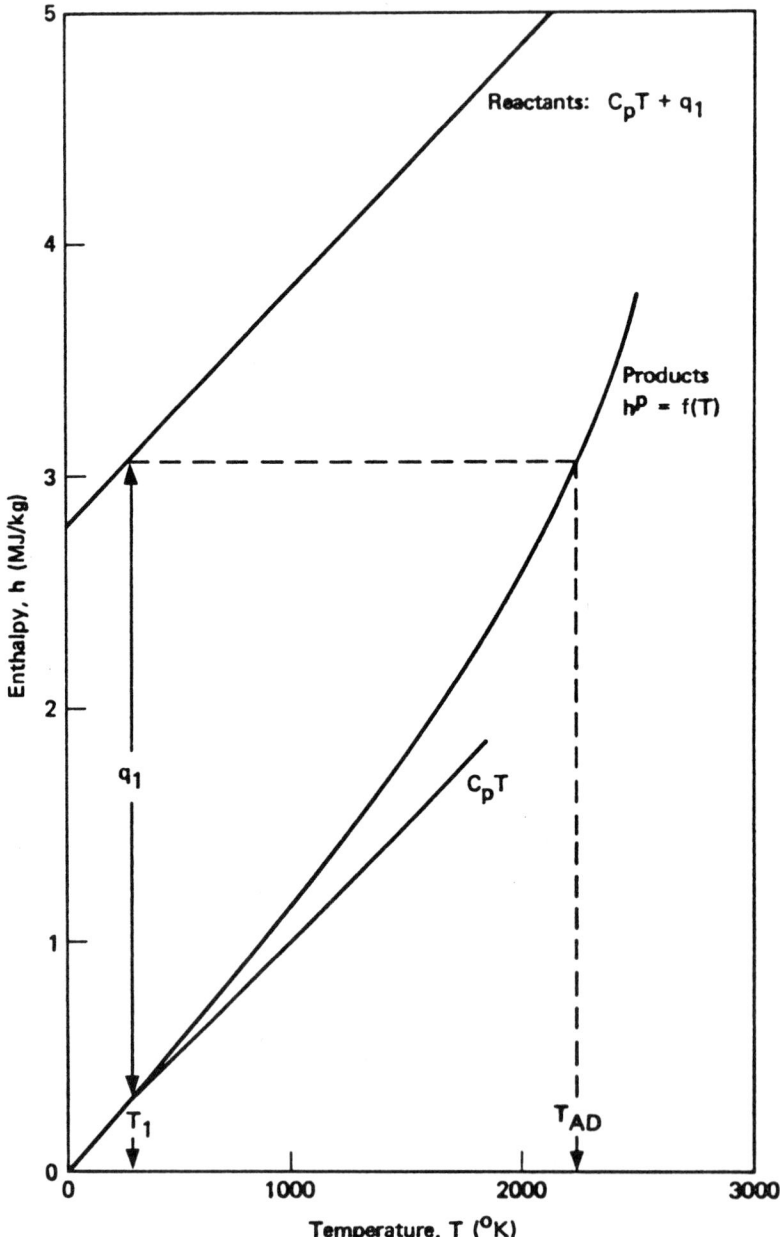

Fig. 2 The temperature-enthalpy diagram for a representative hydrocarbon/air mixture (a stoichiometric mixture of 90 mole% methane and 10 mole% ethane with air).

(16) can be solved for the dimensionless heat additions

$$Q_3 \equiv \frac{q_3}{RT_1} = \frac{T_3}{T_1}\left(\frac{h^P(T_3 P\nu)}{RT_3} \frac{\gamma}{\gamma-1} - \frac{(P-1)(1+\nu)}{2}\right) \qquad (19)$$

$$Q_1 \equiv \frac{q_1}{RT_1} = Q_3 + \left[\frac{h^P(T_3) - h^P(T_1)}{RT_1} - \frac{\gamma}{\gamma-1}\left(\frac{T_3}{T_1} - 1\right)\right] \qquad (20)$$

where we have used the relation $T_4 = T_3 P\nu$ to eliminate temperature T_4.

Examination of Eqs. (19) and (20) indicates that the dimensionless heat releases Q_1 and Q_3 depend on 1) the temperature ratio ahead of the flame, $T_3/T_1 = f(F,Z)$; 2) the pressure ratio $P(F,Z,\gamma)$ and specific volume ratio ν (F) across the flame; and 3) the equation-of-state parameters γ, R, $h^P(T)$.

For similarity analysis of constant velocity flames (where $F_4 \equiv 0$), these variables along with the burning velocity are known functions of F, Z, and x ahead of the flame; and therefore Q_1 and S can be evaluated at each point along the integral curve. When Q_1 and S match the assumed values, this determines the flame front location. This flame model can probably also be used for blast waves driven by variable velocity flames (where $F_4 \neq 0$) if one can specify F_4 in terms of known quantities.

The most important contribution of this paper is the generalized flame model described above. As shown here, the Hugoniot relation can be interpreted as giving the heat release as a function of pressure and specific volume ratios across the flame even when fairly general equations of state are used for the combustion products. Note that we do not have have to solve analytically the Hugoniot relation [e.g., Eq. (19), which is a nonlinear relation for generalized equations of state] for pressures or specific volume ratios across the flame as is usually done in combustion theory. Instead, the mechanical jump conditions alone are sufficient to determine both P and ν (a consequence of specifying $F_4 = 0$), and these values can be used in the Hugoniot relation to determine the flame location for a given Q_1.

Methane/Air Calculations

Analysis has been performed for a stoichiometric mixture of 90 mole% methane and 10 mole% ethane with air. Parameters for the reactants are

$$M = 27.94 \text{ moles/mole-g}^\dagger$$

$$R = 0.294 \text{ kJ/kg K}^\dagger$$

$$q_1 = 2.755 \text{ MJ/kg at STP}$$

$$\gamma = 1.39$$

Equilibrium thermodynamic calculations were performed for this mixture. It was found that the molecular weight of the products changed by only 1%, and therefore the change in molecular weight due to combustion was neglected. The calculated temperature enthalpy for the combustion products is shown in Fig. 2. The slope of the products curve (i.e., the C_p of the products) increases as temperature increases. This curvature of the $h^p(T)$ curve causes the heat release $q(T_i) \equiv h^p(T_i) - h^r(T_i)$ to decrease at elevated temperatures. The temperature-enthalpy curve for the combustion products was fit with a tenth-order polynomial in temperature

$$h^p(T) = \sum_{i=0}^{10} b_i T^i \qquad (21)$$

The coefficients b_i were determined by a least-squares routine, yielding

$$b_0 = 5.965467E-02 \quad b_1 = -1.924248E-02$$
$$b_2 = 8.826051E-02 \quad b_3 = -3.199495E-02$$
$$b_4 = 6.757593E-03 \quad b_5 = -8.739552E-04$$
$$b_6 = 7.168432E-05 \quad b_7 = -3.747641E-06$$
$$b_8 = 1.210761E-07 \quad b_9 = -2.203909E-09$$
$$b_{10} = 1.729367E-11$$

---

†The same value was used for reactants and products.

where $[h]$ = MJ/kg and $[T]$ = 100 K. This fit is valid for the temperature range 250 K < T < 2500 K; it is accurate to better than 0.1% above 1000 K and has a maximum error of 3% at about 300 K. The ambient conditions ahead of the shock were taken as the following:

$$P_1 = 1.01325 \text{ bars} \qquad a_1 = 345 \text{ m/s}$$
$$\rho_1 = 1.1832 \text{ kg/m}^3 \qquad u_1 = 0$$
$$T_1 = 288 \text{ K} \qquad \gamma_1 = 1.39$$

To begin the analysis, a series of values was assumed for the piston parameter Z_p, and Eqs. (2) and (3) were integrated with a fourth-order Runge-Kutta scheme from the piston (F=1, $Z=Z_p$) to the shock front. At each step of the integration, the nondimensional heat addition Q_1 and the burning velocity S were evaluated for each integral curve, according to the generalized flame model [i.e., Eqs. (19) and (20)]. It was found that values of Q_1 decreased and S increased as one traveled along an integral curve from the piston toward the shock. The flame location was identified by requiring Q_1 = 32.170 (corresponding to a heat addition of 2.755 MJ/kg for the methane/air mixture). In general, the values of the burning velocity determined for each of these integral curves did not equal the value desired, so new values were chosen for Z_p and the procedure was repeated. After several iterations on Z_p, the solutions corresponding to burning velocities of S = 1, 5, 10, 20, 50, and 100 m/s were determined. The values of Z_p are listed in Table 1. The integral curves for the spherical case are shown in Fig. 3. The flame locus is shown as a chain-dashed curve.

Table 1 Iterated values of the piston parameter Z_p vs burning velocity S

S, m/s	Z_p		
	j=0	j=1	j=2
100	0.83	0.595	0.535
50	2.16	1.51	1.34
20	9.4	6.9	6.2
10	32.	24.4	23.
5	117.3	94.	88.5
1	2750.	2225.	2300.

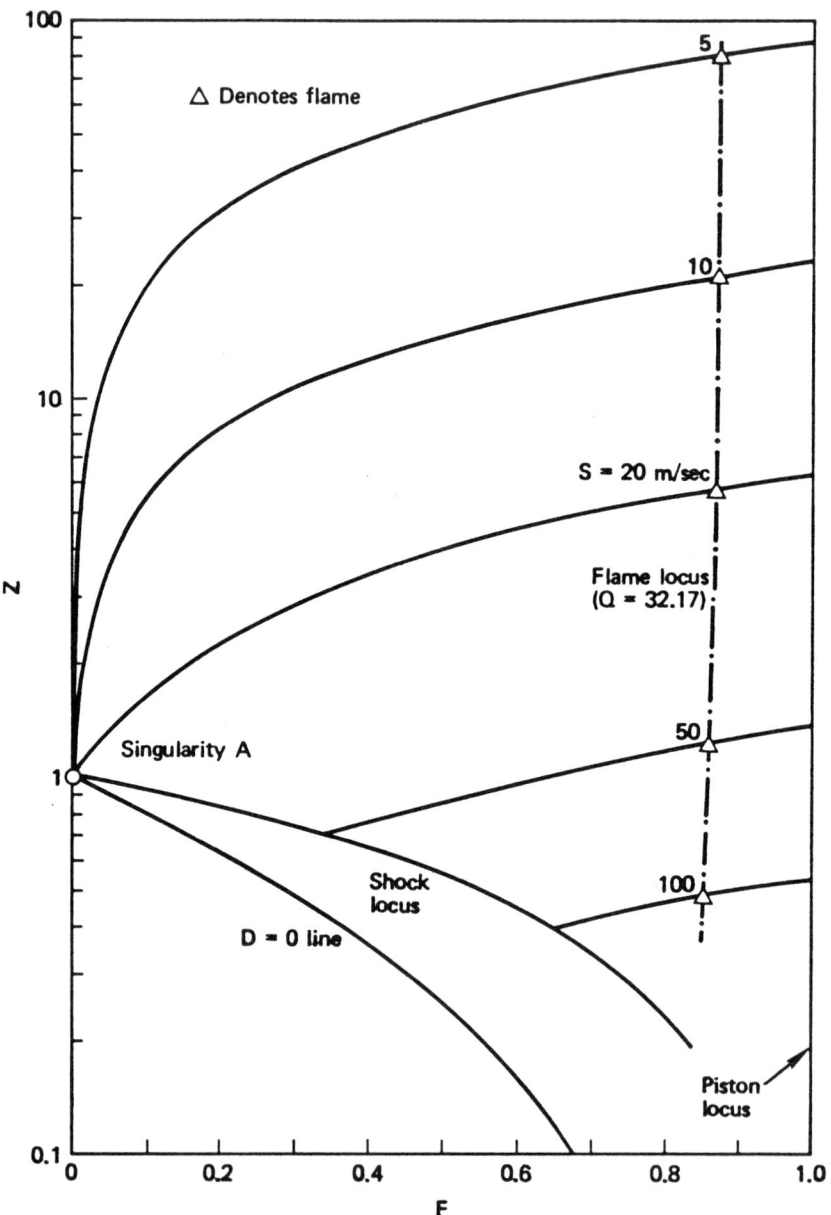

Fig. 3 Integral curves in the phase plane corresponding to flame- and piston-driven spherical blast waves ($j=2$, $\gamma=1.39$).

The accuracy of the numerical integration was checked by reducing the step size, recalculating, and observing the change in the solution. It was thus determined that at or below a step size of $F = 10^{-3}$, for $0.1 \leq F \leq 1$, and $\Delta F = F/10^3$, for $0 < F < 0.1$, changes in the solution were imperceptible (e.g., below 0.1%). The global accuracy of the solutions has been verified by calculating the nondimensional mass integral:

$$J_1 = \int_0^1 (\rho/\rho_1) \, x^j \, dx \tag{22}$$

which should satisfy the theoretical relation $J_1 = 1/(j+1)$. Typical errors in global mass conservation are less than 10^{-4}% for the piston solutions and 0.1% for the flame solutions.

The resulting flowfields for planar, cylindrical, and spherical flame-driven blast waves propagating in the methane/air mixture are shown in Figs. 4-7. The solutions have been continued past the flame to the piston, so these figures include the complete results for piston-supported blast waves also. Both solutions display the typical characteristics of piston-driven waves, with all gasdynamic variables increasing toward the interior of the wave. Across the flame, the values of pressure and density drop discontinuously, with the magnitude of the drop increasing as the burning velocity is increased. Temperatures increase across the flame to values slightly above the adiabatic flame temperature of 2224 K; the maximum calculated value is 2262 K for the planar case with S = 100 m/s. This small increase in temperature is a consequence of the fact that the slope of the temperature-enthalpy curve for the combustion products increases at high temperatures, so that large increases in the temperature of the reactants (due to compressive heating) cause smaller increases in the temperature of the products (see Fig. 2). Inside the flame, the thermodynamic variables are constant and the velocity is zero.

Two key parameters related to the potential mechanical damage from such blast waves are the overpressure at the shock, Δp_2, and at the flame, Δp_3. Figure 8 shows that these overpressures increase as the burning velocity of the

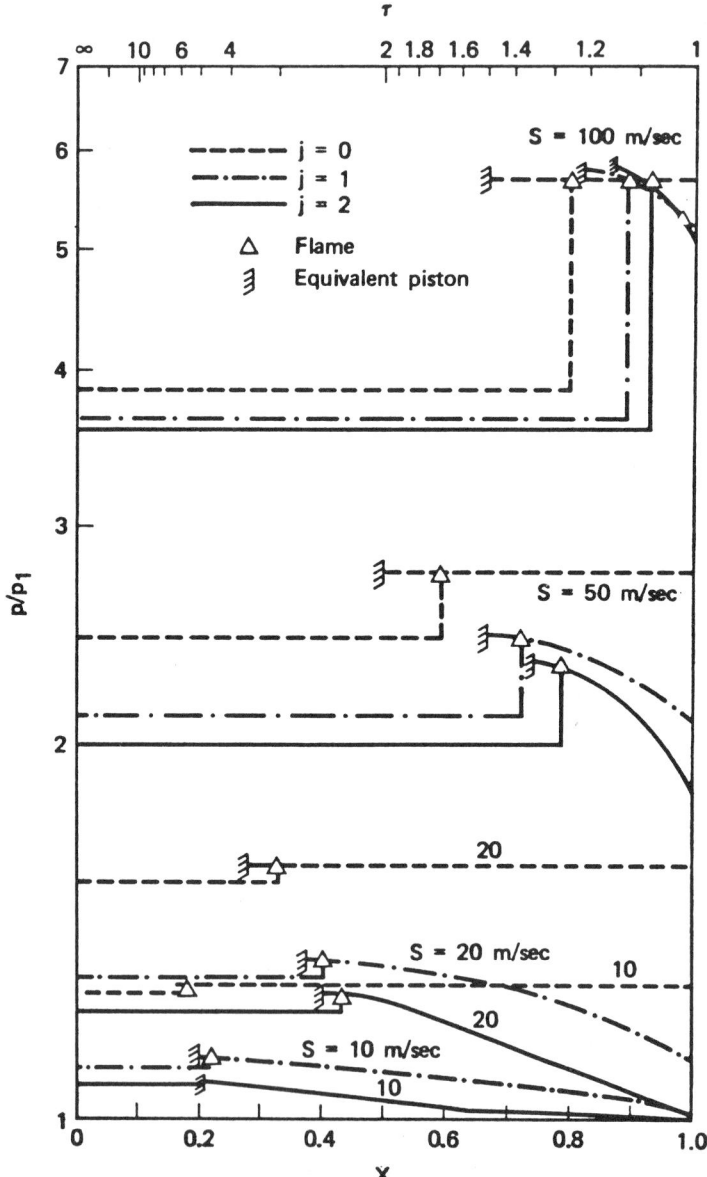

Fig. 4 Blast wave pressure distributions corresponding to flames with various burning velocities ($Q_1 = 32.170$).

SIMILARITY ANALYSIS OF FLAME-DRIVEN BLAST WAVES

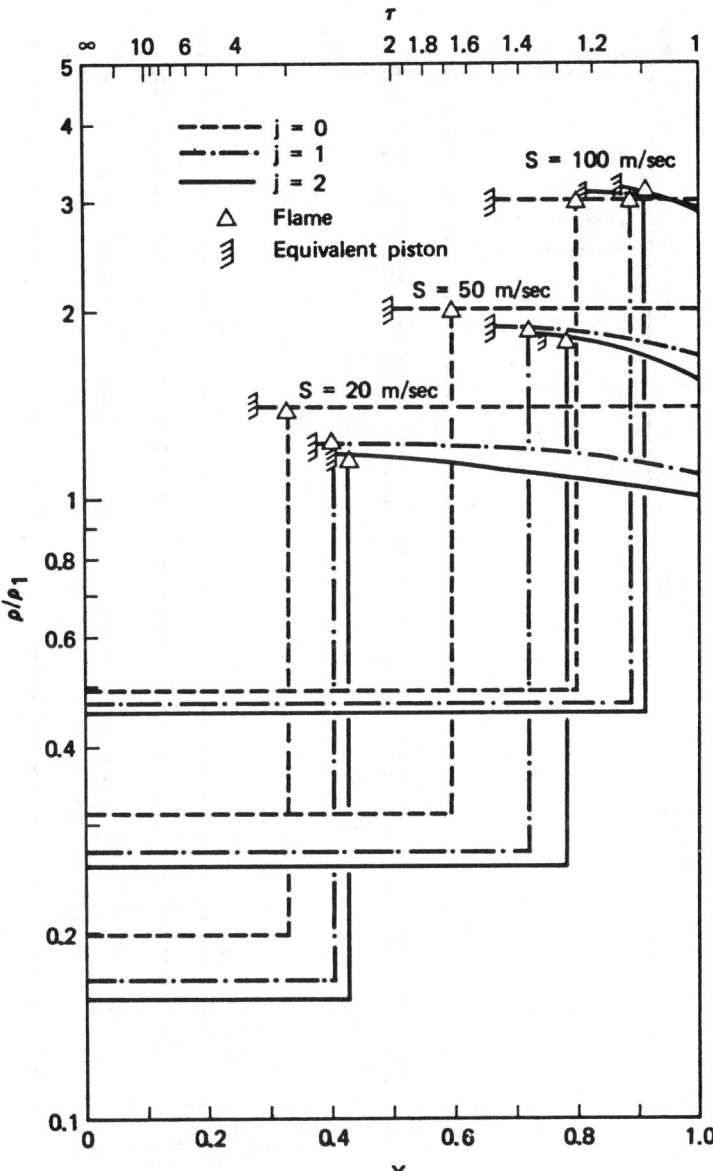

Fig. 5 Blast wave density distributions corresponding to flames with various burning velocities (Q_1=32.170).

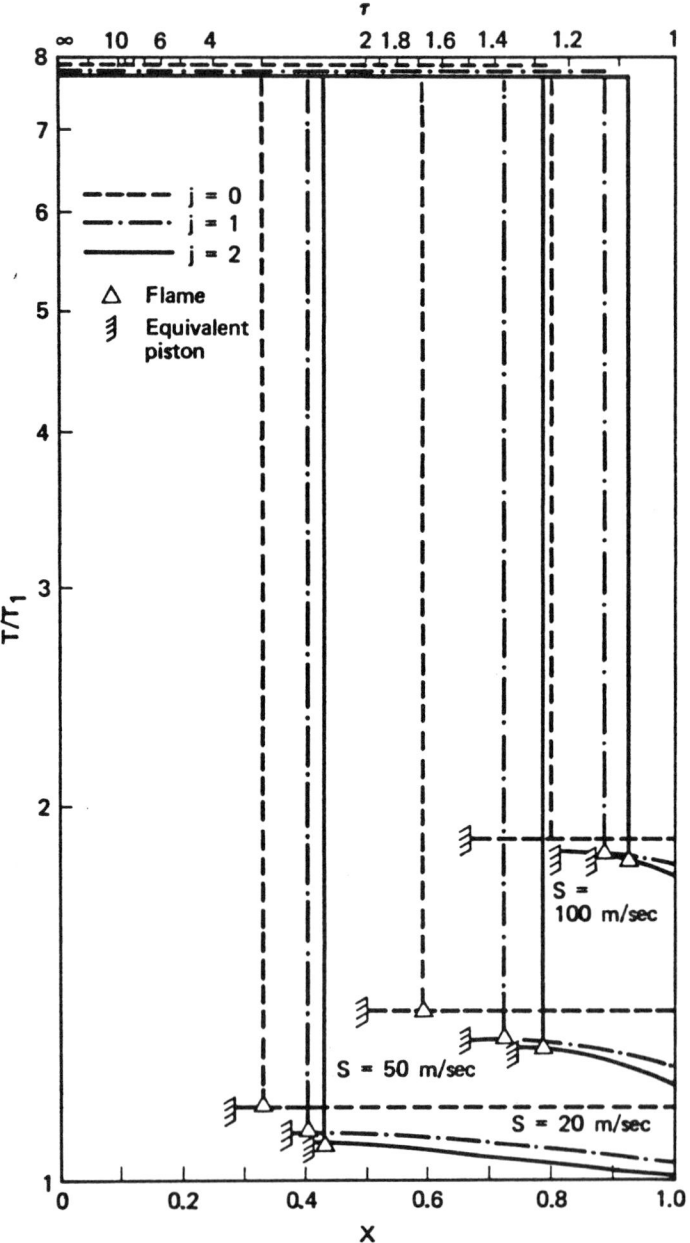

Fig. 6 Blast wave temperature distributions corresponding to flames with various burning velocities ($Q_1=32.170$).

SIMILARITY ANALYSIS OF FLAME-DRIVEN BLAST WAVES 191

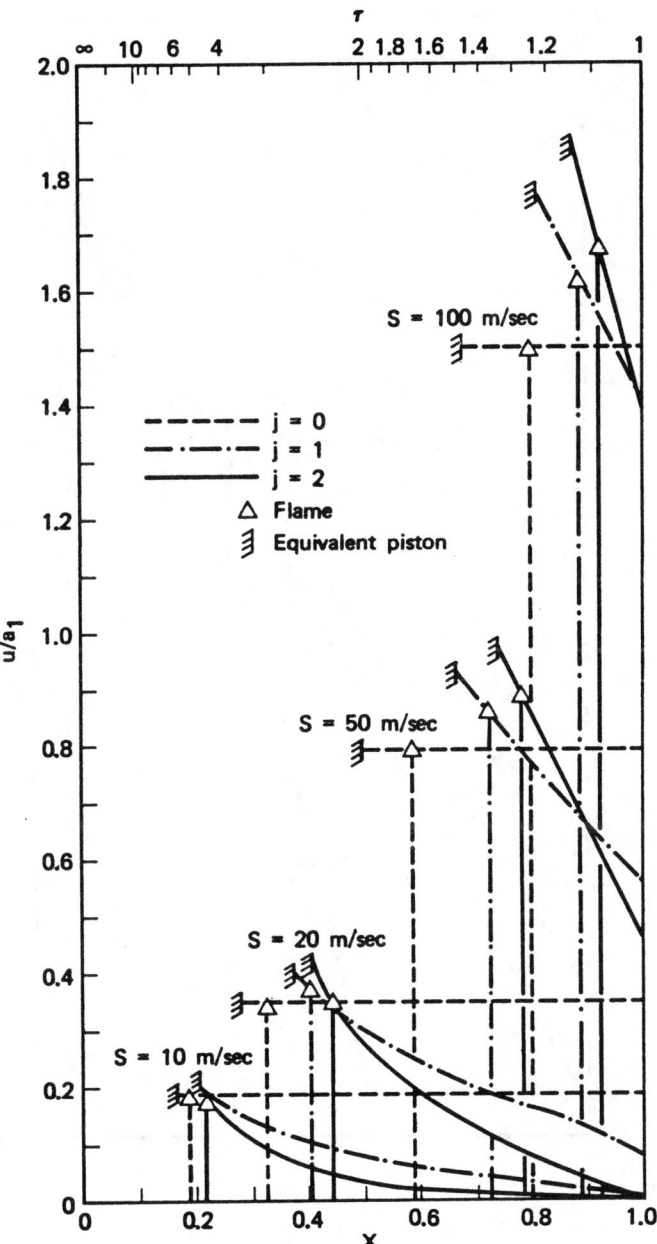

Fig. 7 Blast wave velocity distributions corresponding to flames with various burning velocities (Q_1=32.170).

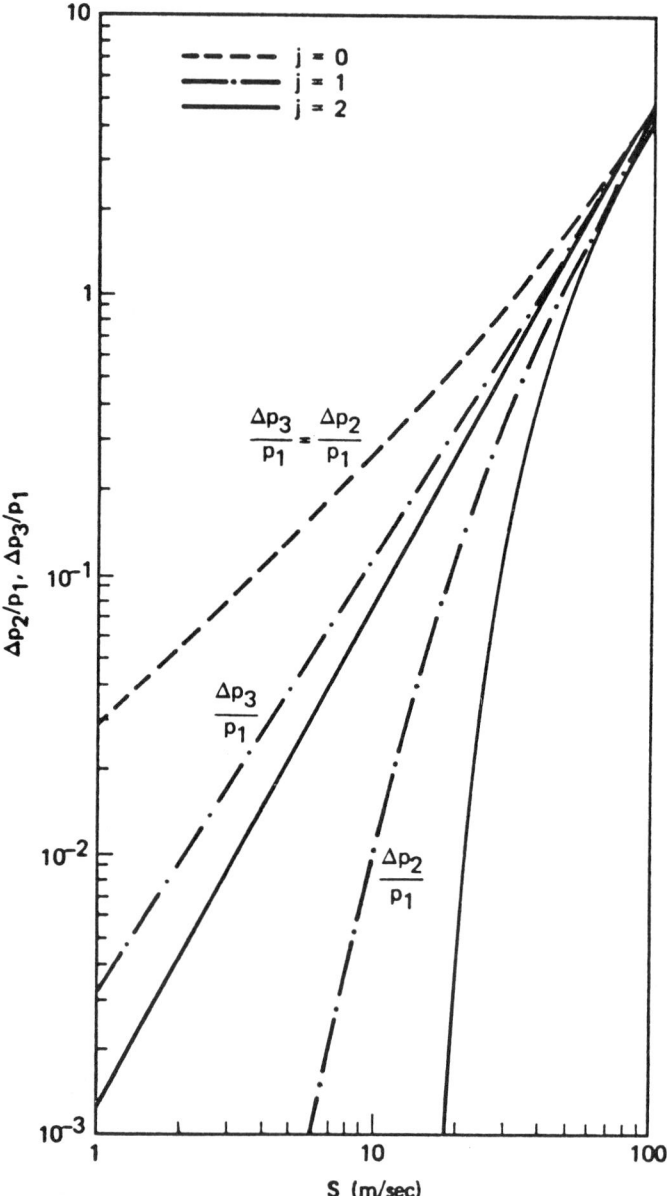

Fig. 8 Overpressure at the shock and in front of the flame as a function of the burning velocity of the flame ($Q_1 = 32.170$).

flame increases. Typical values of laminar burning velocities are around 1 m/s, and blast waves driven by such flames are very weak (being essentially acoustic waves). Turbulence can substantially increase the burning velocity. The turbulence can be self-generated by the flame, or it can be induced mechanically by placing obstacles in the flow. There is some experimental evidence (Moen et al. 1983) that such obstacle-generated turbulence can increase the relative burning velocity to perhaps 10-20 m/s under strongly confined conditions. According to the present analysis, flame propagation with a burning velocity of 20 m/s would produce a shock overpressure of 0.113 and 0.0039 atm for cylindrical and spherical waves, respectively; the corresponding overpressure at the flame would be 0.339 and 0.261 atm, respectively. The accuracy of these values is estimated to be about +2%. The present overpressure results for $j = 1$ and $S = 20$ m/s are from 30 to 50% higher than our previous work (Kuhl et al. 1973), which used a constant gamma flame model and somewhat more generic thermodynamic properties for the vapor cloud ($\gamma_3 = 1.3$, $\gamma_4 = 1.2$, and $T_4/T_1 = 7.0$). These differences are a consequence of improvements in the flame model, equations of state, and heat release employed.

The heat addition at the flame front, Q_3, varies with flame velocity owing to preheating of the unburned gases by compression. Values of Q_3 were also computed as part of the present analyses, and it was found that Q_3 differed from Q_1 by much less than 0.1%; i.e., in the present calculations there was no significant change in heat addition due to compressional heating, even for burning velocities as high as 100 m/s. This can be explained as follows. For the present solutions, the maximum temperature just ahead of the flame was 539 K ($j=0$, $S=100$ m/s). Figure 2 shows that at this temperature, the enthalpy of the combustion products is negligibly different from $C_p T$. Consequently, for this combustible mixture one can assume $Q_3 = Q_1$ in Eq. (19), and Eq. (20) becomes superfluous. Note that if one uses the constant gamma flame model i.e., $h^p = [\gamma_4/(\gamma_4 - 1)] RT$ with a gamma value that fits the combustion products near the flame temperature, the opposite conclusion would be reached. For example, assuming $T_3/T_1 = 1.873$ and $\gamma_4 = 1.158$ at 2224 K, the constant gamma verison of Eq. (20) gives $Q_3 = Q_1 - 3.29$, and there would be about a 10% decrease in heat addition for the higher flame velocities. To obtain the most realistic results, the new flame model is recommended.

It should be pointed out that these similarity results are essentially equivalent to closed-form solutions. Since the flowfield depends on an ordinary differential equation which can be integrated as accurately as desired by simply reducing the integration step, the solutions can be made arbitrarily accurate within the framework of assumptions employed (e.g., equations-of-state, etc.). We believe that the results given herein represent benchmark calculations that can be used, for example, to verify finite-difference codes which can then be applied to situations requiring more general geometries.

Summary and Conclusions

A new flame model is proposed for similarity analysis of flame-driven blast waves. It is shown that the Hugoniot relation can be solved to give the heat release in terms of the pressure and specific volume ratios across the flame for fairly general equations of state for the combustion products. The mechanical conditions are sufficient to specify the pressure and specific volume ratios across the flame in terms of known parameters just ahead of the flame. Hence this flame model allows one to evaluate the heat release and burning velocity throughout the blast wave flowfield, and thereby identify the flame location.

The generalized flame model was used in a similarity analysis of constant velocity, flame-driven blast waves in a stoichiometric methane/air mixture. Planar, cylindrical, and spherical wave results were obtained for a range in burning velocities from 1 to 100 m/s. The results from this generalized flame model give from 30 to 50% higher overpressure than the previous calculations (Kuhl et al. 1973), which were based on a constant gamma flame model. These differences are a consequence of improvements in the flame model, equations of state, and heat release employed.

It should be pointed out that these similarity results are essentially equivalent to closed-form analytic solutions. Since they are exact, the similarity solutions can serve as benchmark calculations which can be used to validate finite-difference codes which can then be applied to physical problems requiring more general geometries.

References

Guirguis, R. H., Kamel, M. M., and Oppenheim, A. K. (1983), Self-similar blast waves incorporating deflagrations of variable speed. Shock Waves, Explosions, and Detonations:

AIAA Progress in Astronautics and Aeronautics (edited by Bowen, Manson, Oppenheim, and Soloukhin),Vol. 87, pp. 121-156. AIAA, New York.

Kuhl, A. L., Kamel, M. M., and Oppenheim, A. K. (1973) 14th International Symposium on Combustion, pp. 1201-1215. The Combustion Institute, Pittsburgh, Pa.

Moen, I. O., Donato, M., Knystauvas, R., and Lee, J. H. (1981) Turbulent, flame propagation and acceleration in the presence of obstacles. Gasdynamics of Detonations and Explosions: AIAA Progress in Astronautics and Aeronautics (edited by Bowen, Manson, Oppenheim, and Soloukhin), Vol. 75, pp. 33-47. AIAA, New York.

Sedov, L.I. (1959) Similarity and Diminsional Methods in Mechanics (4th ed.), Gostekhizdat, Moscow. (English translation edited by M. Holt, Academic Press, New York, 1959).

Taylor, G. I. (1946) Proc. Roy. Soc. Ser. A, 186, 273.

Optical Interferometry of Spherical Shock Waves

V.F. Klimkin* and V.V. Pickalov†
Academy of Sciences, Novosibirsk, USSR
and
R.I. Soloukhin‡
Academy of Sciences, Minsk, USSR

Abstract

When interferometry is used to study unsteady processes in gases, liquids, and plasma, interference records of finite-width fringes can be easily interpreted only in planar or cylindrical disturbances. In this work the interferometric method for analysis of spherical phase disturbances is developed. With this method one can reconstruct the radial distribution of the refractive index of an object in question when the shifts of a small number of the interference fringes are measured. The integral equation is derived for the local values of the refractive index as a distance-sphere center function for the spherical symmetry. As a result of the numerical and physical experiments, the method is shown to be sufficiently reliable that it can be used in the high-speed interferometric studies.

Introduction

The interferometric recording, the optical picture of the transmitted light intensity distribution, depends on the phase difference $\Delta\phi(x,y)$ between the reference and probing

---

Presented at the 8th ICOGER, Minsk, USSR, Aug. 23-26, 1981. Copyright © American Institute of Aeronautics and Astronautics, Inc., 1982. All rights reserved.
*Senior Scientist, Institute of Pure and Applied Mechanics.
†Research Scientist, Institute of Pure and Applied Mechanics.
‡Professor and Director, Heat and Mass Transfer Institute.

waves passing through the inhomogeneity under study. A wave front retardation $\Delta\phi(x,y)$ (in the geometrical optics approach and at small deviation angles) is equal to (Landenburg and Bershader 1954)

$$\Delta\phi(x,y) = (2\pi/\lambda)\int_{z_1}^{z_2}[n(x,y,z)-n_0]dz \quad (1)$$

where $n(x,y,z)$ is the inhomogeneity refractive index at the point (x,y,z); n_0 is the refractive index of an undisturbed medium; and z_1 and z_2 are, respectively, the incidence and emergence points for a beam entering the inhomogeneity. The retardation $\Delta\phi(x,y)$ is related to the relative number of fringes $k(x,y)$ by which the interference picture has been shifted by

$$k(x,y) = \Delta\phi(x,y)/2\pi \quad (2)$$

As $\Delta\phi(x,y)$ and $k(x,y)$ are the integral characteristics of the optical inhomogeneity, interpretation of the interferograms in terms of the fringes of a finite width is straightforward only for the planar or cylindrical inhomogeneities. In such cases the absence of the refractive index gradients along one of the axes x,y leads to the unambiguous determination of the relative retardation in the chosen cross section of an object through measurement of the fringe shift in terms of the undisturbed fringe period parts. Thus the problem is reduced to measurements of the value k as a function of only one variable, say, x. The local values of the object refractive index in the given cross section are determined either by algebraic calculations (a plane one-dimensional layer) (Ladenburg and Bershader 1954)

$$n(x) = n_0 \pm [k(x)\lambda/L]$$

where L is the geometrical beam path in the inhomogeneity, or by solution of the Abel integral equation (cylindrical symmetry) (Ladenburg and Bershader 1954)

$$k(x) = \frac{2}{\lambda}\int_x^R [n(r) - n_0]\frac{rdr}{\sqrt{r^2 - x^2}} \quad (3)$$

where R is the inhomogeneity radius in the cross section, y = const, and x is the distance from the chord to the symmetry axis. When the optical inhomogeneity (with refractive index) varies also in the y direction, the local values of the refractive index are given for a planar two-dimensional layer by

$$n(x,y) = n_0 + [k(x,y)\lambda/L]$$

and for an axisymmetric inhomogeneity by

$$k^*(x,y) = \frac{2}{\lambda} \int_x^R [n(y,r) - n_0] \frac{r\,dr}{\sqrt{r^2 - x^2}} \qquad (4)$$

Then the problem is reduced to interferogram measurements of the value k^* as a function of x,y and to the two-dimensional interpolation of the data on a relative retardation $k^*(x,y)$ in the given cross section y = const. Since many interference fringes are required for such a study, essential difficulties arise when the micro-objects with a typical size ∼ 100 μm are studied. In this case there appears a sufficiently complicated problem of computational mathematics on the approximation and smoothing of a two-variable function prescribed in a tabular form on irregular mesh points grid (Marchuck 1980).

The present paper develops an interference technique suitable for the study of spherical phase microinhomogeneities. This technique requires a small number of interference fringes (even one fringe) to reconstruct the radial distribution of the refractive index in the object under study.

Results and Discussion

In the case of a spherical symmetry the local change in the refractive index $\Delta n(\rho)$ is determined from the measured relative shift of the m-th interference fringe $k_m(x)$ by

$$k_m(x) = \frac{2}{\lambda} \int_{\rho_m(x)}^R \frac{\Delta n(\rho)\rho\,d\rho}{\sqrt{\rho^2 - \rho_m^2(x)}} \qquad (5)$$

where ρ is the spherical variable, R is the inhomogeneity radius, and $\rho_m(x)$ is the distance from the inhomogeneity

image center to the points chosen on the m-th dispaced
interference fringe in the image plane. Unlike the conventional Abel equation (4), a two-dimensional interpolation over a large number of interference fringes is not
needed to solve the integral equation (5). The radial
distribution of the refractive index $\Delta n(\rho)$ is reconstructed
within the whole spherical inhomogeneity. Equation (5) was
solved with the use of the statistical regularization
methods (Turchin et al. 1970; Klimkin and Picklov 1979a).

Numerical experiments with a number of optical
inhomogeneity models have been used to verify that reliable
results are obtainable. As an example, $\Delta n(\rho)$ was
reconstructed for an optical inhomogeneity model with a
refractive index jump characteristic of a spherical shock
wave:

$$\Delta n(\rho) = \exp(1 - R/\rho) \quad 0 \leq \rho \leq R$$
$$= 0 \quad \rho > R \quad (6)$$

The values $k(x)$ were determined from Eqs. (5) and (6). A
physical experiment was modeled through the introduction of
random normal errors into the corresponding $k(x)$. The
results of the $\Delta n(\rho)$ reconstruction by the statistical
regularization method for the zero and third fringes are
represented in Figs. 1a and 1b, respectively. The zero
undisturbed fringe passes through a central cross section
and the distance between the undisturbed fringes equals
0.1R. The error inherent in the reconstruction of the jump
and adjacent regions (the solid line in Fig. 3) does not
exceed 2-4% for the 5% measurement error k of the maximum
value (the broken line in Fig. 3). There is some smoothing
of the solution in the central region which is due to the
boundary condition introduced into the algorithm and the
decreasing one for the zero fringe.

At conditions of a real physical experiment, the
reconstruction accuracy was determined with a model device
including the Mach-Zehnder interferometer and the
helium-neon laser which served as a light source and a
special cuvette (Klimkin and Pickalov 1979b). A sphere of
melted quartz with a radius R = 480 μm was the object in
question. A sample was placed into a cuvette filled with a
liquid whose refractive index could be controlled. The
value of the refractive index jump was $\Delta n = n_\ell - n_q$. A
typical interferogram of the object (Fig. 2) shows the shift
of the interference fringes within the disturbance. This
shift has been measured with the help of a comparator. The

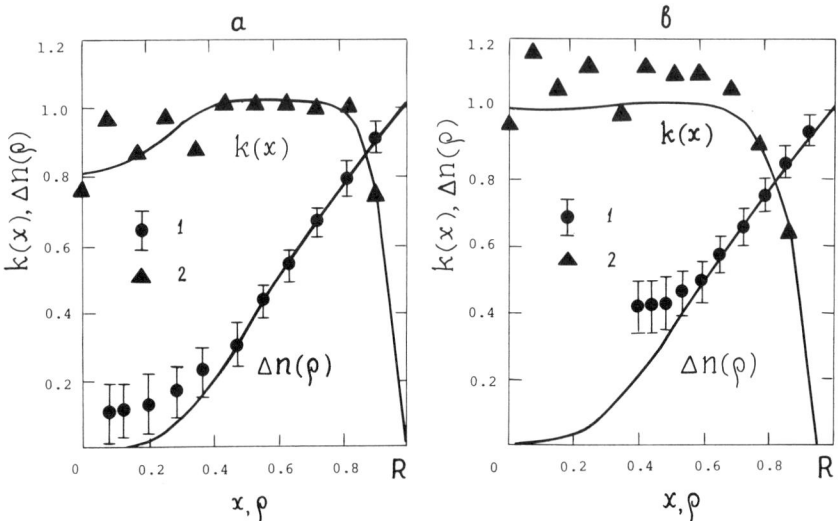

Fig. 1 Reconstruction of the model $\Delta n(\rho)$ by the statistical regularization method.

results of the sphere refractive index profile reconstruction for the zero interference fringe are shown by the closed circles in Fig. 3.

The deviation from the condition Δn = const can be attributed to a fine structure of the disturbance arising in the process of manufacturing a microsample. However, a mean of the reconstructed values of the refractive index change (the solid line in Fig. 3) is in a good agreement with the one measured using a conventional refractometer design (the broken line in Fig. 3), the measuring accuracy being about 10^{-4}. The open circles in Fig. 3 represent the results of the $\Delta n(\rho)$ reconstruction obtained from the Abel integral equation (3) at the same initial conditions. The qualitative coincidence of results indicates that satisfactory data, with an accuracy within $\lesssim 6\%$, can be obtained from the Abel integral equation if a cylindrical symmetry for the sphere cross sections close to the central one is assumed.

The results of the mathematical and physical modeling lead to the conclusion that the technique developed has proven to be sufficiently reliable. As a consequence the method was used in the interferometric studies of microdisturbances produced at the initial stage of the electric discharge in the liquid dielectrics. In these experiments a ruby laser with a pulse duration of about 5 ns served as a light source. As has previously been shown (Klimkin and Ponomarenko 1979), the development of a

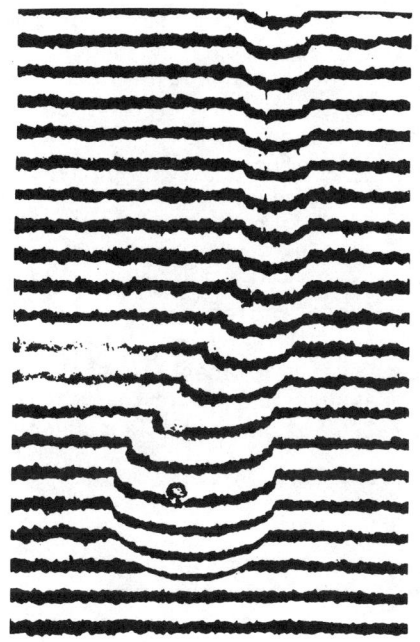

Fig. 2 Interferogram of the model micro-object of melted quartz. Radius of the spherical part is 480 μm.

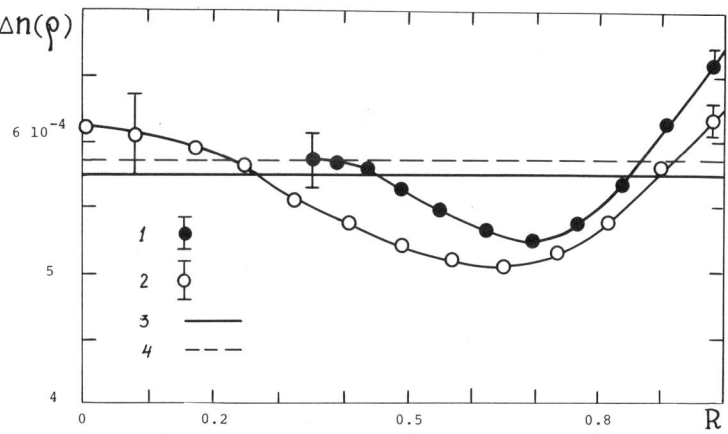

Fig. 3 Results of the refractive index profile reconstruction for the object shown in Fig. 2.

discharge in a liquid dielectric in the uniform electric field has two stages. At the first stage, near the electrode surface, spherical shock waves of about 10^{-2} cm in radius are produced as a result of the local energy release. These shocks favor the formation of the initial ionization channel. Klimkin and Ponomarenko (1979) proposed a new approach to the problem of studying the initial stage of the electric discharge in the liquid dielectrics. The shock waves were investigated with the help of the optical interferometry at a sufficiently large distance from the point of energy release so that it was possible to follow the fringe shift behind the shock wave front. The pressure dependence at the shock wave front upon the radius for disturbances that do not cause a breakdown, has been used to estimate a characteristic channel radius and a pressure in the energy release region from an approximate hydrodynamic model. The size of the energy release region may also be estimated from the short wave theory of Yakovlev (1961) with the pressure profile in a narrow region adjacent to the shock wave front. Usually the pressure change in the shock wave in time is characterized by the exponential dependence

$$p = p_0 \exp(-t/t_0)$$

where the constant t_0 corresponds to the time during which the pressure falls as much as e times, while the time t is measured from the moment of the wave arrival at the given point. If the pressure profile deformation over distance in the region $P \leq 300$ atm is neglected, the following expression for the pressure change behind the shock wave front holds:

$$p = p_0 \exp(-\rho/\rho_0)$$

where ρ_0 is the distance at which the pressure falls e times, while ρ is measured from the shock wave front R. In the case of the spherically symmetric flow in water there exists the following expression (Yakovlev 1961):

$$\frac{\rho_0}{r_0} \simeq \frac{t_0 C_0}{r_0} = \frac{1.92 \ln(17{,}000/P_0) - 1.11}{\sqrt{\ln(17{,}000/P_0) - 0.5}}$$

where C_0 is the sound velocity, and r_0 is the characteristic channel radius. Equation (7) yields the values of a pressure P_0 at the shock wave front and of a decay time constant t_0 close to the experimental ones at $P_0 \leq 250$ atm.

Thus one can attempt to estimate a characteristic channel size for a given disturbance from Eq. (7) and the pressure profile parameters P_0, r_0.

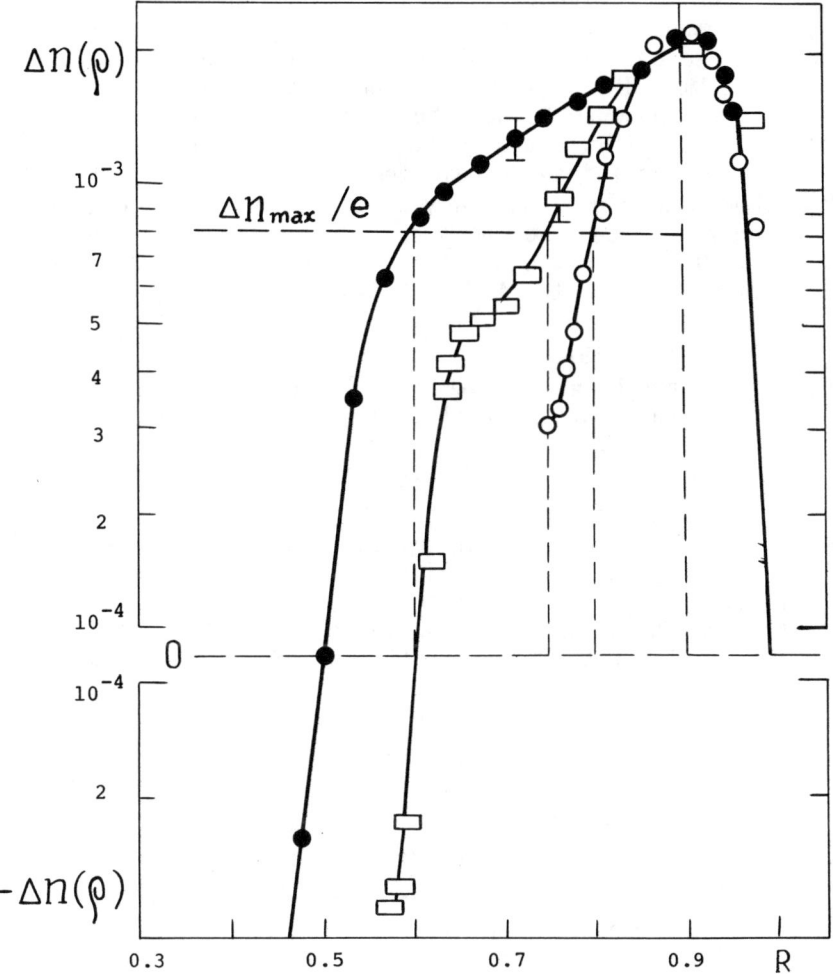

Fig. 4 Reconstruction of the refractive index distribution for a shock wave in distilled water.

The reconstruction of the refractive index change behind the spherical shock wave front in water with a radius $R \simeq 170$ µm is shown in Fig. 4. Only three fringes were used. Both the pressure from the water isentropicity condition and the refractive index dependence upon the pressure can be estimated (Klimkin and Ponomarenko 1979). At the pressure P $\lesssim 10^3$ atm, when the density-temperature dependence is negligible, the pressure profiles reproduce, with an accuracy up to constant, the corresponding profiles $\Delta n(\rho)$. Good agreement of results in the maximum pressure region should be noted. An estimate of the maximum pressure yields $P_0 \simeq 140$ atm. The profile difference observed in the other pressure regions can be attributed to a violation of the spherical symmetry in the development of initial ionization channels behind the shock wave front. The upper profile is described with a good accuracy by the exponential dependence in a region adjacent to the shock front. The characteristic channel size r_0 for the given disturbance at the pressure $P_0 \simeq 140$ atm was estimated from Eq. (7) to be $\rho_0/r_0 \simeq 4$. For ρ_0, from the different profiles for a radius of the energy release region (Fig. 4) $r_0 \simeq 5\text{-}13$ μm. This result is in qualitative agreement with the results of the shock wave dynamics measurements (Klimkin and Ponomarenko 1979).

Conclusions

In this paper a method of the interferometric measurements of spherical microdisturbances is proposed. This technique was verified in numerical and physical modeling. Its use in the experimental studies of the electric discharge in water enables one to obtain the results dealing with the initial stage of its development.

References

Klimkin, V. F. and Pickalov, V. V. (1979a) On microinterferometry capabilities for studying nonstationary processes. Zh. Prikl. Mekh. Tekh. Fiz. 3, 14-26.

Klimkin, V. F. and Pickalov, V. V. (1979b) Optical interferometry of pulsed micro-discharges. J. Phys. (Paris) 40, C7-329-330.

Klimkin, V. F. and Ponomarenko, A. G. (1979) Investigation of pulsed electric discharge in liquids by means of optical interferometry. Zh. Tekhn. Fiz. 49, 1896-1904.

Ladenburg, R. and Bershader, D. (1954) Interferometry, in Physical Measurements in Gas Dynamics and Combustion (ed. R. W. Ladenburg, et al.) Vol. IX, p. 47. Princeton University Press, Princeton, N. J.

Marchuck, G. I. (1980) Metody Vychislitel'noi Matematiki. Nauka, Moscow, USSR.

Turchin, V. F., Kozlov, V. P., and Malkevich, M. S. (1970) The use of mathematical statistics methods for solving ill-posed problems. Phiz. Usp. 102, 345-386.

Yakovlev, Y. S.(1961) Gidrodinamika Vzryva (Explosion Hydrodynamics). Leningrad, USSR.

A Study of Explosive Shock Tubes

G.A. Shvetsov,* and V.M. Titov,* V.P. Chistyakov,* and I.A. Stadnichenko*
USSR Academy of Sciences, Siberian Division, Novosibirsk, USSR

Abstract

Experimental results on high-velocity gas flows generated in explosive shock tubes with single and tubular gas-cumulative explosive charges are presented. Shock wave velocity and conversion efficiency of explosive energy to gas flow kinetic energy, as a function of the geometric sizes and number of the tubular explosive charges, have been measured. The cellular structure of an explosive charge produces a shock wave velocity of about 10^4 m/s in air under normal pressure for a minimal weight of an explosive, while the shock tube cross-section area is some hundreds or thousands square centimeters. This charge also gives a 2.5- to threefold increase in the conversion efficiency of explosive energy to gas kinetic energy over that of a single tubular explosive charge. Shock tubes with tubular explosive charges have been employed in experiments on conversion of explosive to electromagnetic energy in pulsed-MHD generators. They were also used for the investigation of structural and phase transitions on the surface of steel specimens affected by high-enthalpy gas flow. The cellular explosive charges have been used to produce powerful light pulses within the visible and ultraviolet regions of the spectrum and cumulative barium plasma accelerators for the investigation of near-Earth space.

Presented at the 8th ICOGER, Minsk, USSR, Aug. 23-26, 1981. Copyright © American Institute of Aeronautics and Astronautics, Inc., 1983. All rights reserved.
*Scientist, Lavrentyev Institute of Hydrodynamics.

Introduction

The use of high explosives to generate strong shock waves and high-velocity gas flows extends the possible applications of the shock tube for investigation of a large number of problems on high-velocity gasdynamics and nonstationary magnetic hydrodynamics. At present some explosive shock tube designs have produced shock wave velocities above 10 km/s at high densities of gas kinetic energy. These velocities have been obtained in air under normal conditions. Typical cross sections of such explosive shock tubes are shown in Fig. 1.

In the Voitenko compressor (Fig. 1a) (Voitenko 1966), an explosion drives a metal plate into a rapidly converging chamber containing a driver gas. The motion of the plate produces a strong wave when compressed and forces gas into the shock tube channel. Shock wave velocities of 40 km/s are achieved in air under normal conditions.

In the implosion-driven shock tube (Fig. 1b) (Glass et al. 1974) a detonation is generated by explosion of a thin copper wire placed in the geometrical center of a hemispherical chamber filled with a stoichiometric hydrogen-oxygen mixture. The detonation wave velocity is about 2.9 km/s;

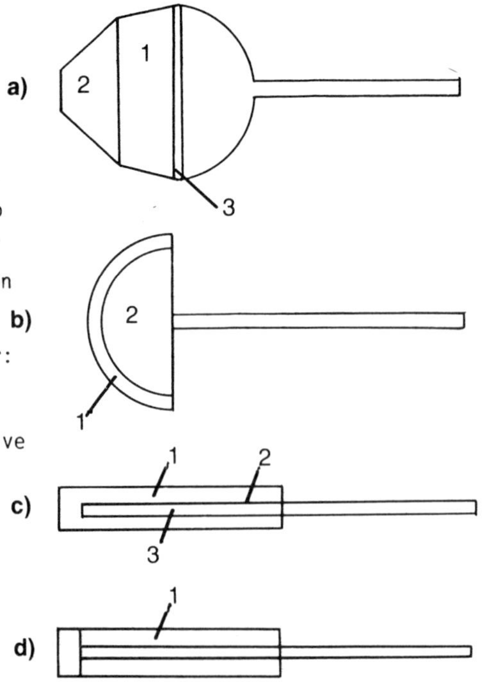

Fig. 1 Schematics of explosive driven shock tubes. a) Voitenko compressor: 1) high explosive, 2) plane wave generator, 3) metal driver plate; b) implosion driven: 1) high explosive, 2) explosive gaseous mixture ($2H_2+O_2$); c) linear accelerator: 1) high explosive, 2) thin-walled metal container, 3) He, 42 atm; d) tubular gas-cumulative charge: 1) high explosive.

impact of the detonation on a hemispherical shell of PETN leads to an explosion. A strong converging shock compresses the detonation products and forces compressed gas into the shock tube channel. The characteristic shock wave velocities in the channel are about 10-20 km/s.

In the explosion-driven shock tube (linear accelerator) (Fig. 1c) (Gill and Simpkinson 1969), a metallic container filled with high-pressure gas (usually helium) is placed in the cavity of the explosive cylindrical charge. Upon detonation of the explosive charge, the container is compressed and acts as a high-speed piston on the gas in the container. The best results are obtained in the shock tube filled with helium at a pressure of 42 atm. In the experiments the shock wave velocities of 14-16 km/s are obtained at air pressures of 0.25-0.4 Torr in the shock tube channel.

In Fig. 1d a tubular gas-cumulative explosive charge is shown. Detonation of a tubular explosive charge generates a high-enthalpy detonation product flow, i.e., a gas-cumulative jet, in its cavity at the expense of the cumulative effects. Shock wave velocities of 12-14 km/s are achieved in air at normal pressure. The tubular explosive charge detonation and gas flow in the charge cavity have been investigated by many authors (Woodhead 1959; Titov et al. 1968; Zharikov et al. 1967; Zagumennov et al. 1969; Titov and Shvetsov 1979).

A discussion of the operating characteristics of the above-mentioned explosive shock tubes and their advantages and disadvantages is beyond the scope of the present paper. This work is concerned with the requirement that a high explosive shock tube efficiently convert explosive energy to directional gas kinetic energy and produce a shock wave velocity of 10-15 km/s in the channels of large cross sections (some hundreds or thousands of square centimeters). This is particularly important for the explosive shock tubes used in the pulsed-MHD generators and for the explosive plasma injectors used in space research (Ruzhin et al. 1981).

The conversion efficiency of explosive energy into plasma energy (including internal and kinetic energy) in the Voitenko compressor is about 6% (Voitenko et al. 1975). The linear accelerator efficiency is 7% (Gill et al. 1969). Comparison of the conversion efficiency for the shock tubes in Fig. 1a-c has been made by Glass et al. (1974). The authors consider the characteristics of the UTIAS implosion-driven shock tube (Fig. 1b) to be superior to those of the linear accelerator (Fig. 1c), but inferior to the characteristics of the Voitenko compressor (Fig. 1a).

The conversion efficiency of chemical energy of a gas-cumulative explosive charge to directed kinetic energy of a gas flow is estimated as 8-10% at $d_2/d_1 = 2.4$, where d_2 is the external diameter and d_1 is the internal diameter of and explosive charge (Titov and Shvetsov 1979). Teno and Sonju (1974) estimate that the conversion efficiency of a gas-cumulative jet is 13.6% when a tubular explosive charge of $d_2/d_1 = 3.3$ detonates in air under normal conditions.

While the above investigations were not intended to produce the maximum conversion efficiency, it is not possible to compare with certainty the efficiency of these explosive shock tubes. However, the conversion efficiency of explosive energy into gas flow kinetic energy is apparently higher in the case of tubular gas-cumulative explosive charge. The simple design of the explosive shock tube with the tubular explosive charge is an obvious advantage.

Operation of the explosive shock tubes with the large cross-section channels is difficult. For effective operation the Voitenko and UTIAS shock tubes must have a hemisphere diameter 8-10 times greater than the channel diameter (Voitenko 1966; Glass 1974). For a channel diameter of 30 cm, the hemisphere diameter (as the explosive charge diameter) should be 2.5-3 m. As a consequence, it is practically impossible to use such shock tubes under laboratory conditions. With linear systems (Fig. 1c, d), the experimental results cited later in this paper indicate that for effective operation the charge length $l_0 \geq 5d_1$. For $d_1 = 30$ cm, the explosive charge length should be longer than 1.5 m. Use of linear devices in the laboratory is impracticable because of the large weight of the explosive.

The aim of the present study is to design a large-diameter explosive shock tube which can be used under laboratory conditions.

The Experimental Results

The design of an explosive shock tube is based on the experimental insight that a certain active explosive layer exists in the tubular explosive charge. This active layer determines the formation of a gas-cumulative jet and its energy (Titov et al. 1968; Titovard Shvetsov 1979). An external explosive layer is essentially an inert shell with respect to this active layer. The substitution of an inert shell for this passive explosive layer gives a 40% increase in the conversion efficiency of explosive energy to directed kinetic energy of a gas-cumulative jet (Titov and

Shvetsov 1979). A decrease in the quantity of the explosive passive shell increases the conversion efficiency of a gas-cumulative jet. This may be accomplished with a cellular explosive charge (Fig. 2). An increase in the number n of single tubular explosive charges makes it possible to produce shock tube channels of any cross section. The explosive charge length is determined by the diameter d_1 of a single explosive charge. In addition, when the tubular explosive charges are arranged as in Fig. 2, detonation product flows will appear in the gaps between the charges and increase the kinetic energy of the gas flow.

The operation of the explosive cellular charge tube depends both on the explosive charge length l_0 and on the d_1, d_2, and n. The shock wave velocity in the channel will also depend on the ratio of the tube channel area to the hole area of the explosive charge and on other parameters.

To solve this multiparameter problem, a simple method of gas flow kinetic energy estimation by the volume of a target crater formed in collision of a high-velocity gas flow with a solid target has been chosen. Numerous investigations of high-velocity impact and interaction of cumulative jets with targets indicate that the crater volume is proportional to the kinetic energy of a particle or cumulative jet (Eichelberger and Kineke 1967). If the crater volume is measured, the influence of each parameter on the kinetic energy of the gas flow can be studied. An absolute conversion efficiency η can be estimated by normalization of the results for the volume of a target crater due to a tubular charge with the known mass velocity, density, and conversion efficiency. The characteristic values of the gas-cumulative jet parameters for the charge at $d_2/d_1 = 2.4$ and $l_0/d_1 = 25$ follow: 1) the mass flow velocity is 10 km/s (Pryakhin et al. 1971); the density range is 0.17-0.20 g/cm^3 (Silvestrov and Urushkin 1971; Baum et al. 1975); the conversion efficiency is 8-10% (Titov et al. 1968; Titov and Shvetsov 1979). The results of conversion efficiency for single and cellular charges have

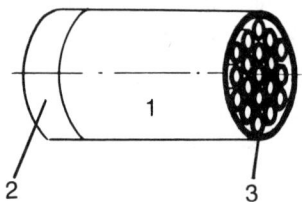

Fig. 2 Schematic of cellular explosive charge. 1) High explosive, 2) plane wave generator, and 3) η - single tubular charges.

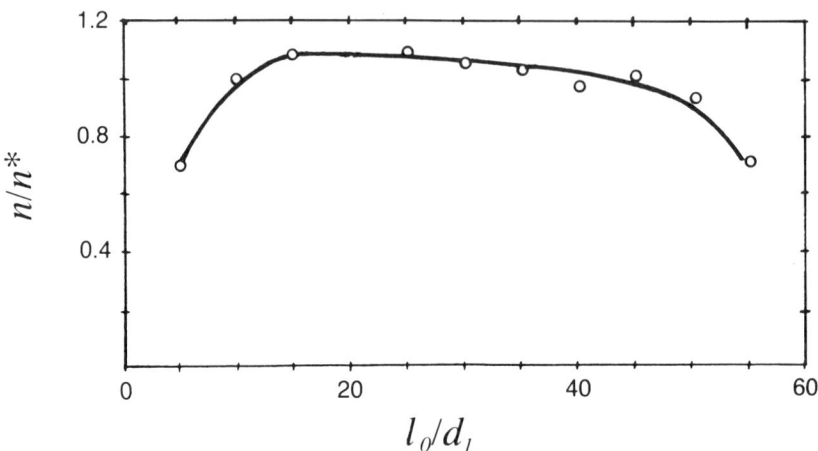

Fig. 3 Conversion efficiency of shock tube driven by cellular explosives charge: d_2 = 15 mm, d_1 = 5 mm.

been normalized by n^*. Explosive charges of TNT-50%, RDX-50% have been used in all the experiments, and the target material hardness was the same in all the experiments.

In Fig. 3 the dependence of the conversion efficiency on the explosive charge length is shown. The initial segment of the curve indicates the gas flow development in the channel. When l_0/d_1 varies from 10 to 45, the ratio n/n^* remains approximately constant. The decrease in n at $l_0/d_1 > 45$ is apparently due to jet deceleration on the channel wall and to the influence of the added mass of air (Zagumennov et al. 1969).

In Fig. 4 the dependence of the shock wave velocity on the distance X from the charge end is shown. This figure may be used to select d_1 to obtain a predetermined shock wave velocity for a given length of the shock tube channel.

The dependence of the shock wave velocity on the distance X for cellular charges, when the tube is absolutely filled with charges, is shown in Fig. 5.

The dependence n/n^* for a cellular charge with n single charges is presented in Fig. 6, while Fig. 7 reports dependence of n/n^* on d_2/d_1. The experimental results demonstrate that with a cellular explosive charge the conversion efficiency of explosive energy to gas flow kinetic energy can be considerably increased over the operation with n single explosive charges in series.

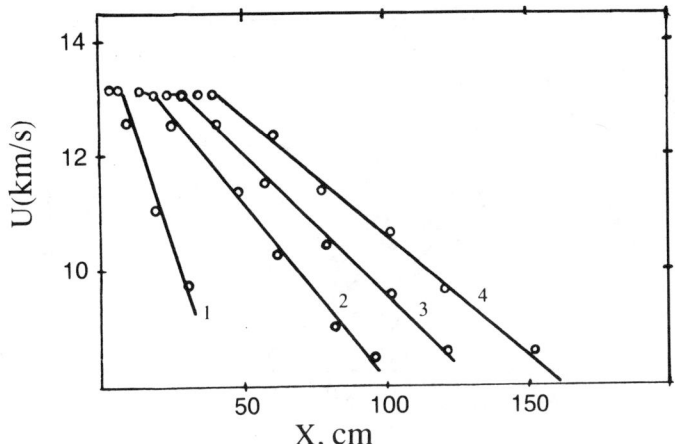

Fig. 4 Shock wave decay in shock tube driven by cellular explosives. Curves: 1) d_1 = 5 mm; 2) d_1 = 10 mm; 3) d_1 = 15 mm; and 4) d_1 = 20 mm. d_1/d_1 = 2.4; l_0/d_1 = 15.

Discussion

The experimental results show that the cellular structure of the explosive charge can lead to a 2.5- to threefold increase in the conversion efficiency over that of a single charge. This observation can be explained in an approximate fashion as follows. The volume of the formed crater, V, is $nV_1 + mV_2$. Here V_1 is the volume of a crater due to a single tubular explosive charge; V_2 is the volume of a crater due to a gas flow in the gap between tubular explosive charges; and m is the number of gaps.

Within the investigated range of d_1, d_2, l_0, n, and m, the measured values of V and those calculated from the foregoing definition of V coincide with an acccuracy higher than 20%. If $\eta = V/nE_{exp}$, and $\eta_1 = V_1/E_{exp}$, then it follows from the foregoing definition of V that

$$\eta = \eta_1[1 + (m/n)(V_2/V_1)]$$

Here E_{exp} is the explosive energy of a single tubular charge; η is the conversion efficiency of cellular explosive charge energy to kinetic gas flow energy; η_1 is the conversion efficiency of explosive single tubular charge energy to the kinetic gas-cumulative jet energy. The volumes V_1 and V_2 are measured directly, and m = $6k^2$ and n

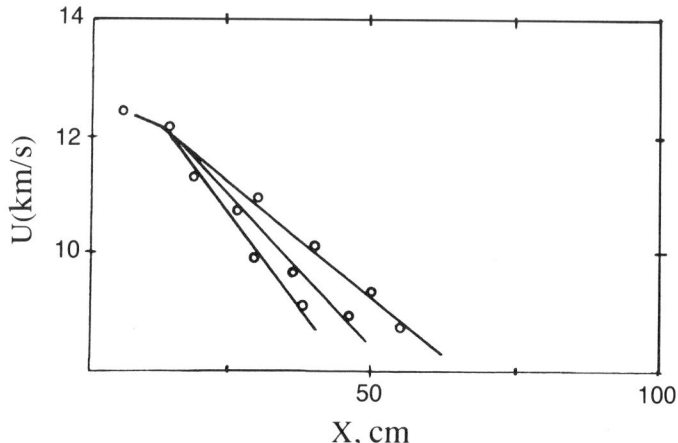

Fig. 5 Shock wave decay in shock tube driven by cellular explosives. Curves: 1) $\eta = 7$, 2) $\eta = 19$, and 3) $\eta = 37$. $d_1 = 5$ mm; $d_2 = 15$ mm; $d_2 = 15$ mm; $l_0/d_1 = 15$.

$= 3k(k+1) + 1$ for the close-packed triangle arrangement. Here k is the order of arrangement (k = 0, 1, 2, 3, ...). For the explosive charge, $d_2 = 15$ mm, $d_1 = 5$ mm, $l_0/d_1 = 15$, and $V_2/V_1 = 1.22$. For k = 2, the foregoing formula predicts that $\eta/\eta_1 = 2.54$. This is in good agreement with the experimental results: $\eta/\eta_1 = 2.5$ (from Fig. 6). For k = 5, $\eta/\eta_1 \approx 3$; and at $k \to \infty$, $\eta/\eta_1 \to 3.45$.

This study indicates that an explosive shock tube driven by a cellular explosive charge can meet the requirements specified above.

Application of Explosive Shock Tubes

The explosive shock tube with tubular explosive charges has been used: 1) to study explosive to electromagnetic energy conversion in pulsed-MHD generators, 2) to investigate structural and phase transitions at the surface of steel specimens exposed to a high-enthalpy gas flow, 3) to develop a pulsed light source of high power density within the visible and ultraviolet regions of the spectrum, and 4) to power cumulative barium plasma accelerators for investigation of near-Earth space.

In the experiments with the model pulsed-MHD generators emphasis was placed on the determination of the limiting electric characteristics for a linear MHD generator and on the maximum ratio of energy released in the generator load to ne energy, ω_0, of the initial magnetic field in the MHD channel. The experiments indicate that the tubular explosive charges produce an energy 1.5-2 times that of the ω_0 obtained in the load (Titov et al. 1968; Titov and Shvetsov 1979). This is apparently the greatest achievement of the explosive MHD generator with a gaseous working body. For comparison it should be noted that when the MHD

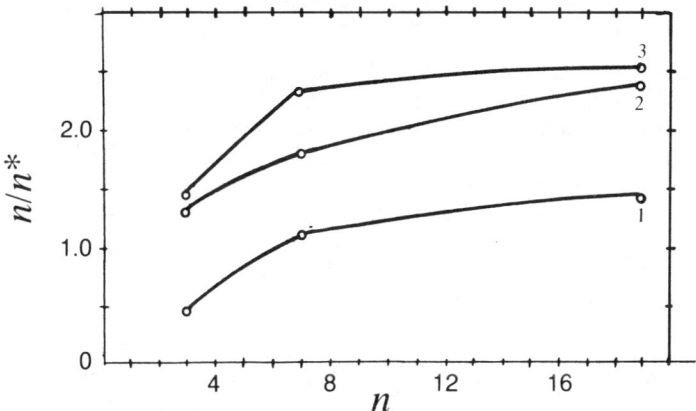

Fig. 6 Influence of the number of single charges on the conversion efficiency. Curves: 1) d_2 = 20 mm, d_1 = 12.5 mm; 2) d_2 = 12 mm, d_1 = 5 mm; and 3) d_2 = 15 mm, d_1 = 5 mm, l_0/d_1 = 15.

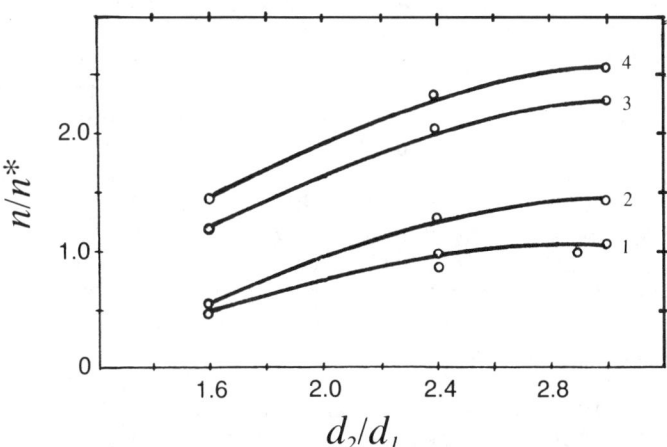

Fig. 7 Influence of explosive charge diameter ratios on conversion efficiency. Curves: 1) n = 1, 2) n = 3, 3) n = 7, and 4) n = 19.

generator uses a plasma formed at the exit of a detonation wave at a plane end of a charge, the energy transmitted to the ohmic load does not exceed 0.25 ω_0, the MHD channel dimensions being comparable (Jones and McKinnon 1965). The conversion efficiency of explosive to electromagnetic energy was 5%.

The possibility of the application of explosive shock tubes with tubular explosive charges for surface hardening of steel work-pieces has been studied experimentally. Heat fluxes to the work-piece surface, which is exposed to a high-enthalpy gas flow, are 10^5-10^6 W/cm^2. At these heat flux densities irreversible structural and phase ($\alpha \rightarrow \gamma$Fe) transitions harden the surface layers of the work-piece of ferrous alloys to a depth of 10-100 μm. In these experiments, the tubular work-pieces of steel 3 and 45 were used. For steel 3, the microhardness H_v was increased to 700-900 kg/mm^2 from initial values in the range 100-230 kg/mm^2. For steel 45, H_v was increased from 200-250 to 800-1000 kg/mm^2. An increase in microhardness has been achieved when steel work-pieces are exposed to high-power laser radiation (Rykalin et al. 1965). The use of an explosion-generated plasma for hardening steel work-pieces has an advantage over the use of lasers because large surfaces can be treated without the motion of the work-piece and internal surfaces inaccessible to a straight laser ray can be hardened.

Shock-wave-generated cellular explosive charges produce powerful pulsed light sources within the visible and ultraviolet regions of the spectrum. The total radiation energy measured by a calorimetric transducer within the wavelength spectrum of 380-2600 Nm ranged from several to tens of kilojoules. The radiation energy essentially depends on n and the shock tube length before the luminescence bulb. The pulsed light source brightness temperature, measured by the method described by Model (1957), varied by 18,000 to 6000 k. As the radiation maximum lay withi- the nearest ultra violet spectrum region, a considerable part of the radiation energy of the explosive lamp was not registered by the calorimetric transducer. The total duration of luminescence was about 600 μs. However, the main energy part (∼ 60%) occurs in the first 200 μs. The light sources have been used for high-speed reflected light photography of the explosive destruction of metal structures.

Shock waves driven by a cellular explosive charge have produced barium plasmas which are used for investigation of near-Earth space. Haerendal (1973) outlined the main requirements for explosive barium plasma injectors: The jet velocity must exceed 10 km/s; the barium vapor output in ratio to the injector weight and its volume must be a maximum; and the jet must have a minimum velocity gradient along its length.

Hollow-charge projectiles with a conic cavity have been used to produce barium plasmas (Haerendal 1973; Wescott et al. 1972). With these injectors an effective barium evaporation of 12% at a cone angle of 30 deg was achieved with a maximum jet velocity of 15 km/s. As the jet has a large velocity gradient along its length, the jet front brightness attenuated rapidly and, consequently, the luminescent length of the magnetic force line decreased. As noted by Haerendal (1973), only 2% of barium vapor moves at a velocity above 10 km/s. In Fig. 8 the oscillogram of mass velocity profile of the gas-cumulative jet generated by the detonation of a tubular explosive charge is presented. The jet mass velocity has been determined from the emf induced by the ionized gas moving in a transverse uniform magnetic field (Pryakhin et al. 1971). The jet front velocity is 16.5 km/s (the ambient air pressure being 1 mm Hg), and the velocity at the plateau is 10.9 km/s. The bulk of the

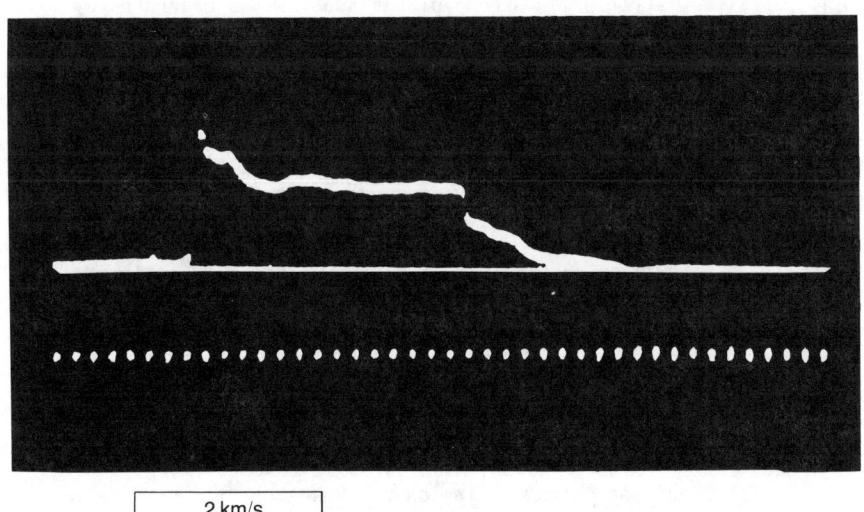

Fig. 8 Oscillogram of mass velocity of a gas-cumulative jet generated by detonation of a tubular explosive charge.

jet mass ("plateau") has a small velocity gradient. The cellular cumulative injector was first used in "Spolokh-2" experiment (Zhulin et al. 1979; Ruzhin et al. 1981). It contained 3.5 kg of explosive and 0.6 kg of metallic barium as six thin-walled cylindric inserts of a cellular explosive charge. About 8% of the barium mass had a velocity above 10 km/s.

One of the important findings of these experiments was the discovery (Zhulin et al. 1979) of periodic electron precipitations caused by cumulative barium plasma injection in the internal radiation zone of the Earth. One of the factors which causes such precipitation is the use of a cellular cumulative injector which possesses a nearly rectangular distribution of mass velocity.

References

Baum, F. A. et al. (1975) Fizika Vzryva (Physics of Explosion). Nauka, Moscow, USSR.

Eichelberger, R. J. and Kineke, J. H. (1967) Hypervelocity Impact High-Speed Physics (edited by K. Vollrath and G. Thomer). Springer-Verlag, Vienna.

Gill, S. P. and Simpkinson, W. L. (1969) Analysis and performance of an explosively driven shock tube. Proc. of the Seventh Int. Shock Tube Symposium, Toronto, Canada, pp. 366-396.

Glass, I. I. et al. (1974) Strong planar shock waves generated by explosively driven spherical implosions. AIAA J. 12, 367-374.

Haerendal, G. (1973) Results from barium cloud releases in the ionosphere and magnetosphere. Space Res. XIII, 601-617.

Jones, M. S. and McKinnon, C. N. (1965) Explosive driven linear MHD generator. Proc. of the Conf. on Magagauss Magnetic Field Generation by Explosives and Related Experiments, Frascati, Italy.

Model, I. S. (1957) High-temperature measurements in strong shock waves in gases. Zh. Eksper. Teor. Fiz. 32, 714-718.

Pryakhin, G. V., Titov, V. M., and Shvetsov, G. A. (1971) Electromagnetic study of high-velocity gas flows. Prikl. Mekh. Tekh. Fiz. 3, 137-140.

Ruzhin, Y. Y., Skomarovsky, V. S., and Shvetsov, G. A. (1981) Cumulative plasma injectors used in space research. Proc. of XV Int. Conf. on Phenomena in Ionized Gases, Part II, Minsk, USSR, 905-906.

Rykalin, N. N. (1965) Laser Material Treatment. Mashinostroyenie, Moscow, USSR.

Silvestrov, V. V. and Urushkin, V. P. (1971) Method of density measurements for high-velocity gas jets. Contin. Dyn. 7, 125-128.

Teno, J. and Sonju, O.K. (1974) Development of explosively driven MHD generator for short pulse aircraft high power. Technical Report AF APL-TR-74-48.

Titov, V. M., Fadeenko, Y. I., and Titova, N. S. (1968) Acceleration of solid particles with cumulative explosion. Dokl. Acad. Nauk SSSR 180, 1051-1052.

Titov, V. M. and Shvetsov, G. A.(1979) High energy electric pulse generation by cumulative explosion. Magagauss Physics and Technology. Plenum Press, New York.

Voitenko, A. E. (1966) Strong shock waves in air. Zh. Tekh. Fiz. 36, 178-189.

Wescott, E. M. et al. (1972) Two successful geomagnetic-field-line tracing experiments. J. Geophys. Res. 77, 2982-2984.

Woodhead, D. W. (1959) Advance detonation in a tubular charge of explosive. Nature London 183, 1756-1759.

Zagumennov, A. S. et al. (1969) Detonation of extended charges with cavities. Prikl. Mekh. Tekh. Fiz. 2, 79-83.

Zharikov, I. F. et al. (1967) Effect of light radiation from an explosion-driven source on a solid substance. Prikl. Mekh. Tech. Fiz. 1, 31-44.

Zhulin, I. A., Titov, V. M., Shvetsov, G. A., et al. (1979) Experimental study of disturbance caused by cumulative barium injection in the ionosphere. Proc. VI All-Union Conference of Rarified Gasdynamics, Part III. Novosibirsk, USSR, pp. 114-119.

The Taylor Instability of Contact Boundary Between Expanding Detonation Products and a Surrounding Gas

S.I. Anisimov,* Y.B. Zel'dovich,† N.A. Inogamov,* and M.F. Ivanov*

L.D. Landau Institute for Theoretical Physics, Moscow, USSR

Abstract

An analysis of the Rayleigh-Taylor turbulence which arises during the expansion of a spherically/cylindrically symmetrical charge of a condensed explosive into a low-density gas is reviewed. Two limiting cases have been identified. In the first case, the mean spatial scale of the Rayleigh-Taylor modes $\bar{\lambda}$ is much less than the thickness of the shock-compressed layer Δr of detonation products. This case is called free Rayleigh-Taylor turbulence. In the second case, after some period of time when the condition $\bar{\lambda} \ll \Delta r$ is fulfilled, a motion exists in which the Rayleigh-Taylor modes achieve the size $\bar{\lambda} \sim \Delta r$. At this stage of motion the turbulence properties have been considered.

Introduction

In this paper the explosion of a spherical (cylindrical) charge of a high explosive in a gaseous medium is considered. The analysis is restricted by the case when the pressure and density of the gas surrounding a charge are much less than those of detonation products (DP). In this case the expansion of DP is affected by the gas only in a thin layer adjacent to the contact boundary. During the initial stage of expansion, when the condition holds:

$$r_c(t) - r_c(0) \ll r_c(0)$$

Presented at the 8th ICOGER, Minsk, USSR, Aug. 23-26, 1981. Copyright © 1983 by S. I. Anisimov, Y. B. Zel'dovich, N. A. Inogamov, M. F. Ivanov. Published by the American Institute of Aeronautics and Astronautics, Inc. with permission.

*Scientist.
†Academician.

this layer is limited from the inner side of a characteristic; here, $r_c(t)$ is the radical position of the contact boundary. At later times a shock wave (SW) plays the role of the inner boundary of this layer. Thus the region where the flow is perturbed owing to the interaction of DP with a gas is limited by two shock waves (see Fig. 1). The first (external) SW is well recognized, while the second (internal) one is only briefly mentioned in relevant literature (Stanijukovich 1955). It is the internal shock that plays a very important part in the problem under consideration, because its position controls the spatial scale of Taylor modes and, hence, the rate of intermixing of DP and shock-compressed gas.

Unidimensional Flow Properties

In this investigation the coordinate frame fixed on the contact boundary is used. Since in the course of DP expansion the velocity of the boundary decreases, this frame is not inertial. Changes in the velocity are equivalent to the presence of a gravitational field directed along the radius. The order of magnitude of gravity acceleration can be estimated from the equation

$$g \simeq (v_0^2/r_0)(\rho_g/\rho_d)^{(1/n)}$$

where v_0 is the initial sound speed in DP, $r_0 = r_c(0)$, ρ_g and ρ_d are initial densities of the gas and DP, respectively, n = 2 for a cylindrical flow, and n = 3 for a spherical flow. This estimation follows from

$$g \simeq r_T/t_T^2 \quad t_T \simeq r_T/v_0 \quad \rho_g r_T^n \simeq \rho_d r_0^n$$

where r_T equals the maximum radius of the contact boundary. At the maximum radius the mass of the gas involved in motion is comparable to the mass of DP.

In the presence of a gravity field the directions "up" and "down" should be defined. For this purpose, the axis x in the coordinate system under consideration is in the "up" direction if it is directed against the gravity force vector. The region of mutual influence confined between two pressure jumps is termed the region of deceleration (see Figs. 1 and 2). Indeed, the freely expanding detonation products are first decelerated sharply as they pass through SW, which confines the deceleration region from above, and are then decelerated gradually together with the remaining

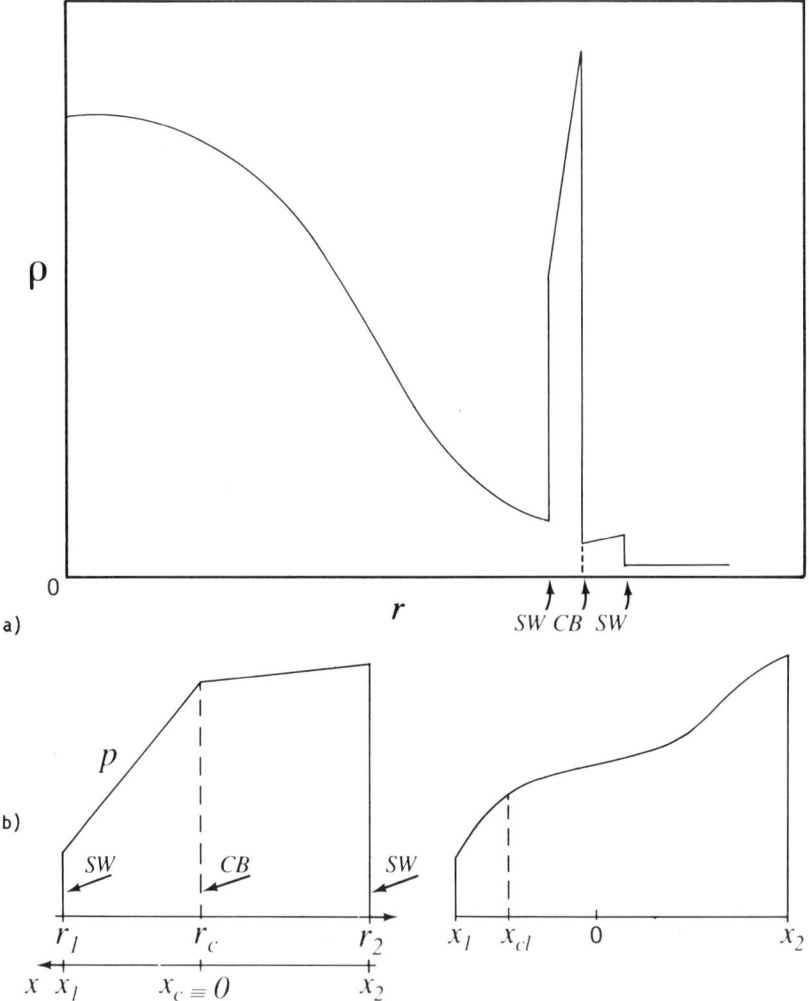

Fig. 1 Distribution of density in an unperturbed, exactly one-dimensional flow (top), and distribution of pressure in the deceleration region (bottom). On the left: one-dimensional flow. Axes correspond to the coordinate system connected with the one-dimensional contact boundary (CB). On the right: angle-averaged instantaneous pressure field in real flow.

mass confined in the region of deceleration. In turn, a gas inserted into the deceleration region from below is also gradually decelerated after a sharp initial acceleration connected with passing through the lower SW. An approximate instantaneous distribution of pressure and density over the "height" in the region of deceleration can be found by

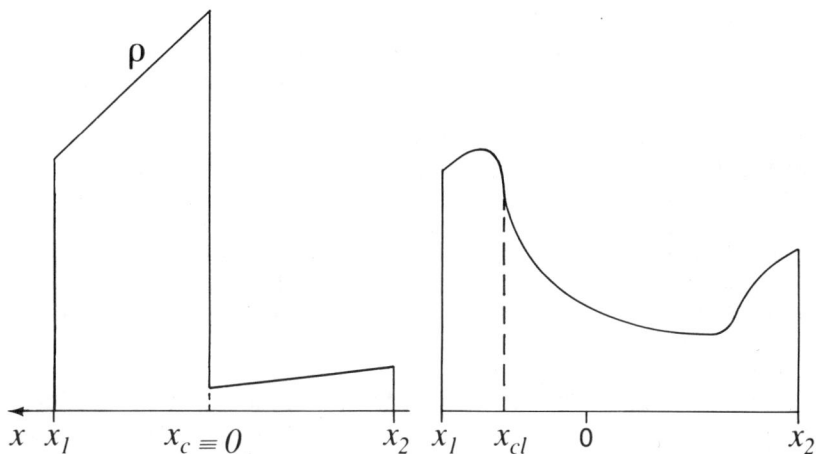

Fig. 2 Deceleration region. On the left: distribution of density in one-dimensional flow. On the right: angle-averaged distribution of density in real flow.

integrating the equation of hydrostatics, if the distribution of entropy over the mass and the instantaneous value of pressure beyond the front of the lower SW are known. The accuracy of a hydrostatic approximation increases as the relative thickness of a shock-compressed layer of DP, $(r_c - r_1)/r_c$, decreases. Here $r_1(t)$ is the radius of inner SW. It follows from one-dimensional calculation that $H_1 = c_1^2/g$ is comparable to the thickness of the shock-compressed layer of DP:

$$r_1 \simeq r_c - r_1$$

It follows from calculation that $H_1 \ll r_c$. The velocity of sound in detonation products at the contact boundary is denoted by c_1; g is deceleration of the contact boundary.

The region of deceleration is divided by the contact boundary into two regions: upper and lower. The calculation shows that there is a density jump at the contact boundary, the ratio of densities being high and slowly varying with motion. This fact defines the character of real flow, in which three-dimensional effects are very essential.

Interface Instability

Thus far the properties of a one-dimensional flow have been discussed. The most important property of this flow is its instability, which violates the rigorously one-dimensional character of flow near the contact boundary. The source of instability arises from the fact that the expanding DP have a higher density than a shock-compressed gas (the pressure is continuous while passing through the contact boundary) and are located above in the "gravitational field" (see Figs. 1 and 2) in this analogy. It is well known that the equilibrium between heavy and light liquids in this case is unstable. A slight distortion of the plane boundary disrupts the equilibrium, and the heavy upper liquid sinks in the light lower one, whereas the resulting vacant volume is filled with the ascending light liquid.

The instability of the contact boundary in explosion was first mentioned in a paper of Anisimov and Zel'dovich (1977). Somewhat later there appeared independently the experimental papers of Davydov et al. (1978, 1979) in which this instability was observed in a cylindrical explosion. In the following, a nonlinear development of the Taylor (interchange) modes is considered. The character of the development depends essentially on the singularities of one-dimensional flow discussed previously: existence of the upper shock wave in DP, and 2) a small density ratio, $\mu = \rho_g/\rho_d$, in the vicinity of the contact boundary.

The one-dimensional calculations indicate that the deceleration region is thin when compared to the contact boundary radius. One can show that the shock wave (SW) velocity and the pressure behind it are not sensitive to the flow modifications due to interchange instability. The law of motion of the deceleration region depends on the total mass confined in the region and does not practically depend on the thermodynamic properties of fluids intermixed inside the region.

Nonlinear evolution of Taylor modes leads to the turbulent mixing of upper and lower fluids. The structure of the intermixed layer in the case of free Taylor turbulence is described in Appendix A. The motion of detonation products downwards is stopped by the lower SW. Therefore the concentration of products should be higher near the shock wave (see Fig. 1). It should be noted that under free turbulence without a lower boundary this is not so: The concentration of a heavy component increases monotonously with height.

If $x_{cl}(t)$ is the lower boundary of nonmixed products of detonation in a real flow (see Figs. 1 and 2), the mass of DP penetrated into the mixed layer and the shift of the lower boundary of DP $x_+ = x_{cl}(t) - x_c(t)$, as compared to the

TAYLOR INSTABILITY OF CONTACT BOUNDARY

results of a one-dimensional calculation, are interrelated. Here, $x_c(t)$ is the coordinate of the contact boundary in one-dimensional calculation, $x_c \equiv 0$, because the coordinate frame connected with the one-dimensional contact boundary.

In the case of an infinitely thick atmosphere ($H_1 \gg x_+$), i.e., of free Taylor turbulence, the shift of the boundary is $x_+ = \alpha_+ (gt^2/2)$; the mean scale of interchange modes is $\bar{\lambda} = (\alpha_+/Fr)^2 gt^2$; the speed of the boundary is $\dot{x}_+ = \alpha_+ gt$ (see Appendix A); and compressibility may be neglected (Inogamov 1980). The distortion of free regime of intermixing occurs at the value of the shift x_+ equal to H_1, when the mean scale $\bar{\lambda}$ is of the order of H_1 and $\dot{x}_+ \simeq c_1^1$. The time t_s necessary for the sound to pass the DP atmosphere is small when compared to r_c/r_c. For $r_c \lesssim r_T$, for example, $t_s \simeq t_T(H_1/r_T)^{1/2}$. The characteristic time of intermixing, $\bar{\lambda}/\dot{x}_+$, is of the order of t_s. If $\bar{\lambda} \simeq H_1$, then the quasistationary regime of motion is established under which self-regulation of the mean scale of interchange modes takes place: $\bar{\lambda}_* \simeq v_{in}^2/g$. In this case the thickness of the upper atmosphere turns out to be $\simeq \bar{\lambda}_*$, and the mass influx into the atmosphere through the upper SW and the mass efflux through the atmosphere lower boundary are balanced. Here v_{in} is the velocity of the detonation products flowing into SW, and the asterisk denotes the parameters of flow in the regime of self-regulation.

Two-dimensional numerical calculations have been performed in order to answer the question: At what moment does the saturation of free turbulence take place? (See Appendix B.) For low back pressure and an initial density ratio (DP to gas) of 10^4 to 10^5, a small initial amplitude perturbation (of the order of 1% of the initial wavelength), with the wavelength several times that of the DP atmosphere thickness, produces nonlinear behavior in times which are small compared to the stop time. For these conditions when the CB has moved of the order of the stop radius, the Taylor turbulence is in the nonfree regime. Hence, one may conclude that 1) the mass of the detonation products penetrated into the shock-compressed gas is of the order of that penetrated into the deceleration region; 2) in the case of a wide spectrum of initial perturbations the perturbations with the wavelength $\lambda > H_1$ turned out to be unimportant for the motion under consideration; and 3) the spectrum of modes is restricted from both sides and appears to be rather narrow, the characteristic number of the harmonics being $l \sim 2\pi r_c/\bar{\lambda}_*$.

Appendix A: Free Taylor Turbulence

Two incompressible liquids with different densities are assumed to fill the upper and lower half-spaces. At the initial moment of time, the plane $x = 0$ serves as a boundary between the two liquids. At the initial moment a wide spectrum of perturbations (i.e., without any separated modes) exists. As a result of development of interchange instability there appears a region of a "two-flux" motion in which the heavy and light liquids flow to meet one another.

The upper and lower boundaries of the domain involved in intermixing are denoted by $x_+(t)$ and $x_-(t)$, respectively. When the density ratio is high, the main contribution to the shift of the center of masses of the intermixed layer comes from sedimentation of the heavy liquid. Since the speed of sinking at $\mu \ll 1$ is high, $|x_-(t)|$ exceeds $|x_+(t)|$ by several times. This means that the volume concentration (not the weight one) of heavy liquid in the layer is not large. While the heavy liquid sinks, large streams disperse gradually. As a result there arises a mass flow into the small-scale region. It should be noted that intermixing can in no way be assumed spatially homogeneous.

If $\mu \ll 1$, then at the upper boundary of the domain involved in the motion there arises a flow with a known structure, typical for the Taylor instability. The structure is of a mean scale, which we denote by $\bar{\lambda}$. As is well known (Inogamov 1978), the interaction of harmonics results in increasing $\bar{\lambda}$. No intermixing in the upper layer is practically observed, the upper boundary of the intermixed layer is sharply defined.

The speed of motion of the boundary is determined by that of bubble rising:

$$\dot{x}_+ = Fr(g\bar{\lambda})^{1/2} \qquad (A1)$$

The Froude number at $\mu = 0$ is approximately equal to 0.23-0.40. The value $Fr = 0.23$ corresponds to two-dimensional periodic flow in the form of rolls, and the value 0.40 corresponds to the square lattice of bubbles. The Froude number slowly decreases with increasing μ. At $\mu \ll 1$ the approximate formula is valid:

$$Fr(\mu) \simeq Fr(0)(1-\mu)^{1/2}$$

The factor $(1-\mu)^{1/2}$ arises owing to evident renormalization of acceleration.

From simple dimensional considerations

$$x_+ = \alpha_+ (gt^2/2) \qquad (A2)$$

where $\alpha = \alpha_+(\mu)$ is some dimensionless function. At $\mu \ll 1$, we put $\alpha_+(\mu) \simeq \alpha_+(0)(1-\mu)$. From Eqs. (A1) and (A2) it follows that

$$\bar{\lambda} = 2(\alpha_+/Fr)^2 (gt^2/2)$$

The value $\alpha_+(0)$ is determined from the results of experiments. In experiments with a low level of noise, $\alpha_+(0)$ can be lowered down to the order of 0.1. In this case

$$\bar{\lambda}/x_+ \sim 1$$

With increasing noise level, $\alpha_+(0)$ increases (Inogamov 1978) in proportion:

$$[\log(\frac{1}{k\alpha|(k)|})]^{-1}$$

where $\alpha(k)$ is the amplitude of the k-th harmonics in the spectrum of the primary noise. The ratio $\bar{\lambda}/x_+$ may characterize the noise.

Under the same dimensional consideration for the lower boundary,

$$x_- = \alpha_- (gt^2/2)$$

The value α_- is very sensitive to variation of μ and is less sensitive to the noise. At $\mu = 0$, we apparently have $\alpha_-(0) = 1$. At $\mu = 0.05$, Anuchina and Kucherenko (1978) report that $\alpha_-(0.05) \simeq 0.2$. To avoid a misunderstanding it should be noted that Anuchina and Kucherenko (1978) present the values of $(\alpha_+ + \alpha_-)$ and the graphs of density profile. To find α_+ and α_- separately it is necessary to determine, by a simple integration, the position of initial plane.

Appendix B: Two-Dimensional Numerical Calculation

The nonlinear growth of perturbations has been investigated through the solution of two-dimensional

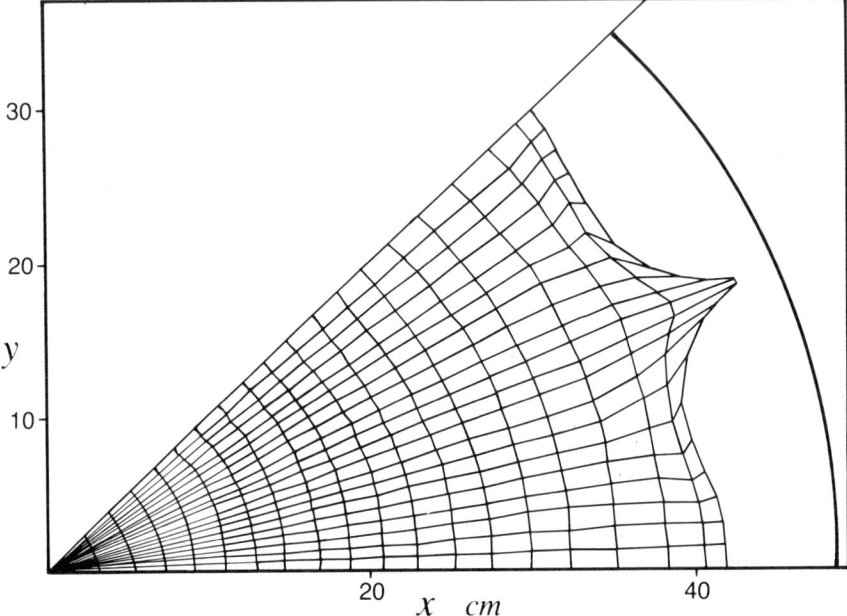

Fig. 3 Typical structure of flow resulting from two-dimensional numerical calculation. Two layers of Lagrangian cells adjacent to the contact boundary form the atmosphere. The perturbation wavelength exceeds approximately by an order of magnitude the thickness of the atmosphere.

nonstationary equations of gasdynamics for the expansion of detonation products into a gaseous medium. For this purpose, a totally conservative Lagrangean-Eulerian scheme (Hirt et al. 1974) was applied. For both media, equations of state of an ideal gas with different adiabate indices were adapted. Initial values of density and temperature corresponded to the instantaneous detonation of high explosives. Small perturbations of the boundary shape, or the density profile in Lagrangean cells near the boundary, were predetermined at the initial moment of time. Evolution of perturbations with wavelengths of the order of 0.1-0.05 was studied at the cylindrical symmetry of the main flow. Since artificial viscosity was used in the finite-difference scheme, the acceleration of the contact surface in the calculation continued during some finite time. When the boundary remained stable, the amplitude of perturbations did not increase, whereas their phases varied. At the subsequent stage of deceleration, the amplitude of perturbations increased with time, first by the exponential and then approximately by the linear law. Formation of "jets" of heavy medium was observed, and their penetration into the light medium became apparent. Figure 3 shows the shape of the boundary for the time instant of ~ 250 μs. The initial radius of the cylindrical charge was 5 cm; the initial temperature of the detonation products was 3×10^3 K; the initial perturbation of the boundary for the eight harmonics had the amplitude of $10^{-2}\, r_0$.

References

Anisimov, S. I. and Zel'dovich, Y. B. (1977) Rayleigh-Taylor instability of boundary between detonation products and gas in spherical explosion. Pis'ma Zh. Eksp. Teor. Fiz. 3, 1081-1084.

Anuchina, N. N. and Kucherenko, Y. A. (1978) Turbulent mixing on accelerated boundary between different density fluids. Izv. Akad. Nauk SSSR. Mekh. Zhidk. Gaza 6, 157-160.

Davydov, A. N., Lebedev, E. F. and Perkov, S. A. (1978) Gasdynamic instability in propagation of cylindrical shock waves. Detonation. Critical Phenomena. Physico-Chemical Transformations in Shock Waves, pp. 76-79. Chernogolovka, OIKhF AN SSSR.

Davydov, A. N., Lebedev, E. F., and Perkov, S. A. (1979) Experimental investigation of gas-dynamic instability in the plasma flow following the cylindrical shock wave. Preprint N 1-40, Institute of High Temperatures, USSR, p. 26.

Hirt, C. W., Amsdem, A. A., and Cook, J. L. (1974) An arbitrary Lagrangian-Eulerian computing method for all flow speeds. J. Comput. Phys. 14, 227-253.

Inogamov, N. A. (1978) Turbulent stage of the Taylor instability Pis'ma Zh. Eksp. Teor. Fiz. 4, 743-747.

Inogamov, N. A. (1980) Rayleigh-Taylor instability in a compressible medium. Preprint ITF AN SSSR, Chernogolovka, USSR.

Stanijukovich, K. P. (1955) Nonstationary Motion of Continuous Medium, p. 729. Gostechizdat, Moscow, USSR.

Chapter III. Gaseous Detonations

Properties of Detonation Waves in Hydrocarbon-Oxygen-Nitrogen Mixtures at High Initial Pressures

P. Bauer* and C. Brochet†

Laboratoire d'Energétique et de Détonique, Poitiers, France

Abstract

On the basis of brightness temperature and detonation velocity measurements, a study of the detonation characteristics of several fuel-oxygen-Nitrogen mixtures at a high initial pressure (up to 50 bar) has been undertaken. A comparison between experimental values and results of computations based on two equations of state has been done. It appears that a calculation using a virial equation of state as proposed by Boltzmann, assuming that the products behave as ideal mixtures, yields detonation velocities in good agreement (less than 2%) with the experimental ones. The brightness temperatures recorded with a four colors pyrometer ($\lambda = 0.657$ μm, $\lambda = 0.783$ μm, $\lambda = 0.915$ μm and $\lambda = 1.008$ μm, respectively) are strongly wavelength dependent which does not allow any comparison with a calculated equilibrium temperature.

Presented at the 8th ICOGER, Minsk, USSR, Aug. 23-26, 1981.
Copyright © American Institute of Aeronautics and Astronautics, Inc., 1983. All rights reserved.
*Assistant Professor.
†Maitre de Recherche.

Introduction

In a previous investigation (Bauer et al. 1979) on the theoretical and experimental determination of detonation characteristics of C_2H_4-O_2-N_2 fuel rich mixtures, at initial pressures ranging from 2 to 45 bars, we pointed out that 1) the detonation velocity was a very sensitive parameter to check the validity of the equations of state; 2) the temperature was less sensitive to the choice of an equation of state, but remained an important parameter which had to be taken into account to confirm the validity of any equation of state; 3) of the equations of state which could describe the thermodynamic behavior of the detonation products at such high pressures, the ideal gas EOS and the empirical Becker, Kistiakowsky, and Wilson (B.K.W) EOS yielded velocities that were in poor agreement with experimental data; 4) computations based on the Boltzmann EOS were inconclusive because of the necessity of introducing too many simplifying assumptions; and 5) detonation velocities which were derived from measurements in two tubes of different sizes, at initial pressures of from 3 to 7 bars, were not dependent upon their diameter.

The computation code based on the Boltzmann EOS has been improved, and a second code based on the Carnahan and Starling EOS (Carnahan and Starling 1969) has been developed. These two EOS have been tested to determine which best fitted the properties of detonation waves we measured.

To extend this investigation to other hydrocarbons, the detonation velocity fuel (propane or methane) oxygen-nitrogen nixtures has been measured. The effect of pressure on the structure of the detonation front and the detonability limits of methane fuels were investigated.

Experimental Setup

Detonation velocity measurements were performed in a stainless steel, 6.5-m-long cylindrical tube with an inside diameter of 15 mm. Mixtures can be detonated in this tube at initial pressures of up to 50 bars without any damage. The complete setup (see Fig. 1) has already been described in detail in a previous study (Bauer et al. 1979).

To cancel the mechanical effects of the detonation wave, a stainless steel chamber of 120 mm i.d. was set up at

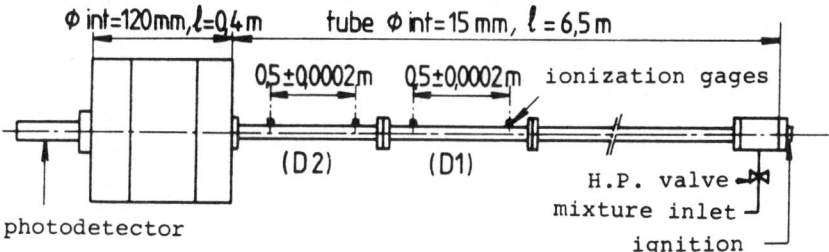

Fig. 1 Experimental setup.

one end of the tube. A glass diaphragm separates the chamber from the tube and breaks as soon as the detonation front hits it. Thus the detonation products rapidly expand in the chamber which had been previously evacuated. A pyrex porthole 20-mm thick with a diameter of 76 mm placed perpendicular to the tube axis allows direct optical measurement of the forthcoming detonation front. For this purpose, a multiwavelength pyrometer (λ = 0.657, 0.783, 0.915, and 1.008 µm, respectively) is used. Its response time is shorter than 50 ns (Bouriannes et al. 1977), and it is focused on the internal side of the glass diaphragm. The resolving power is $\sim$ 3 mm and the depth of focus is close to 20 cm. The brightness temperatures are thus recorded for each wavelength.

All the mixtures were prepared from pure compounds at a pressure of 65 bars. The Redlich-Kwong EOS (Kemp et al. 1975) was used to correct compositions for compressibility effects.

Equation of State of the Detonation Products

The Chapman-Jouguet (CJ) calculations, performed on an HP 9845 desk computer for mixtures described in Tables 1 and 2, are based on two real gas equations of state and on the following main assumptions: 1) the detonation yields entirely gaseous products, and the involved species are CO_2, CO, H_2O, H_2, O_2, N_2, OH, O, N, NO, NO_2, and H; 2) fundamental thermodynamic parameters (i.e., enthalpy, Gibbs free energy, etc.) can be derived from the temperature dependent polynomial development proposed by Gordon and McBride (1971).

Boltzmann Equation of State

This equation of state in the virial form was proposed by Boltzmann (Taylor 1952):

Table 1 Methane and propane - O_2-N_2 mixture parameters for CJ calculations

Mixture	Equivalence ratio	$Z = n_{N_2}/n_{O_2}$	Composition
A	1.18	0.96	CH_4 + 1.70 O_2 + 1.63 N_2
B	1.18	2.00	CH_4 + 1.69 O_2 + 3.38 N_2
C	1.18	3.01	CH_4 + 1.69 O_2 + 5.09 N_2
D	1.18	3.20	CH_4 + 1.70 O_2 + 5.44 N_2
E	1.18	3.21	CH_4 + 1.70 O_2 + 5.46 N_2
F	1.18	3.23	CH_4 + 1.70 O_2 + 5.49 N_2
G	1.14	2.00	C_3H_8 + 4.37 O_2 + 8.72 N_2

Table 2 Ethylene-O_2-N_2 mixture parameters for CJ calculations calculations

Mixture	Equivalence ratio	$Z = n_{N_2}/n_{O_2}$	Composition
H	1.05	0.87	C_2H_4 + 2.85 O_2 + 2.48 N_2
I	1.38	1.77	C_2H_4 + 2.17 O_2 + 3.85 N_2
J	1.29	1.92	C_2H_4 + 2.32 O_2 + 4.46 N_2

Table 3 Coefficients a_i for the expansion of $A(T*)$

	a_0	a_1	a_2	a_3
2<T*<10	-8.6996E-2	4.1051E-3	-1.5938E-5	3.4847E-8
10<T*<100	-1.9898	1.7258E-1	-6.8717E-3	1.5202E-4
100<T*<1000	-8.4872	3.5784	-7.4843E-1	8.4737E-2

	a_4	a_5	a_6
2<T*<10	-4.2580E-11	2.7120E-14	-7.0073E-18
10<T*<100	-1.8694E-6	1.1955E-8	-3.0971E-11
100<T*<1000	-5.2794E-3	1.7020E-4	-2.2168E-6

DETONATIONS IN HYDROCARBON-OXYGEN-NITROGEN MIXTURES

$$pV/nRT = 1 + x + 0.625\, x^2 + 0.287\, x^3 + 0.193\, x^4$$

with $x = b/V$, where $b = \Sigma n_i b_i$ represents the second virial coefficient (Pujol 1968). The mixture, which is supposed to be ideal, contains n_i moles of each specie i.

Carnahan and Starling Equation of State

For nonattracting rigid spheres, Carnahan and Starling (CS) (1969) proposed the following equation of state:

$$pV/nRT = (1 + y + y^2 - y^3)/(1-y)^3$$

where y is a function of the reduced temperature $T^* = kT/\varepsilon$ and can be derived from the Rowlinson relation (Barrere 1976; Rowlinson 1964):

$$y = B(T^*)\, N\, n\, \sigma^3/V$$

where

$$B(T^*) = (\pi\sqrt{2}/6)(T^*)^{3/\alpha}[1 + A(T^*)/\alpha]^3$$

and k, N, n, and α are respectively the Boltzmann constant, the Avogadro number, the global number of moles in the mixtures, and the Lennard-Jones potential exponent (Fickett 1962).

The tabulated values of the function $A(T^*)$ were put in the following polynomial form:

$$A(T^*) = \sum_{i=0}^{6} a_i\, T^{*i}$$

where the coefficients from a_0 to a_6 are given in Table 3. Moreover, for real mixtures the following interaction laws were used:

$$\sigma = (\Sigma_i \Sigma_j \sigma_{ij}^3\, X_i X_j)^{1/3}$$

and

$$\varepsilon = (\Sigma_i \Sigma_j \varepsilon_{ij}^3\, \sigma_{ij}\, X_i X_j)/\sigma^3$$

where σ_{ij} and ε_{ij} are, respectively, the diameter and the interaction energy of two molecules which belong to species i and j. The respective molar fractions of these latter are X_i and X_j. Values of σ_{jj} and ε_{jj} are derived from empirical expressions (Barrère 1976; Fickett 1962):

$$\sigma_{ij} = (\sigma_i + \sigma_j)/2 \quad \varepsilon_{ij} = (\varepsilon_i \varepsilon_j)^{1/2}$$

Experimental Results and Discussion

Comparison Between Measured and Computed Velocities

Influence of Inert Additives. Given the nitrogen-to-oxygen ratios $Z = nN_2/nO_2$, we can split the mixtures into the following two groups:

1) $Z < 3$: For these mixtures (A, B, G) a very good agreement between the velocities measured on the two sections of the tube was obtained. This had been observed for mixtures H, I, and J in a previous study (Bauer et al. 1979). A comparison between the experimental values and results of a computation based on the above equations of state reveals the validity of the Boltzmann equation of state in this range of pressures (see Fig. 2). Indeed, for almost all these mixtures, measured values agree to within 1% with the calculations. The highest average discrepancy was obtained for mixture H and remains smaller than 2%. Values computed from the Carnahan and Starling EOS are 3% smaller than the experimental results (4% for mixture H).

2) $Z > 3$: For mixtures C, D, E, and F, a systematic discrepancy between the velocities recorded on the two sections of the tube was observed. This dispersion of the results is depicted in Fig. 3. It reveals the uncertainty in the assumption of a self-sustained detonation front. In that case a conclusion on the validity of any of these two equations of state is baseless.

Influence of the Various Parameters of the Carnahan and Starling Equation of State. The parameters involved in the Carnahan and Starling EOS can be discussed on the basis of the experimental results (see Fig. 4). Given an increase of the σ_i value, calculations performed for mixture A show little improvement on the agreement with the experimental results. This increase was chosen to be equal to 20%, which is consistent with the uncertainty on the knowledge of this parameter (see, for example, Hirschfelder et al. 1954).

The same variation of ε_i has no effect on the calculated values. Finally, the value $\alpha = 9$, proposed by Pralin and Giddings (1955), does not suit our experimental

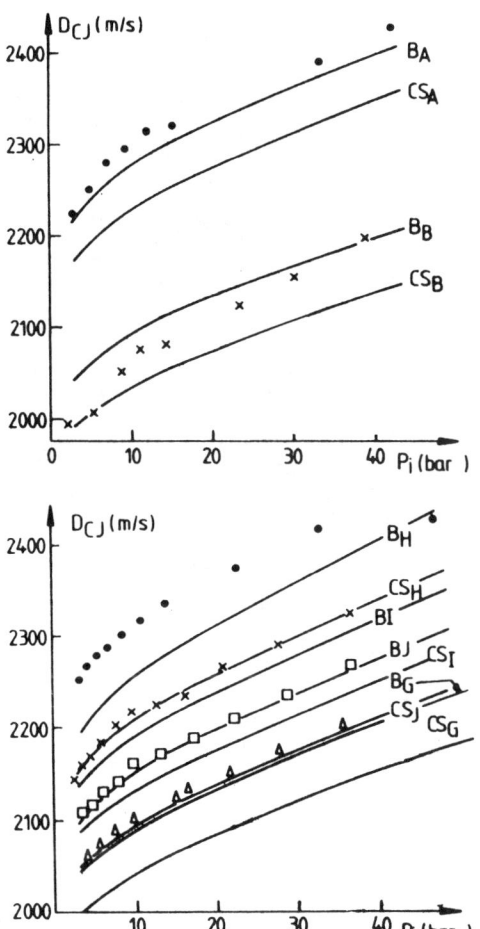

Fig. 2 Detonation velocities. a) Mixture A: $CH_4 + 1.7\ O_2 + 1.63\ N_2$; •, experimental values; B_A, CS_A calculated values derived from the Boltzmann and CS EOS, respectively; Mixture B: $CH_4 + 1.69\ O_2 + 3.38\ N_2$; x, experimental values; B_B, CS_B, computed values using the Boltzmann and CS EOS. b) Mixture G: $C_3H_8 + 4.37\ O_2 + 8.72\ N_2$; △, experimental; B_G, CS_G, calculated. Mixture H: $C_2H_4 + 2.85\ O_2 + 2.48\ N_2$; •, experimental; B_H, CS_H, calculated. Mixture I: $C_2H_4 + 2.17\ O_2 + 3.85\ N_2$; x, experimental; B_I, CS_I, calculated. Mixture J: $C_2H_4 + 2.32\ O_2 + 4.46\ N_2$; □, experimental; B_J, CS_J, calculated.

conditions. These results show that the use of the Carnahan and Starling EOS depends strongly upon the value given to the parameters which are involved, mainly the value of σ_i. Therefore before asserting that the CS EOS is inappropriate, a more thorough parametric study should be undertaken.

Discussion of the Values of Brightness Temperatures and Computed CJ Temperatures

The radiative energy emitted by the detonation wave was recorded for mixtures B and G. On the basis of a calibration of the pyrometer, performed using two different sources which can be considered as a blackbody (Kato et al. 1981),

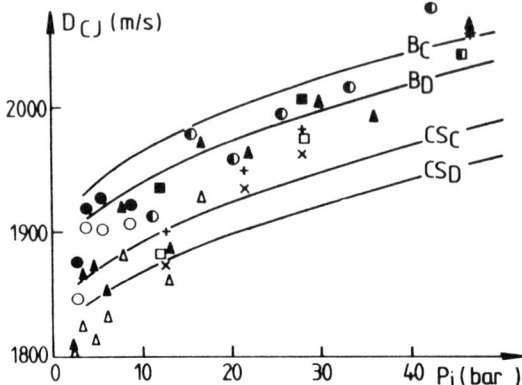

Fig. 3 Detonation velocities. Mixture C: $CH_4 + 1.69\ O_2 + 5.09\ N_2$; •, o, values measured on two different sections of the tube; B_C, CS_C, calculated on the basis of the Boltzmann and CS EOS, respectively. Mixture D: $CH_4 + 1.7\ O_2 + 5.44\ N_2$; ▲, △, experimental; B_D, CS_D, calculated. Mixture E: $CH_4 + 1.7\ O_2 + 5.46\ N_2$; +, x, experimental. Mixture F: $CH_4 + 1.7\ O_2 + 5.49\ N_2$; ■, □, experimental

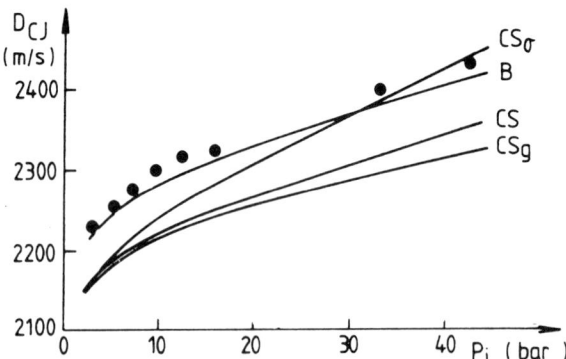

Fig. 4. Influence of the parameters involved in the CS EOS on the detonation velocity. Mixture: $CH_4 + 1.7\ O_2 + 1.63\ N_2$; •, experimental; B computed with the Boltzmann EOS; CS computed with the CS EOS taking $\alpha = 12$; CS_g computed with the CS EOS taking $\alpha = 9$; CS_σ computed with the CS EOS with an increase of 20% of all the σ_i.

values of the brightness temperature of the detonation wave were thus derived. These values are reported in Fig. 5. This temperature is strongly wavelength dependent, particularly for initial pressures lower than 10 bars, but also for higher pressures. The discrepancy between the brightness temperatures corresponding to each wavelength and their mean value is from 200 to 300 K. This wavelength dependence means that, in this range of densities, the equilibrium temperature cannot be compared to the brightness temperature. Increasing the density of the products should have led to similar results, as in condensed explosives where the brightness temperature is no longer wavelength dependent (Kato et al. 1979). Nevertheless, these conditions can also be reached with gaseous explosives, but for much higher initial pressures or for very rich mixtures yielding solid carbon particles.

The values are nearly in a line in a semilogarithmic diagram (Fig. 5). The highest brightness temperatures were

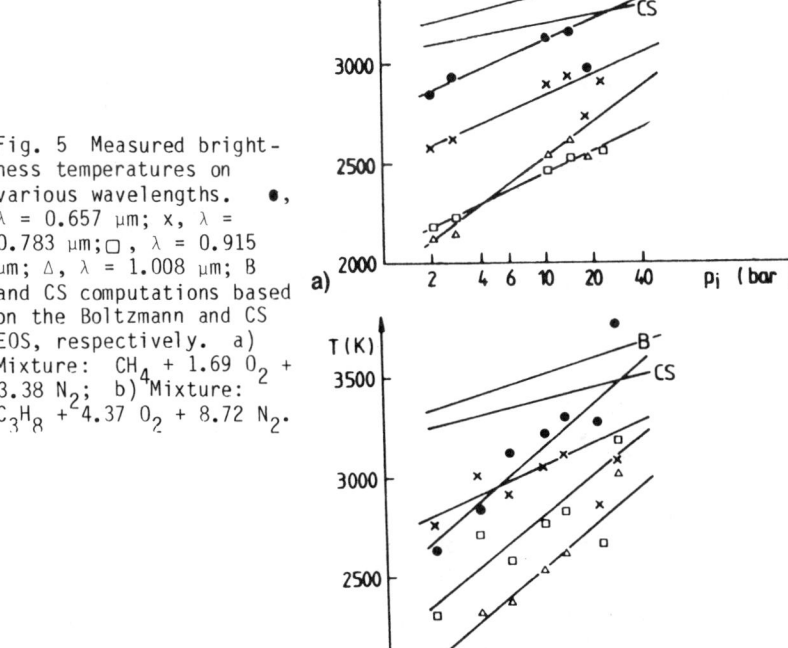

Fig. 5 Measured brightness temperatures on various wavelengths. ●, $\lambda = 0.657$ μm; x, $\lambda = 0.783$ μm; □, $\lambda = 0.915$ μm; △, $\lambda = 1.008$ μm; B and CS computations based on the Boltzmann and CS EOS, respectively. a) Mixture: $CH_4 + 1.69\ O_2 + 3.38\ N_2$; b) Mixture: $C_3H_8 + 4.37\ O_2 + 8.72\ N_2$.

obtained for $\lambda = 0.657$ μm and, for a given pressure, the values decrease as λ rises.

Even though the comparison between the computed detonation temperature and the measured brightness temperature is unreliable, we have plotted on the same figure the results of the calculations based on both equations of state. Computations based on the Boltzmann EOS give higher values (200 K) than the calculations based on the CS EOS. The tendency of these two sets of results shows that some agreement between calculation and experiments could be obtained for a higher range of initial pressures which cannot be investigated with this experimental device. But, actually, it appears that these measurements are not sufficient to confirm the validity of the Boltzmann EOS, on the basis of another characteristic parameter such as temperature, at least in this range of densities.

Influence of the Initial Pressure on the Structure of the Detonations and on the Detonability Limits

To correct the experimental detonation velocity in the comparison with the computed detonation velocity, the effect of the diameter should have been taken into account. In the

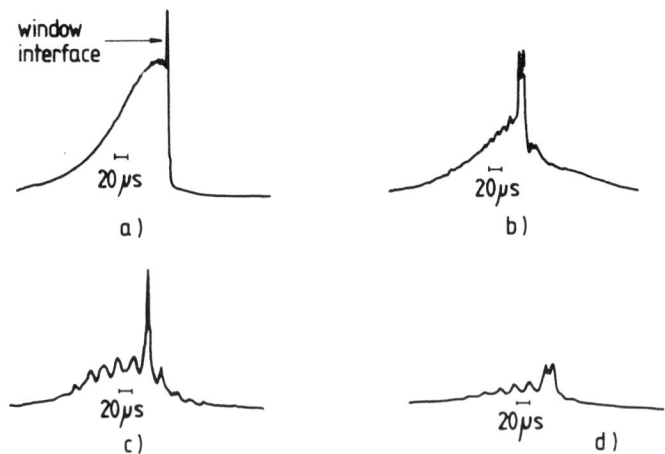

Fig. 6 Recordings of the brightness temperature on a single wavelength, $\lambda = 0.657$ μm, for various initial pressures; a) $p_i = 35.85$ bars; b) $p_i = 12.83$ bars; c) $p_i = 4.59$ bars; d) $p_i = 2.31$ bars, giving a scale of the structure of the detonation front for mixture $CH_4 + 1.7\ O_2 + 5.44\ N_2$.

C_2H_4-O_2-N_2 mixtures in such proportions that $Z < 3$, the detonation velocities obtained in a tube with an inside diameter of 15 mm, at initial pressures higher than 5 bars, can be considered as the ideal ones (with an uncertainty of less than 1%).

The latter correspond to an infinite diameter load. But, in the mixtures more diluted with N_2 ($Z > 3$) and more particularly in those containing CH_4, this effect of diameter cannot be neglected anymore. In such mixtures, a spinning detonation was observed as the initial pressure decreases. This is shown in Fig. 6, for mixture D. When $p_i > 12$ bars, the frequency is roughly 100 kHz and tends to become the characteristic frequency of the spinning detonation ($\cong 42$ kHz) for $p_i < 5$ bars. In this case, the experimental conditions are much lower than the critical conditions (pressure and diameter) as defined by Matsui and Lee (1979). Indeed, the critical diameter derived from these authors' study for such mixtures would be 3 m and 4-5 cm only for $p_i = 1$ bar and $p_i = 50$ bars, respectively.

We also checked that no detonation could be obtained in the whole range of our initial pressures for mixtures having a Z greater than 3.23.

For $3 < Z < 3.2$, the conditions of firing appeared to be very sensitive to the initial pressure.

Summary and Conclusion

The comparison between measured and computed detonation velocities of CH_4-O_2-N_2, C_2H_4-O_2-N_2, and C_3H_8-O_2-N_2 mixtures at initial pressures ranging from 2 to 45 bars has shown: 1) the validity (within 1%) of the Boltzmann virial equation of state, based on the assumption of an ideal mixture (this validity only holds for mixtures in which $Z < 3$); and 2) a disagreement attaining 5% between experimental values and those calculated on the basis of the Carnahan and Starling equation of state, for a Lennard-Jones potential equal to 12. However, this discrepancy could be due to a difficult choice of the parameters involved in this equation of state.

The validity of the Boltzmann equation of state was not confirmed by the investigation on the detonation temperature. Indeed, the brightness temperatures which were recorded for mixtures B and G are strongly wavelength dependent, even at initial pressures up to 40 bars, and do not allow any conclusion on the true temperature of the products.

For CH_4-O_2-N_2 and C_2H_4-O_2-N_2 mixtures, we observed that the velocities derived from the Boltzmann equation of state were respectively lower than the experimental values when $Z < 1$ or 2, and higher when Z was increasing. This might result from the influence of the diameter, which becomes more sensitive as Z rises. As shown previously, this influence can be considered as negligible for a C_2H_4-O_2-N_2 mixture ($Z = 2.6$) at initial pressures $3 < p_i < 7$ bars. In that case, the critical diameter defined by Matsui and Lee (1979) is slightly smaller than the inside diameter of the detonation tube. In the CH_4-O_2-N_2 mixture in which $Z > 3$, a self-sustained detonation was no longer observed and no detonation at all occurred for $Z > 3.23$. Thus a CH_4-air mixture cannot be detonated in the tube for this range of initial pressures. In these boundary conditions we obtained a spinning detonation with the characteristic frequency of 42 kHz. These detonability limits, observed in our experimental conditions, are very sensitive to the initial pressure and these results show that an increase of this pressure leads to a higher stability of the detonation and, as a matter of fact, to a decrease of the critical diameter.

Acknowledgments

The authors are grateful to M. Y. Sarrazin and to M. M. Ferdjouni for their assistance in carrying out the experiments.

References

Barrère, M. (1976) private communication.

Bauer, P., Krishnan, S., and Brochet, C. (1981) Detonation characteristics of gaseous ethylene, oxygen, and nitrogen mixtures at high initial pressures. Gasdynamics of Detonations and Explosions: Progress in Astronautics and Aeronautics (edited by Bowen, Manson, Oppenheim, and Soloukhin), Vol. 75, pp. 408-422. AIAA, New York.

Bouriannes, R., Moreau, M., and Martinet, J. (1977) Un pyromètre rapide à plusieurs couleurs. Rev. Phys. Appl. 12, 893-899.

Carnahan, N. F. and Starling, K. E. (1969) Equation of state for non-attracting rigid spheres. J. Chem. Phys. 51, 635-636.

Fickett, W. (1962) Detonation properties of condensed explosives calculated with an equation of state based on intermolecular potentials. Rept. LA-2712, Los Alamos Scientific Laboratory, Los Alamos, N. Mex.

Gordon, S. and McBride, B. J. (1971) Computer program for calculation of complex chemical equilibrium compositions, rocket performance, incident and reflected shocks and Chapman-Jouguet detonations. NASA SP-273.

Hirschfelder, J. O., Curtiss, C. F., and Bird, R. B. (1954) Molecular Theory of Gases and Liquids. John Wiley and Sons, New York.

Kato, Y., Bouriannes, R., and Brochet, C. (1979) Mesure de la température de luminance des détonations d'explosifs transparents et opaques. Actes du H.P.D. Symposium, pp. 439-449. Ed. du C.E.A., Saclay, France.

Kato, Y., Bauer, P., Brochet, C., and Bouriannes, R. (1981) Brightness temperature of detonation wave in nitromethane-tetranitromethane mixtures and in gaseous mixtures at a high initial pressure. Seventh International Symposium on Detonation, pp. 403-308. U.S. Naval Academy, Annapolis, Md.

Kemp, M. K., Thompson, R. E., and Zigrang, D. J. (1975) Equations of state with two constants. J. Chem. Educ. 52, 802.

Matsui, H. and Lee, J. H. (1979) On the measure of the relative detonation hazards of gaseous fuel oxygen and air mixtures. 17th International Symposium on Combustion, pp. 1269-1280. The Combustion Institute, Pittsburgh, Pa.

Parlin, R. B. and Giddings, J. C. (1955) A solid state model for detonations. Rept. XVI. Institute for the Study of Rate Processes, University of Utah, Salt Lake City, Utah.

Pujol, Y. (1968) Détermination de la composition et de l'enthalpie de mélanges gazeux à haute température compte tenu de leur second coefficient de viriel. Entropie 20, 39-47.

Rowlinson, J. S. (1964) The virial expansion in two dimensions. Mol. Phys. 7, 593-594.

Taylor, J. (1952) Detonation in Condensed Explosives. Oxford Press, Oxford, England.

Overdriven Gaseous Detonations

T.P. Gavrilenko* and E.S. Prokhorov†
Academy of Sciences, Novosibirsk, USSR

Abstract

The overdriven detonation waves, generated by the interaction of a Chapman-Jouguet (CJ) detonation with a conical insert in a circular tube and a wedge inserted into a flat channel, have been studied experimentally. In the case of an irregular (Mach) CJ wave reflection from the surface of the wedge or the conical channel, the Mach stem was an overdriven wave, whose velocity was 1.3 times the CJ velocity. The critical wedge angle and the critical angle between the conical channel and the axis of the tube were found to be 40 ± 1 and 45 ± 1 deg, respectively. As these angles were decreased, the Mach stem size increased, and its degree of overdrive decreased. The locus of the triple point in the flat channel was not self-similar, while it was nearly self-similar in the circular tube. The results indicated a direct correlation between the Mach stem size and the cell size in the CJ wave. For the mixtures: C_2H_2 + $2.5O_2$, $2H_2 + O_2$, and $CH_4 + 2O_2$ at initial pressures from 0.05 to 1 atm, the pictures of an irregular reflection were identical when the cell size was the same. In the case of the secondary irregular reflection of the Mach stems at the centerline of the tube the generated wave velocity was about 1.6 times the CJ velocity and remained constant along the narrow channel for a distance of about two or three narrow channel widths.

Presented at the 8th ICOGER, Minsk, USSR, Aug. 23-36, 1981. Copyright © American Institute of Aeronautics and Astronautics, Inc. 1983. All rights reserved.
*Senior Researcher, Institute of Hydrodynamics.
†Researcher, Institute of Hydrodynamics.

OVERDRIVEN GASEOUS DETONATIONS

Introduction

This paper presents an experimental investigation in a flat channel or a circular tube of overdriven detonation waves arising from the interaction of a Chapman-Jouguet (CJ) detonation with a wedge inserted into the channel or with a tube of conically narrowing cross section. The angle between a conical surface and centerline of the tube is called the cone angle. If the angle between the wedge surface and the incident detonation wave front exceeds a critical value, an irregular Mach reflection of detonation takes place. In this case a Mach stem is an overdriven detonation wave.

The overdriven wave parameters were investigated with the smoked foil method as well as Schlieren and streak photographic methods. Such parameters as the critical angle, the dependence of the degree of overdriving α ($\alpha = D_1/D_0$, where D_1 is the overdriven wave velocity, and D_0 is the CJ velocity) on the cone angle, the dependence of the Mach stem size on the wedge angle (or cone angle) and on the initial pressure of the mixture were found experimentally. The smoked foil method was used to observe the symmetric intersection of Mach stems (secondary Mach reflection). Also the behavior of the doubly overdriven wave in the channel with a constant cross section that follows the conical section has been investigated.

Experimental System

A rectangular 10x70-mm channel, used in the experiments, was equipped with glass windows located 700 mm from an initiation point. The wedge, whose angle could be changed from 5 to 45 deg through 5-deg intervals and near to 40 deg through 1-deg intervals, was inserted into the chanel against the windows.

A circular tube, 80 mm in diameter and 1000 mm long was cut along its axis and mounted on a 1000x120x15-mm flat plate. At one end, initiation was realized; at the other, an axially cut cylinder with a conically narrowing cross section was mounted. Its maximum diameter of 80 mm was narrowed down to 20 or 10 mm. The wedge angle was changed from 30 to 45 deg through 5-deg intervals and near to 45 deg through 1-deg intervals. With this construction it was easy to obtain information with the smoked foil method. Stoichiometric mixtures of $C_2H_2 + 2.5O_2$ and $2H_2 + O_2$ under an initial pressure of from 0.05 to 1 atm were used.

Results

The critical wedge angle for both mixtures was found to be equal to 40 ± 1 deg. This result differs from the calculated value of 34 ± 0.4 deg obtained under the assumption of chemical equilibrium among the detonation products (Gavrilenko et al. 1979). In a circular tube the critical angle of a cone was greater than that of a wedge and equals 45 ± deg. The Schlieren photographs show that for all wedge angles the Mach stem was straight and normal to the wedge surface.

The triple point trajectory was determined from smoked foils. The cell size strongly depended on the degree of overdrive α, and the Mach stem was an overdriven wave in both the circular tube and the flat channel. Therefore the size of the cells behind its front was always less than that behind the CJ wave front. The dependence between the cell size in an overdriven wave and the degree of overdrive α is shown in Fig. 1 (Gavrilenko and Prokhorov 1980). Here a_0 denotes the cell size in the CJ wave. From the observed dependence the detonation wave velocity in channels with variable cross sections can be determined with a high accuracy. Presented in Figs. 2 and 3 are the smoked foil records (a flat channel in Fig. 2; a circular tube in Fig. 3). The region with the large-scale cells corresponds to an incident wave, that with the small-scale cells corresponds to a Mach stem, and the interface between the regions is the triple point trajectory.

Discussion

In contrast with the irregular reflection of strong shock waves, the triple point propagation of an irregular CJ detonation reflection is not self-similar. This may be explained by the complex interaction that occurs when the structure of the large-scale cells interacts with the triple point. The triple point trajectory configuration (h, the length of the Mach stem) is strongly affected by the initial pressure of a mixture. If these trajectories are constructed in dimensionless coordinates with respect to the cell size a_0 of the incident CJ wave (where L is the length along the surface of the wedge or cone), at the same wedge angle the trajectories for different mixtures and at different initial pressures are coincident with a "roughness" curve. This roughness does not exceed the cell size a_0 (Gavrilenko and Prokhorov 1980). The cell size a_0 in these processes can be regarded as a scale which determines geometrical dimensions and other important characteristics. So far this explicit correlation has not been explained.

OVERDRIVEN GASEOUS DETONATIONS

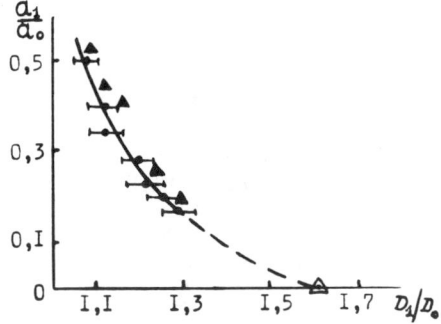

Fig. 1 Dependence of the cell size in an overdriven wave on the degree of overdrive. ▲ - $2H_2 + O_2$; ● - $C_2H_2 + 2.5O_2$; ▵ - experimental data of Gordeyev (1976).

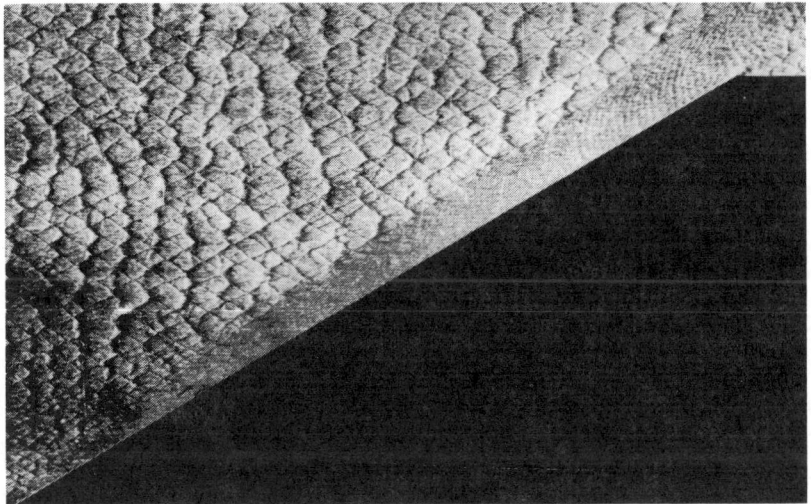

Fig. 2 Smoked foil record obtained in a flat channel.

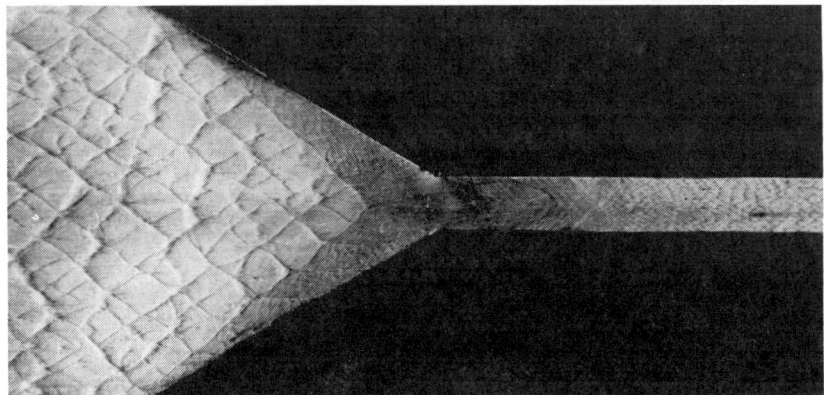

Fig. 3 Smoked foil record obtained in a circular tube.

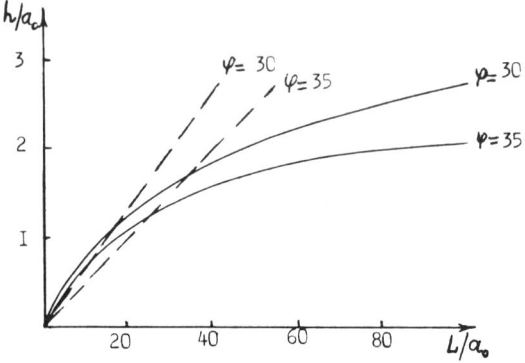

Fig. 4 Dependence of h/a_0 vs L/a_0 for two wedge or cone angles. The solid lines are for the flat channel; the dashed lines illustrate data obtained in the circular tube.

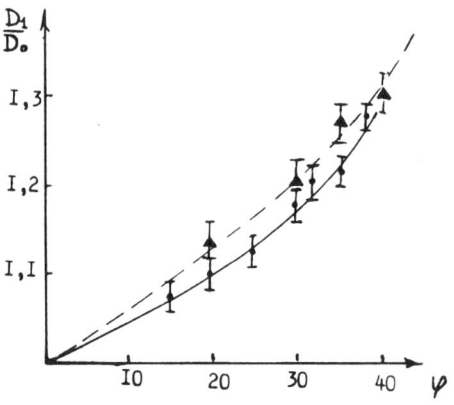

Fig. 5 Dependence of degree of overdrive D_1/D_0 vs the angle of a wedge or cone. The solid lines are for the flat channel; the dashed lines illustrate data obtained in the circular tube.

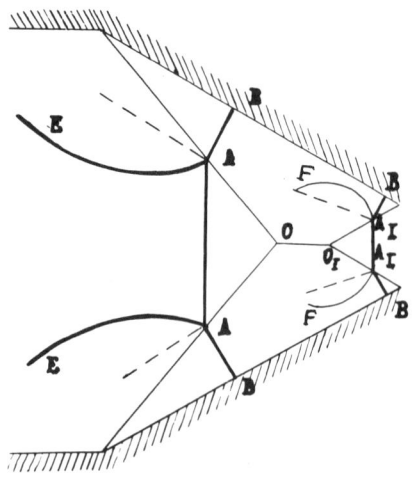

Fig. 6 Scheme of the second irregular Mach stem reflection. The drawing also illustrates growth before intersection at the centerline.

OVERDRIVEN GASEOUS DETONATIONS

The Mach stem size (dimensionless in a_0) increases with decreasing the wedge (cone) angle. In this case the degree of overdrive decreases. Presented in Figs. 4 and 5 are the dependences of h/a_0 vs L/a_0 and D_1/D_0 vs the angle of a wedge or cone, respectively. Solid lines are related to the values in a flat channel; dashed lines illustrate data obtained in a circular tube.

The maximum degree of overdrive is 1.30 in a flat channel at a wedge angle of 38 deg. At the same cone angle the overdrive is higher. The amount of overdrive is further increased by a latter irregular Mach stem reflection. Such a reflection picture is shown in Fig. 3. The angle formed by the tube axis and stem front is supercritical; therefore at the intersection of the Mach stems irregular reflections occur. Such a reflection with a previous stem growth is shown in Fig. 6. Along the line OO_1, the Mach stem reflection is regular and then, later, an irregular reflection is observed (after O_1). The regular reflection phenomenon is caused by the Mach stem curvature in the vicinity of the triple point, even though near the wedge surface the stem is normal to it. As a result, the angle between its front and the tube axis becomes subcritical. The A_1A_1 wave formed under a secondary irregular reflection is overdriven. Its velocity, determined from streak photography records through a slit parallel to the axis, is equal to 1.5 to 1.7 D_0. A detonation wave propagating with such a velocity has a smooth front (Manjaley and Subbotin 1976; Gordeyev 1976), as verified experimentally, because within this region cells are not observed on the smoked foil records.

If the walls of the channel of constant cross section are placed exactly on the trajectory of the A_1A_1 wave so that the A_1A_1 front is normal to the walls when it enters the channel, the A_1A_1 wave velocity is constant for distances of about one channel width. Then it decreases until at about from 10 to 15 channel widths it coincides with the CJ velocity.

In this case, the reflected waves A_1F generated at the secondary irregular reflection propagate into the gaseous state formed by the AE waves and can be considered to be the piston which sustains a quasistationary A_1A_q wave propagation. The AE waves can also serve as a piston for the CJ wave front when the reflection from the wedge is regular. This phenomenon was observed (Gavrilenko, Nikolaev, and Topchian 1979) when the wedge angle was subcritical.

Further investigations of dynamical wave characteristics are needed if detonation wave propagation through channels with variable cross sections is to be predicted.

Conclusions

As a result of the above-mentioned observations, the following conclusions have been made:

1) The critical wedge angle of irregular reflection of a CJ detonation in $2H_2 + O_2$ and $C_2H_2 + 2.5O_2$ is equal to 40 ± 1 deg. The critical cone angle for the same mixtures is 45 ± 1 deg.

2) The Mach stem is an overdriven wave normal to the wedge surface. The maximum stem overdriving in a flat channel is 1.3, in a circular tube it is about 1.4. With increasing distance from the surface of the wedge (cone), the stem becomes curved.

3) The triple stem motion along the surface of the cone or wedge is not self-similar. There is a unique correlation between the dimensions of the Mach stem and the cell size in the incident wave.

4) An irregular Mach stem reflection at the centerline can generate a wave with an overdrive of from 1.5 to 1.7 which propagates as a quasistationary wave for distances equal to, approximately, one channel width of the small channel.

References

Gavrilenko, T. P., Nikolaev, Y. A., and Topchian, M. E. (1979) Investigation of overdriven detonation waves. Fiz. Goreniya Vzryva 15, 119-123.

Gavrilenko, T. P. and Prokhorov, E. S. (1980) Experimental investigation of irregular reflection from the wedge. Proceedings of the Sixth All-Union Symposium on Combustion and Explosion, Chernogolovka, USSR, pp. 103-106.

Gordeyev, V. E. (1976) Limiting velocity of overdriven detonation and the stability of shocks in the detonation wave spin. Dokl. Adad. Nauk SSSR 226, 619-662.

Manjaley, V. I. and Subbotin, V. A. (1976) Experimental investigation of overdriven detonation waves stability in gas. Fiz. Goreniya Vzryva 12, 935-942

Motion of Solid Bodies in Combustible Gas Mixtures

M.M. Gilinsky,* L.I. Zak,* and T.S. Novikova*
Moscow State University, Moscow, USSR

Abstract

Results of a theoretical study of solid bodies moving at high speeds in detonating gas mixtures are reported. An analysis of earlier experimental results and theories describing various combustion regimes near sharp and blunt bodies is presented. New two-dimensional computation results for low-frequency head shocks oscillating in a different mode than previously observed are presented. Changes in global aerodynamic properties of such flows are investigated, and the reason why the observed phenomena decrease the blunt body drag is explained.

Introduction

Intensive experimental studies of ballistic nature, where models are shot into various gas mixtures in which exothermic reactions can be initiated brought about a number of technical applications of combustion in hypersonic flows. Such experiments were carried out in the U.S., France, Federal Republic of Germany, and the USSR [see for example, Ruegg and Dorsey 1962; Behrens et al. 1965; Lehr 1972; Toong 1974; Chernyi 1969; Soloukhin 1961] in a number of combustion regimes with smooth shock waves and a combustion front, unsteady regimes with regular and irregular oscillations of the gas flow, and steady detonation waves with a small cellular structure of the flow.

The theoretical foundation of individual combustion regimes has been developed in several publications (Toong 1974; Chernyi and Gilinsky 1969; Levin 1965; Gilinsky 1974; Gilinsky et al. 1980). New results in the study of this problem are given below. Two questions are considered: 1)

Paper presented at the 8th ICOGER, Minsk, USSR, Aug. 23-26, 1981. Copyright © American Institute of Aeronautics and Astronautics, Inc., 1983. All rights reserved.
*Senior Scientist, Institute of Mechanics.

the magnitude of the wave drag of solid bodies in detonating gases; and 2) a possible mechanism of flow self-oscillation in front of a blunt body.

Wave Drag

The problem of the influence of heat transfer on the wave drag of bodies has received relatively little attention. Behrens et al. (1965) and Ruegg and Dorsey (1962) found a substantial decrease (up to about 50%) in the general drag of blunt bodies due to heat release. The pioneer estimation of the sphere and circular cylinder drag in such flows was given by Schneider (1968), who analyzed hypersonic stream flow around a body with the formation of an infinite thin detonation wave and with an approximation of constant gas density in the shock layer. The latter assumption requires the heat supply in the wave to decrease as $q = q_0 \cos_0^2 \theta$ (Fig. 1). Schneider (1968) found that the

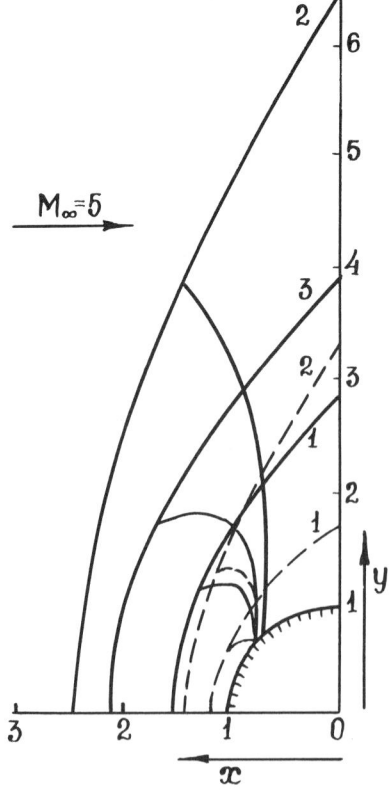

Fig. 1 Supersonic flow around sphere or cylinder with a detonation front. 1) $q = Q/V_\infty^2 = 0$; 2) $q = 0.35$; 3) $q = 0.335 \cos^2\theta$; $q_{max} = 0.48$.

sphere drag reduction can reach about 40% while that of a circular cylinder, about 36%. The theoretical analysis shows that these assumptions (Schneider 1968) are too severe and result in an overestimation of drag reduction coefficients.

Indeed, Prandtl's formula for an exothermic flow yields

$$X(\gamma, M_\infty, F) = \frac{p_{oq}}{p_{oa}}$$

$$= \frac{(1 + \frac{\gamma-1}{2} M_\infty^2 + \frac{M_\infty^2 - 1}{2} F)(\frac{2\gamma}{\gamma+1} M_\infty^2 - \frac{\gamma-1}{\gamma+1})^{\gamma/(\gamma-1)}}{\frac{\gamma+1}{2} M_\infty^2 [1 + \frac{\gamma}{\gamma+1}(M_\infty^2-1)F][1 + \frac{2\gamma}{\gamma+1}(M_\infty^2-1)]}$$

$$\times \frac{1 + \frac{2\gamma}{\gamma+1}(M_\infty^2-1)F}{1 + \frac{2\gamma}{\gamma+1}(M_\infty^2-1)} \quad (1)$$

where F is a parameter characterizing the degree of the detonation wave overpressure; $F = 1$ corresponds to the Chapman-Jouguet (CJ) wave; while $F = 2$ is an ordinary shock wave without heat release ($1 \leq F \leq 2$).

The limit expressions at $M_\infty \to \infty$ and $\gamma \to 1$ for the function X are as follows:

$$X(\gamma, \infty, F) = \frac{F}{2} \frac{2(\gamma-1) + F}{(\gamma+1)F}^{\gamma/(\gamma-1)} \qquad X(1, \infty, F) = \frac{F}{2} \exp\frac{2-F}{2F} \quad (2)$$

Equations (1) and (2) show that the minimum value of X is reached in the case of the CJ wave, i.e., X min = $X(1, \infty, 1) = 0.824$. Thus the decrease of the stagnation pressure as a result of heat transfer cannot exceed 18%. The use of a modified Newton's theory for the pressure distribution gives the same value of 18% for the drag reduction due to heat supply.

More exact estimates must be obtained by numerical methods. Such computations were made with the unsteady finite-difference method of Babenko and Pusanon (1965). The models of an infinitely thin detonation wave with three different models of heat release were used: 1) $q = $ const, 2) $q = q_0\cos^2\theta$, 3) $q = q_0\sin^2\theta$. The first model corresponds to the case which results in the stabilization of a CJ detonation at some distance from the body. The second model corresponds to the heat supply decrease resulting from the finite rate exothermic reactions and the splitting of the wave into an adiabatic shock and the combustion front. The

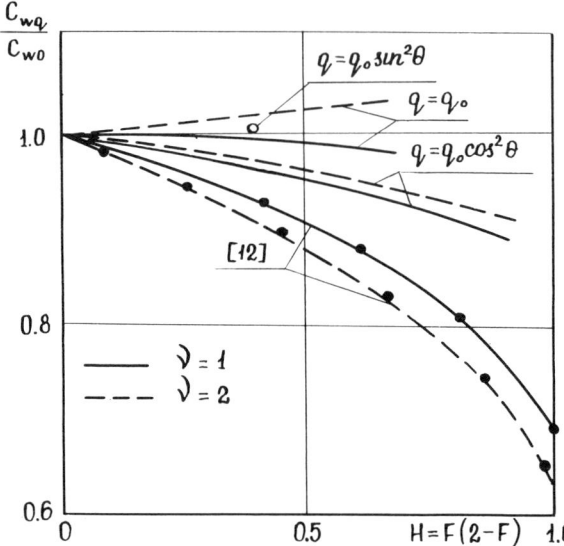

Fig. 2 Influence of heat transfer on the drag coefficient of bodies.

third model corresponds to the hypersonic flying model velocities when, in the vicinity of the stagnation line dissociation, reactions are dominating and the process is endothermic on the whole, while it is exothermic in the periphery.

The computation domain included sub- and transonic flow regions up to the midsection of the body ($\theta = \pi/2$). The grid contained 41 longitudinal and 21 transverse modes. The calculation error was about 0.5%. Along with integral checks of the Bernoulli integral, isentropic conditions, and conservation of the flow rate through various cross sections, the correctness of determining the local characteristics of the flow was verified. These included the angles of the detonation wave inclination at the sonic point, the sonic line inclination toward the incident flow, and the ratio of the curvatures of the shock wave and the streamline. The appropriate exact relations were written out. Their analysis allows the study of symptotic properties of the flow behind the attenuating detonation wave.

Figure 1 shows the flow geometry past a sphere (dashed line) and circular cylinder (solid line) for different models of heat supply. In Fig. 2 the ratio of the drag coefficients with and without heat supply is plotted. The maximum drag reduction occurs for model 2 and reaches about 10%. It is interesting to note that for a sphere with model 1 the drag increases slightly. This is attributed to the dominant influence of the pressure due to heat supply in the hypersonic part of the flow ahead of the sphere in comparison with its decrease in the subsonic region in the vicinity of the stagnation point. The dark circles in Fig. 2 show the results of Schneider (1968).

Thus it is found that the decrease of the wave drag of a blunt body and a profile as a result of heat release in the shock layer cannot reach the values of 40 and 50% observed in the experiments. It is necessary to look for other additional reasons to explain this fact, namely, the change of friction during burning, base pressure, unsteady effects, etc.

In addition to previous calculations, the computations of the detonation gas flow around segment-conic bodies and wedgelike profiles with a blunt and sharp front parts were performed. In the first case, the calculations were made from the midsection according to the stationary scheme of Godunov (1976); in the second, the whole flow was calculated according the the same scheme. The ranges of the wedge (cone) angles β were taken as $0 \leq \beta \leq 0.5$ to avoid noticeable influence of the flow separation from the rear surface. The calculations show that heat supply caused a slight pressure rise on the rear surface of segment-conic bodies and the drag coefficient decreased by no more than 2-8%. A quick rise in the drag of sharp bodies in combustion flow of about 50% and more was noted.

Flow Pulsations

The accuracy of measurement of aerodynamic characteristics of bodies in the detonative flow is less than in usual flows, which may be due to unsteady pulsations caused by burning (Chernyi and Chernjavsky 1973; Ruegg and Dorsey 1962). As a consequence, the direct numerical modeling of two-dimensional unsteady blunt-body flow is of great interest.

The model considered is a gas mixture flow with nonequilibrium exothermic reactions which arises when an axisymmetrical blunt body (face-end cylinder) instantaneously acquires a supersonic velocity V_∞. Following Levin (1965) and Taki and Fujiwara (1978), a two-stage kinetic model of combustion characterized by two reactions is used:

$$\frac{d\phi}{dt} = K_\phi \alpha^l \rho^n p^m \exp(-E_\phi/RT) = \omega_1 \qquad (3)$$

$$\frac{d\beta}{dt} = -K_\beta p^2 \{\beta^2 \exp(-E_\beta/RT) - (1-\beta)^2 \exp[-(E_\beta + q)/RT]\} = \omega_2 \qquad (4)$$

where $0 \leq \phi \leq 1$ is equal to unity at the end of the induction period and $\bar{\beta}$ changes in the course of entire particle combustion from 1 to its equilibrium value. Dimensionless variables as defined by Taki and Fujiwara (1978), are introduced, with the only difference being the choice of the body midsection radius as a characteristic linear dimension L_*. The numerical values of the constants K_ϕ, K_β, I, m, n, E_ϕ, and E_β are also those of Taki and Fujiwara (1978). Among the dimensionless variables determining the flow, M_∞ and K_1 are the most important ($K_1 = K_\phi P_\infty L_*/RT_\infty \sqrt{q}$, where R is the gas constant).

Gilinsky et al. (1980) performed a numerical solution with an extension of the "through computation scheme" of Godunov (1976) to the case of nonequilibrium flow. When Eq. (3) is integrated, the term $d\phi/dt$ was neglected, i.e., the induction time was determined in quasisteady approximation. Equations (3) and (4) were integrated by implicit difference scheme, the first one being integrated along quasisteady trajectories of particles.

The results of parametric investigations made on the network $M = 30$, $N = 60$, $M_1 = 17$, $N_1 = 15$ (Fig. 3) were represented as isobars, Mach number profiles, stagnation pressure oscillograms, positions of a shock wave and the

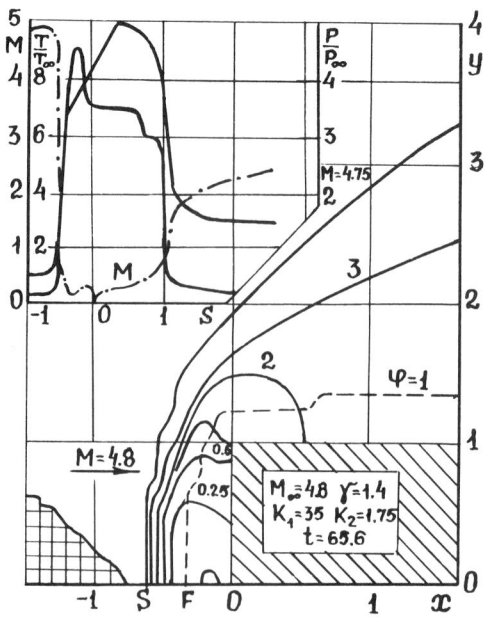

Fig. 3 Supersonic motion of cylinder in combustible gas mixture: lines of M = const; pressure P, temperature T, and Mach number M profiles along the stagnation streamline.

combustion front along the flow symmetry line as functions of time.

Three series of calculations were made for $M_\infty = 4$, 4.8 and 6 with a variation of K_1 over a wide range. A typical picture of constant Mach numbers ("isomach") is presented in Fig. 3, where the numbers above the lines show the value of this magnitude. The dashed line shows the boundary of ignition, line $\phi = 1$.

In Fig. 4 the dark lines are the stagnation pressure oscillograms at $M_\infty = 4.8$, $K = 35$; the dotted and dashed lines present the dependence for the wave detachment from the body; the dashed lines show the position of the combustion front along the stagnation line (points S and F in Fig. 3).

Discussion

The analysis of the obtained solutions allows us to reach the following conclusions. For each value of the freestream Mach number M_∞ there are two critical parameters, $K_1^{(1)}$ and $K_1^{(2)}$. For $K_1 \leq K_1^{(1)}$, ignition of the mixtures occurs at the initial stage of the unsteady flow but combustion is not sustained and an adiabatic flow around the

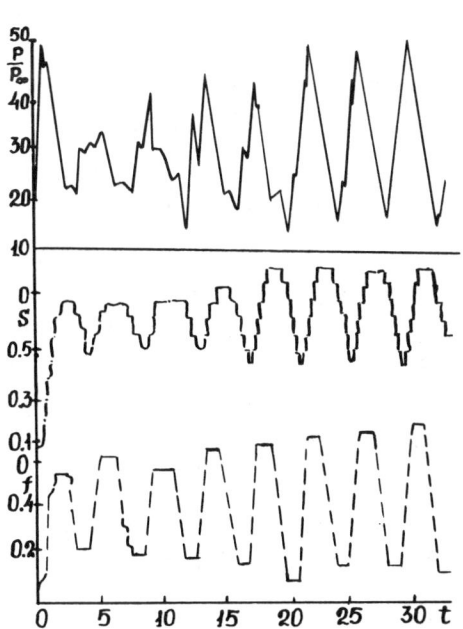

Fig. 4 Stagnation pressure oscillogram (P), position of shock wave (S), and the combustion front (f).

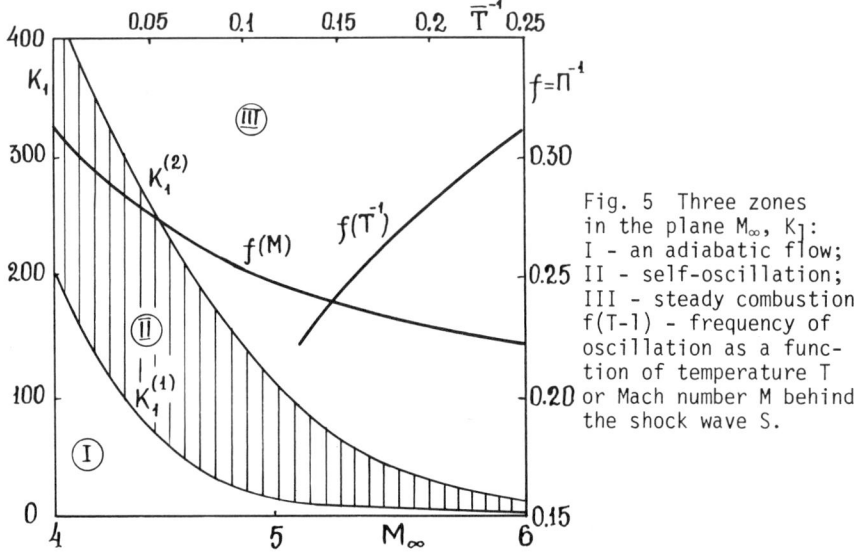

Fig. 5 Three zones in the plane M_∞, K_1:
I - an adiabatic flow;
II - self-oscillation;
III - steady combustion
$f(T^{-1})$ - frequency of oscillation as a function of temperature T or Mach number M behind the shock wave S.

face-end cylinder becomes steady. At $K_1 \geq K_1^{(2)}$, the gas mixture ignites and a steady combustion zone is established. In the case of $K_1^{(1)} < K^1 < K_1^{(2)}$, continuous flow oscillations appear.

Thus, in the plane of M_∞, K_1, the limiting curves $K_1^{(1)}(M_\infty)$ and $K_1^{(2)}(M_\infty)$ divide the entire region into three zones (Fig. 5): adiabatic flow, self-oscillations, and steady combustion.

The frequency of oscillations decreases as M_∞ or the temperature behind the shock wave T_s increases, the dependence of T_s^{-1} being close to linear one as observed by Behrens et al. (1965) and Lehr (1972). Comparison of the maximum oscillation frequency with experimental results due to Soloukhin (1961) indicates pure divergent oscillations are similar. So, for the stoichiometric hydrogen-oxygen mixture at P_∞ = 1 atm, D = 15 mm, the frequency ν is equal to about 0.052 MHz and is twice as low as the frequency obtained by Lehr (1972) for a cylindrical model with a spherical blunt nose.

To understand better the mechanism self-oscillations, more detailed information of one oscillation cycle was printed by the computer. Figure 6 shows the pressure and temperature distributions along the symmetry axis for every

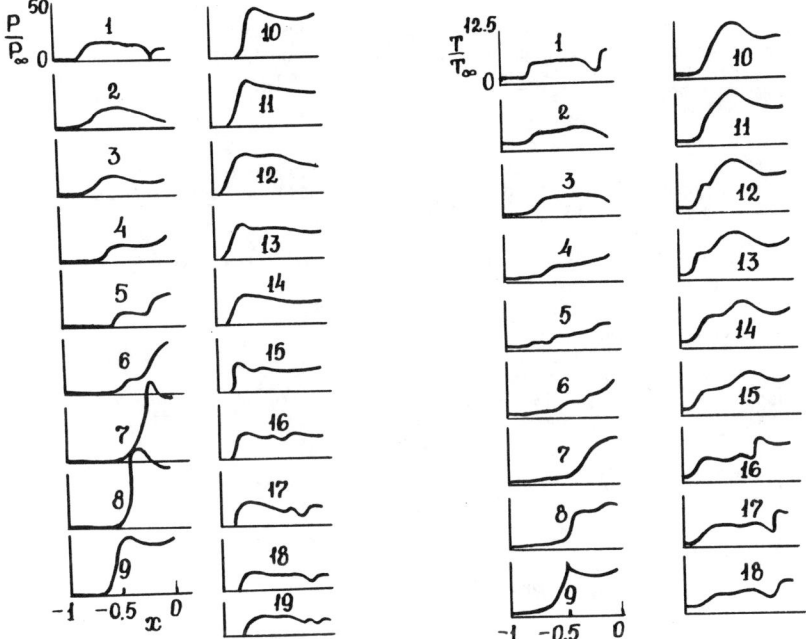

Fig. 6 Pressure and temperature profiles along the stagnation line over one cycle of self-oscillation.

ten computation steps. The process of oscillation proceeds in the following way. As a shock wave travels to the body, the combustion front also moves to the body so that the compression wave runs on its wall (points 2 and 4 in Fig. 6). The wave is then reflected and the pressure and temperature rise (points 4 and 5). Intensive burning initiated near the wall causes the inside wave to collide with the head wave, producing a strong detonation wave (points 7 and 8) that propagates upstream. Subsequently, the wave is damped as a consequence of gas expansion behind it, but is still moving forward (points 9-16). After this, the wave comes back to the body, the combustion attenuates (points 16-18), then the compression wave collides again with the wall and so on.

The oscillations depend drastically on the intensity of the compression wave incident on the wall. It must be sufficient for the ignition in the reflected wave to occur for a definte time. If this is not the case, either because of small K_1 (the zone of heat release is "broad," $K_1 = 20 K_2$) or because of large K_1 (the compression wave has time to reduce without reaching the end wall), the flow is stabilized.

References

Babenko, K. I. and Pusanov, V. V. (1965) Difference method of solution of three-dimensional problems of gas dynamics. <u>Transactions of the Second All-Union Conference on Mechanics</u>, Moscow, USSR.

Behrens, H., Struth, W., and Wecken, F. (1965) Studies of hypervelocity firings into mixtures of hydrogen with air or with oxygen. <u>Tenth Symposium (International) on Combustion</u>. The Combustion Institute, Pittsburgh, Pa.

Chernyi, G. G. (1969) Supersonic flow past bodies with formation of detonation and combustion fronts. <u>Problems of Hydrodynamics and Continuum Mechanics</u>. USSR Academy of Sciences, Moscow, USSR.

Chernyi, G. G. and Chernjavsky, S. U. (1973) Motion of blunt bodies at high velocity in hydrogen-oxygen mixture. <u>Rep. Acad. Sci. USSR</u>, 212.

Chernyi, G. G. and Gilinsky, S. M. (1969) High speed motion of a body in chemically active gases. <u>Fluid Dyn. Trans.</u> 4.

Gilinsky, M. M. (1974) Unsteady regimes of supersonic combustible flow around blunt bodies. <u>Trans. Inst. Mech. USSR</u> 32.

Gilinsky, M. M., Zak, L. I., and Simanovsky, G. P. (1980) Numerical calculation of self-oscillation regimes of combustion in a blunt-body flow. <u>Sixth All-Union Symposium on Combustion and Explosion</u>, Alma-Ata, USSR Academy of Sciences, Moscow, USSR.

Godunov, S. K. et al. (1976) Numerical solution of multi-dimensional problems of gas dynamics. Moscow, USSR.

Lehr, H. F. (1972) Experiment on shock induced combustion. <u>Astronaut. Acta</u> 17.

Levin, V. A. (1965) Motion of shock and detonation waves in combustible gas mixtures. Ph.D. Thesis, Moscow State University, Moscow, USSR.

Maurer, F. and Brings, W. (1968) Been/flussung des Widerstands und der Kopfwelle durch Wärmezufuhr im Staupunktbereich stumpfer Körper bei Überschallanströmung. Vortrag auf dem VII GAS-Kongress.

Ruegg, F. W. and Dorsey, W. W. (1962) A missile technique for the study of detonation waves. <u>T. Res. Nat. Bur. Stand.</u> 66, 51-58.

Schneider, W. (1968) Über den einfluss von warmezufuhr bei hyperschallströmung um kugel und kreiszylinder. <u>Z. Flugwiss.</u> 16.

Soloukhin, R. I. (1961) Pulsating combustion behind shock wave in supersonic flow. Appl. Mech. Theor. Phys. (5).

Taki, S. and Fujiwara, T. (1978) Numerical analysis of two-dimensional nonsteady detonations. AIAA J. 16.

Toong, T.-Y. (1974) Instabilities in reaching flows. Astronaut. Acta 1.

Direct Initiation of Detonation in LNG/Air Clouds

J. Kurylo,* J.M. Thomsen,† and F.M. Sauer*
Physics International Company, San Leandro, Calif.

Abstract

A detailed numerical simulation of the direct initiation of detonation of a typical LNG/air mixture by a subcritical spherical energy source has been carried out. In the simulation, the chemical reaction is modeled by Westbrook's 26 species, 75 step methane-ethane oxidation mechanism. The gasdynamic flowfield is computed using Oppenheim's CLOUD CODE. Analysis of the simulation indicates that during the process of initiating detonation, the shock and reaction fronts experience three distinct modes of propagation: the strongly coupled, weakly coupled, and uncoupled modes. The strongly coupled mode is characterized by a very strong shock front ($M_n > 8$), extremely short postshock induction times ($\tau < 0.1$ μs), production of large quantities of H, O, and OH, and by an endothermic gas reaction. The weakly coupled mode is characterized by a strong shock front ($M_{CJ} < M_n < 8$), production of H_2O, CO, and CO_2, and significant chemical heat release (>95% Q_{CJ}). Short postshock induction times ($0.1 < \tau < 2.5$ μs) promote, via a cell explosion mechanism, the cyclic generation of pressure peaks that sustain the detonation. The uncoupled mode is characterized by the propagation of a shock-reaction front complex. During its propagation, the reaction front intermittently generates pressure waves that substantially increase the pressure and temperature of large portions of the induction zone as well as maintain the strength of the

---
Presented at the 8th ICOGER, Minsk, USSR, Aug. 23-26, 1981. Copyright © American Institute of Aeronautics and Astronautics, Inc., 1982. All rights reserved.
*Staff Physicist, Shock Simulation and Reactive Systems Department.
†Staff Scientist, Nuclear Effects Division.

lead shock. These intermittent waves are thought to be the mechanism by which detonation is reinitiated.

Introduction

Liquefied natural gas (LNG) is a stable liquid mixture at atmospheric pressure and a temperature of -260°F, occupy-

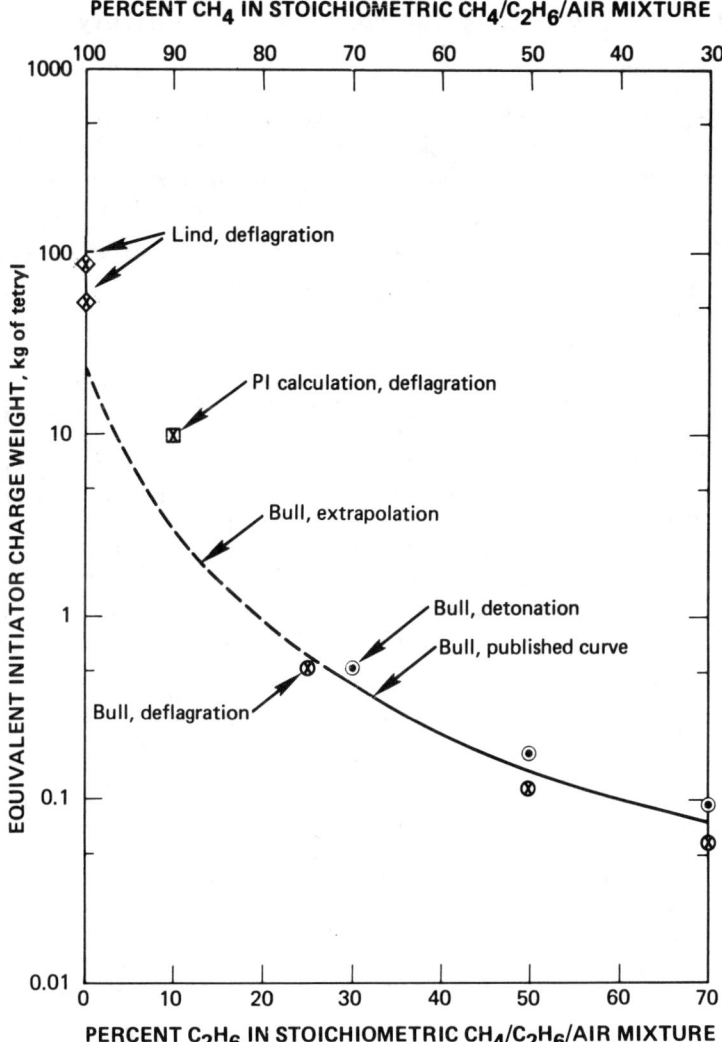

Fig. 1 Critical spherical tetryl charge weights for sustained detonation in stoichiometric methane/ethane/air mixtures.

ing only 1/600 of its volume as a gas (Weast 1978). LNG is composed predominately of methane (93% by volume), ethane (4.5%), propane (2.3%), and negligible amounts (0.2%) of higher hydrocarbon fuels (Westbrook 1981a). When exposed to the atmosphere, LNG rapidly vaporizes while mixing with air (U.S. DOE 1980). The resulting low-lying pancake-shaped fuel/air cloud encompasses a wide range of fuel/air compositions (lean at the edge of the cloud to rich at the LNG source site) and fuel chemical compositions ($CH_4/C_2H_6/C_3H_8$/higher hydrocarbons). In the likely event that the cloud comes into contact with a suitable ignition source, the gas cloud will combust.

Detonation of the gas cloud can either be self-initiated, via deflagration to detonation transition (DDT), or directly initiated. DDT requires an extreme degree of flame acceleration [U_{lam}=45 cm/s (Andrews and Bradley 1972), $U_{turb}/U_{lam} \geq 685$ (Gordon and McBride 1971)] by turbulence and/or pressure wave-flame interaction mechanisms. Although DDT has been observed under confined situations, it is unlikely that under unconfined situations the necessary acceleration mechanisms, and therefore burning speeds, can be achieved or maintained to support self-initiation (Moen et al. 1980; Urtiew 1980; Parnarouskis et al. 1980). In DDT, the ignition source characteristics are unimportant.

Direct detonation initiation (Lee 1977; Westbrook and Haselman 1979) occurs if the source generated blast front maintains a critical shocked gas state for a certain minimum duration, a period of time at least on the order of the chemical induction time of the shocked gas. If the duration is too short, the expansion wave associated with the blast wave decay cools the gas below its critical limit. Direct initiation of detonation depends crucially on the energy and power characteristics of the initiation source.

Figure 1 shows the experimental results of Bull et al. (1976, 1979) and Parnarouskis et al. (1980)[‡] for the equivalent critical spherical tetryl charge weight [tetryl high explosive characterized by Bull et al. (1976)] required to directly initiate detonation in an unconfined stoichiometric methane/ethane/air mixture as a function of the ethane content (percent by volume) in the fuel. Bull's extrapolation of his experimental data to the lower ethane concentrations, typical of LNG clouds, is included in Fig.

[‡]Hemispherical sources assumed, charge equivalence (energy basis) 1 kg tetryl = 0.8626 kg Composition B.

1. Bull predicts that 22 kg of tetryl would be required to detonate an unconfined pure methane/air mixture. The shape of the extrapolated curve suggests that the critical tetryl charge weight has a greater than exponential sensitivity to small variations in the ethane content of the fuel. While Bull's experiments clearly established that direct initiation of detonation is achieved in certain methane/ethane/air mixtures, analysis of the experiments did not yield detailed information about the dynamic processes occurring between the lead shock and the reacting fluid behind it. Bull and Elsworth (1978) did, however, provide experimental evidence of the decay of the initially overdriven detonation to an uncoupled shock-reaction front complex that later reaccelerated to a Chapman-Jouguet (CJ) detonation.

As a result of the experimental observations by Bach et al. (1969), Soloukhin and Ragland (1969), and Kamel and Oppenheim (1973), the following dependence of the direct initiation of detonation on the initiating energy of the source has evolved. For sufficiently large (supercritical) initiation energies, the initially overdriven detonation decays continuously to a multiheaded CJ detonation. Throughout the decay process, the blast front and reaction front are closely coupled. For sufficiently low (subcritical) initiation energies, a progressive decoupling of the reaction front from the blast front occurs, culminating in the decay of the initially overdriven detonation to a deflagration. The source generated blast front continuously decays to a sound wave. When the initiation energy is at the critical value, the overdriven detonation wave first decays, but the blast and reaction fronts remain closely coupled. As the wave expands, decoupling begins as the reaction front recedes from the blast front. However, unlike the subcritical regime, the decoupling process terminates when the total chemical energy released by combustion begins to contribute significantly to the blast motion. The blast and reaction fronts then propagate as a "quasisteady" complex with a constant sub-CJ (0.7-0.9 D_{CJ}) speed. During the quiasisteady period, instabilities develop leading to a highly asymmetrical multiheaded CJ detonation.

Several models (Zeldovich et al. 1956; Korbeinikov et al. 1972; Lee and Ramamurthi 1976; Edwards et al. 1976; Strehlow and Salm 1976; Sichel 1977; Westbrook and Haselman 1979; Bull and Elsworth 1979; Atkinson et al. 1980) incorporating the essential features of the events characteristic of the critical initiation energy regime have been used to estimate the critical initiation energy. These models do not provide the time-dependent details of the interaction between the gasdynamic processes and chemical reaction leading to the establishment of the critical states necessary for detonation initiation. These models generally require specification of the critical energy for one mixture concentration or specification of the Mach number of the shock-reaction complex or its postshock temperature and density. The sensitivity of the predicted critical energy to small variations in the estimate of postshock conditions is quite significant.

A number of analytic models that simultaneously account for the time-dependent gasdynamic and chemical kinetic processes have been attempted (Kyong et al. 1972; Levin and Markov 1975; Oran et al. 1978; Boni et al. 1978a). For example, Boni et al. (1978a, 1978b) developed a comprehensive finite-difference model to simulate the critical initiation energy experiments of Bull et al. (1976, 1979). Their results, extrapolated to pure methane/air mixtures, predict that over 4,000 kg of tetryl are required to directly initiate detonation. A possible explanation for the large discrepancy between their prediction and that of Bull et al. (1976) and Westbrook and Haselman (1979) may lie in the two-step oxidation mechanism used by Boni et al. (1978b) to simulate the chemical reaction of methane/oxygen/nitrogen mixtures. Their two-step mechanism predicts chemical induction times, throughout the temperature range of interest, which are more than one order-of-magnitude less than that determined experimentally by Burcat et al. (1971). Unfortunately, in the presentation of their results, they also did not include the details of the gasdynamic-chemical kinetic processes occurring in the region bounded by the shock front and the reacting zones for the cases which did and did not culminate in detonation.

Therefore we have developed a finite-difference code called IGNITION, which for the first time blends a validated complex reaction mechanism for LNG/air mixtures with a time-proved gasdynamic finite-difference code. The goals of performing very detailed numerical calculations of the direct initation event are 1) to demonstrate the feasibility of performing calculations using a very detailed chemical reaction mechanism; 2) to present detailed profile snapshots and histories of the relevant species and gasdynamic properties for cases which did and did not culminate in detonation; 3) to provide an estimate of the critical energy required to directly initate detonation in typical LNG/air cloud mixtures, which could then be experimentally verified; and 4) to identify those gasdynamic and chemical kinetic processes and critical states primarily responsible for initiating detonation.

Two simulations were performed: 1) verification of the LNG/air chemical reaction mechanism; and 2) direct initiation of detonation of a stoichiometric 90% (by volume) methane/10% ethane/air mixture by a 7.6 kg spherical C-4 charge, equivalent in energy to 10 kg to tetryl.

Model Equations

The flowfield is considered to be symmetric and one dimensional so that the independent variables are time t and position r. The effects of transport phenomena are neglected. Under these circumstances, the conservation equations of mass, momentum, species, and energy can be expressed, respectively, in the following Lagrangian form:

$$\text{(Mass)} \quad \frac{\partial v}{\partial t} = \frac{v}{r^j}\frac{\partial}{\partial r}(r^j u)$$

$$\text{(Momentum)} \quad \frac{\partial u}{\partial t} = -v\frac{\partial p}{\partial r}$$

$$\text{(Species)} \quad \frac{\partial y_k}{\partial t} = v\omega_k \quad k = 1,\ldots,n$$

$$\text{(Energy)} \quad \frac{\partial e}{\partial t} = -p\frac{\partial v}{\partial t}$$

where v is the specific volume, p the pressure, u the mass velocity, e the specific internal energy of the mixture, y_k and ω_k the k-th species concentration and production rate [for details see Williams (1965)], respectively, while j = 0, 1, and 2 denotes geometries with plane, line, and point symmetry, respectively.

The mass velocity is given by the kinematic relation

$$u = \frac{\partial r}{\partial t}$$

The high-energy solid initiators are characterized by the Jones-Wilkins-Lee (JWL) (Lee et al. 1968, 1973) equation of state. Table 1 presents the JWL equation, its coefficients, and the gasdynamic data required to describe C-4 and tetryl solid explosives.

An equation of state relating specific internal energy to temperature and an equation of state expressing the perfect gas behavior of the combustible gas mixture are

Table 1 The JWL equation of state for C-4 and tetryl solid explosives:

$$P = A\left(1 - \frac{\omega}{R_1} \frac{\rho_o}{\rho}\right) e^{-(R_1 \rho_o/\rho)} + B\left(1 - \frac{\omega}{R_2} \frac{\rho_o}{\rho}\right) e^{-(R_2 \rho_o/\rho)} + \frac{\omega \rho_o}{\rho} E$$

Explosive	A, Mbar	B, Mbar	R_1	R_2	ω	ρ_o, gm/cc	E_o, Mbar-cc/cc	P_{CJ}, Mbar	D_{CJ}, km/s	Γ_{CJ}
C-4	6.0977	0.12950	4.50	1.40	0.25	1.601	0.09000	0.2798	8.193	2.8380
Tetryl	5.8680	0.10671	4.40	1.20	0.28	1.500	0.06405	0.2000	7.300	2.9968

INITIATION OF DETONATION IN LNG/AIR CLOUDS

given by

$$\text{(State)} \quad e = \sum_{k=1}^{n} x_k e_{0_k} + x_k \int_{T_0}^{T} c_{v_k} \, dT$$

$$\text{(State)} \quad pv = R_u T / \overline{M}$$

where

$$x_k = M_k v y_k$$

$$\overline{M} = 1 / \sum_{k=1}^{n} v y_k$$

where e_{0_k} is the specific internal energy of formation at the reference temperature T_0; R_u is the universal gas constant; $\overline{M}$ is the average molecular weight of the gas mixture; and c_{v_k}, x_k, and M_k are the k-th species specific heat at constant volume, mass fraction, and molecular weight, respectively.

The Numerical Model and Verification of the Reaction Mechanism

Our numerical model uses Westbrook's (1979) 75 step, 26 species reaction mechanism to simulate the reaction of CH_4/C_2H_6/air mixtures. Table 2 contains the detailed mechanism, the forward and reverse Arrhenius coefficients for each elementary step, and a list of the participating species. The preexponential factors of elementary steps 54 and 75 have been increased and decreased by factors of 4-1/2 and 2/3, respectively, over those reported by Westbrook (1979). The first modification is required to accurately model pure C_2H_6/air mixture induction times (Burcat et al. 1971). It, however, very slightly affects computed induction times for fuels containing less than 20% C_2H_6. The second modification (Westbrook, 1981b) also has a negligible effect. The enthalpy and specific heat at constant pressure of the species are modeled in the conventional manner (Gordon and McBride 1971), that is, by a pair of fourth-order polynomials in temperature covering the range

Table 2 Westbrook's methane-ethane oxidation mechanism

Reaction rates in cm^3-mole-s-cal, $k = AT^n \exp(-E_a/RT)$ [a]

	Reaction		Forward			Reverse		
			A	n	E_a	A	n	E_a
1.	O_2+M	$= O+O+M$	5.100E+15	0.00	1.150E+05	4.700E+15	-0.28	0.000E+00
2.	H_2+M	$= H+H+M$	2.200E+14	0.00	9.600E+04	3.000E+15	0.00	0.000E+00
3.	H_2+O_2	$= H+HO_2$	5.500E+13	0.00	5.780E+04	2.500E+13	0.00	7.000E+02
4.	$H+O_2$	$= O+OH$	2.200E+14	0.00	1.679E+04	1.740E+13	0.00	6.770E+02
5.	$H+O_2+M$	$= HO_2+M$	1.650E+15	0.00	-1.000E+03	2.310E+15	0.00	4.590E+04
6.	$H+HO_2$	$= OH+OH$	2.500E+14	0.00	1.900E+03	1.200E+13	0.00	4.010E+04
7.	$OH+M$	$= H+O+M$	8.000E+19	-1.00	1.037E+05	1.000E+16	0.00	0.000E+00
8.	H_2+O	$= H+OH$	1.800E+10	1.00	8.900E+03	8.300E+09	1.00	6.950E+03
9.	O_2+OH	$= HO_2+O$	6.420E+13	0.00	5.661E+04	5.000E+13	0.00	1.000E+03
10.	H_2+OH	$= H+H_2O$	2.200E+13	0.00	5.146E+03	9.500E+13	0.00	2.030E+04
11.	H_2+HO_2	$= H_2O_2+H$	7.300E+11	0.00	1.870E+04	1.700E+12	0.00	3.750E+03
12.	H_2O+M	$= H+OH+M$	2.200E+16	0.00	1.050E+05	1.400E+23	-2.00	0.000E+00
13.	H_2O+O_2	$= OH+HO_2$	6.330E+14	0.00	7.386E+04	5.000E+13	0.00	1.000E+03
14.	H_2O+O	$= OH+OH$	3.400E+13	0.00	1.835E+04	3.450E+12	0.00	1.101E+03
15.	H_2O+HO_2	$= H_2O_2+OH$	2.800E+13	0.00	3.279E+04	1.000E+13	0.00	1.800E+03
16.	H_2O_2+M	$= OH+OH+M$	1.200E+17	0.00	4.550E+04	9.100E+14	0.00	-5.070E+03

(Table continued next page)

Table 2 (cont.) Westbrook's methane-ethane oxidation mechanism.
Reaction rates in cm^3-mole-s-cal, $k = AT^n \exp(-E_a/RT)$ [a]

	Reaction		Forward			Reverse		
			A	n	E_a	A	n	E_a
17.	$H_2O_2 + O_2$	= $HO_2 + HO_2$	4.000E+13	0.00	4.264E+04	1.000E+13	0.00	1.000E+03
18.	$CO + OH$	= $CO_2 + H$	1.275E+07	1.30	-7.650E+02	1.430E+09	1.30	2.158E+04
19.	$CO + O + M$	= $CO_2 + M$	5.900E+15	0.00	4.100E+03	5.500E+21	-1.00	1.318E+05
20.	$CO + HO_2$	= $CO_2 + OH$	1.000E+14	0.00	2.300E+04	1.100E+15	0.00	8.484E+04
21.	$CO + O$	= $CO_2 + O$	3.140E+11	0.00	3.760E+04	2.780E+12	0.00	4.383E+04
22.	$CH_4 + M$	= $CH_3 + H + M$	1.400E+17	0.00	8.840E+04	2.840E+11	1.00	-1.951E+04
23.	$CH_4 + H$	= $CH_3 + H_2$	1.250E+14	0.00	1.190E+04	4.800E+12	0.00	1.143E+04
24.	$CH_4 + O$	= $CH_3 + OH$	1.600E+13	0.00	9.200E+03	2.672E+11	0.00	6.640E+03
25.	$CH_4 + OH$	= $CH_3 + H_2O$	3.470E+03	3.08	2.007E+03	5.750E+02	3.08	1.668E+04
26.	$CH_4 + HO_2$	= $CH_3 + H_2O_2$	2.000E+13	0.00	1.800E+04	1.050E+12	0.00	1.448E+03
27.	$CH_4 + O_2$	= $CH_3 + HO_2$	7.630E+13	0.00	5.859E+04	1.000E+12	0.00	4.000E+02
28.	$CH_2 + H$	= $CH + H_2$	2.700E+11	0.67	2.570E+04	1.897E+11	0.67	2.873E+04
29.	$CH_2 + O$	= $CH + OH$	1.900E+11	0.68	2.500E+04	5.863E+10	0.68	2.593E+04
30.	$CH_2 + OH$	= $CH + H_2O$	2.700E+11	0.67	2.570E+04	8.213E+11	0.67	4.388E+04
31.	$CH_3 + O$	= $CH_2O + H$	1.300E+14	0.00	2.000E+03	1.700E+15	0.00	7.163E+04
32.	$CH_3 + OH$	= $CH_2O + H_2$	4.000E+12	0.00	0.000E+00	1.200E+14	0.00	7.172E+04

(Table continued next page)

Table 2 (cont.) Westbrook's methane-ethane oxidation mechanism.

Reaction rates in cm^3-mole-s-cal, $k = AT^n \exp(-E_a/RT)^a$

	Reaction		Forward A	n	E_a	Reverse A	n	E_a
33.	CH_3+O_2	$= CH_3O+O$	2.400E+13	0.00	2.900E+04	1.520E+14	0.00	7.330E+02
34.	CH_3+HO_2	$= CH_3O+OH$	1.600E+13	0.00	0.000E+00	1.000E+00	0.00	0.000E+00
35.	CH_3O+M	$= CH_2O+H+M$	5.000E+13	0.00	2.100E+04	9.910E+08	1.00	-2.563E+03
36.	CH_3O+O_2	$= CH_2O+HO_2$	1.000E+12	0.00	6.000E+03	1.280E+11	0.00	3.217E+04
37.	CH_2O+M	$= HCO+H+M$	5.000E+16	0.00	7.200E+04	2.120E+11	1.00	-2.077E+04
38.	CH_2O+H	$= HCO+H_2$	3.981E+12	0.00	3.760E+03	3.155E+11	0.00	1.843E+04
39.	CH_2O+O	$= HCO+OH$	5.000E+13	0.00	4.600E+03	1.750E+12	0.00	1.717E+04
40.	CH_2O+OH	$= HCO+H_2O$	5.400E+14	0.00	6.300E+03	1.870E+14	0.00	3.612E+04
41.	CH_2O+HO_2	$= HCO+H_2O_2$	1.000E+12	0.00	8.000E+03	1.090E+11	0.00	6.593E+03
42.	CH_2O+O_2	$= HCO+HO_2$	3.660E+15	0.00	4.604E+04	1.000E+14	0.00	3.000E+03
43.	CH_2O+CH_3	$= HCO+CH_4$	1.000E+10	0.50	6.000E+03	2.090E+10	0.50	2.114E+04
44.	$HCO+M$	$= H+CO+M$	1.450E+14	0.00	1.900E+04	5.050E+11	1.00	1.553E+03
45.	$HCO+H$	$= CO+H_2$	2.000E+14	0.00	0.000E+00	1.310E+15	0.00	9.000E+04
46.	$HCO+O$	$= CO+OH$	1.000E+14	0.00	0.000E+00	2.880E+14	0.00	8.790E+04
47.	$HCO+OH$	$= CO+H_2O$	1.000E+14	0.00	0.000E+00	2.800E+15	0.00	1.051E+05
48.	$HCO+O_2$	$= CO+HO_2$	3.300E+12	0.00	7.000E+03	7.400E+12	0.00	3.929E+04

(Table continued next page)

INITIATION OF DETONATION IN LNG/AIR CLOUDS 273

Table 2 (cont.) Westbrook's methane-ethane oxidation mechanism.

Reaction rates in cm^3-mole-s-cal, $k = AT^n exp(-E_a/RT)$ [a]

	Reaction		Forward			Reverse		
			A	n	E_a	A	n	E_a
49.	$HCO+CH_3$	$= CO+CH_4$	3.000E+11	0.50	0.000E+00	5.140E+13	0.50	9.047E+04
50.	$CH+O_2$	$= CO+OH$	1.350E+11	0.67	2.570E+04	5.187E+11	0.67	1.856E+05
51.	$CH+O_2$	$= HCO+O$	1.000E+13	0.00	0.000E+00	1.334E+13	0.00	7.195E+04
52.	CH_2+O_2	$= HCO+OH$	1.000E+14	0.00	3.700E+03	4.117E+13	0.00	7.658E+04
53.	C_2H_6	$= CH_3+CH_3$	2.500E+19	-1.00	8.831E+04	1.000E+13	0.00	0.000E+00
54.	C_2H_6+H	$= C_2H_5+H_2$	2.417E+03	3.50	5.200E+03	4.374E+03	3.50	2.730E+04
55.	C_2H_6+O	$= C_2H_5+OH$	2.500E+13	0.00	6.360E+03	4.600E+12	0.00	1.123E+04
56.	C_2H_6+OH	$= C_2H_5+H_2O$	6.720E+13	0.00	2.447E+03	1.200E+14	0.00	2.457E+04
57.	$C_2H_6+CH_3$	$= C_2H_5+CH_4$	5.500E-01	4.00	8.280E+03	3.000E+10	0.00	1.250E+04
58.	C_2H_5	$= C_2H_4+H$	3.800E+13	0.00	3.800E+04	7.800E+08	1.00	-3.000E+03
59.	$C_2H_5+O_2$	$= C_2H_4+HO_2$	1.000E+12	0.00	5.000E+03	1.330E+11	0.00	1.370E+04
60.	$C_2H_5+C_2H_3$	$= C_2H_4+C_2H_4$	3.244E+17	0.00	3.557E+04	1.000E+18	0.00	1.029E+05
61.	C_2H_4+M	$= C_2H_3+H+M$	3.800E+17	0.00	9.816E+04	2.337E+12	1.00	-1.180E+04
62.	C_2H_4+H	$= C_2H_3+H_2$	6.300E+13	0.00	6.000E+03	7.900E+12	0.00	7.400E+03
63.	C_2H_4+O	$= CH_3+HCO$	1.000E+13	0.00	1.130E+03	4.770E+11	0.00	3.118E+04
64.	C_2H_4+O	$= CH_2O+CH_2$	2.500E+13	0.00	5.000E+03	3.016E+12	0.00	1.568E+04

(Table continued next page)

Table 2 (cont.) Westbrook's methane-ethane oxidation mechanism.
Reaction rates in cm^3-mole-s-cal, $k = AT^n \exp(-E_a/RT)^a$

	Reaction		A	Forward n	E_a	A	Reverse n	E_a
65.	C_2H_4+OH	$= C_2H_3+H_2O$	1.000E+14	0.00	3.500E+03	5.022E+13	0.00	1.776E+04
66.	C_2H_3+M	$= C_2H_2+H+M$	3.000E+16	0.00	4.050E+04	4.600E+12	1.00	-1.360E+03
67.	C_2H_2+M	$= C_2H+H+M$	1.000E+14	0.00	1.140E+05	1.108E+09	1.00	7.670E+02
68.	C_2H_2+H	$= C_2H+H_2$	2.000E+14	0.00	1.900E+04	4.179E+13	0.00	1.321E+04
69.	C_2H_2+O	$= C_2H+OH$	3.200E+15	-0.60	1.700E+04	2.937E+14	-0.60	9.112E+02
70.	C_2H_2+O	$= CH_2+CO$	6.700E+13	0.00	4.000E+03	1.260E+13	0.00	5.467E+04
71.	C_2H_2+OH	$= C_2H+H_2O$	6.000E+12	0.00	7.000E+03	5.428E+12	0.00	1.636E+04
72.	$C_2H_2+O_2$	$= HCO+HCO$	4.000E+12	0.00	2.800E+04	1.000E+11	0.00	6.365E+04
73.	C_2H+O	$= CH+CO$	5.000E+13	0.00	0.000E+00	3.160E+13	0.00	5.943E+04
74.	C_2H+O_2	$= HCO+CO$	1.000E+13	0.00	7.000E+03	8.444E+12	0.00	1.384E+05
75.	$CH3OH+M$	$= CH_3+OH+M$	2.000E+18	0.00	8.000E+04	9.660E+12	1.00	-1.098E+04

[a]Species: A_r, CH, CH_2, CH_2O, CH_3, CH_3O, CH_3OH, CH_4, CO, CO_2, C_2H, C_2H_2, C_2H_3, C_2H_4, C_2H_5, C_2H_6, H, HCO, HO_2, H_2, H_2O, H_2O_2, N_2, M, O, OH, O_2.

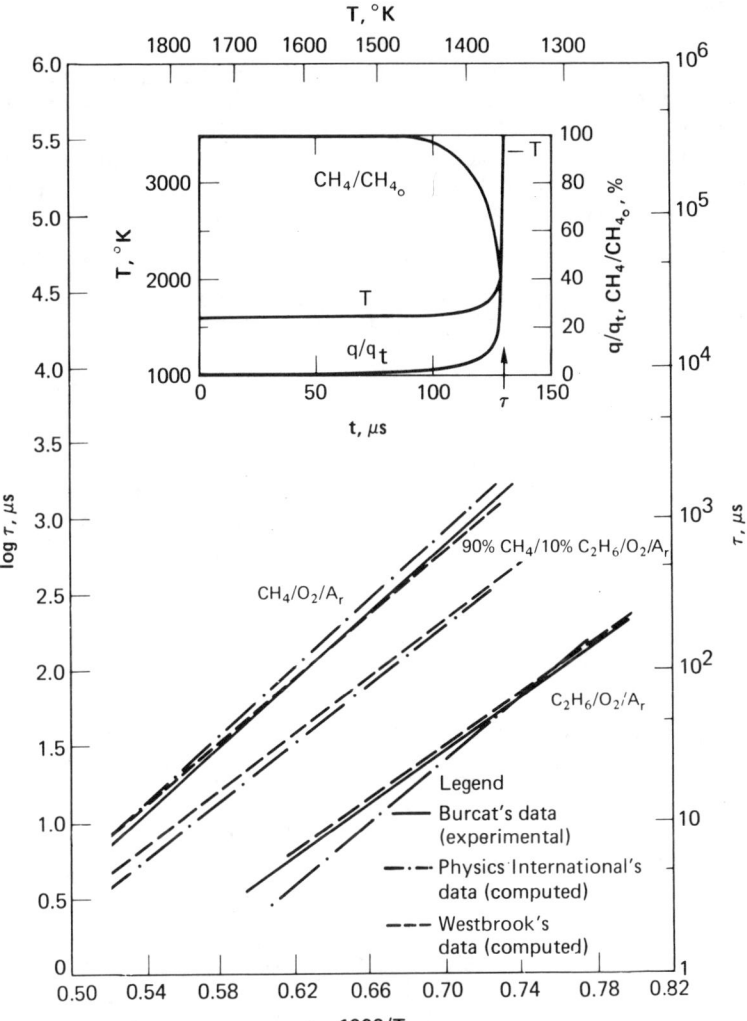

Fig. 2 Experimental and computed induction times of stoichiometric $CH_4/C_2H_6/O_2/A_r$ mixtures under-going an adiabatic constant volume chemical reaction agree closely over the experimental range of initial mixture temperatures. The insert depicts details of the $CH_4/O_2/A_r$ reaction.

300<T<5000 K. The numerical model uses Hindmarsh's (1972a, 1972b, 1974) stiff ordinary differential equation (ODE) integration technique, based on the work of Gear (1971), to integrate the system of stiff ODEs derived from the chemical mechanism.

The purpose of the first simulation is to verify the accuracy of the LNG/air reaction mechanism. Its accuracy is determined by comparing the inverse temperature dependence of the induction delay time τ computed using the reaction mechanism with that calculated by Westbrook and Haselman (1979) and that experimentally determined by Burcat et al. (1971). The calculations were carried out in stoichiometric $CH_4/C_2H_6/O_2/A_r$ mixtures in which the partial volume of argon was five times that of the oxygen and the mixture density was 3.4078×10^{-3} gm/cc [Westbrook and Haselman (1979) used a density of 2.75×10^{-3} gm/cc]. The insert in Fig. 2 shows

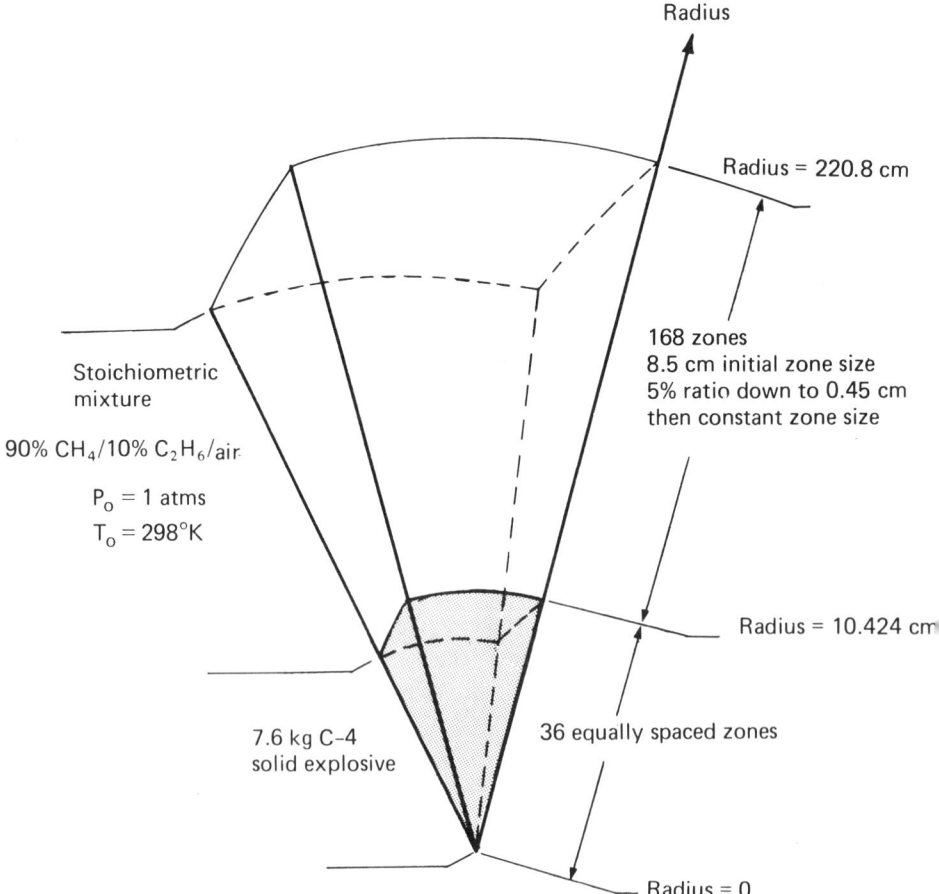

Fig. 3 Schematic of the initial flowfield used in the simulation of direct initiation of detonation of a stoichiometric 90% CH_4/10% C_2H_6/air mixture by 7.6 kg of spherical solid explosive.

the detailed results of a calculation for pure $CH_4/O_2/A_r$ at an initial temperature of 1600 K. The induction delay time was defined to be the time at which the CO times the O concentration reaches a maximum.

Figure 2 compares the computed induction delay time of our reaction mechanism (chain dotted line) for three stoichiometric gas mixtures, 100% $CH_4/O_2/A_r$, 90% CH_4/10% $C_2H_6/O_2/A_r$, and 100% $C_2H_6/O_2/A_r$, with that calculated by Westbrook and Haselman (1979) (dashed line) and that experimentally determined by Burcat et al. (1971) (solid line). All three results agree closely over the experimental range of initial mixture temperatures, though our calculated results have a slightly higher overall activation energy than do the others. The complete reaction mechanism allows us to accurately model the dependence of τ on 1/T as well as the dependence of τ on the ethane content of the fuel. This latter dependence is especially significant given the range of ethane content of LNG (<30%). For example (Westbrook and Haselman 1979), a 7% addition of ethane to a pure CH_4 fuel at 1500 K decreases τ from 270 to 135 s, a factor of 2, while an increase in the ethane content of the fuel from 30 to 100% is required to achieve a similar factor of 2. The ethane content insignificantly affects the CJ detonation properties and the heat released.

Wilkins' (1969) first-order accurate Lagrangian finite-difference scheme, as implemented in the CLOUD CODE by Cohen et al (1975), is used to integrate the governing time-dependent gasdynamic equations. Explicit artificial viscosity is used to stably compute the motion and/or development of shock fronts within the flowfield. Artificial viscosity smears shock fronts from over 3 to 5 Langrangian cells.

Program IGNITION uses the numerical technique of operator splitting (Yanenko 1971) to simultaneously compute the effects due to gasdynamics and chemical reaction. The time step used to advance the calculations is chosen to be the minimum between the maximum allowable gasdynamic time step (approximately 2/3 of the Courant time step) and an estimated overall chemical time step. The chemical time step is estimated such that temperature change in each gaseous cell resulting from chemical reaction during the time step does not exceed 3% of the temperature in the cell at the beginning of the timestep, up to a maximum of 75 K.

Figure 3 presents a detailed schematic of the second numerical simulation: an unconfined spherical homogeneous stoichiometric 90% CH_4/10% C_2H_6/air mixture surrounds a centered 7.6 kg spherically symmetric C-4 solid explosive (shaded region). The solid explosive of radius 10.424 cm is divided into 36 equally spaced cells. The gas cell adjacent to the solid explosive has an initial 8.5 cm zone size. The gas cells extending from the solid initiator decrease in size at a 5% rate until a 0.45 cm zone size is attained. Thereafter the zone size remains constant. The first 220.8 cm of the flowfield requires 198 zones (see Table 3). The size of the gas cells adjacent to the solid igniter is quite adequate for modeling the impulse transmitted to the gas by the detonation of the solid explosive and for modeling the highly overdriven detonation wave transmitted from the solid explosive. During the simulation, the gas cell adjacent to the solid explosive is compressed by a factor of 18. Smaller cell sizes are used well before that point in the flowfield at which the energy released by the solid explosive equals the heat of combustion released by the gas mixture. These cells are then further compressed by the lead shock to approximately 0.1 cm, a size much smaller than the length of the induction zone.

In the course of carrying out the simulation, the flowfield was rezoned to increase the maximum allowable gasdynamic time step. Rezoning blends pairs of highly compressed cells into a single, larger, more manageable cell. Cell rezoning was restricted to the solid explosive and the gas cell adjacent to the solid igniter-gas interface. Rezoning was only carried out after the detonation front was 12.6 charge radii from the source. Table 3 contains the details of the rezoning.

Throughout the simulation, chemical kinetic calculations were restricted to those cells whose temperature was between 800 and 6000 K. At the time of the rezoning, there existed an extended region of combusted gas behind the reaction front that was quite uniform in species concentration, temperature, and particle velocity. Given this distribution, the chemical kinetic calculations were then further restricted to the 12 cells behind the reaction front. The reaction front was defined as that location at which the temperature first exceeds 2250 K. A description of the flowfield when this modification was implemented is also contained in Table 3.

Results for Simulation No. 2

Simulation No. 2 predicts that 7.6 kg of solid explosive C-4 is insufficient to cause direct initiation of detonation in a 90% CH_4/10% C_2H_6/air mixture initially at

Table 3 Details of the initial zoning and rezoning of the flowfield and flowfield conditions supporting the restriction of chemical kinetics calculation near the reaction front

Explosive				Initial Zoning		Combustible mixture		
Type	Weight kg	Radius cm	No. of zones	Radius cm	No. of zones	First cell zone size cm	Cell size reduction %	Constant zone size cm
C-4	7.596	10.424	36	220.80	162	8.5	5	0.45

Rezone

Time s	Detonation front position cm	Detonation front position charge radii	No. of zones rezoned	Rezoned cell numbers
433.9	142.0	13.62	5	29,31,33,35,37

Chemical Kinetics Restriction

Solid-gas interface		Reaction front						Temperature		Particle velocity		Maximum species deviation,a %	No. of cells behind reaction front b
Position, cm	Cell no.	Position, cm	Cell no.	No. of cells	Length, cm	End position, cm	End cell no.	Average, K	Maximum deviation, %	Average, km/s	Maximum deviation, %		
117.7	36	142.0	64	11	10.0	139.0	60	3053	0.8	1.283	3.0	18.6	12

a For species concentration $\geq 10^{-8}$ moles/g-mix. b Chemical reaction restricted beyond this location.

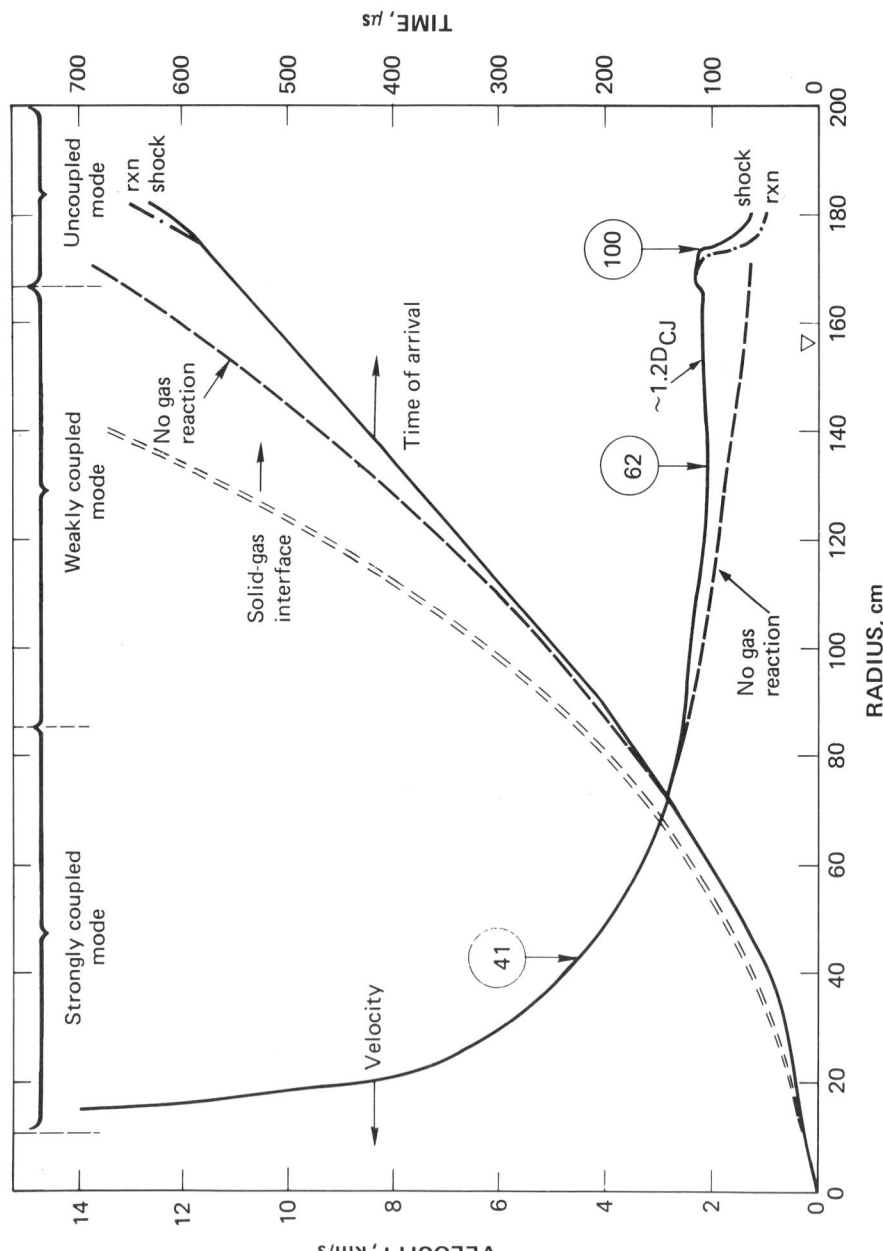

Fig. 4 Shock and reaction front trajectories and velocity profiles illustrate the three distinct modes of propagation of the shock-reaction front complex.

298 K and atmospheric pressure. The simulation also predicts that during the latter stages of propagation the reaction front becomes significantly detached from the shock front. Evidence of this event is depicted by the separation of the shock (solid line) and reaction (chain dotted line) front trajectories, and their front velocity profiles presented in Fig. 4.

Analysis of the computational output reveals that the shock-reaction front complex undergoes three distinct modes of propagation during the course of its evolution (see Fig. 4). The first mode, the strongly coupled mode, corresponds to the highly overdriven detonation state which is achieved when the source generated 7.3-km/s blast wave first begins to propagate into the gas mixture. This mode is characterized by a very strong shock front ($M_n > 8$) and postshock gas temperatures on the order of 2800 K. These conditions result in postshock induction delay times on the order of 0.1 μs (Fig. 7 in Westbrook and Haselman 1979).

Figure 5a presents the concentration histories of the fuel gases (CH_4, C_2H_6), oxygen (O_2), major product species (H_2O, CO, CO_2), selected radicals (H, OH, O, CH_3), and hydrogen (H_2) within Lagrange cell 41 (see Fig. 4). Cell 41, located 43.1 cm ($\approx$ 4 charge radii) from the source center, undergoes a process characteristic of the strongly coupled mode. The consumption of the fuel and formation of the product species occur very rapidly, because radical generation is instantaneous. Figure 5b presents the pressure and temperature histories of cell 41. The rapid heat release due to combustion manifests itself as a distinct increase in the thermodynamic properties. Figures 6a and 6b present the species and pressure and temperature profiles at 125.6 μs after source ignition. At this time the detonation wave is highly overdriven. Its propagation is predominantly determined by the detonation of the solid explosive and not by the gaseous combustion process. Complete reaction occurs well within a single cell. In the strongly coupled mode of propagation, the reaction front is closely coupled to the motion of the lead shock front.

As the overdriven detonation front expands, it continues to decay. Figure 4 indicates that once the detonation front decays to a velocity of approximately 1.2 D_{CJ}, it maintains this velocity for an extended period of time ($\approx$ 80 μs). This is the weakly coupled mode of propagation. Figures 7a and 7b present the species and pressure and temperature histories, respectively, of Lagrange cell 62 (see Fig. 4). Cell 62 was initially

282   J. KURYLO, J.M. THOMSEN, AND F.M. SAUER

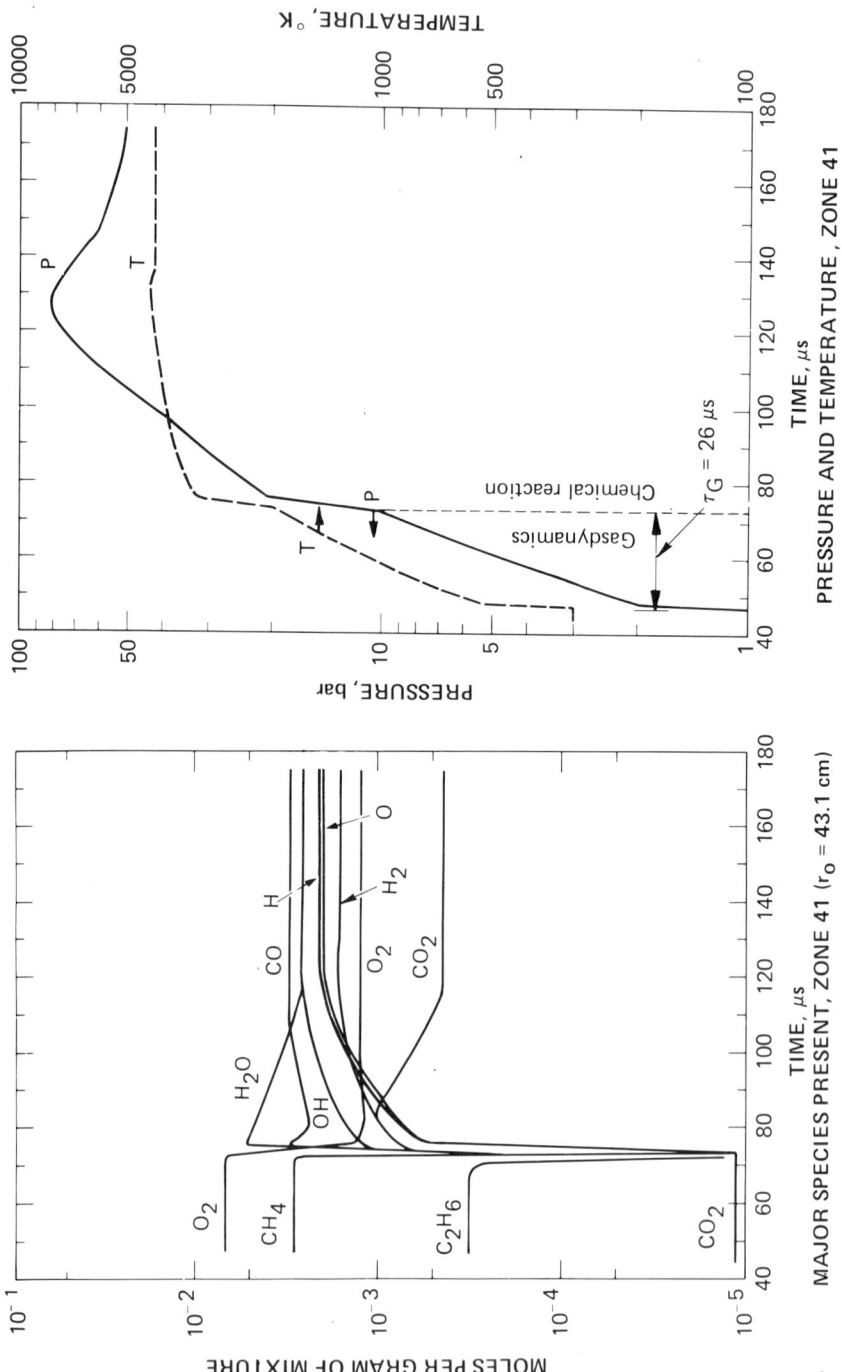

Fig. 5 Strongly coupled region: rapid fuel consumption at the leading edge of the shock – zone 41 (r_o = 43.2 cm). a) Major species concentration histories, b) pressure and temperature histories.

INITIATION OF DETONATION IN LNG/AIR CLOUDS 283

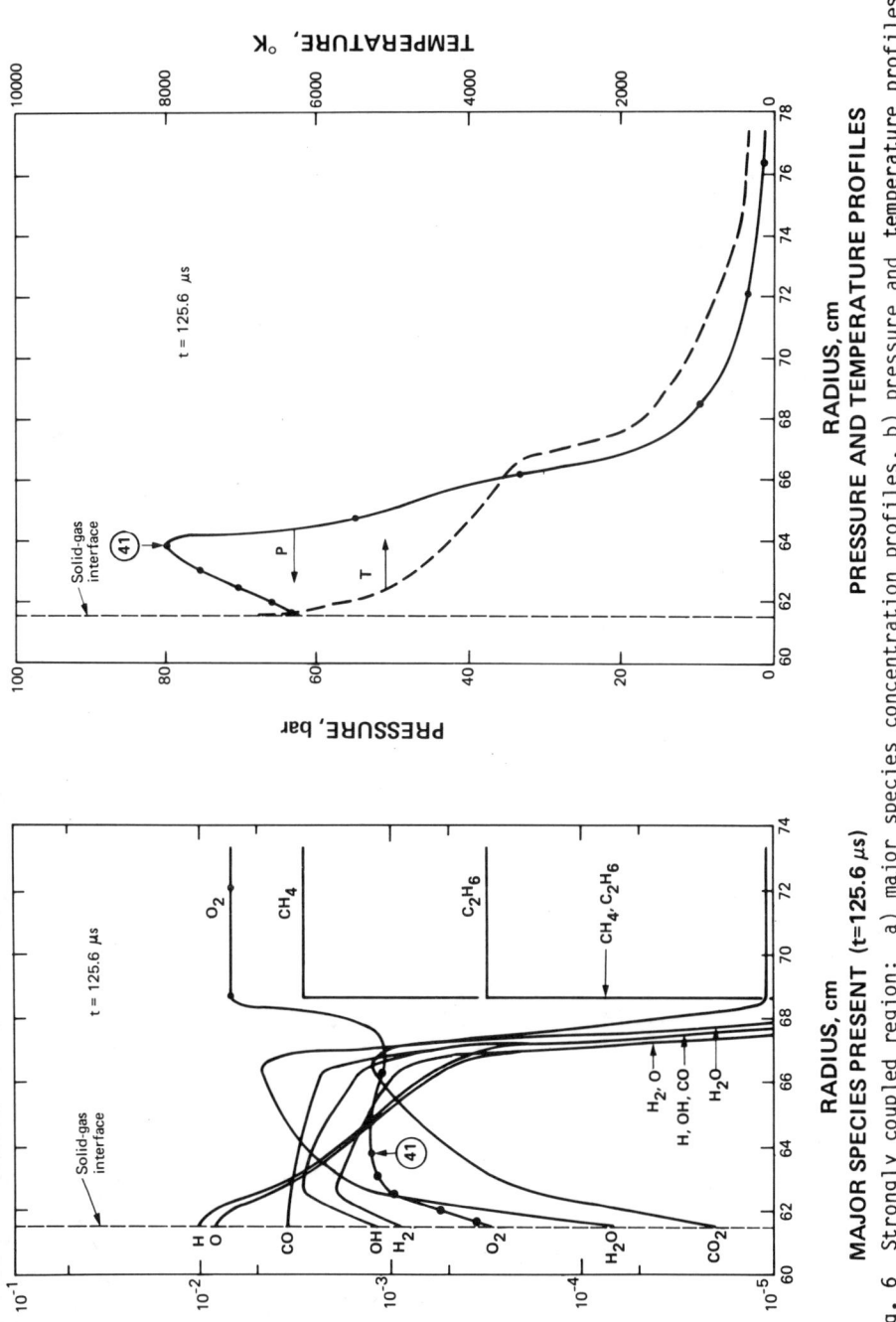

Fig. 6 Strongly coupled region: a) major species concentration profiles, b) pressure and temperature profiles at 125.6 μs after ignition (o denotes cell center).

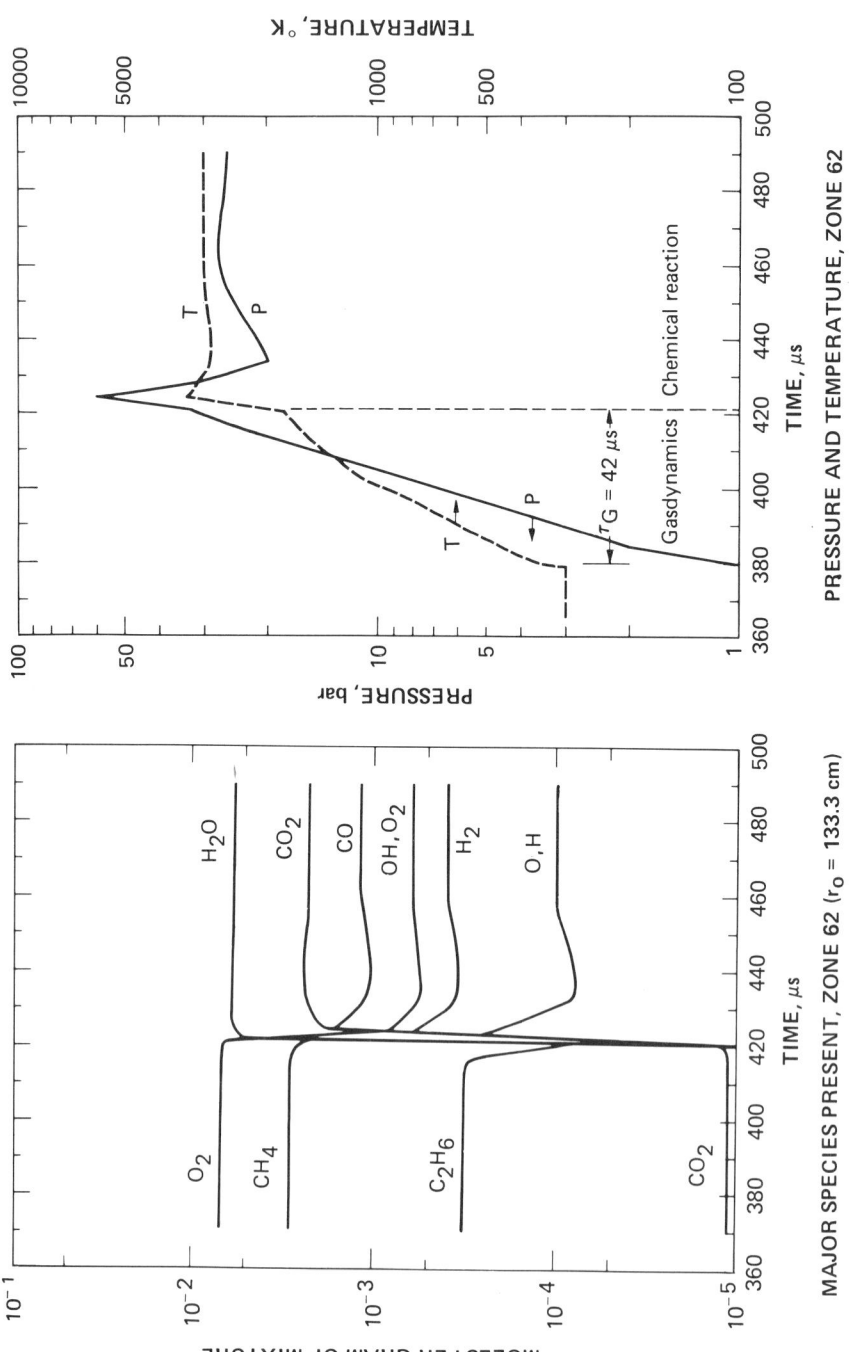

Fig. 7 Weakly coupled region: fast fuel consumption at, or slightly behind, the maximum shock pressure - zone 62 (r_o = 133.3 cm). a) Major species concentration histories, b) pressure and temperature histories.

located 133.3 cm ($\approx$ 13 charge radii) from the source center. During this mode of propagation, the postshock temperatures are reduced to approximately 1900 K. Induction delay times therefore increase by as much as a factor of 25 over that characteristic of the strongly coupled mode, causing a definite curvature to the C_2H_6 and CH_4 concentration histories.

Figures 8a and 8b display the species and pressure and temperature profiles at 466.2 μs. They show that complete combustion occurs at or one to two cells downstream of the lead shock front. Therefore, in this mode of propagation, the shock front is followed closely by the reaction front. The length of the region between the shock and reaction fronts (the induction zone thickness) is determined by product of the induction time and the shock velocity relative to the postshock particle speed. If a further decrease in the shock strength should occur, the induction time would sharply increase causing a similar increase in the induction zone thickness. The effects of artificial viscosity on the shock-reaction coupling is minimal. Because chemical reaction is significantly only at temperatures above 1500 K, reaction occurs only in the last cell of the smeared shock front, that is, at the back surface of the shock.

Throughout this mode of propagation, combustion within individual cells occurs explosively, that is, the cell experiences a very rapid and large pressure increase similar to that achieved by combustion under constant volume, constant internal energy, adiabatic conditions. The large pressure spike followed by the strong rarefaction wave recorded in Fig. 8b dramatically illustrates the explosive nature of the combustion process. The cyclic generation of pressure peaks, generated as successive cells combust explosively, sustains the approximately constant velocity of propagation of the detonation. The explosive combustion phenomenon is a physical reality. Cells explode because of the very large differences between the rates at which the heat of combustion is released at the end of the induction period (see the insert in Fig. 2) and the characteristic rate at which that energy can be carried away by gasdynamic processes. The effects displayed in Fig. 8b are for finite size cells. If an infinitesimally fine mesh had instead been used in this simulation, the numerous cells forming the front of the detonation would have had induction times (time remaining before the release of the heat of combustion) that continuously spanned the range from zero at the reaction front up to the postshock induction time. As the detonation

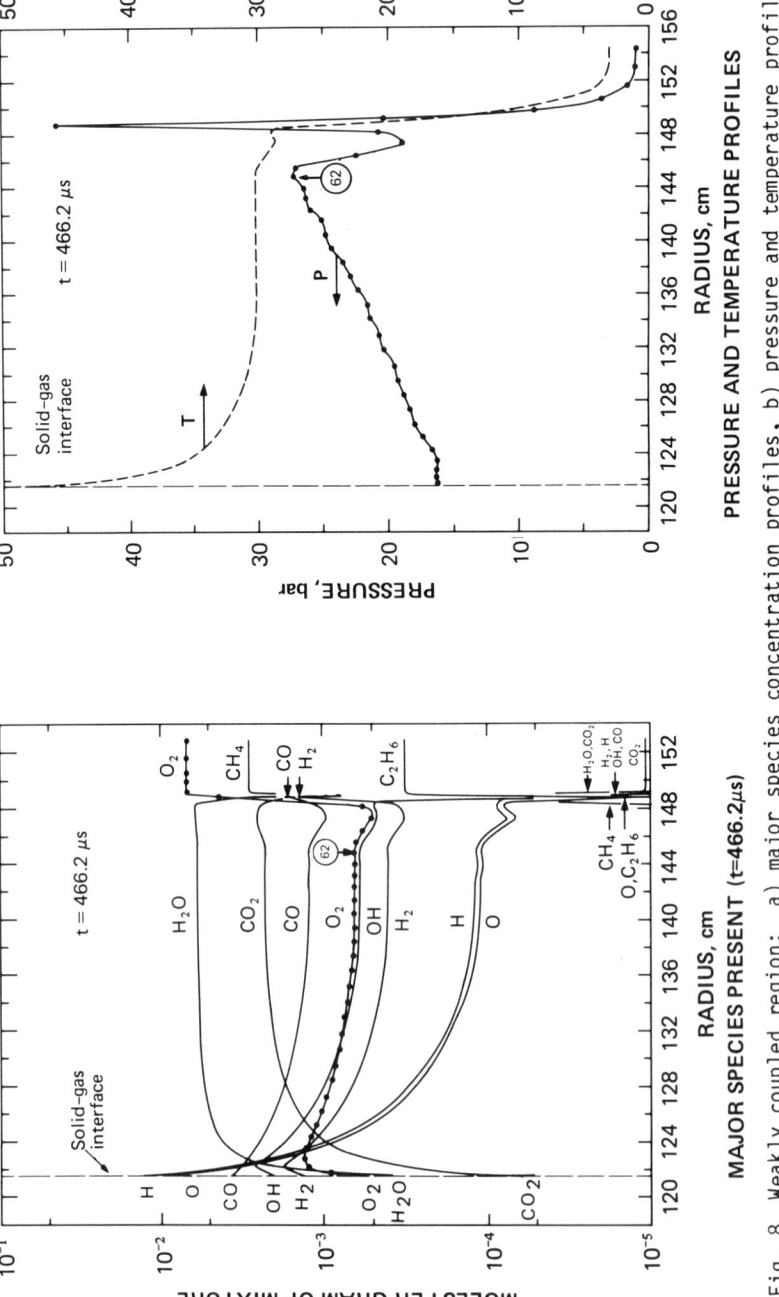

Fig. 8 Weakly coupled region: a) major species concentration profiles, b) pressure and temperature profiles at 466.2 μs after ignition (o deontes cell center).

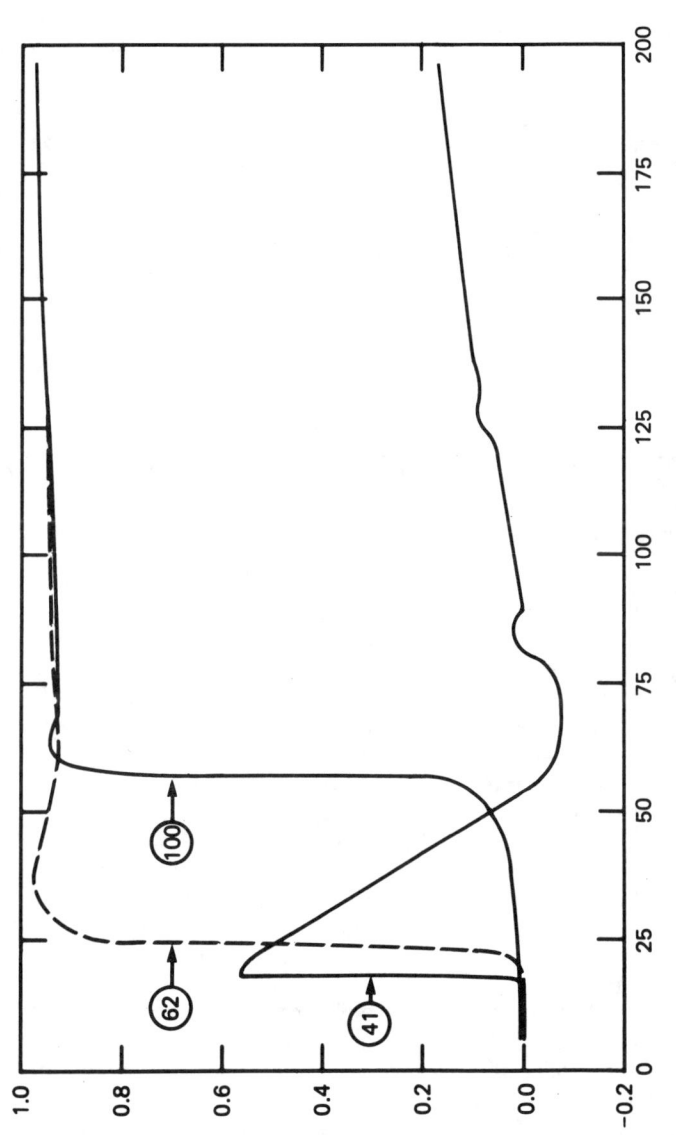

Fig. 9 Postshock heat release histories of zones 41, 62, and 100 expressed as a fraction of the chemical heat released at the Chapman-Jouget state, $q_{CJ} = -553.6$ cal/g ($\tau = 0$ when $T = 900$ K; $t_{41} = 56.9$, $t_{62} = 399.7$, $t_{100} = 579.0$ μs).

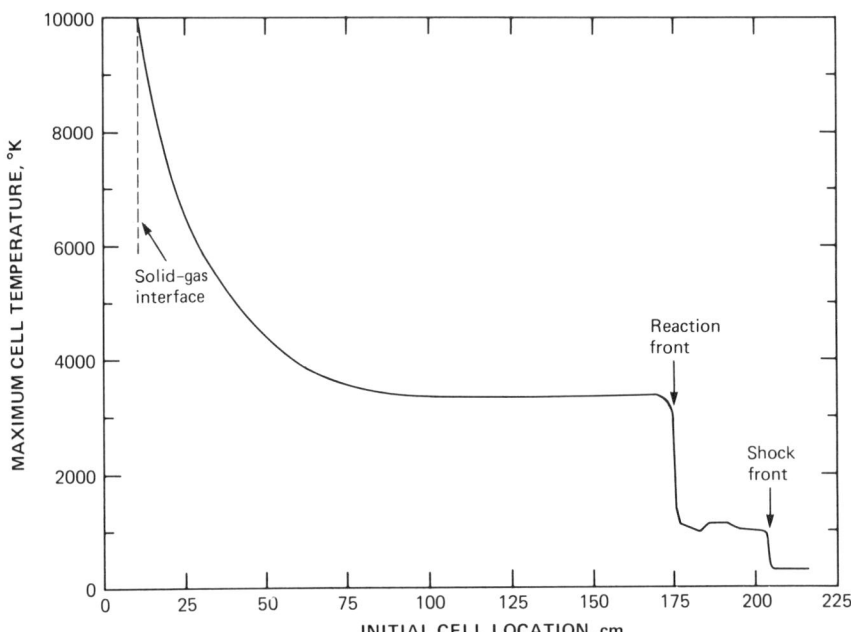

Fig. 10 Maximum cell temperature vs initial cell center location (t = 800.4 μs).

front progressed, there would have been a continuous succession of cells which combusted explosively, thereby maintaining the peak detonation pressure.

The inverted triangle in Fig. 4, located 156.7 cm (≈ 15 charge radii) from the source center, identifies the point in the initial flowfield at which the energy contained within the source theoretically equals the heat of combustion released by the gas mixture. This location approximately corresponds to the extent of the weakly coupled mode. The shock front trajectory and velocity profile, denoted by the single dashed lines, represent the results of a calculation carried out under the same set of initial conditions as in the present calculation, but without chemical reaction. The first noticeable departure of the "reaction" and "no reaction" front trajectories and velocity profiles roughly corresponds to the end of the strongly coupled mode of propagation and the beginning of the weakly coupled mode. As the energy released by the combusted gas approaches that of the source energy, a greater and greater deviation of the front trajectories should occur. However, the computed deviation appears to be much greater than expected because of the early leveling off of the detonation front velocity.

Analysis of the calculation indicates that the heat release history of individual cells may be a mechanism that decreases the rate of decay of the overdriven detonation and aids in maintaining the approximately constant propagation velocity of the detonation. Figure 9 displays the postshock heat release histories of Lagrange zones 41, 62, and 100. The heat released is expressed as a fraction of the chemical heat released at the Chapman-Jouguet state, where q_{CJ} = -553.6 cal/g (Gordon and McBride 1971). Overall, very little net heat is released by the gas cells detonated during the strongly coupled mode. It is only during the mid- to latter stages of the weakly coupled mode that these cells release a significant amount of the theoretical heat of combustion. Very little heat is released because the heat of combustion, which is initially released, quickly gets chemically reabsorbed by the disassociation of the major product species into H and O atoms and OH and CO molecules. The cells detonated during the weakly coupled and uncoupled modes release heat in excess of 95% q_{CJ}.

Figure 10 is a plot of the maximum temperature attained in each Lagrange cell vs the initial location of the cell. As the very high temperatures achieved in the zones near the solid initiator gas interface slowly decay, the highly dissociated products of combustion begin to recombine. The heat of combustion which is then rereleased contributes to the motion of the detonation front.

The uncoupled mode of propagation begins when the shock front first starts to separate from the reaction front. Shortly after separation, the postshock temperature is on the order of 1500 K. Though this temperature is not significantly different from the temperature characteristic of the weakly coupled mode, this decrease in temperature causes a significant increase in induction delay time and induction zone thickness (see Fig. 2). Once the shock front separates, it quickly decelerates to a velocity of approximately 0.8 D_{CJ}, with a concomitant drop in the postshock temperature to approximately 1100 K. At this temperature, induction times are on the order of 3000 μs, which effectively rules out the reestablishment of detonation by chemical induction-time-related processes. The separation process is clearly portrayed in Fig. 4 by the trajectories and velocity curves of the fronts.

Figure 4 shows that by the time the shock-reaction front complex reaches Lagrange cell 100, located 174.0 cm ($\approx$ 17 source radii) from the source center, the shock and reaction fronts have clearly separated. The species

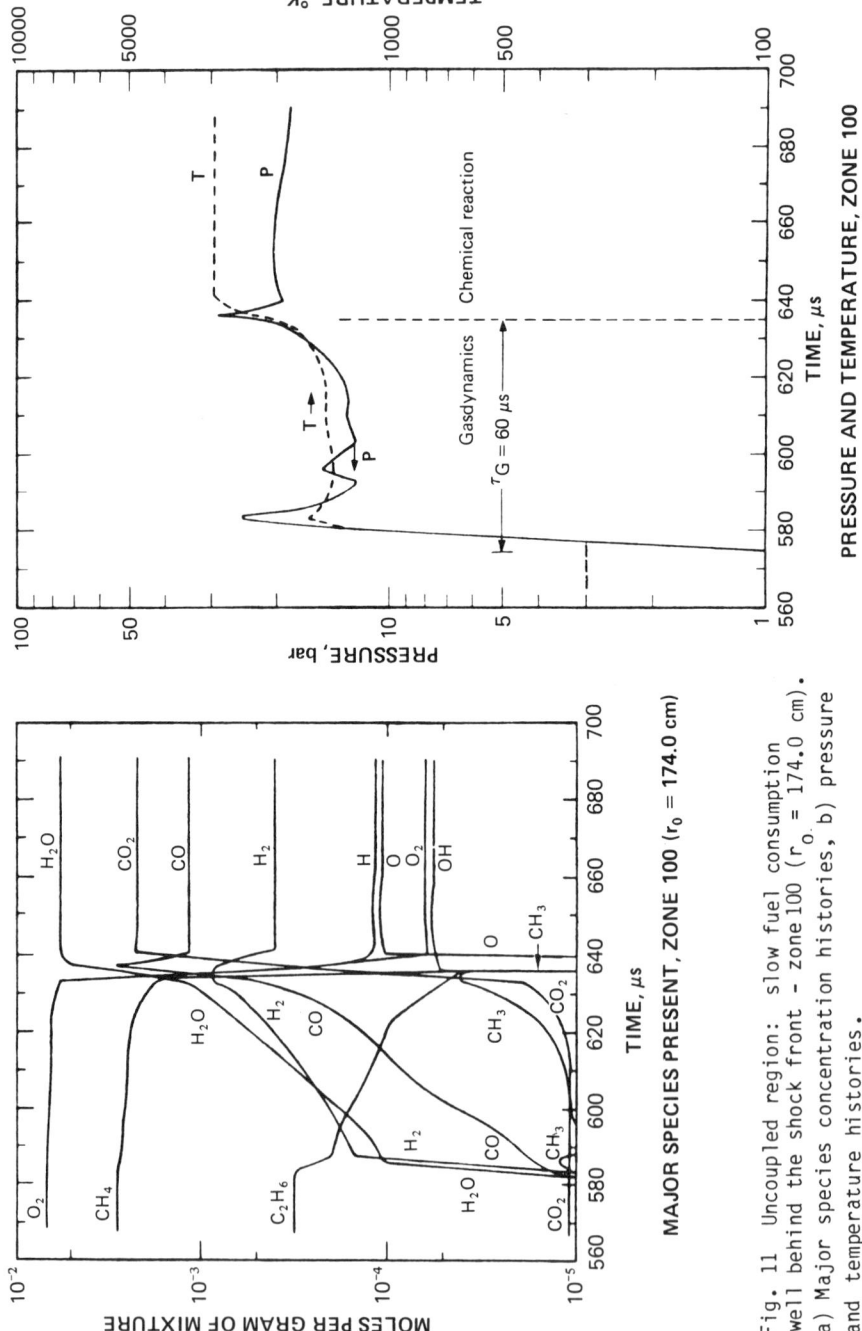

Fig. 11 Uncoupled region: slow fuel consumption well behind the shock front - zone 100 ($r_0 = 174.0$ cm). a) Major species concentration histories, b) pressure and temperature histories.

histories of cell 100 depicted in Fig. 11a reflect the uncoupled nature of the shock reaction front. Upon passage of the lead front, H_2O, CO, and H_2 are immediately formed, while significant quantities of C_2H_6 begin to disassociate into CH_3. This process agrees with Westbrook's (1979) postulated mechanism for the sensitization of CH_4/air mixtures by C_2H_6. The profiles depicting the consumption of fuel and production of major product species and radicals have a noticeable curvature to them. Figure 11b presents the pressure and temperature histories. The first major peak in the pressure curve represents the passage of the lead shock front. The second major peak represents combustion of the gas some 60 μs later. In between these peaks is a record of the passage of a forward propagating shock front produced by the explosion of a cell. The temperature and pressure histories are quite similar. Figures 12a and 12b display the species and pressure and temperature profiles at 749.2 μs after ignition. It can be seen that combustion occurs rapidly at the reaction front. Figure 12b indicates that an elevated temperature plateau lies within the reaction zone. Artificial viscosity has no effect on the induction zone thickness. Throughout this mode of propagation, the distribution of cells from the lead shock to the reaction front and beyond was very fine, so fine that it appears as a solid line in Figs. 12a and 12b.

Figures 13a-13d display the pressure, temperature, and density profiles at four instants during the uncoupled mode of propagation. These profiles present the details of the intensification and subsequent decay of an extended elevated temperature plateau located within the induction zone.

Figure 13a shows that at 632.9 μs after detonation initiation there already exists a short ($\ell_p \cong 0.25$ cm) elevated temperature (T_p = 920 K) plateau behind the lead shock front. Figure 13b shows that by 690.3 μs, the plateau has been extended to 0.4 cm and the temperature elevated to 1105 K. This is an increase in plateau temperature of 185 K despite the spherical expansion of its constituent cells. The mechanism responsible for these changes is also depicted in Fig. 13b. Cells which combust at the reaction front do so explosively, but far less rapidly and intensively than in the uncoupled mode. Each explosion sends out a forward and backward propagating pressure pulse. The forward pulse interacts with the induction zone, reintensifying the elevated temperature portion, and interacts with the lead

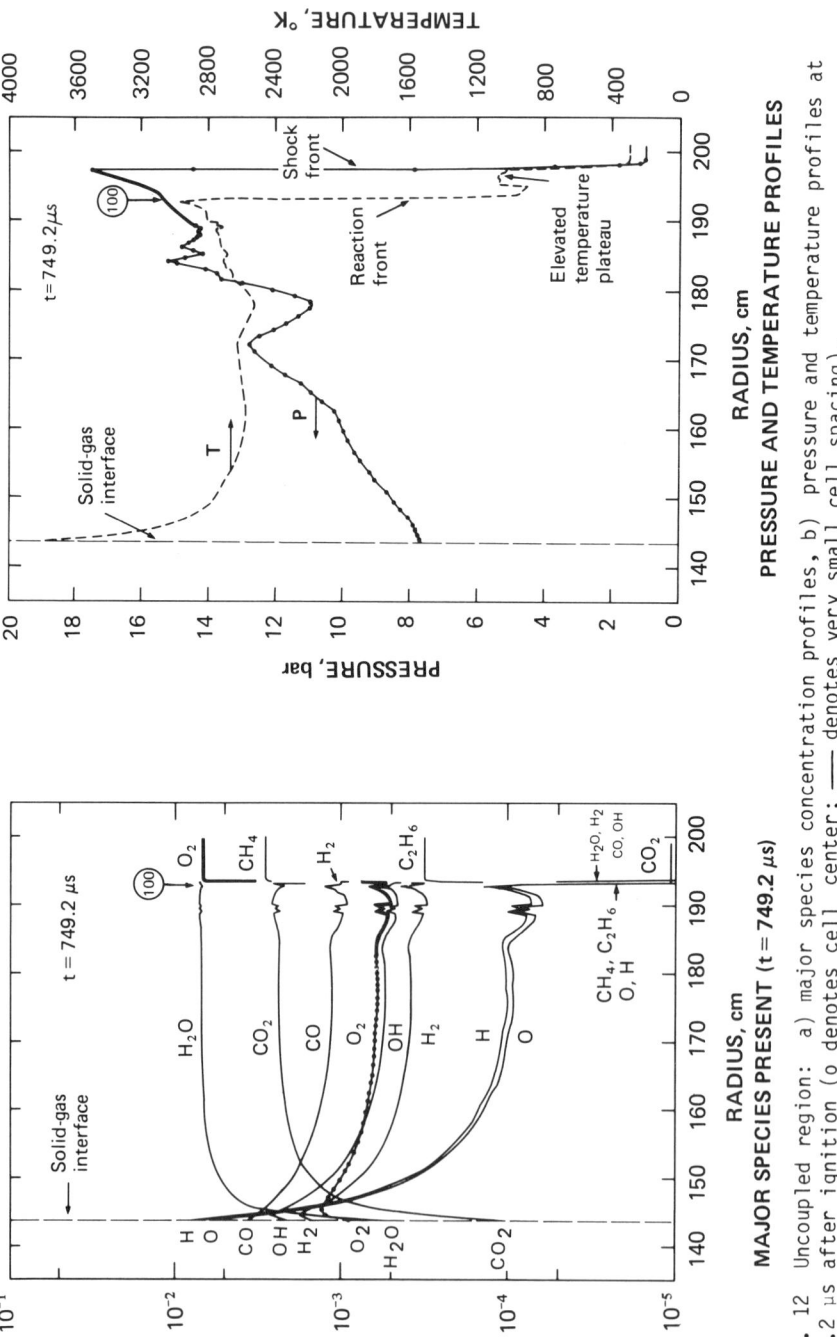

Fig. 12 Uncoupled region: a) major species concentration profiles, b) pressure and temperature profiles at 749.2 μs after ignition (o denotes cell center; ——— denotes very small cell spacing).

shock front, thereby extending the length of the plateau. Cell 96 is shown to be on the verge of interacting with the lead shock front. Figure 13c shows that by 743.1 μs the plateau has been extended to 2.0 cm. However, the intensified shock front has not been supported during the last 50 μs, and therefore has begun to decay (T_p = 1070 K). Figure 13d shows that at 800.4 μs, the temperature of the plateau ($\ell_p \cong 3.5$ cm) has further decayed to 1010 K. This temperature is sufficiently low to cause the complete disintegration of the shock-reaction complex and thus failure of the 7.6 kg charge of C-4 (equivalent to 10 kg of tetryl) to directly initiate detonation in a stoichiometric 90% CH_4/10% C_2H_6/air mixture. If an infinitesimally fine

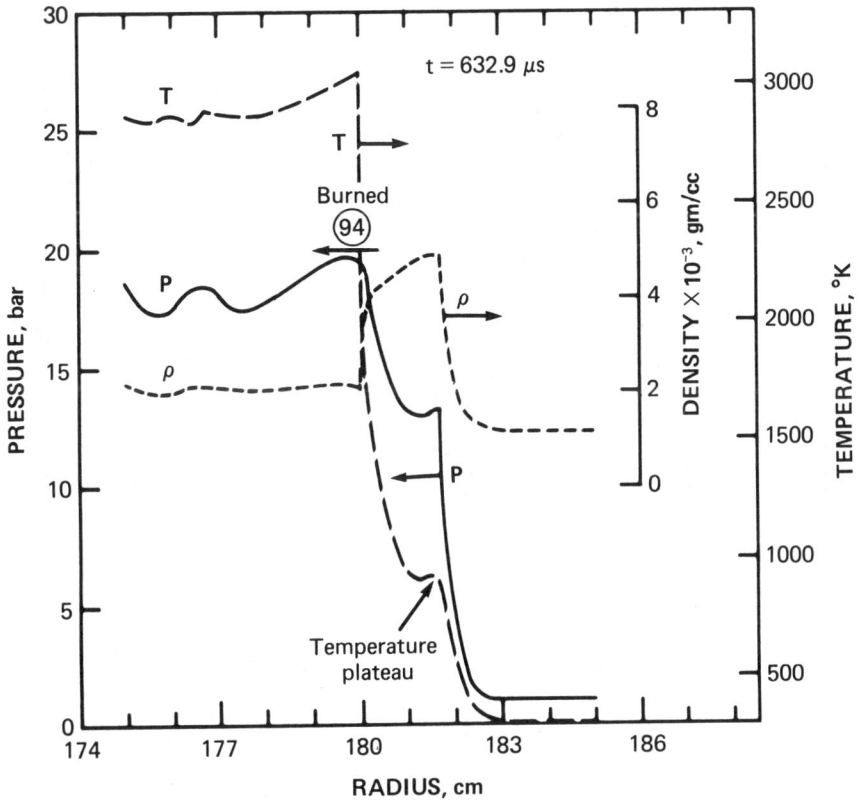

a) PRESSURE, TEMPERATURE AND DENSITY PROFILES (t = 632.9 μs)

Fig. 13 Pressure, temperature, and density profiles at a) 632.9, b) 690.3, c) 743.1, and d) 800.4 μs after ignition illustrate the formation of an elevated temperature plateau behind the shock front.

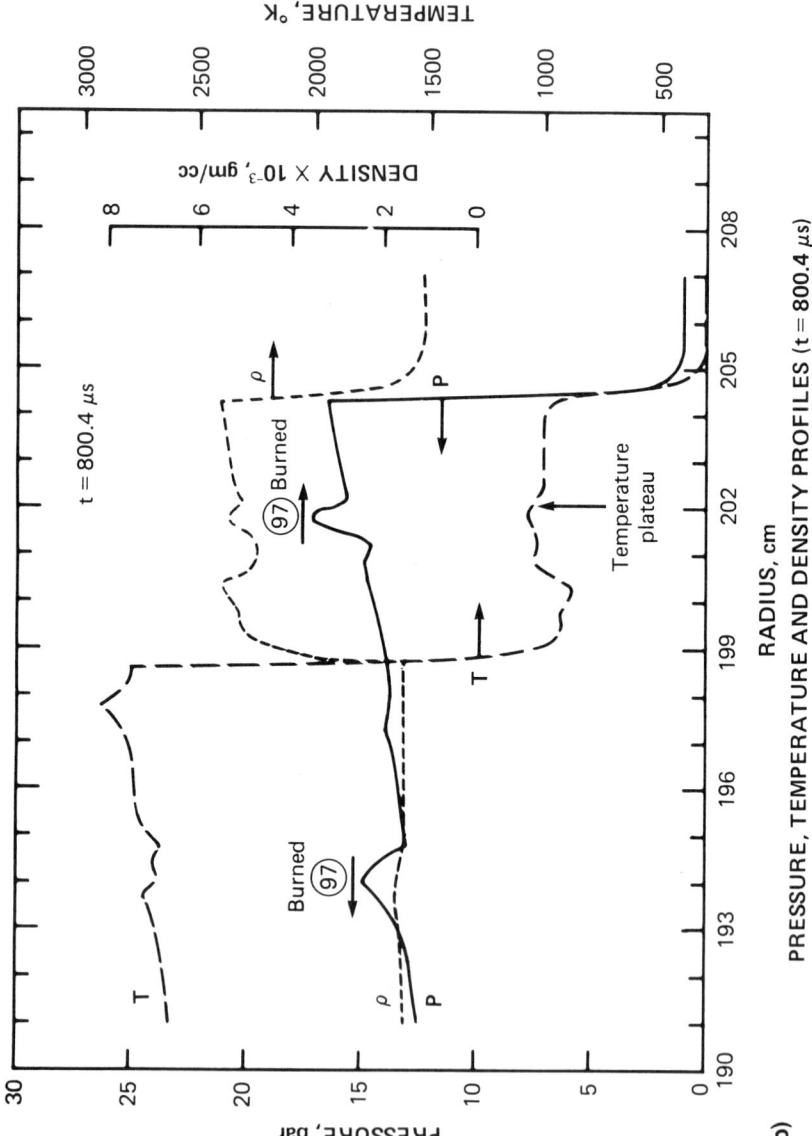

Fig. 13 (cont.) Pressure, temperature, and density profiles at a) 632.9, b) 690.3, c) 743.1, and d) 800.4 μs after ignition illustrate the formation of an elevated temperature plateau behind the shock front.

INITIATION OF DETONATION IN LNG/AIR CLOUDS

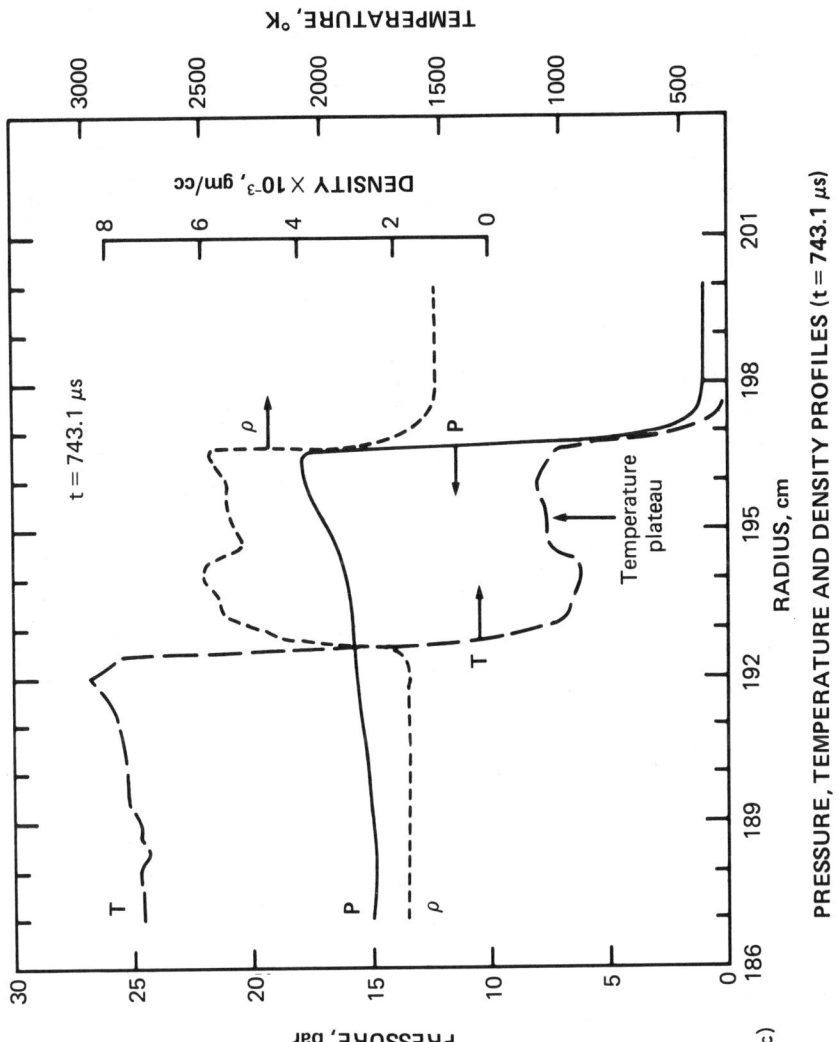

Fig. 13 (cont.) Pressure, temperature, and density profiles at a) 632.9, b) 690.3, c) 743.1, and d) 800.4 μs after ignition illustrate the formation of an elevated temperature plateau behind the shock front.

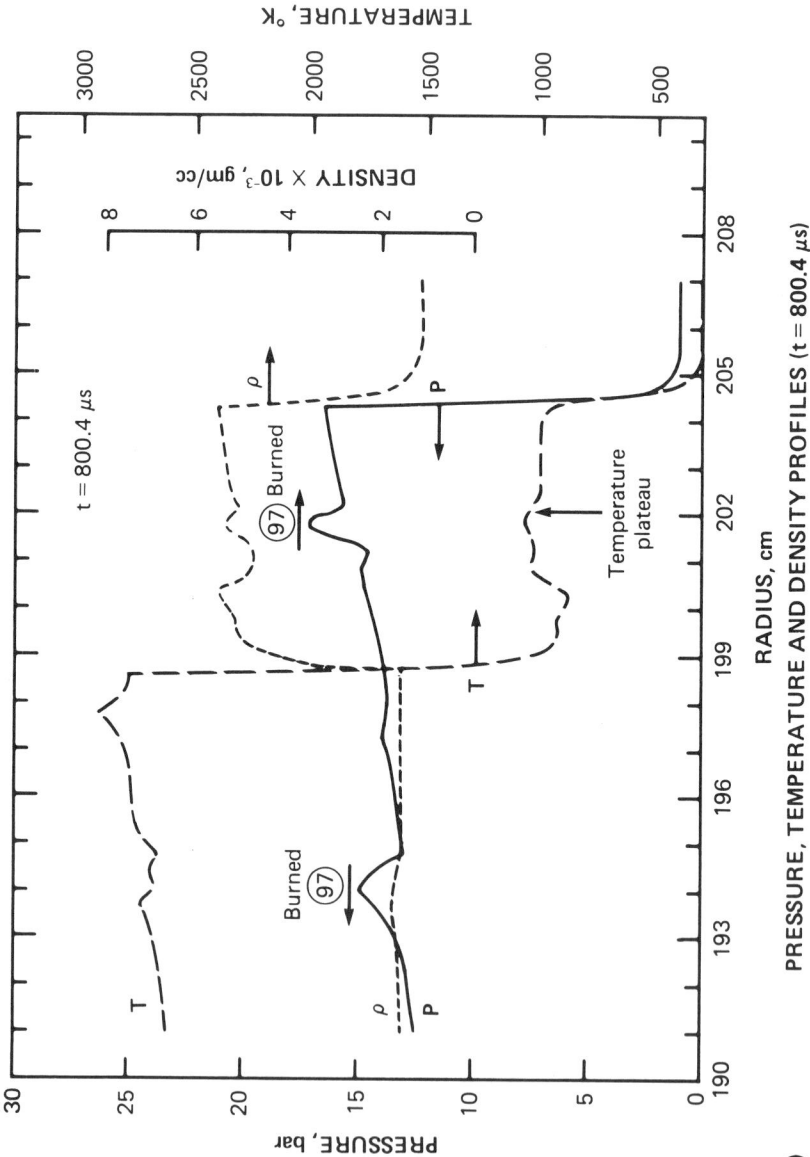

Fig. 13 (cont.) Pressure, temperature, and density profiles at a) 632.9, b) 690.3, c) 743.1, and d) 800.4 μs after ignition illustrate the formation of an elevated temperature plateau behind the shock front.

mesh had instead been used in this simulation, we expect that the reaction front would have continuously rather than intermittently emitted pressure disturbances. These disturbances, though far weaker than the intermittent pulse, should interact with the lead shock and induction zone in a manner similar to the finite-size mesh simulation, that is, it should culminate in the intensification of a portion of the induction zone -- the elevated temperature plateau.

The significance of the predicted elevated temperature plateau is that if the temperature had been raised to 1370 K instead of 1105 K, the plateau induction times would have been sufficiently short ($<100\ \mu s$) to expect an event equivalent to an "explosion within an explosion," which was observed (Urtiew and Oppenheim 1966) to cause transition to detonation in tubes. Or, if the reaction front had reached the temperature plateau, the significantly elevated plateau temperatures would have drastically increased the rate of cell explosions, further intensifying the extended plateau, and possibly leading to transition to detonation.

The cell explosion mechanism which produces the intermittent waves is thought to be a mechanism by which detonation can be reinitiated.

Conclusions

A detailed calculation of the chemical kinetic and gasdynamic processes occurring during the direct initiation of detonation in LNG/air mixtures has been carried out. Analysis of the computed results indicates that in the process of initiating detonation, the shock-reaction complex passes through three distinct modes of propagation: the strongly coupled, weakly coupled, and uncoupled modes of propagation. The strongly coupled mode is characterized by a high Mach number shock front and very short postshock induction times. During this mode of propagation, the combustion of the gas mixture near the explosive source produces large quantities of radical species (H, OH, O) and thus very little net heat is released. Further from the initiating source, in the weakly coupled region, the high postshock temperature promotes rapid chemical reaction, resulting in superimposed pressure spikes that tend to sustain the overdriven detonation. The major products of combustion are H_2O, CO, and CO_2, and the heat released is approximately 95% of the theoretical amount. The shock and reaction fronts are slightly separated.

When the shock front separates from the reaction front (the uncoupled region), the moderate postshock temperature promotes the decomposition of C_2H_6 into CH_3, which in turn produces the H, O, OH, and HO_2 radicals which efficiently attack the remaining fuel. Westbrook's chemical induction time hypothesis appears to work well at this time, i.e., an induction zone is formed between the shock and reaction front. A potentially important effect of the interaction of the reaction front on the shock front occurs during the latter stages of the uncoupled mode of propagation. Pressure signals emanating from the reaction plane temporarily raise the temperature behind the shock front from 920 to 1105 K. Though this increase in temperature is insufficient to cause transition to detonation, the cell explosion/intermittent wave mechanism which caused the increase has significant implications for the mechanism by which detonation is reinitiated.

Two additional simulations of the direct initiation of detonation are currently in progress: 1) direct initiation of a 90% CH_4/10% C_2H_6/air mixture by a 30.4 kg spherical charge of C-4 (equivalent in energy to 40 kg of tetryl solid explosive); and 2) direct initiation of a 70% CH_4/30% C_2H_6/air mixture by a 0.675 kg spherical tetryl charge. The first simulation is expected to detail the processes occurring during the direct initiation of detonation by a supercritical initiator. The second simulation is expected to verify that our model predicts detonation initiation for a solid initiator/gas mixture for which experimental evidence of detonation initiation exists (Bull et al. 1979). The results of these simulations will be reported later.

References

Andrews, G. E. and Bradley, D. (1972) The burning velocity of methane-air mixtures. Combust. Flame 19, 275-288.

Atkinson, R., Bull, D. C. and Shuff, P. J. (1980) Initiation of spherical detonation in hydrogen/air. Combust. Flame 39, 287-300.

Bach, G. G., Kynstautas, R., and Lee, J. H. (1969) Direct initiation of spherical detonations in gaseous explosives. 12th Symposium (International) on Combustion, pp. 853-864. The Combustion Institute, Pittsburgh, Pa.

Boni, A. A., Su, F. Y., and Wilson, C. W. (1978a) Numerical simulation of unsteady combustion and detonation phenomena. AIAA Paper 78-947 presented at AIAA/SAE 14th Joint Propulsion Conference, Los Vegas, Nev., July 25-27.

Boni, A. A., Wilson, C. W., Chapman, J., and Cook, J. L. (1978b) A study of detonation in methane/air clouds. Acta Astron. 5, 1153-1169.

Bull, D. C., Elsworth, J. E., Hooper, G., and Quinn, C. P. (1976) A study of spherical detonation in mixtures of methane and oxygen diluted by nitrogen. J. Phys. D. Appl. Phys. 9, 1191-2000.

Bull, D. C. and Elsworth, J. E. (1978) Initiation of spherical detonation in hydrocarbon/air mixtures. Acta Astron. 5, 997-1008.

Bull, D. C. and Elsworth, J. E. (1979) Concentration limits to unconfined detonation of ethane-air. Combust. Flame 35, 27-40.

Bull, D. C., Elsworth, J. E., and Hooper, G. (1979) Susceptibility of methane-ethane mixures to gaseous detonation in air. Combust. Flame 34, 327-330.

Burcat, A., Scheller, K., and Lifshitz, A. (1971) Shock tube investigation of comparative ignition delay times for C_1-C_5 alkanes. Combust. Flame 16, 29-33.

Cohen, L. M., Short, J. M., and Oppenheim, A. K. (1975) A computational technique for the evaluation of dynamic effects of exothermic reactions. Combust. Flame 24, 319-334.

Edwards, E. H., Hooper, G., and Morgan, J. M. (1976) The experimental investigation of the direct initiation of spherical detonations. Acta Astron. 3, 117-130.

Gear, C. W.(1971) Numerical Initial Value Problems in Ordinary Differential Equations. Prentice-Hall, Englewood Cliffs, N. J.

Gordon, S. and McBride, B. J. (1971) Computer program for calculation of complex chemical equilibrium compositions, rocket performance, incident and reflected shocks and Chapman-Jouguet detonations. NASA SP-273.

Hindmarsh, A. C. (1972a) Linear multistep methods for ordinary differential equations: method formulations, stability, and the methods of Nordsieck and Gear. UCRL-51186, Lawrence Livermore Laboratory, Livermore, Calif.

Hindmarsh, A. C. (1972b) Construction of mathematical software - part III: the control of error in the Gear package for ordinary differential equations. UCID-30050, Lawrence Livermore Laboratory, Livermore, Calif.

Hindmarsh, A. C. (1974) Gear: ordinary differential equation system solver. UCRL-30001, Rev. 3, Lawrence Livermore Laboratory, Livermore, Calif.

Kamel, M. M. and Oppenheim, A. K. (1973) Photographic laboratory studies of explosions. L'Aerotecnica Missilie Spazio, 2: Rivista dell'Associazione Itaniana di Aeronautica e Astronautica, pp. 122-134, Tamburini, Milano.

Korbeinikov, V. P., Levin, V. A., Markov, V. V., and Chernyi, G. G. (1972) Propagation of blast waves in a combustible gas. Astron. Acta 17, 529-537.

Kyong, W. H., Bach, G. G., Lee, J. H., and Knystautas, R. (1972) A theoretical study of spherical gaseous detonation waves. MERL-72-7, McGill University, Canada.

Lee, E. L., Hornig, H. C., and Kury, J. W. (1968) Adiabatic expansion of high explosive detonation products. UCRL-50422, Lawrence Livermore Laboratory, Livermore, Calif.

Lee, E. L., Finger, M., and Collins, W. (1973) JWL equation of state constants for high explosives. UCID-16189, Lawrence Livermore Laboratory, Livermore, Calif.

Lee, J. H. and Ramamurthi, K. (1976) On the concept of the critical size of a detonation kernal. Combust. Flame 27, 331-340.

Lee, J. H. S. (1977) Initiation of gaseous detonations. Ann. Rev. Phys. Chem. 28, 75-104.

Levin, V. A. and Markov, V. v. (1975) Initiation of detonation by concentrated release of energy. Combustion, Explosion, Shock Waves 11, 429-536.

Moen, I. O., Donato, M., Knystautas, R., and Lee, J. H. (1980) Flame acceleration due to turbulence produced by obstacles. Combust. Flame 39, 21-32.

Oran, E., Young, T., and Boris, J. (1978) Application of time-dependent numerical methods to the description of reactive shocks. 17th Symposium (International) on Combustion, pp. 43-54. The Combustion Institute, Pittsburgh, Pa.

Parnarouskis, M. C., Lind, C. D., Raj, P. P. K., and Cece, J.M. (1980) Vapor cloud explosion study. Paper 12 presented at the Sixth International Conference and Exhibition on Liquefied Natural Gas, Kyoto, Japan.

Sichel, M. (1977) A simple analysis of the blast initiation of detonations. Acta Astron. 4, 409-424.

Soloukhin, R. I. and Ragland, K. W. (1969) Ignition processes in expanding detonations. Combust. Flame 13, 295-302.

Strehlow, R. A. and Salm, R. J. (1976) The failure of marginal detonations in expanding channels. Acta Astron. 3, 983-994.

Urtiew, P. A. and Oppenheim, A. K. (1966) Experimental observations of the transition to detonation in an explosive gas. Proc. R. Soc. (London) Ser. A 295, 13-28.

Urtiew, P. A. (1980) private communication. Lawrence Livermore Laboratory, Livermore, Calif.

U. S. Department of Energy (1980) Liquefied gaseous fuels safety and environmental control assessment program: second status report. DOE/EV-0085, Vols. 1 and 2.

Weast, R. C. (1978) CRC Handbook of Chemistry and Physics, 59th ed. CRC Press Inc., Boca Raton, Fla.

Westbrook, C. K. (1979) An analytic study of the shock tube ignition of mixtures of methane and ethane. Combust. Sci. Tech. 20, 5-17.

Westbrook, C. K. and Haselman, L. C. (1979) Chemical kinetics in LNG detonations. UCRL-8229, Rev. 1, Lawrence Livermore Laboratory, Livermore, Calif.

Westbrook, C. K. (1981a) private commmunication. Lawrence Livermore Laboratory, Livermore, Calif.

Westbrook, C. K. (1981b) private communication. Lawrence Livermore Laboratory, Livermore Calif.

Wilkins, M. L. (1969) Calculation of elastic-plastic flow. UCRL-7322, Rev. 1. Lawrence Livermore Laboratory, Livermore, Calif.

Williams, F. A. (1965) Combustion theory. Addison-Wesley Publishing Co., Reading, Mass.

Yanenko, N. N. (1971) The Method of Fractional Steps. Springer-Verlag, New York.

Zeldovich, Y. B., Kogarko, S. M. and Simonov, N. N. (1956) An experimental investigation of spherical detonation in gases. Zh. Tekh. Fiz. 26(8), 1744-1768 (in Russian); translated in Sov. Phys. Tech. Phys. 1(8), 1689-1713 (1957).

Influence of Walls on Pressure Behind Self-Sustained Expanding Cylindrical and Plane Detonations in Gases

D. Desbordes* and N. Manson†
Université de Poitiers, Poitiers, France
and
J. Brossard‡
Université d'Orléans, Bourges, France

Abstract

This investigation deals with the wall effects on the detonation front velocity and on the pressure field behind, in C_2H_2-O_2 and C_3H_8-O_2 mixtures, confined in a cylindrical sector (h = 2.5 cm thickness) and in a tube (ϕ = 10.7 cm i.d.). It is known that the steady detonation velocity D_p in tubes varies with the ϕ i.d., and that the cylindrical expanding detonation velocity D_c varies with the thickness h of the confinement. In the present work, the $D_p(D_c)$ value is lower than the value extrapolated from the $D_p[\phi^{-1}]$ ($D_c[h^{-1}]$) relation obtained from tubes (sectors) of smaller diameters (thicknesses). In expanding cylindrical detonations, the pressure evolution is governed by the Zeldovitch-Taylor (ZT) ideal wave theory. However, in the tube, the expansion of the detonation products just behind the front is stronger than predicted by the plane ZT theory. The pressure decrease is correlated with the measured curvature radius of the front ($\sim$ 0.7 m). Near it, the flow is two dimensional; it becomes one dimensional in the rear. Then, after a quasi-self-similar distance from the front, the expansion conforms to the plane ZT model.

Presented at the 8th ICOGER, Minsk, USSR, Aug. 23-26, 1981. Copyright © American Institute of Aeronautics and Astronautics, Inc., 1982. All rights reserved.

*Assistant Professor, Laboratoire d'Energétique et de Détonique.
†Professor, Laboratoire d'Energétique et de Détonique.
‡Professor, Laboratoire de Recherche Universitaire.

Introduction

In a previous study (Desbordes, Manson, and Brossard 1981) we have shown that, in gaseous mixtures, the pressure evolution in the wake of self-sustained spherical and hemispherical detonation waves, instantaneously created, is quite well predicted by the ideal one-dimensional model of Zeldovitch (1942) and Taylor (1950) (ZT model).

In self-sustained "plane" detonation waves propagating in a tube, it is usually assumed that the properties of the detonation front depend on wall effects, and the results of the experiments performed in tubes with several diameters are extrapolated to the infinite diameter (Brochet et al. 1963; Getzinger et al. 1966; Veyssière 1971; and Renault 1972). It is to be noticed that, in most of these experiments, the tubes have been chosen long enough to consider that the plane rarefaction wave is negligible. Thus such obtained values of pressure, density, and Mach number may be assumed constant in the wake near the reaction zone and therefore may be considered as the values associated with the front.

As the front has a tridimensional structure and the flow is not strictly one dimensional, the comparison between measurements and results of the plane theory is only approximate.

Indeed, experiments show that the rarefaction behind the front is stronger than expected and depends on the tube (Edwards et al. 1963, 1970). X-shaped waves are observed by Fay and Opel (1958) and Edwards et al. (1963). These waves arise from the interaction between the detonation front and the wall; Fay (1962) explains their existence as a result, in the supersonic region of the flow, of the growth of a boundary layer just behind the detonation front (shock).

To obtain more insight into these phenomena, new experiments in a large-diameter (ϕ = 107 mm) cylindrical detonation tube have been performed. The pressure measurements were done on the wall and on the centerline of the tube; and the values of self-sustained detonation velocity were compared with those obtained in tubes of smaller diameter for the same mixture.

The results of this comparison and also of the pressure evolution with the ZT model are presented. The results obtained in large tubes are explained on the basis of the front curvature and of the existence of the X-shaped waves (oblique waves) in the expansion.

Some investigations on cylindrical detonation have also been done in a sector of h = 2.5 cm thickness. Although the results obtained by Brochet et al. (1970) have shown that the self-sustained detonation velocity was dependent on the wall effects, the pressure profiles, in this case, are in good agreement with the ZT model.

Experimental Details

The Detonation Vessels

The cylindrical diverging detonations are performed in a steel equilateral sector of H = 60 cm height and h = 2.5 cm thickness (see Fig. 1). The tube used for the plane detonation is a steel pipe having a circular section (ϕ_0 = 107 mm i.d.) and L_0 = 4.8 m length (see Fig. 2).

In both devices, the detonation was quasi-instantaneously initiated on the bottom of the tube and on one of the corners of the equilateral sector by means of a capacitive discharge in an exploding wire F [$W = (1/2)\ CV^2$ = 150 J].

Measurements

The tube and the sector are equipped with side ports which accommodate the pressure gages (KISTLER 603 B) and the ionization probes, triggering the 0.1-µs chronometers used for the front velocity measurements: 1) on the expanding sector, the thermally insulated pressure gages are positioned from the initiation position at C_1, R_1 = 16.6 cm, C_2, R_2 = 29.6 cm, C_3, R_3 = 40.6 cm, and C_4, R_4 = 53.6 cm; 2) on the tube, the pressure gages are i) on the wall at distances L_1 = .03 m, L_2 = 1.3 m, L_3 = 2.3 m, L_4 = 3.3 m, and L_5 = 4.3 m from the exploding wire F; ii) on the axis of

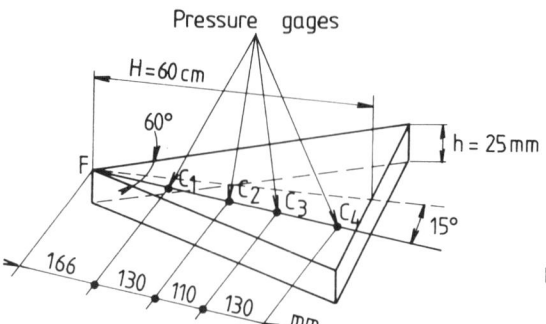

Fig. 1 Detonation tube.

the tube at $L_5' = 4.3$ m on cylindrical support of $\phi = 22$ mm (Desbordes, Manson, and Brossard 1981); and iii) at the end of the tube at $L_6 = 4.8$ m, normally to the axis.

Two ionization probes, separated by a distance of 50 cm, are set at 1.3 and 3.3 m from the end F of the tube, and give the mean velocities $\bar{D}_{p1}$ and $\bar{D}_{p2}$, respectively, with an uncertainty of about 0.2%.

The uncertainty of the pressure gages is 2% on the pressure levels and 1% on the mean velocity measurements.

Mixtures

The following mixtures were detonated at room temperature $T_f = 293 \pm 2$ K and pressure p_f below 400 mbars: $C_2H_2 + 2.5\ O_2$ (mixture A) and $C_3H_8 + 5\ O_2$ (mixture B). These mixtures are prepared in accordance with the usual laboratory techniques (Brossard 1971).

Results

Detonation Velocities

The measured values of the self-sustained detonations velocities (see Fig. 3 for mixture A) of expanding cylindrical D_c and plane D_p are lower than the calculated Chapman-Jouguet (CJ) values. The smaller p_f is, the greater is the discrepancy. However, we remark that 1) the cylindrical detonations are always strong near the initiation line F and on a maximum radius of 30 cm for $p_f \simeq 14$ mbars (i.e., $\bar{D}_{c1} \geq \bar{D}_{c2} \geq \bar{D}_{c3} \geq \bar{D}_{c4} = \bar{D}_c$); however, for $p_f \geq 200$ mbars, the discrepancy between $\bar{D}_{c1}$ and D_c is very small (∼1%); and 2) in the tube, with mixtures A and B, the value of mean detonation velocity $\bar{D}_p$ is always greater ($\bar{D}_{p1} - \bar{D}_{p2}$

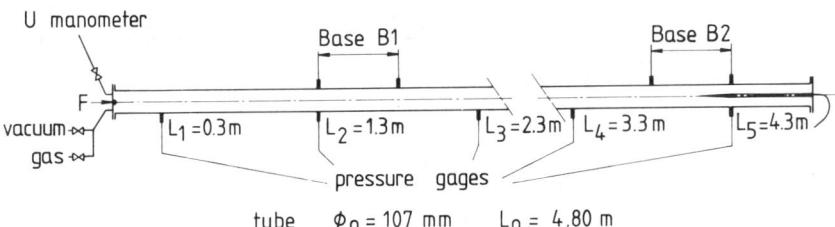

Fig. 2 Detonation sector (cylindrical propagation).

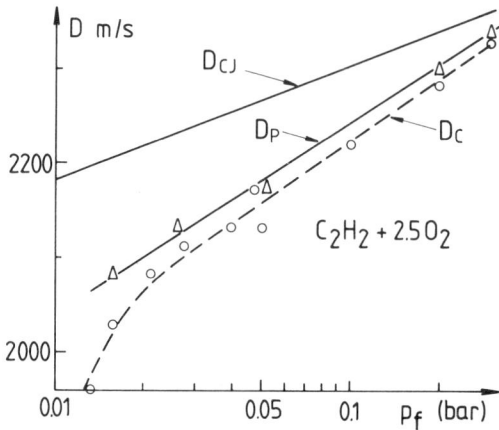

Fig. 3 Variation of detonation velocities D_p (plane), D_c (expanding cylindrical), and D_{CJ} vs p_f, in mixtures $C_2H_2 + 2.5\ O_2$ at $T_f = 293$ K.

< 1%) than $\bar{D}_{p2}$; this latter value is the one which can be expected in self-sustained and stable detonation (D_p).

The D_p values are plotted as a function of the inverse of the tube diameter ϕ^{-1} on Figs. 4a and 4b, respectively, for mixture A (300 mbars $\geq p_f \geq$ 100 mbars) and for mixture B (p_f = 300 and 400 mbars). Values of spherical detonation velocities $D_{s\infty}$ (Desbordes, Manson, and Brossard 1981) are also given for these mixtures. In all cases, the same behavior of the D_p values is apparent: a linear dependence of D_p on ϕ^{-1} until $\phi \sim$ 50 mm and for $\phi >$ 50 mm observed value of D_p and $D_{s\infty}$ (ϕ^{-1} = 0) fall below the extrapolation of the linear curve.

Thus the values of $D_{p\infty}$ obtained by the extrapolation to infinite diameter (ϕ^{-1} = 0) of the straight line $D_p(\phi^{-1})$ for tubes of internal diameters $\phi <$ 50 mm do not have a physical significance, such as, a plane detonation propagating without losses. This fact is in agreement with observations of Brochet et al. (1970) of cylindrical expanding detonation velocities D_c, strongly influenced by the thickness h of the sector: $D_c(h^{-1})$ reaches a maximum and tends to $D_{s\infty}$ when h^{-1} tends to zero.

Pressure

Cylindrical Detonations. The pressure records (see Fig. 6) indicate the following:

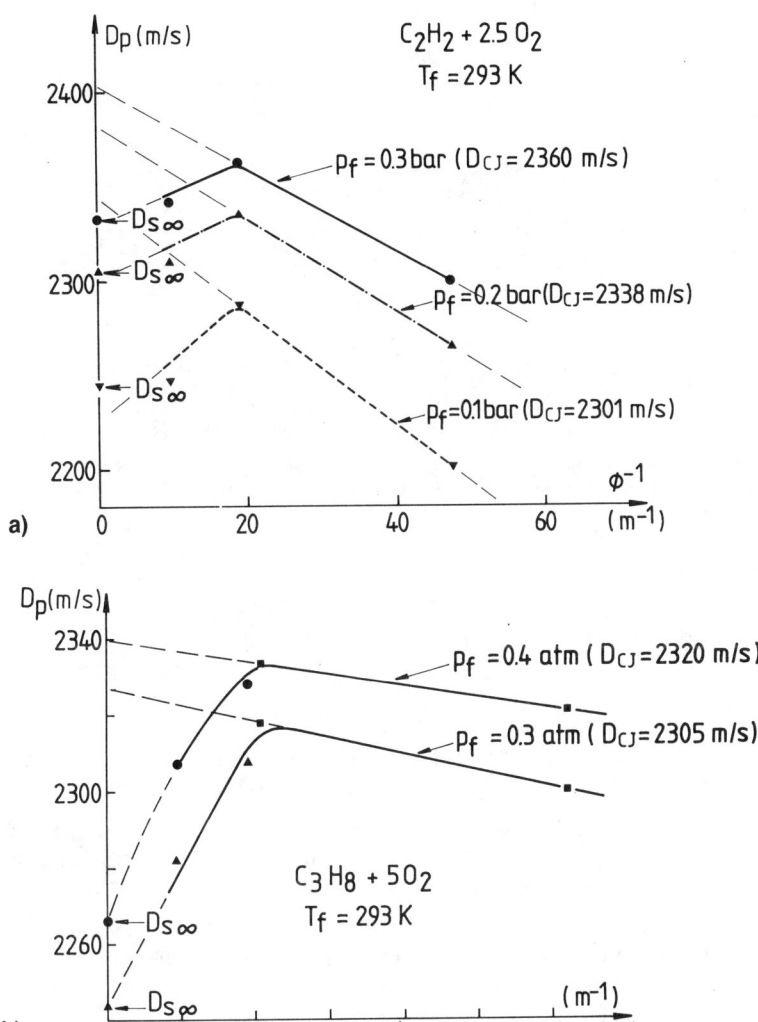

Fig. 4 Variation of detonation velocity D_p vs the inverse diameter tube ϕ^{-1}, in mixtures a) $C_2H_2 + 2.5\ O_2$ and b) $C_3H_8 + 5\ O_2$ at $T_f = 293$ K.

1) The ratio $\pi = p_{CM}/p_{CJ}$ of the maximum pressure p_{CM} normalized by the p_{CJ} value is related to R, but this correlation also depends on p_f (see Fig. 5 for mixture A). We notice that i) when the detonations are stable (i.e.,

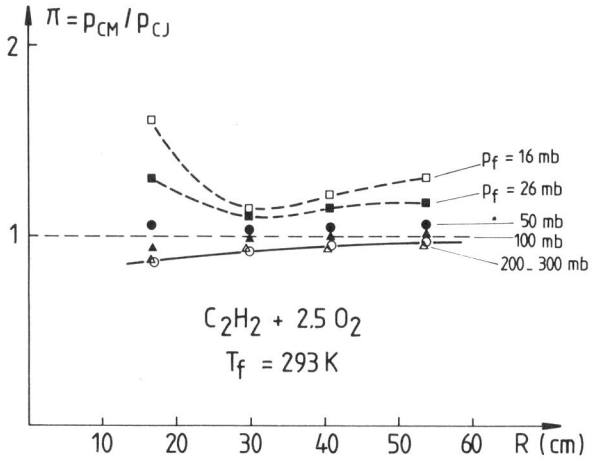

Fig. 5 Variation of π ratio vs distance of propagation R for cylindrical expanding detonation in $C_2H_2 + 2.5\ O_2$ at $T_f = 293$ K.

with a constant detonation velocity) even near the origin ($p_f \geq 200$ mbars), π grows with R independently of p_f and reaches 0.96 - 1 at R ≃ 0.5 m; and ii) when the detonation is strong ($p_f < 200$ mbars), π is generally greater than 1 in the first position (R_1); this value decreases in R_2 and when, in all cases, the detonation has reached its constant velocity regime, π (generally > 1) grows slowly with R.

2) In the case of stable detonations, and 4-5 μs after the peak, the rarefaction pressure p(t) is in agreement with the ZT model (Figs. 6a and 6b). For very low-pressure p_f, the model does not give good agreement with theory. In that case, the multidimensional structure becomes important and of the same order as the sector thickness, and therefore noticeably modifies the pressure profiles (Fig. 6c). When the detonation is strong, the decrease in the rarefaction wave is sharper than in stable detonations.

All these facts agree with those previously observed in cylindrical detonations by Lee et al. (1966) and in spherical expanding detonations by Desbordes, Manson, and Brossard (1981).

<u>Plane Detonations</u>. The pressure peaks p_{PM} obtained on the tube wall are quasi-independent of the distance of pressure gages from the point of initiation (see Fig. 7).

The ratio $\pi = p_{PM}/p_{CJ}$, for $p_f \geq 100$ mbars, on the wall is about 1.1 and on the centerline of the tube is equal to

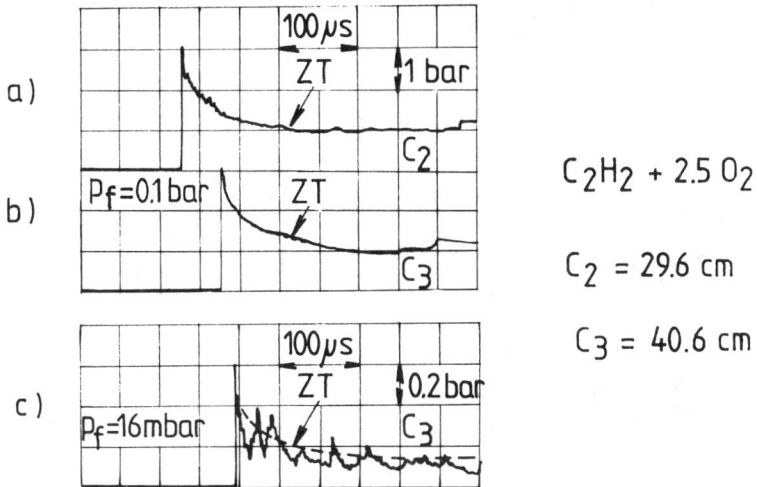

Fig. 6 Typical pressure recordings in cylindrical expanding detonations and corresponding ZT profiles.

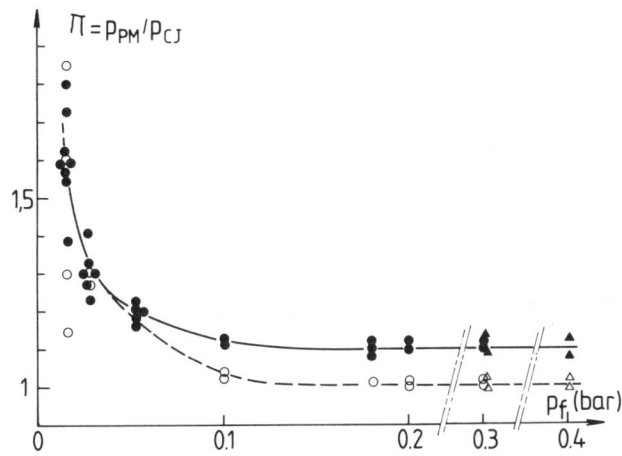

Fig. 7 Variation of π ratio vs p_f for plane detonation in $C_2H_2 + 2.5\ O_2$ (O, ●) and $C_3H_8 + 5\ O_2$ (△, ▲) mixtures, measured on the wall (black) and on the centerline of the tube (white).

1. If $p_f < 100$ mbars, on the centerline and on the wall, π grows with p_f^{-1}; for example, the mean value of π for $p_f = 16$ mbars is 1.6 but the dispersion is very important. Thus, below $p_f \sim 100$ mbars, the influence of the three-dimensional structure increases and the distance between the generalized CJ plane and the front grows.

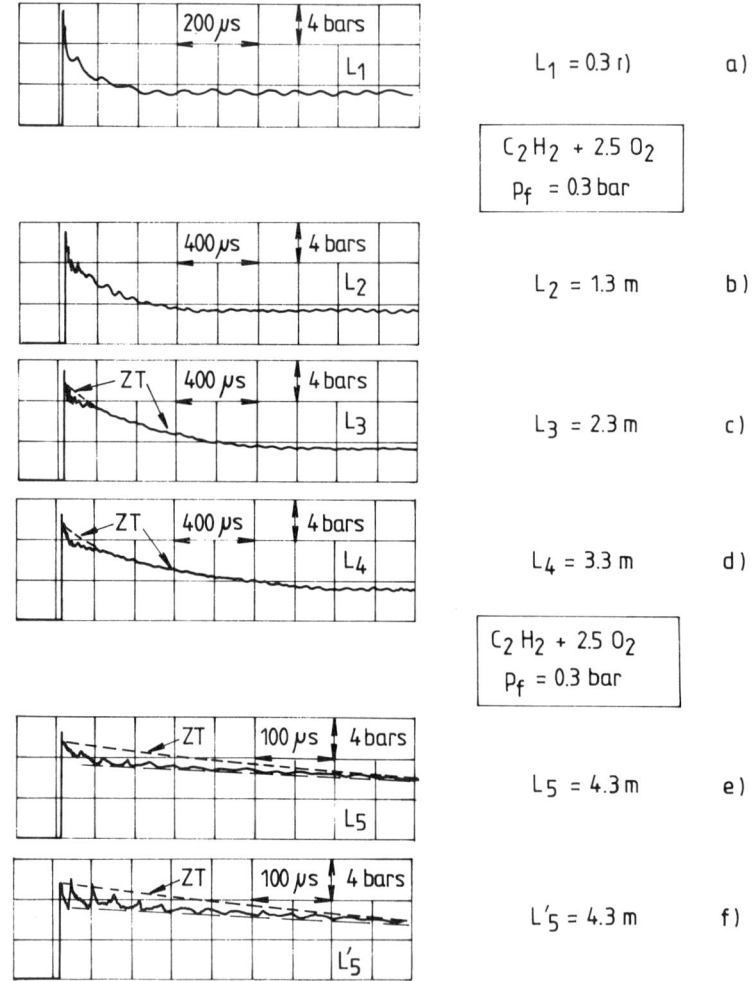

Fig. 8 Typical pressure recordings in plane detonations and corresponding ZT profiles.

If the pressure profile at different gage positions (see Fig. 8), and more precisely on the wall and on the centerline, at a given distance L_5 = 4.3 m, are examined, the following features are significant (see Fig. 9): 1) the presence of periodic pressure peaks whose amplitude diminishes with time suggests the existence of X-shaped waves (oblique waves) in the wake, as already observed by Fay and Opel (1958) and Edwards et al. (1963); 2) the pressure decay in the rarefaction wave starts from the quasi-p_{CJ} value; and 3) during the first 30 to 40 µs this decay is sharper on the centerline of the tube than on the wall and below the value predicted by the ZT plane model. Later, both records agree and the pressure decreases very slowly, reaches a quasi-constant value, and follows closely the ZT profile for the distance ℓ_1 from the front corresponding to a ratio $\ell_1/L \sim 0.13 \pm 0.1$ (where L is the distance traveled by the detonation front from the initiation point).

Discussion of the Flowfield

The X-shaped wave, starting from the interaction between the detonation front (shock) and the wall, is created by the reflection on the wall of the detonation front, which propagates spherically near its initiation point. Then, when the propagation is long enough, the phenomenon becomes stationary relative to the front. A 10% overestimation (if $p_f > 100$ mbars) of the pressure is observed just before the CJ state has been attained on the wall. It is not a simple Mach wave near the front (the Mach number M > 1 relative to the flow ahead), and the modification of the pressure of the flow behind it is not

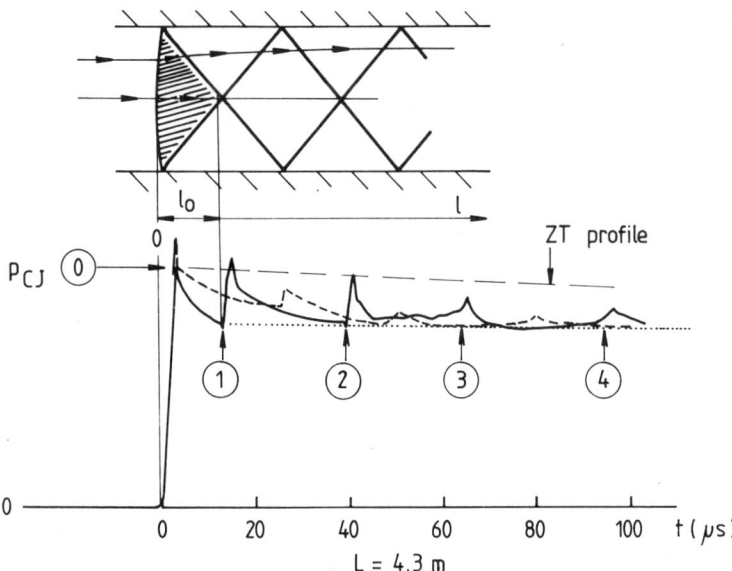

Fig. 9 Pressure time profiles observed on the wall (- - -) and on the centerline of the tube (——), and scheme of the bidimensional flow behind the front.

negligible, as we can observe on the pressure recordings (see Figs. 8 and 9).

The flow, just behind the detonation front of the hatched zone of Fig. 9, is not perturbed by the oblique wave. In this first region the pressure decays on the centerline from p_{CJ} at the front to 0.77 p_{CJ} or so after a lapse of time of 12-13 µs or a distance ℓ_o of 28-30 mm (with a D_p velocity of about 2.3 mm/µs). Such a pressure decrease in the supersonic region is due, as suggested by Fay (1959) and Edwards et al. (1975), to the divergence of the flow behind the front. This divergence is expected to be set by the tube characteristics (inner diameter, nature of the wall, roughness, etc.), and its origin relative to the front is constant after the detonation has been propagating over a sufficiently long distance from its initiation point.

To estimate the radius of curvature R_F of the front, the particle path in the first region is assumed to diverge linearly with distance ℓ, measured along the tube axis from the front. With the CJ condition (the distance between the shock and the end of the reaction zone is assumed small compared with the length ℓ_o of the first region) taken normally to the front, the area increase is deduced in the divergence (see Fig. 10a). Therefore the isentropic evolution of the flow parameters (M, p/p_{CJ}, a/D_{CJ}, V/D_{CJ}) is obtained vs the adimensional distance λ [= $(R_F - \ell)/R_F$] for ℓ and ϕ_o << R_F (see Fig. 10b). If $1 \leq \lambda \leq 0.95$, these profiles are in good agreement with the spherical ZT ones obtained with the R_F radius.

Thus, if the observed pressure profile in the first region of Fig. 9 is compared with the computed one of Fig. 10b, the value ℓ_o, corresponding to $\lambda = 0.972$, yields a curvature radius of the front R_F of 1-1.1 m.

Systematic measurements of the mean curvature radius of the front are performed in mixture A for $26 \leq p_f \leq 300$ mbars, using two pressure gages set on the end wall of the tube, perpendicularly to the flow (one, C_o, on the centerline of the tube; the other, C_5, near the wall at 5 cm from the center). An example of pressure recordings is given in Fig. 11. We can deduce that 1) the maximum overpressure of both signals is about 2.5 times the CJ values for $p_f \geq 100$ mbars; 2) for lower values of p_f, the pressure jump is higher then 2.5 p_{CJ}; this observation corroborates

PRESSURE BEHIND SELF-SUSTAINED DETONATIONS 313

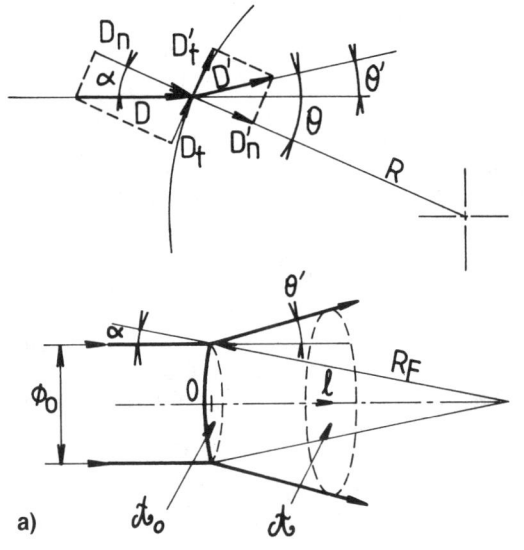
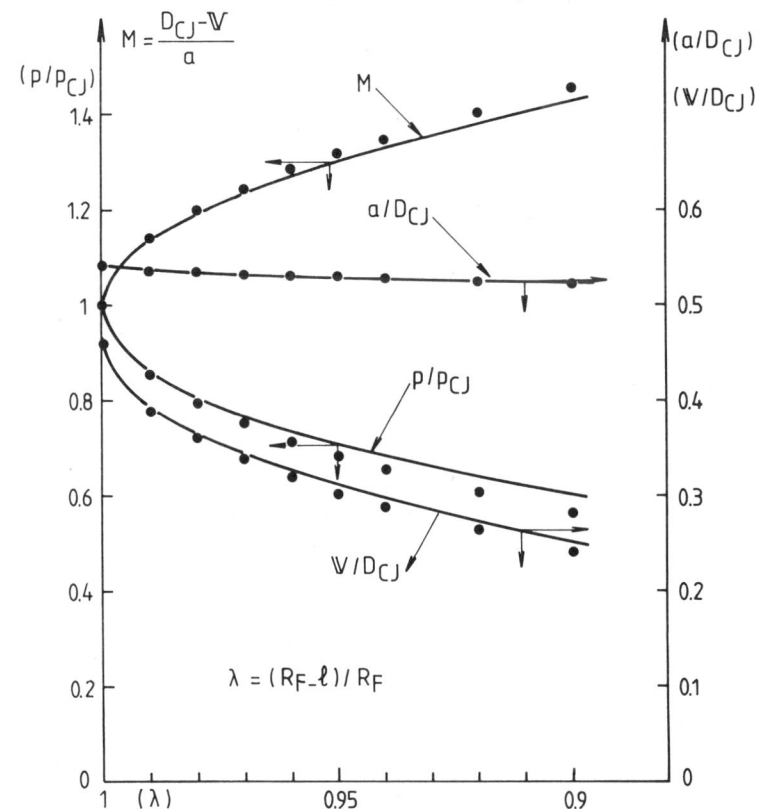

Fig. 10a) Scheme of the diverhent flow behind the curved front and of the growth of area Λ. b) Calculated variation of reduced parameters (Mach number M, pressure p/p_{CJ}, sound velocity a/D_{CJ}, and particle speed V/D_{CJ}) of the flow near the detonation front vs the adimensional radius λ, for ZT spherical (—) and for linear bidimensional (●) expansions.

those on static pressure measurements performed on the centerline which show that π is greater than 1 and increases, whereas $p_f < 100$ mbars; and 3) the delay time t, between the two pressure rises, is 1.0 ± 0.2 µs, except at very low pressure ($p_f < 50$ mbars) where the dispersion is very important. For $D_p \sim 2.3$ mm/µs, $R_F = 1.1 \pm 0.2$ m.

This value agrees with the one deduced from the pressure decay in the supersonic flow of the first region. Also, the isentropic expansion in this region corresponds to an increase of the Mach number of the flow (see Fig. 10b). Furthermore, as the temperature (also the sound velocity a) remains quasi-constant and poorly dependent on the expansion geometry, the growth of material velocity, as well as the decay of density in the flow, are proportional to the Mach number and to the pressure, respectively.

On the wall and just behind the front, the flow is modified by the oblique wave. The pressure sharply decreases after the peak p_{PM} to p_{CJ}, as can be observed on the pressure recordings. The expansion starts and the pressure falls from the p_{CJ} value to $0.91\ p_{CJ}$ at a distance ℓ of 30 mm and to $0.87\ p_{CJ}$ at 60 mm. A curvature radius of the front of 6 m can be derived from this decay. This means that, if the particle path diverges just after the curved

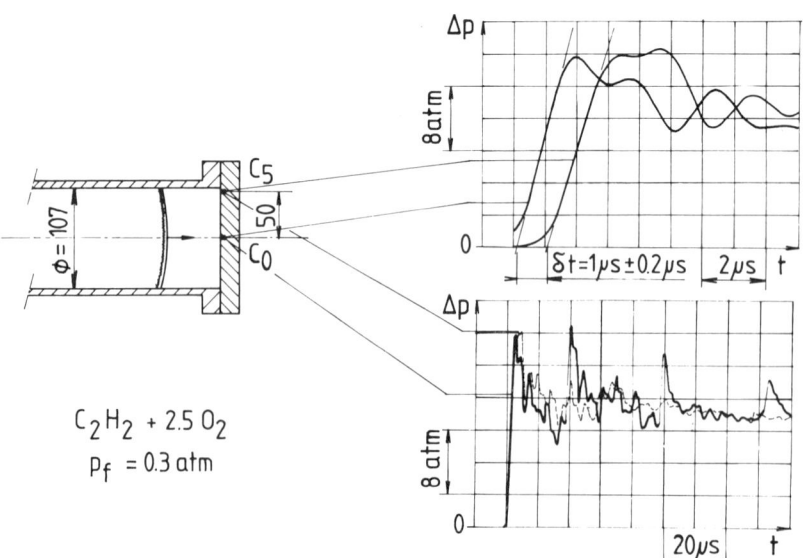

Fig. 11 Pressure time recordings of the reflection pressure for a detonation wave in $C_2H_2 + 2.5\ O_2$ at $p_f = 0.3$ bar, $T_f = 293$ K.

front, the first oblique wave straightens (not completely) the path with respect to the wall; so do the following waves with a smaller amplitude. Finally, the particle path becomes parallel to the wall and the pressure profile is the ZT one for the distance from the front $\ell \geq \ell_1 = 0.13\ L$, showing that the friction and heat losses at the wall remain small.

In the region between the diverging spherical flow near the front and the plane flow far from it ($\ell_0 < \ell < \ell_1$), the assumption of isentropic flow holds. The Mach number of the flow (derived from the Mach angle of oblique waves) and the pressure evolution are in good agreement (Desbordes et al. 1981).

Conclusion

From these experiments, the steady detonation velocity in tubes varies with the ϕ i.d., as does the cylindrical expanding detonation velocity with the thickness h of the confinement. It tends to the $D_{s\infty}$ (steady spherical detonation velocity) value when ϕ^{-1} (or h^{-1}) tends to zero.

The pressure evolution of burnt gases in stable cylindrical expanding detonations, like in spherical propagation, is well predicted by the ideal wave model (ZT).

However, the experimental results obtained for plane detonations in tubes of $\phi \sim 0.1$ m, i.e., the curvature of the front with mean radius of R_F (~ 1 m) and the corresponding decay (on the centerline of the tube) of the pressure in the expansion flow just behind the front, confirm that the flow divergence occurs in confined detonations, as suggested by Fay (1959) and Edwards et al. (1975). This divergence yields very important (infinite) gradients just behind the front.

Far from the front, at a distance of 0.13 L (where L is the whole length of the detonation propagation), the expansion of the detonation products corresponds to the plane ZT model. Therefore there is a transition region between the diverging flow (quasi-spherical with radius R_F) just behind the curved front and the plane flow far from it. In this region, the flow (quasi-isentropic) is modified by the existence of oblique waves, whose influence decreases in the rear. The mean pressure and Mach number of the flow are quasi-constant but differ from those of the front and agree with the previous observations given by numerous workers.

Acknowledgments

The authors wish to express their gratitude to C. Guerraud and S. Sec for their technical assistance.

References

Brochet, C., Manson, N., Rouze, M., and Struck, W. (1963) Influence of the initial pressure on the velocity of stable detonations in stoichiometric mixtures of propane-oxygen and acetylene-oxygen. C.R.A.S. 257, 2412.

Brochet, C., Brossard, J., Manson, N., Cheret, R., and Verdes, G. (1970) A comparison of spherical, cylindrical and plane detonation velocities in some condensed and gaseous mixtures. Proceedings of the Fifth (Int.) Symposium on Detonation, Pasadena, Calif., p. 41.

Brossard, J. (1971) Etude experimentale des detonations et des deflagrations spheriques divergentes dans quelques melanges gazeux. Revue IFP, 26, 11.

Desbordes, D., Manson, N., and Brossard, J. (1981) Pressure evolution behind spherical and hemispherical detonations in gases. Gasdynamics of Detonations and Explosions: Progress in Astronautics and Aeronautics (edited by Bowen, Manson, Oppenheim, and Soloukhin), Vol. 75, pp. 150-165, AIAA, New York.

Desbordes, D., Souletis, J., Manson, N., and Brossard, J. (1981) Evolution de la pression dans les produits des detonations autonomes "cylindriques divergentes" et "planes" des melanges gazeux. Colloq. Int. Berthelot-Vieille-Mallard et Le Chatelier, Bordeaux, II, 443-448.

Edwards, D. H., Jones, T. G., and Price, B. (1963) Observations on oblique shock waves in gaseous detonations. J. Fluid Mech. 17, 21-32.

Edwards, D. H., Brown, D. R., Hooper, G., and Jones, A. T. (1970) The influence of wall heat transfer on the expansion following a C.J. detonation wave. J. Phys. D 3, 365-376.

Edwards, D. H., Jones, A. T., and Phillips, D. E. (1975) The secondary shock wave in supported detonations with flow divergence. J. Phys. D 8, 891-901.

Fay, J. A. and Opel, G. (1958) Two-dimensional effects in gaseous detonation waves. J. Chem. 29, 955-956.

Fay, J. A. (1959) Two-dimensional gaseous detonations: Velocity deficit. Phys. Fluids 2, 283-289.

Fay, J. A. (1962) The structure of gaseous detonation waves. Eighth Symposium (Int.) on Combustion, pp. 30-40, Williams and Wilkins, Baltimore, Md.

Getzinger, R. W., Bowen, J. R., Oppenheim, A. K., and Boudart, M. (1966) Steady detonations in gaseous ozone. Tenth Symposium (Int.) on Combustion, pp. 779-784. The Combustion Institute, Pittsburgh, Pa.

Lee, J. H., Lee, B. H. K., and Shanfiels, I. (1966) Two-dimensional unconfined gaseous detonation waves. <u>Tenth Symposium (Int.) on Combustion</u>, pp. 805-815. The Combustion Institute, Pittsburgh, Pa.

Renault, G. (1972) Propagation des détonations dans des mélanges gazeux contenus dans des tubes de section circulaire et rectangulaire: influence de l'état de la surface interne des tubes. Thèse de 3è Cycle, ENSMA, Poitiers, France.

Taylor, G. I. (1950) The dynamics of the combustion products behind plane and spherical detonation fronts in explosives. <u>Proc. R. Soc., Ser. A</u> 200, 235-247.

Veyssière, M. (1971) Contribution à l'étude des caractéristiques physiques des produits de détonation dans les mélanges gazeux. Thèse de Doctorate d'Etat, ENSMA, Poitiers, France.

Zeldovitch, Y. (1942) Distribution de la pression et de la vitesse dans les produits de détonation; cas d'une onde sphérique divergente. <u>Zh. E.T.P.</u> (USSR) 12, 389-406.

Kinetic Modeling of Ethane/Air Detonability

R. Atkinson* and D.C. Bull†
Shell Research Ltd., Thorton Research Center, Chester, England

Abstract

A detailed chemical kinetic model of ethane oxidation incorporating 64 elementary reactions has been used to predict the detonability of ethane/air mixtures as a function of ethane concentration. Using a constant of proportionality for the Zeldovich criterion ($E \propto \Delta^3$, where E is the detonation initiation energy and Δ an induction length) derived from the hydrogen/air detonation system, it is possible to predict on an absolute basis the detonability of ethane/air mixtures to solid explosive charges. Comparison between predicted and experimental data for the ethane/air system shows good agreement over a wide range of ethane concentrations. Analogous approaches can, in principle, be used to predict the detonability behavior by solid explosive charges for any fuel/air system whose detailed chemical kinetics are accurately known. A lack of accurately known rate and mechanistic data for some of the elementary reactions over the appropriate temperature range places very real limitations upon further exploitation of this method at the present time. However, it is demonstrated that only a complete chemical kinetic scheme can correctly predict the existence of absolute concentration limits to detonation. The implications of the simplifying assumptions made concerning the gasdynamic features of the model are discussed and prospects for a model incorporating

---
Presented at the 8th ICOGER, Minsk, USSR, Aug. 23-26, 1981. Copyright © 1983 by Shell Research LTD. Published by the American Institute of Aeronautics and Astronautics with permission.
*Senior Scientist; present address: Statewide Air Pollution Research Center, University of California, Riverside, Calif.
†Senior Scientist; present address: Schneggenstrasse 24, 8620 Wetzikon, Switzerland.

both gasdynamic and chemical kinetic elements are considered.

Introduction

Numerous attempts have been made to assess the concentration limits to detonability for a variety of fuel/air mixtures. However, most of these experimental studies have been concerned with planar detonations in tubes, and as such are always limited by enthalpy loss mechanisms associated with the confining walls and complicated by shock-wall interactions, which in some circumstances, e.g., for some cylindrical systems (Brossard et al. 1967), appear to lead to artificial support of a marginal wave. The validity of earlier attempts (Pusch and Wagner 1965) to predict absolute concentration limits by extrapolation of data obtained from experiments in tubes of varying diameter to that of an imaginary infinite diameter is at least open to question. Experiments with unconfined waves are much more difficult to perform and always appear to show an extremely steep dependence of initiation energy requirements upon shifting to weak or rich mixtures. It has been shown possible to model the experimental observations very closely with a simple model (Bull et al. 1979a; Bull 1979) based on "global" kinetics, but the enigma remains that so long as a simple Arrhenius-type expression is used to predict the oxidation chemistry there will be no ultimate concentration limits to detonability. However, the "feel" resulting from detonation tube experiments and from other combustion properties in general is that some, possibly reasonably sharply defined, concentration limits will probably exist.

This paper presents the second part of a study aimed at determining how the reaction kinetics in a gas detonation vary as the rich and lean concentrations are approached and in what way, if any, limit phenomena arise.

In the first part of this study (Atkinson et al. 1980), it was shown from a detailed chemical kinetic computer model of the hydrogen/air system under detonation conditions that the variation with stoichiometry of the calculated induction time to ignition, τ, was in good accord with experimental detonation data. This work also showed that the detonation limits are principally governed by the gas temperature behind the lead shock in a detonation and are associated with the transition from ignition behavior [with a rapid increase to high levels of the radical species H, $O(^3P)$, and

Table 1 The reaction mechanism for ethane/air detonation

Reaction						
1	CH_4 + O_2	→	CH_3 + HO_2	
2	CH_4 + N_2	→	CH_3 + H	+ N_2
3	H + CH_4	→	CH_3 + H_2	
4	HO + CH_4	→	CH_3 + H_2O	
5	O + CH_4	→	CH_3 + HO	
6	CH_3 + HO_2	→	CH_3O + HO	
7	CH_3 + HO	→	HCHO + H_2	
8	CH_3 + O	→	HCHO + H	
9	CH_3 + O_2	→	HCHO + HO	
10	CH_3 + CH_3	→	C_2H_6	
11	CH_3O + N_2	→	HCHO + H	+ N_2
12	CH_3O + O_2	→	HCHO + HO_2	
13	C_2H_6	→	CH_3 + CH_3	
14	HO + C_2H_6	→	H_2O + C_2H_5	
15	H + C_2H_6	→	H_2 + C_2H_5	
16	O + C_2H_6	→	HO + C_2H_5	
17	C_2H_5 + O_2	→	C_2H_4 + HO_2	
18	C_2H_5	→	C_2H_4 + H	
19	O + HCHO	→	HO + HCO	
20	H + HCHO	→	H_2 + HCO	
21	HO + HCHO	→	H_2O + HCO	
22	HCHO + N_2	→	HCO + H	+ N_2
23	HCO + O_2	→	CO + HO_2	
24	HCO + N_2	→	H + CO	+ N_2
25	HO + CO	→	CO_2 + H	
26	C_2H_4 + C_2H_4	→	C_2H_5 + C_2H_3	
27	H + C_2H_4	→	H_2 + C_2H_3	
28	O + C_2H_4	→	CH_3 + HCO	
29	HO + C_2H_4	→	H_2O + C_2H_3	
30	C_2H_4 + N_2	→	C_2H_3 + H	+ N_2
31	C_2H_3 + N	→	C_2H_2 + H	+ N_2
32	C_2H_2 + N_2	→	C_2H + H	+ N_2
33	C_2H_2 + O_2	→	HCO + HCO	
34	H + C_2H_2	→	C_2H + H_2	

(Table continued on next page)

Table 1 (cont.) The reaction mechanism for ethane/air detonation

Reaction									
35	O	+	C_2H_2		$\rightarrow$	CH_2	+	CO	
36	HO	+	C_2H_2		$\rightarrow$	C_2H	+	H_2O	
37	C_2H	+	O_2		$\rightarrow$	HCO	+	CO	
38	C_2H	+	O		$\rightarrow$	CO	+	CH	
39	CH_2	+	O_2		$\rightarrow$	HCO	+	HO	
40	CH	+	O_2		$\rightarrow$	HCO	+	O	
41	H	+	O_2		$\rightarrow$	HO	+	O	
42	O	+	H_2		$\rightarrow$	HO	+	H	
43	HO	+	H_2		$\rightarrow$	H_2O	+	H	
44	H	+	O_2	+ N_2	$\rightarrow$	HO_2	+	N_2	
45	HO_2	+	N_2		$\rightarrow$	H	+	O_2	+ N_2
46	O	+	HO		$\rightarrow$	H	+	O_2	
47	O	+	HO_2		$\rightarrow$	HO	+	O_2	
48	H	+	HO_2		$\rightarrow$	H_2	+	O_2	
49	H	+	HO_2		$\rightarrow$	HO	+	HO	
50	H	+	HO_2		$\rightarrow$	H_2O	+	O	
51	HO_2	+	HO_2		$\rightarrow$	H_2O_2	+	O_2	
52	HO	+	HO_2		$\rightarrow$	H_2O	+	O_2	
53	H_2O_2	+	N_2		$\rightarrow$	HO	+	HO	+ N_2
54	HO	+	H_2O_2		$\rightarrow$	H_2O	+	HO_2	
55	H	+	H_2O_2		$\rightarrow$	H_2	+	HO_2	
56	O	+	H_2O_2		$\rightarrow$	HO	+	HO_2	
57	CH_3	+	C_2H_6		$\rightarrow$	CH_4	+	C_2H_5	
58	HO	+	HO		$\rightarrow$	H_2O	+	O	
59	HO_2	+	C_2H_6		$\rightarrow$	H_2O_2	+	C_2H_5	
60	HO_2	+	$HCHO$		$\rightarrow$	H_2O_2	+	HCO	
61	HO	+	HCO		$\rightarrow$	H_2O	+	CO	
62	H	+	HCO		$\rightarrow$	H_2	+	CO	
63	O	+	HCO		$\rightarrow$	HO	+	CO	
64	HO_2	+	HCO		$\rightarrow$	H_2O_2	+	CO	

OH, and a corresponding rapid loss of H_2 and O_2] to slow combustion (very low radical levels). The transition temperature for the hydrogen/air system was determined to be $\sim$ 1100 K.

In this study an analogous approach has been used to derive the ignition limits of ethane/air mixtures for detonation related conditions and hence to predict concentration limits to unconfined detonation. Since this and the previous modeling studies both use complete reaction kinetic schemes, the detonation related induction period data can also be used to predict the detonability of ethane/air to explosive charges from the known detonability of hydrogen/air to the same initiator. Finally, some consideration is given to modeling other simple systems.

Modeling

The 64-step chemical kinetic model given in Tables 1 and 2 was closely derived from that formulated by Westbrook and Haselman (1981), but used the more up-to-date hydrogen oxidation chemistry reported by Atkinson et al. (1980). As in this previous study, computations were performed that yielded the concentrations of all species in the reaction scheme at selected time intervals after the passage of the lead shock. Again, as reported by Atkinson et al. (1980), the thermal history and density of the gas molecules from the time they encounter the lead shock in an advancing detonation wave is rather crudely approximated. Simply, the starting conditions for the chemical reaction were taken to be the temperature and density that exist behind a lead shock having the identical velocity to a Chapman-Jouguet (CJ) detonation wave appropriate to the gas mixture. Thus no attempt was made to model the proper three-dimensional wave interaction and precise particle history. The implications of this simplification have been discussed elsewhere (Bull et al. 1979a; Bull 1979) and are considered again at the end of this paper. Suffice it at this stage to restate that this simplification has been shown to be sufficiently useful and valid for at least comparative investigation of not unduly dissimilar systems, e.g., comparison between different fuel/air systems. Comparisons between systems of widely dissimilar character, e.g., fuel/air vs fuel/oxygen, certainly require a more carefully described gasdynamic description before meaningful conclusions can be drawn.

Results and Discussion

With the initial conditions set to those behind the lead shock of velocity equal to that of a CJ detonation, as described above, typical calculated concentration histories for selected species [C_2H_6, C_2H_4, C_2H_2, OH, and $\Sigma(O, OH, H)$]

KINETIC MODELING OF ETHANE/AIR DETONABILITY 323

Table 2 Rate constant parameters for the reaction mechanism ($k = AT^n e^{-E/RT}$ in $cm^3 mole^{-1} s^{-1}$ units)

Reaction	A	n	E/R, cal mole^{-1}	Reaction	A	n	E/R, cal mole^{-1}
1	1.3 × 10^{-10}	0.00	28200	33	6.6 × 10^{-12}	0.00	14000
2	2.1 × 10^{-7}	0.00	44200	34	3.3 × 10^{-10}	0.00	9500
3	2.1 × 10^{-10}	0.00	5950	35	1.0 × 10^{-10}	0.00	2000
4	5.3 × 10^{-21}	3.08	1000	36	1.0 × 10^{-11}	0.00	3500
5	2.6 × 10^{-11}	0.00	4600	37	1.7 × 10^{-11}	0.00	3500
6	2.6 × 10^{-11}	0.00	0	38	8.3 × 10^{-10}	0.00	0
7	6.6 × 10^{-12}	0.00	0	39	1.7 × 10^{-11}	0.00	1850
8	2.1 × 10^{-10}	0.00	1000	40	1.7 × 10^{-11}	0.00	0
9	4.2 × 10^{-11}	0.00	14500	41	3.7 × 10^{-14}	0.00	8450
10	4.0 × 10^{-11}	0.00	0	42	3.0 × 10^{-11}	1.00	4480
11	8.3 × 10^{-11}	0.00	10500	43	3.6 × 10^{-11}	0.00	2590
12	1.7 × 10^{-12}	0.00	3000	44	4.1 × 10^{-33}	0.00	-500
13	2.5 × 10^{-19}	-1.00	44200	45	3.5 × 10^{-9}	0.00	23000
14	1.0 × 10^{-10}	0.00	1200	46	3.0 × 10^{-11}	0.00	0
15	2.5 × 10^{-21}	3.50	2600	47	3.0 × 10^{-11}	0.00	0
16	4.2 × 10^{-11}	0.00	3200	48	4.2 × 10^{-10}	0.00	350
17	1.7 × 10^{-12}	0.00	2500	49	4.2 × 10^{-10}	0.00	950
18	4.0 × 10^{-13}	0.00	19000	50	3.0 × 10^{-11}	0.00	0
19	8.3 × 10^{-11}	0.00	2300	51	6.0 × 10^{-14}	0.00	-1200
20	6.6 × 10^{-12}	0.00	1900	52	3.0 × 10^{-11}	0.00	0
21	8.3 × 10^{-10}	0.00	3150	53	2.0 × 10^{-7}	0.00	22900
22	8.3 × 10^{-8}	0.00	36000	54	1.7 × 10^{-11}	0.00	910
23	5.3 × 10^{-12}	0.00	3500	55	5.0 × 10^{-12}	0.00	1400
24	2.6 × 10^{-10}	0.00	9500	56	2.7 × 10^{-12}	0.00	2125
25	2.1 × 10^{-17}	1.30	-400	57	8.3 × 10^{-25}	4.00	4150
26	8.3 × 10^{-12}	0.00	32400	58	1.0 × 10^{-11}	0.00	550
27	1.0 × 10^{-11}	0.00	3000	59	1.7 × 10^{-12}	0.00	7000
28	4.7 × 10^{-11}	0.00	1500	60	1.7 × 10^{-12}	0.00	4000
29	1.7 × 10^{-10}	0.00	1750	61	1.7 × 10^{-10}	0.00	0
30	1.7 × 10^{-10}	0.00	49100	62	3.4 × 10^{-10}	0.00	0
31	5.3 × 10^{-8}	0.00	20300	63	1.7 × 10^{-10}	0.00	0
32	1.7 × 10^{-10}	0.00	57600	64	1.7 × 10^{-10}	0.00	1500

are shown in Fig. 1 for a stoichiometry $\lambda = 1.0$. The course of the combustion is characterized by a slow decomposition of C_2H_6 into C_2H_4 (and H_2) together with a slow rise in the O, H, and OH concentrations, followed by an increasingly rapid conversion of C_2H_4 into C_2H_2, subsequent oxidation of C_2H_2, and a rapid rise in radical levels to a peak at about the time of the maximum C_2H_2 concentration (Fig. 1).

The definition of induction time, which must effectively define the onset of a strongly exothermic reaction (Atkinson et al. 1980), is obviously more ambiguous in this system than in the hydrogen/air system, where the rapid loss of H_2 and O_2 coincided with a rapid increase in the O, H, and OH concentrations. Since the conversion of C_2H_4, at least partially into C_2H_2, and the oxidation of C_2H_2 are due to the reactions with O and H atoms and OH radicals, other possible additional criteria to decide induction times could be the times to achieve the maximum C_2H_4 or C_2H_2 concentrations. To be consistent, however, the same criteria were used in this study as by Atkinson et al. (1980); i.e., τ was defined as the time elapsed until the OH radical concentration reached 10^{-4} x (total pressure) and $\Sigma(O, H, OH)$ reached 3×10^{-4} x (total pressure). These two criteria (arrowed in Fig. 1) always agreed to within 10% and the mean value was used. Use of the other suggested criteria led to different absolute values of τ, but the variation of τ with stoichiometry was not markedly changed.

The induction periods τ obtained from this computer model were found to be in reasonable agreement with those derived from the "global" Arrhenius expression given by Burcat et al. (1972) from shock tube measurements of ignition times, although the computed values of τ showed a greater dependence on the stoichiometry λ than did the global expression times.

The induction times obtained for ethane/air mixtures subjected to the lead shock in a detonation are given in Fig. 2. As in the hydrogen/air system, at both low and high stoichiometries ($\lambda = 0.47$ and 2.51 for the ethane/air system), the radical levels remained low and the reaction behavior assumed that of slow combustion rather than ignition. These cases were characterized by critical shocked gas temperatures of 1270 + 30 K. In a manner totally analogous to the hydrogen/air system, this implies

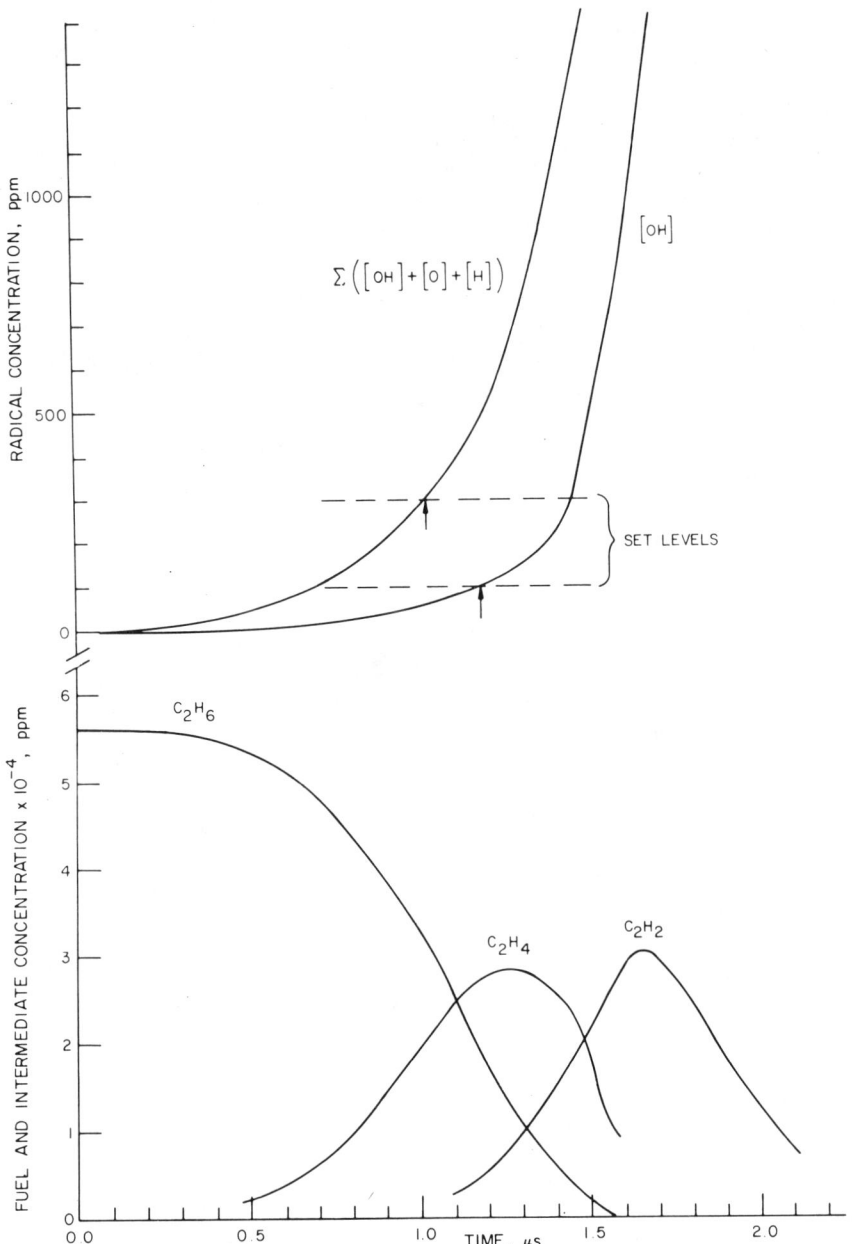

Fig. 1 Species concentration histories for shocked stoichiometric ($\lambda = 1.0$) ethane/air mixture.

that detonation of unconfined ethane/air mixtures cannot be achieved for stoichiometries of $0.47 > \lambda > 2.5$ (i.e., <2.7 or >13.0 Vol.% ethane). These ignition-defined concentration limits to detonability are totally consistent with those determined in a 70-mm tube ($0.47 < \lambda < 2.19$) (Borisov and Loban 1977), since detonation tube limits are expected to be somewhat narrower than those for the unconfined case owing to effects of wall heat losses.

Using the modified Zeldovich criterion (Zeldovich et al. 1956)

$$E \propto \Delta^3 \qquad (1)$$

enables the relationship between ignition kinetics and detonability to be demonstrated. (E is the initiator energy and Δ an induction length which can be roughly approximated

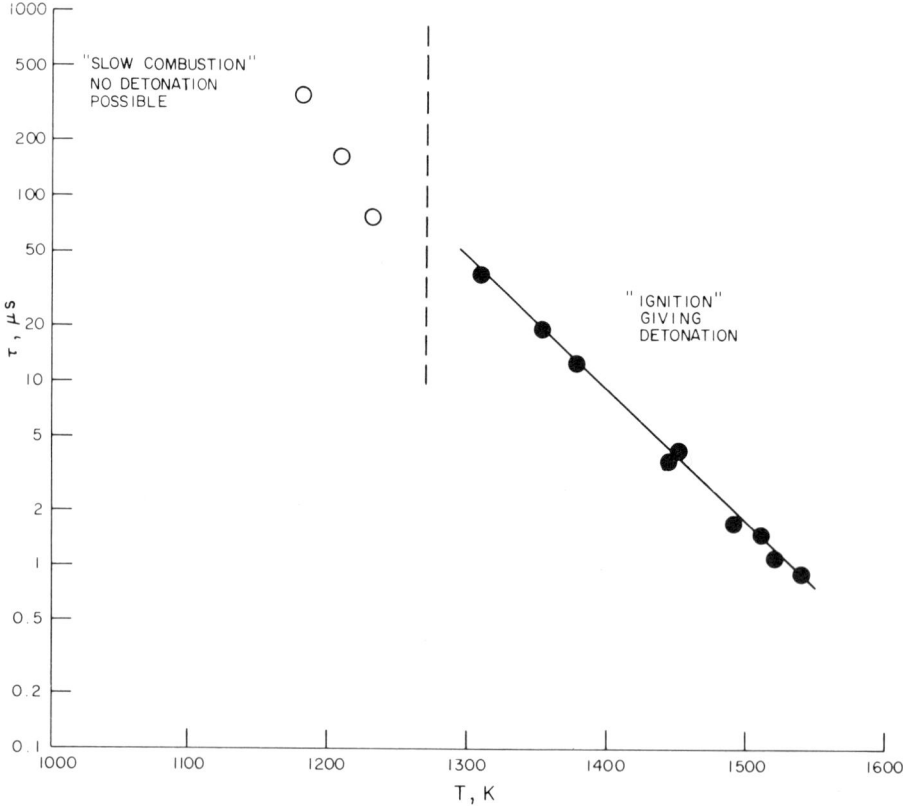

Fig. 2 Computed induction time τ for shocked gas temperatures T, (K) corresponding to a wide range of different concentrations (● = induction period criteria achieved, detonation possible; o - induction period criteria not met, times derived from fuel combustion. Reaction has slow combustion character, detonation not possible).

by $\tau\bar{u}$, where $\bar{u}$ is the mean particle velocity between the shock front and the end of the induction zone.) Since for solid explosive initiators the charge mass m is proportional to E, the expression for the critical mass m_c becomes

$$m_c = k\Delta^3 \qquad (2)$$

where k is a constant of proportionality. Now it would be possible, as we have done previously (Bull et al. 1979a) when we used a crude global kinetic model, to fit the experimental detonation data for unconfined ethane/air detonations with a curve showing the variation of Δ^3 with λ. Such a comparison shows that theory and experiment are in good agreement. Even more instructive, however, is to make a comparison between the absolute values of m_c computed for ethane/air detonation with those for the hydrogen/air system. Thus, since first, the elementary reaction scheme for the hydrogen/air system is now generally accepted to be reliably known, and second, in the course of our hydrogen/air detonability experiments (Atkinson et al. 1980) we determined the critical charge mass just detonating a stoichiometric mixture with some precision, we can compute the constant of proportionality in Eq. (2). If we take (Atkinson et al. 1980) m_c for stoichiometric hydrogen/air as 1.1×10^{-3} kg and $\tau = 0.31 \times 10^{-6}$ s, $\bar{u} = 365$ m s^{-1}, then we find

$$k = 7.6 \times 10^8 \text{ kg m}^{-3} \qquad (3)$$

Applying this same constant of proportionality to the computed kinetic data and particle velocities for the ethane/air system, we derive a curve shown in Fig. 3 together with the experimental detonation data points from Bull et al. (1979a). The fit between the experimental and calculated data is remarkably good, especially considering the many approximations involved and noting that a 10% error in the calculation of τ for each of the hydrogen/air and ethane/air systems can lead to errors of approximately $\pm 60\%$ in the fit of the ethane/air data using the hydrogen/air system as a reference point.

The same modeling technique can, in principle, be extended to predict absolutely the anticipated detonability to solid explosive charge initiators in air of all the lower hydrocarbon gases, either singly or in simple mixtures. The obvious first extension of this methodology is to look at the methane/air system, which is simple on a molecular level, yet of great interest because of its comparative inertness, and also to look at mixtures of ethane and

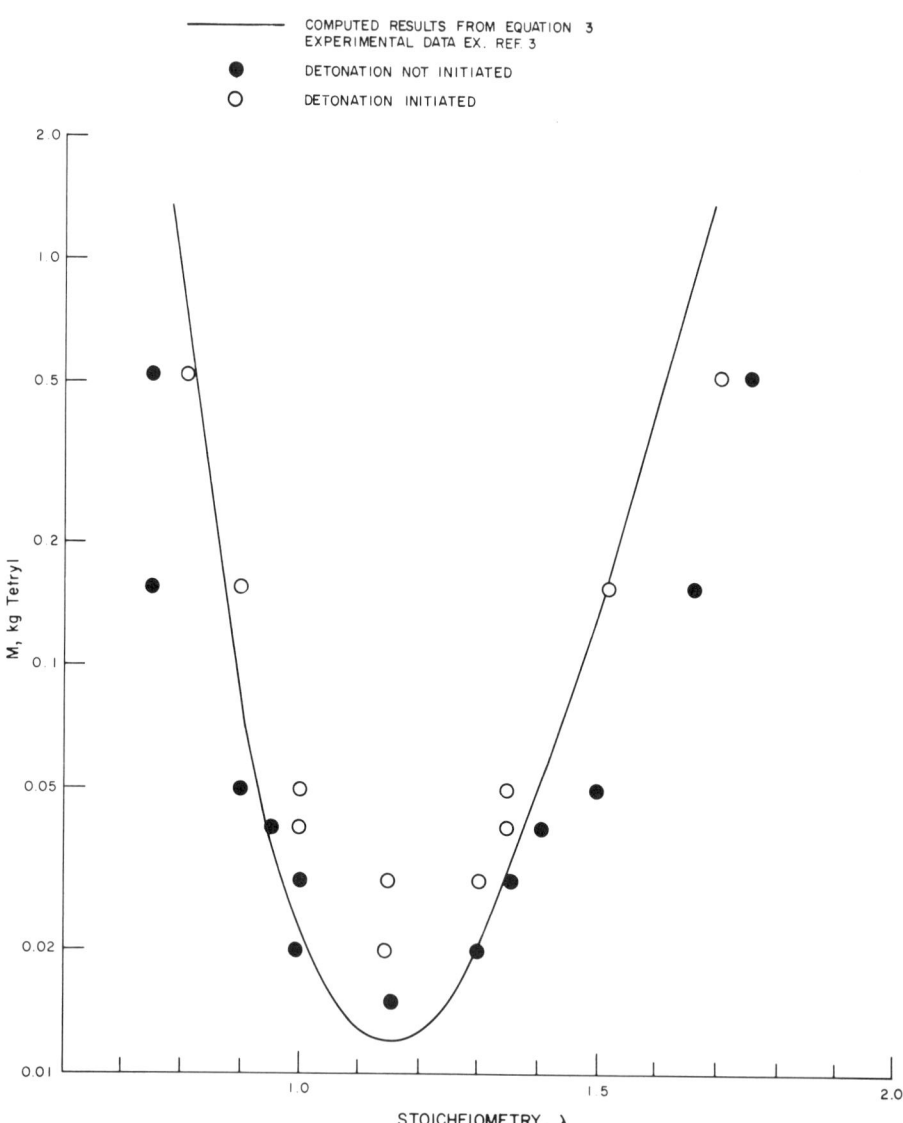

Fig. 3 Calculated (——) and experimental (●,o) values of mass of explosive m against stoichiometry. The calculated line was obtained using Eqs. (2) and (3), and the experimental data are from Bull et al. (1979a) (● - detonation not initiated; o - detonation initiated).

methane, for which experimental detonability data with the same type (Tetryl) of explosive charges exist (Bull et al. 1979b). Further systems which fulfill the criteria of being fairly simple molecules and hence without unduly complex reaction mechanisms, and for which there exist adequate data for experimental comparison (Bull 1979), are ethylene/air and acetylene/air.

Initial extension of the work by considering progressive addition of methane to ethane/air detonation to facilitate comparison with the data of Bull et al. (1979b), and again using Eqs. (2) and (3), led to predictions for critical explosive requirements considerably in excess of those determined by experiment. Thus, for example, for a stoichiometric mixture in air of a gas comprising (0.4 C_2H_6 + 0.6 CH_4) the model predicts 1 kg Tetryl to be the critical initiator--nearly four times the experimentally determined value. These discrepancies suggest that parts of the reaction scheme dealing with CH_4 oxidation, in particular reactions 1-8 in Tables 1 and 2, are somewhat incorrect and lead to excessively long induction times. Further comparison made by running the kinetic model with the initial conditions set to those in the shock tube experiments of Crossley et al. (1972) for ethane/methane oxidation again show an overprediction of induction periods. Finally, sensitivity analysis by slightly varying the rate constants of reactions 1-8 and reexamination of the original literature sources for these data confirm respectively that the model predictions are strongly influenced by the rate constants of these reactions and that the available data are not known with sufficient reliability over the appropriate temperature and density range of these detonation studies. These reactions do not, however, have any strong influence upon the results of the model predictions so long as methane is not present in appreciable amounts, which means that the predictions for ethane/air are essentially totally unaffected by uncertainties from this quarter.

The lack of reliable rate data and the sensitivity of the predictions to the rate constants for the methane reactions does, however, have important consequences and implications. Firstly, there was nothing to be gained from model predictions for methane/air detonations until the ethane/methane mixtures could be successfully paralleled -- they would certainly suggest excessively high values. We wholly resisted the temptation of "fitting" likely values to selected rate constants to achieve this end, since such manipulation of a complex model cannot, we believe, be justified. Rather we chose to point strongly to the important need for more reliable fundamental rate data in this area as an essential prerequisite to future modeling of this type. Examination of the initiation mechanisms for ethylene/air and acetylene/air points, if anywhere, to an even more uncertain picture at the moment for these systems, so that they too must unfortunately await the availability of more reliable fundamental data before sensible model comparisons can be made. The recent model of methanol oxidation by Westbrook and Dryer (1979) is, however, an example of a system for which reliable predictions could probably be made. This would be an interesting topic for further study, since no detonation data for methanol/air have been published as yet.

Finally, it is important to reconsider the influence of the assumptions used in the gasdynamic approximations. The gasdynamic appproach adopted by Bach et al. (1969) and others, with which the authors thoroughly concur, usually makes the important comparison between the initiation of a global detonation as such and the continuous reinitiation process which occurs in a single cell of the detonation frontal structure. Experimental and theoretical studies have shown that a particle entering the lead shock will actually experience a shock somewhat more intense than that having the same velocity as a CJ detonation, and that the density gradient will also change substantially as the particle moves to the reaction zone. However, from a treatment with varying simple assumptions about the appropriate shock strength, it was shown that although assumptions of a stronger shock could lead to a substantially reduced induction period, the influence on the relative values of a self-consistent data set was negligible. The results of the more detailed model study for hydrogen/air and the consistency between it and the present study of ethane/air give confidence to the usefulness of the approximation taken. Since density considerations are far less significant than those associated with the temperature dependence (an exponential one), it is clear that if our assumption underestimates the thermal history of the shocked gas, then the critical temperature which we have found to be asssociated with a change in reaction type and to be responsible for the existence of limit phenomena would be attained first by a slightly weaker or stronger mixture (for the respective limits) than those we predicted.

The difficulties of incorporating correct gasdynamics and complete kinetics are legion, and this paper does not solve them. However, by incorporating a truly detailed model description of hydrocarbon oxidation kinetics with very specific detonation related conditions, it provides new insight into the controlling roles played by specific parts of the reaction scheme on detonability. There would appear to be no reason now why a fuller model of detonation initiation, sophisticated both in kinetics and gasdynamics, should not be constructed both for hydrogen/air and ethane/air. The sensitivity analysis highlights areas where the assumptions inherent in this, and many more simplified approaches, are no longer tenable. The application of such a model to other systems should not be attempted before the chemical soundness of the available rate data has been thoroughly tested. The latter points to the most important current impediment to future progress in theoretical predictions of detonation characteristics.

Conclusions

1) A kinetic model for ethane/air detonation has been described which successfully describes the experimental observations of ethane/air detonability.

2) The model predicts a change in reaction behavior -- under detonation-like initial density conditions there exists a critical temperature of about 1270 K, below which the reaction changes in type from ignition to slow combustion.

3) Using a rather simple approximation concerning the gasdynamic state in detonation initiation this critical change would occur for imaginary detonation in mixtures of concentration less than 2.7 Vol.% and greater than 13.0 Vol.%, and the implication is that these represent absolute limits to the unconfined detonability of ethane/air.

4) Using experimentally computed induction distances from the same simple gasdynamic model it has been possible to compare the well-studied hydrogen/air system with that of ethane/air, and on the basis of detonation data for hydrogen/air alone to predict very closely the detonation initiation requirements for ethane/air.

5) The same method can be used in principle for all systems for which adequate kinetic data are available. Adequate data over the appropriate initial temperature and density range do not, however, yet exist for many of even the simplest molecular systems, and, in particular, methane/air, ethylene/air, and acetylene/air kinetic data are as yet insufficiently precise to permit meaningful detailed kinetic detonation modeling.

References

Atkinson, R., Bull, D. C., and Shuff, P. J. (1980) Initiation of spherical detonation in hydrogen/air. Combust. Flame 39, 287.

Bach, G. C., Knystautas, R., and Lee, J. E. (1969) Direct initiation of spherical detonations in gaseous explosives. Twelfth Symposium (International) on Combustion, pp. 853-864. The Combustion Institute, Pittsburgh, Pa.

Borisov, A. A. and Loban, S. A. (1977) Detonation limits of hydro-carbon-air mixtures in tubes. Fiz. Goreniya Vzryva 13, 729-733.

Brossard, J., Manson, N., and Niollet, M. (1967) Propagation and vibratory phenomena of cylindrical and expanding detonation waves in gases. Eleventh Symposium (International) on Combustion, pp. 623-633. The Combustion Institute, Pittsburgh, Pa.

Bull, D. C. (1979) Concentration limits to the initiation of unconfined detonation in fuel-air mixtures. Trans. Inst. Chem. Eng. 57, 219-227.

Bull, D. C., Elsworth, J. E., and Hooper, G. (1979a) Concentration limits to unconfined detonation of ethane-air. Combust. Flame 35, 27-40.

Bull, D. C., Elsworth, J. E., and Hooper, G. (1979b) Susceptibility of methane-ethane mixtures to gaseous detonation in air. Combust. Flame 34, 327-330.

Burcat, A., Crossley, R. W., Scheller, K., and Skinner, G. B. (1972) Shock-tube investigation of ignition in ethane-oxygen-argon mixtures. Combust. Flame 18, 115-123.

Crossley, R. W., Dorko, E. A., Scheller, K., and Burcat, A. (1972) Effect of higher alkanes on ignition of methane-oxygen-argon mixtures in shock waves. Combust. Flame 19, 373-378.

Pusch, W. and Wagner, H. G. (1965) Effect of the tube cross section on the expansion of an explosion in explosive gas mixtures. I. Effect of rare gases and of the tube cross section on the detonation velocity in explosive gas mixtures. Ber. Bunsen. Phys. Chem. 69, 503-513.

Westbrook, C. K. and Haselman, L. C. (1981) Chemical kinetics in LNG detonations. Gasdynamics of Detonations and Explosions: Progress in Astronautics and Aeronautics (edited by Bowen, Manson, Oppenheim, and Soloukhin), pp. 193-206. AIAA, New York.

Westbrook, C. K. and Dryer, F. L. (1979) A comprehensive mechanism for methanol oxidation. Combust. Sci. Technol. 20, 125-140.

Zeldovich, Y. B., Kogarko, S. M., and Simonov, N. N. (1956) An experimental investigation of spherical detonation of gases. Zhur. Tekh. Fiz. 26, 1744-1768; (1957) Sov. Phys. Tech. Phys. 1, 1689-1713.

Chapter IV.   Heterogeneous Detonations

Detonations Supported by Physical Explosions of Liquefied Gases

Shunichi Tsugé* and Satoshi Kadowaki†
University of Tsukuba, Sakura, Ibaraki, Japan

Abstract

The possibility of detonation of gas/droplets mixtures due to phase change (without chemical reaction) is examined by using one-dimensional analysis. The Chapman-Jouguet point is shown to be reached through volume gain by vaporization of liquid droplets. The existence of a lower liquid temperature limit below which such detonation cannot occur is demonstrated. This temperature, which provides a measure for safety in the transportation or storage of cryogenic liquids, defines a material constant to be compared with the boiling temperature. For most liquids, the temperature is higher than the boiling temperature; they are not hazardous except under superheated conditions. For some liquids, such as benzene, ethylether or freon, it is lower than the boiling temperature, which demonstrates the existence of a detonable temperature range without superheating. The propagation velocity of the Chapman-Jouguet detonation under an idealized condition of sufficiently small droplet size and instant vaporization through leading shock wave is calculated. Factors causing failure in the onset of detonation are discussed.

Introduction

It is well known that superheated liquids often cause explosions while in the process of relaxing to the gaseous

---

Presented at the 8th ICOGER, Minsk, USSR, Aug. 23-26, 1981. Copyright © American Institute of Aeronautics and Astronautics, Inc., 1983. All rights reserved.
*Professor, School of Engineering Sciences
†Graduate Student, School of Engineering Sciences

state. Explosions occurring in melting furnaces are one of the examples where water, upon contact with molten iron, is superheated and explodes. In the same category is the problem of explosions in nuclear reactor (U_2O-Na and Na-water), pulp factories (molten salt-water), and liquid natural gas transport (water-LNG) (the underlined substances are working fluids). For an overview of these problems see Reid (1976) and Schneider (1979).

Among these combinations of heat-bath and working fluid substances, the last one has been investigated most extensively because of its use in industry (Nakanishi and Reid 1971; Katsuyama et al. 1975; Urano et al. 1978). Onset of powerful explosions are reported to take place in a relatively narrow region of a certain temperature difference between the heat-bath (water) and the working fluid (LNG). The shock wave generated in the process has been investigated by Rausch and Levine (1973, 1974).

The possibility is discussed (without analysis) that some of these explosions may be considered detonations (Board et al. 1975); and structures of such detonations have been studied on the basis of a modeled equation of state (Rabie et al. 1979). No undisputed experimental evidence, however, for the existence of such detonations has been reported so far.

The objective of this paper is to show, through a simple formulation of the Rankine-Hugoniot relationship, that a mixture of a gas and evaporating droplets behaves like an exothermic gas under idealized conditions for evaporation and above a certain threshold temperature of the droplets, so that a Chapman-Jouguet (CJ) detonation is theoretically possible. Also, the realization of an idealized condition will be critically discussed as compared with actual situations in laboratories and industrial hazards.

In the present analysis we will limit our considerations to a mixture of a gas and fine-atomized droplets confined in a shock tube generating a one-dimensional wave/flow. The initial temperatures of the gas (T_g) and of the droplets (T_1) are not necessarily equal: The two phases may be in thermal nonequilibrium, since the time needed for equilibration is greater than L/U, where L is the size of the shock tube and U is the shock propagation velocity. We may then consider a gas mixture at room temperature with a cold cryogenic liquid sprayed into the shock tube.

We note at this point that a "confined" space such as the shock tube is the only possible situation in which the onset of detonation can occur. A more realistic example of an industrial hazard such as LNG spilled in an open space (Fig. 1), is ruled out as not capable of causing detonation. In fact, in an analysis of chemical gas detonation in an open space, Tsugé and Fujiwara (1974) derived a necessary condition for detonability of a gaseous explosive with the density ρ in an environment of a gas with the density ρ_0 as

$$\rho_0/\rho > 2/(\gamma + 1) \qquad (1)$$

where γ is the adiabatic index of the environmental gas. This condition, which is subject only to the Newtonian flow approximation of a hypersonic flow theory (Hayes and Probstein 1966), precludes the detonability of the two-phase mixture now under consideration because of the high moisture content,

$$\rho \gg \rho_0$$

of the explosive gas.

Physical Origin of Detonability

Before beginning analysis an intuitive explanation as to how a (rapid) phase transformation can support a detonation is relevant.

Detonation is made self-sustaining by the CJ plane where the local Mach number is unity; in other words, the flow is <u>choked</u>. In the case of chemical detonation propagating in a premixed gas, the sonic flow is achieved by flow acceleration in passing through a flame. Another mechanism generating the same condition is proposed by Fujitsuna and Tsugé (1973) and confirmed by Fujitsuna (1979) to explain the structure of film detonations, where the evaporation and combustion of a condensed-phase explosive deposited on the wall form a throat through their pinching the core flow, thereby reaching the sonic condition (see

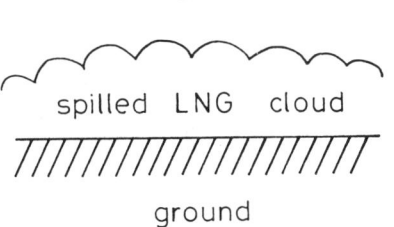

Fig. 1 Schematic figure of the spilled LNG vaporized upon contact with the ground as heat-bath. In the open space detonation does not occur.

Fig. 2). In a gas/droplets mixture in the absence of combustion, a similar Laval nozzle will be formed by the phase transformation, if the "igniting" shock wave is strong enough to shutter and heat the droplets, causing rapid evaporation. Then, a choked flow condition will be realized in view of the enormous volume effect, namely, of the volume ratio of the gas with the liquid of about 10^3 in evaporation (Fig. 3).

One-Dimensional Analysis

Consider a gas and fine droplets mixture of a liquid with the specific volumes v_g and v_l and temperatures T_g and

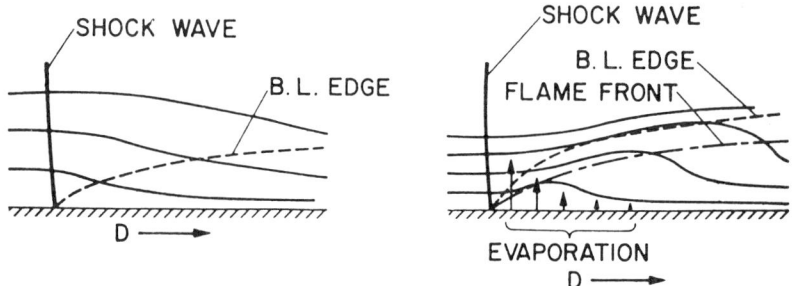

Fig. 2 Pinching effect due to evaporation and combustion in the boundary layer to generate a sonic throat in the film detonation [reproduced from the paper by Fujitsuna and Tsugé (1973)].

Physical Detonation

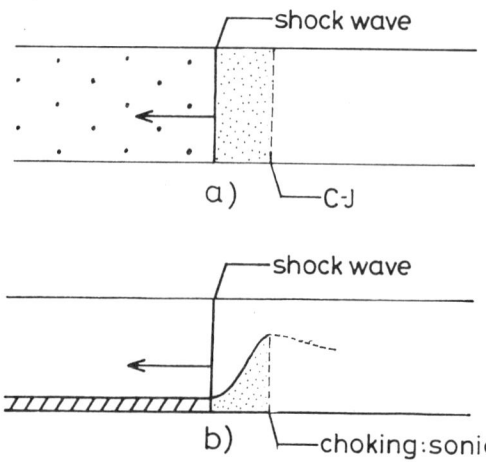

Fig. 3a) Sketch of possible detonation propagating in a homogenized gas/droplets mixture. b) Conceptual equivalent of a) to illustrate the possiblility of forming a sonic throat by volume effects through gasification.

T_1, respectively. As mentioned in the Introduction, the following analysis is not affected by the thermal nonequilibrium ($T_g \neq T_1$). In the formulation of the Rankine-Hugoniot relation, we note the following features characterizing the phenomena under observation: firstly, the specific volume v for the mixture is given by

$$v = Y v_g + (1 - Y) v_1$$

where Y is the mass fraction of the gas. We assume in what follows that the volume occupied by liquid phase is negligible:

$$v \sim Y v_g \qquad (2)$$

Since $v_1/v_g \sim 10^{-3}$, this assumption allows us to cover mixtures with a fairly heavy loading of droplets, for instance, with initial weight fraction of gas being 1%. We also assume that the pressure is given by

$$pv = Y R_M T_g \qquad (3)$$

where v is given by Eq. (2). This means that the liquid phase does not contribute to the pressure. Secondly, the specific enthalpy of the mixture is expressed by

$$h = Y h_g(T_g) + (1 - Y) h_1 (T_1) \qquad (4)$$

with

$$h_g(T) = C_p T$$

$$h_1(T) = C_p T - q(T) \qquad (5)$$

where C_p is the specific heat under constant pressure of the gas phase; h_g and h_1 are the specific enthalpies of the gas and liquid phases, respectively; and $q(T)$ is the specific latent heat at temperature T. We note that material constants specifying the liquid phase appear only through q.

Taking two control surfaces, one just in front of the shock wave and the other at the point where the evaporation has been completed, we write the Rankine-Hugoniot relations as

$$u_0/v_0 = u/v \equiv m$$

$$m^2 v_0 + p_0 = m^2 v + p \qquad (6)$$

$$\tfrac{1}{2} u_0^2 + h_0 = \tfrac{1}{2} u^2 + h$$

where subscript 0 refers to the state before the shock wave. If relations (2), (3), and (4) are substituted into (6) and velocity u is eliminated, we have the following relationship between the pressure and the specific volume nondimensionalized as

$$\xi = v/v_0$$
$$\eta = p/p_0 \tag{7}$$

in the form:

$$\eta = (\frac{\gamma+1}{2\gamma} - \frac{\gamma-1}{2\gamma}\xi + Q)/\frac{\gamma+1}{2\gamma}\xi - \frac{\gamma-1}{2\gamma} \tag{8}$$

with Q defined by

$$Q \equiv \frac{T_1}{T_g} \frac{Y-Y_0}{Y_0} \left(1 - \frac{q(T_1)}{C_p T_1}\right) \tag{9}$$

and γ by

$$\gamma = C_p/(C_p - R_M)$$

assumed as common to shocked and unshocked gas. In the following calculation, γ is set as 1.4. As is readily seen, relationship (8) is nothing but the classical Hugoniot detonation for a gas with nondimensional heat of reaction Q, representing an hyperbola in the $\xi - \eta$ plane (Fig. 4). The

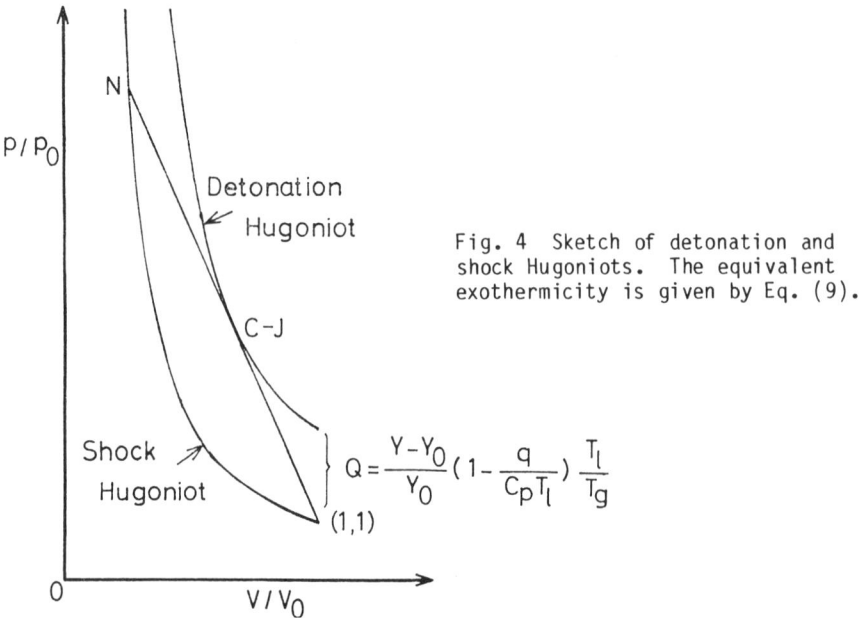

Fig. 4 Sketch of detonation and shock Hugoniots. The equivalent exothermicity is given by Eq. (9).

equivalent exothermicity is now given by Eq. (9) depending on the initial temperature ratio, on the fraction of droplets vaporized within the control surface, and, most crucially, on the liquid temperature. The quantity can be positive or negative depending on whether $C_p T_1$ is greater or less than $q(T_1)$, reflecting the two factors competing in the process: the mass addition in the gas phase due to evaporation and the volume decrement due to latent heat cooling. When the liquid temperature is above critical, namely, $T_1 > T^*$, with T^* defined by

$$f(T^*) \equiv 1 - q(T^*)/C_p T^* = 0 \qquad (10)$$

the mixture behaves like an exothermic gas, and the equivalent exothermicity is greater the greater the initial liquid content is and the higher the liquid temperature is.

Safety Limit Temperature T*

It should be mentioned that the solution of Eq. (10) exists for any gas at a point in the temperature range

$$0 < T^* < T_c$$

where T_c is the critical temperature, since $f(T)$ is negative for sufficiently low temperature, whereas it takes the positive value of unity at $T = T_c$, where

$$q(T_c) = 0$$

The following formula:

$$q(T) = q_0 (1 - T/T_c)^{0.38} \qquad (11)$$

where q_0 is an empirical constant, has been well received as a best fit to the experiment (Sato 1970). Table 1 shows the values of T^* defined by Eq. (10) for various liquids and liquefied gases. The critical temperature T_c at which gaseous and liquid states are no longer discernible and the boiling temperature T_b at which the vapor pressure is 1 atm are also listed for comparison. For the majority of gases we have

$$T^* > T_b \qquad (12)$$

which means that the liquid should be superheated in order to support detonations in the 1 atm environment. According to Table 1, ethylether $(C_2H_5)_2O$, benzene C_6H_6, and freon

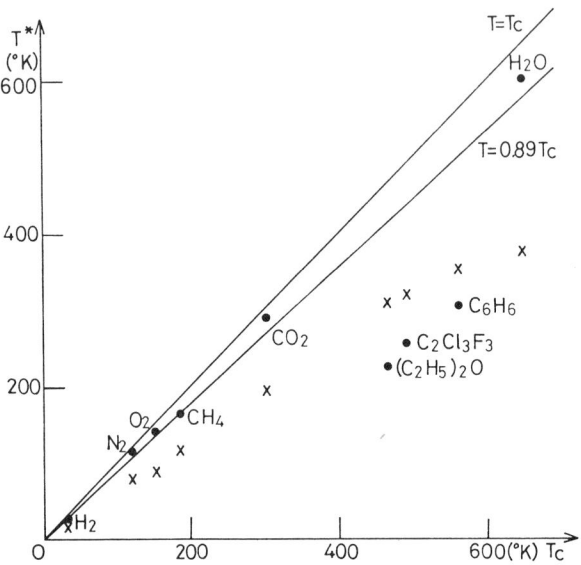

Fig. 5 Safety limit temperature T* (denoted by o) plotted against critical temperature T_c from Table 1. T* is defined by Eq. (10). The boiling temperatures (denoted by x) for respective materials are also shown for comparison.

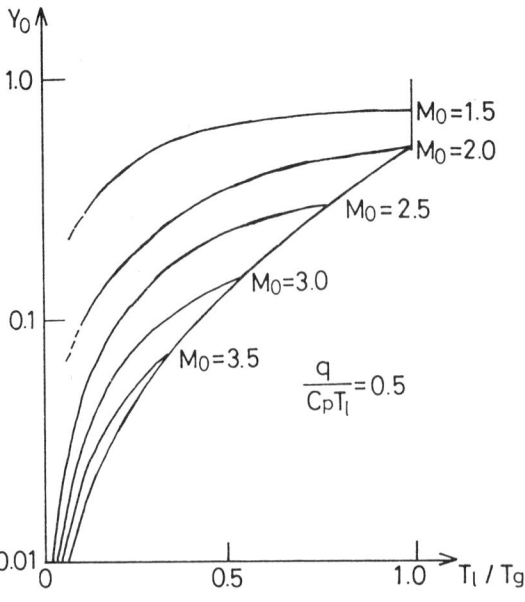

Fig. 6 Propagation Mach numbers of the idealized detonation at liquid temperature $q/C_p T_l$ = 0.5 as a function of initial gas mass fraction Y_0 and liquid/gas temperature ratio. The line designated by $T_{CJ} = T_l$ shows the limit line where the detonated gas reaches the dew point at the CJ point. The dew point is approximated by T_l for simplicity.

Table 1 Safety limit temperature T* calculated for various liquids and liquefied gases on the basis of its definition [Eq. (10)]. The boiling temperature T_b and critical temperature T_c are also listed for comparison.

Gas	T_b, K	T*, K	T_c, K
He	4.2	4.1	5.2
H_2	20.4	24	33.3
N_2	77.3	115	126.2
O_2	90.2	139	154.4
CH_4	111.7	164	190.7
CO_2	194.7	293	305.1
$(C_2H_5)_2O$	307.7	227	467.0
$C_2Cl_3F_3$	320.6	252	487.1
C_6H_6	353.3	308	562.7
H_2O	373.2	605	647.3

R113 $C_2Cl_3F_3$ are exceptional,

$$T^* < T_b \tag{13}$$

so that those liquids have a real detonability temperature range of $T^* < T < T_b$; in other words, no superheating is necessary.

The physical meaning of temperature T* is simply that for T > T* the droplets can act as an equivalent exothermic gas under some idealized conditions. We may say, however, that for liquid temperatures lower than T*, the liquid can never detonate because, then, the mixture behaves as an endothermic gas. In this sense temperature T* may be defined as the safety limit temperature. In fact, if a cryogenic liquid is chilled below this temperature during transportation or storage, safety is fully secured against accidental detonation. The smallness of T* may be looked upon as a qualitative measure for detonability of the liquid upon phase change. To visualize this situation, Table 1 is plotted in the plane T* vs T_c in Fig. 5. Rausch and Levine (1974) have established that the line T= 0.89 T_c represents the upper limit superheating. Points lying above this line correspond to the safe substances in this sense.

Idealized Physical Detonation

Let us assume that momentum and heat losses are negligible between the shock front and the sonic plane, and that the droplets are fine enough to respond quickly to the

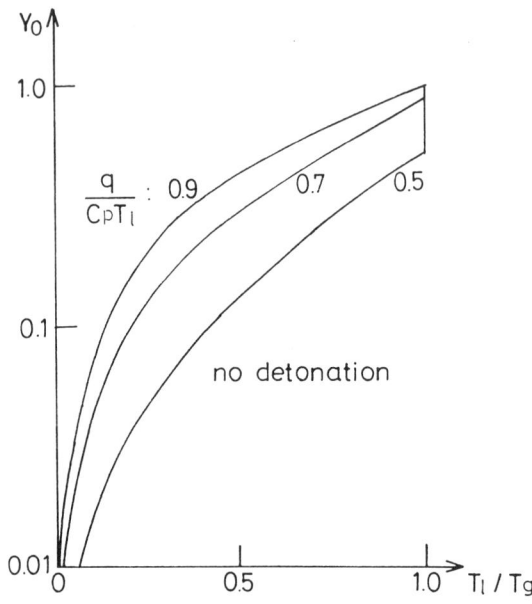

Fig. 7 Dependence of the detonation limit on initial gas mass fraction Y_0 and intial liquid/gas temperature ratio with varied values of $q/C_p T_1$.

motion of the gas without delay. Then, according to the conventional theory of the classical detonation, the CJ condition is expressed by

$$u_0/c = 1 \qquad (14)$$

with

$$c = (\gamma^{1/2}/\beta)c_g \qquad (15)$$

where c is the speed of sound of the mixture, c_g is that of the pure gas, and β is the void fraction

$$\beta = \frac{Y}{Y + (1-Y)(v_1/v_g)} \qquad (16)$$

which is unity within the approximation made at the outset of this analysis. Figure 6 shows the propagation Mach number of the idealized detonation as it depends on the initial mass fraction Y_0 of the gas and the initial temperature ratio T_1/T_g at the fixed value of $q/C_p T = 0.5$. The heavier the droplets loading and the higher the gas temperature compared with the liquid temperature, the higher the Mach number is. The limit line designated by $T_{CJ} = T_1$ is a line such that the accelerated fluid is cooled down to reach the dew point at the CJ plane to be determined by the Clapeyron-Clausius law. Since the dew point differs for

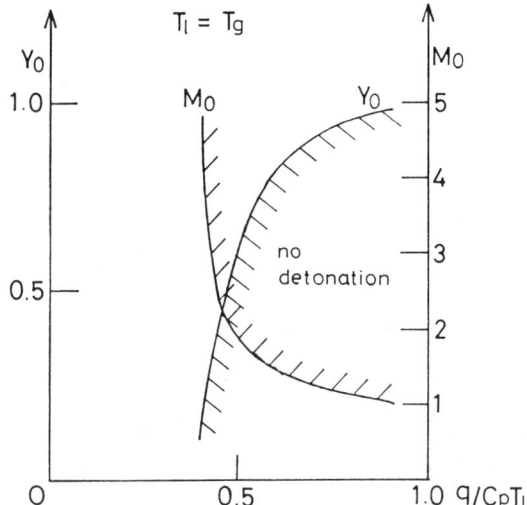

Fig. 8 Limiting Mach number and initial gas mass fraction for varied parameter of $q/C_p T_l$ under thermally equilibrium initial conditions ($T_l = T_g$). The shaded part represents the prohibited zone.

different liquid materials it is replaced by T_l in the figure to get a rough idea of the lower temperature limit required to maintain the gaseous state. Figure 7 shows the limit for different values in the parameter $q/C_p T$. As the equivalent exothermicity diminishes, the limit line recedes to make the detonable region narrower. In Fig. 8, the lower limit of initial gas weight fraction Y_0^*, which corresponds to the upper limit of droplet loading being $1-Y_0^*$, together with the propagation Mach number M_0 are plotted for varied parameter of $q/C_p T$ for the thermally equilibrium initial condition ($T_l = T_g$).

Detonation Failure Due to Finite Evaporation Rate and Finite Momentum Loss

In formulating the Rankine-Hugoniot equations (6) we have assumed no momentum or heat losses and complete gasification of droplets between the two control surfaces. In reality there are finite effects due to these factors by which the detonation failure will take place.

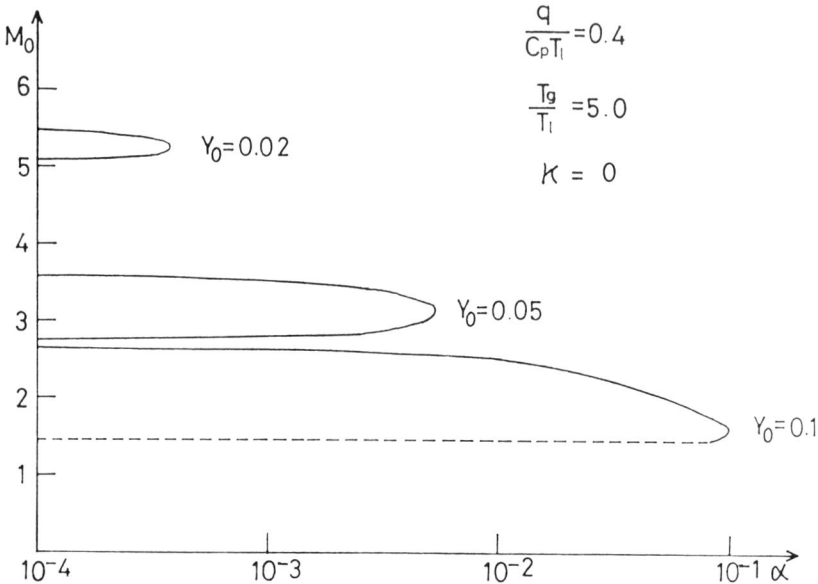

Fig. 9 Effects of momentum loss and finite evaporation rate on the detonation failure. This case ($\kappa = 0$) assumes that the evaporation rate coefficient is independent of the velocity of a droplet relative to the gas.

To investigate these effects we adopt the following finite rate equation for evaporation:

$$\frac{dY}{dx} = A(v_0-v)^{\kappa}(T-T_1)(1-Y) \quad (17)$$

with

$$A = 4\pi r^2 (3v_1/4\pi r^3)k \quad (18)$$

where x is the coordinate directed downstream and r is the mean radius of the droplets. This equation is based on the model that the rate of evaporation is proportional to the total surface area exposed to the gas, and to the temperature difference, with the evaporation coefficient depending on the relative velocity $m(v_0-v)$ of the droplet with the surrounding gas like $k(v_0-v)^{\kappa}$.

We also replace the momentum equation, namely, the second part of Eq. (6), with

$$\frac{d}{dx}(m^2v + p) = B(v_0 - v) \quad (19)$$

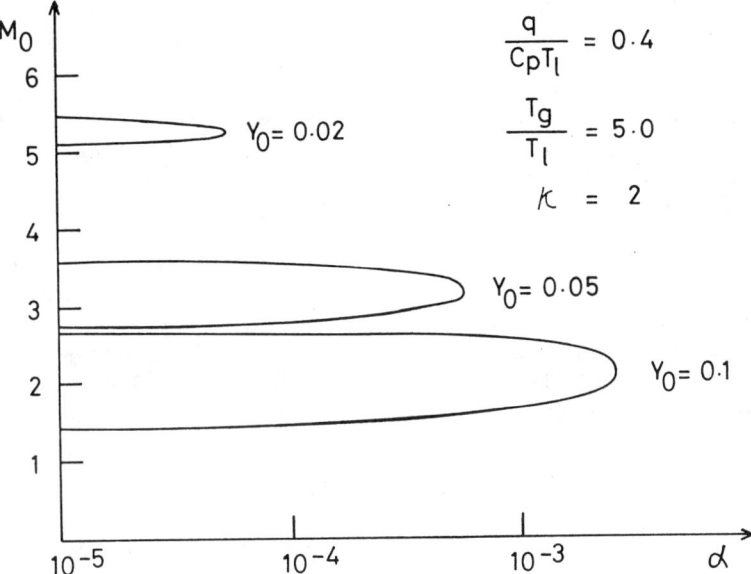

Fig. 10 Effects of momentum loss and finite evaporation rate on the detonation failure. This case ($\kappa = 2$) assumes that the evaporation rate coefficient is proportional to the squared velocity of a droplet relative to the gas.

where the term on the right-hand side represents the momentum loss due to shearing with the sidewalls. In contrast to this term, which acts as a factor causing detonation failure, the thermal interactions with the sidewalls can be heat addition as well as heat loss in the case of cryogenic liquids. Therefore we disregard this ambiguous element in evaluating the failure and assume that the energy equation in the form of Eq. (6) remains valid.

If the independent variable x is eliminated from Eqs. (17) and (19) and the equation is written in the form of $v(Y)$, we have

$$\frac{dv}{dY} = \frac{F(Y,v)}{1-M^2} \qquad (20)$$

with

$$F(Y,v) \equiv \frac{v}{p} \frac{\gamma-1}{\gamma} \frac{C_p T_1 - q}{v} - \frac{B}{A} \frac{(v_0-v)^{1-\kappa}}{(1-Y)(T-T_1)} \qquad (21)$$

where the pressure and the temperature are expressed in terms of v and Y through equation of state (2) and the energy equation [the third equation of Eq. (6)] with enthalpy expression (4) is incorporated. Equation (20) is expected to have a saddle singularity at M = 1 as inferred from the counterpart theory of detonation in premixed gases (Tsuge 1971; Fickett and Davis 1979). This means that only two integral curves can pass through the singular point at M − 1 = 0; in other words, the solution for the CJ detonation can exist only for special values of propagation velocity u_0. The eigenvalue condition may be expressed as

$$F(Y,v) = 0 \quad \text{at } M - 1 = 0 \tag{22}$$

which serves to replace the classical CJ condition (14).

Figures 9 and 10 show the results of the eigenvalue calculation where the propagation Mach number is plotted against friction constant α defined by

$$\alpha = (B/A)(v_0^{2-\kappa}/R_M T_1^2) \tag{23}$$

The only difference between the two is the values of κ specifying the effect of relative velocity on the evaporation coefficient. They are qualitatively similar except that some numerical instability occurs for the case of $\kappa=0$ as the propagation Mach number approches unity. These results demonstrate that detonation failure takes place for above-critical momentum loss in agreement with the case of premixed gas detonation. The only difference is that the lower branch, which is normally considered the unstable branch, does not approach the Mach wave in the limit of vanishing loss, but remains a wave with finite amplitude.

Limitation on Droplet Size

In the foregoing analysis we have idealized reality in discussing the possible detonability of a gas/droplets mixture upon phase transformation. The most, and the only crucial necessity among them is the postulate that the droplets be small enough to move together with gas motion. Theoretically, we may choose a droplet size as small as we wish so that there is no delay in the motion of the droplet relative to the gas. Experimentally, however, this condition is rarely met in two-phase flow experiments designed for industrial problems. Figure 11 shows the speed of sound of the water-air mixture (Hasegawa and Fujiwara 1980) as dependent on air mass fraction Y or void fraction β. The solid curve which represents the idealized circumstance of no relative motion between the two phases

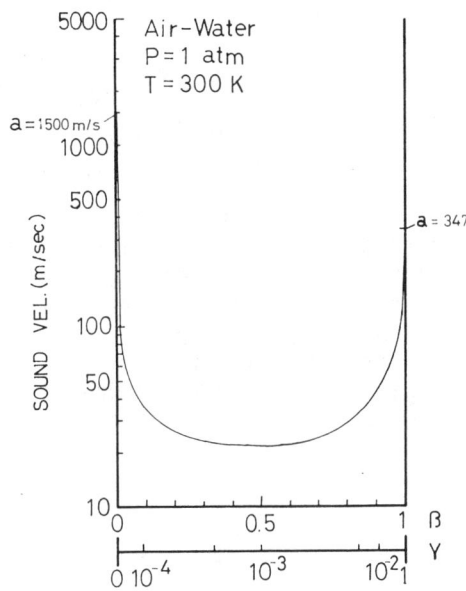

Fig. 11 Speed of sound of an air-water mixture as dependent on void fraction β or the mass fraction of the gas (Hasegawa and Fugiwara 1980). The theoretical curve assumes no delay in the motion of the droplets relative to the motion in the gas phase.

indicates a large discrepancy with the experimental data obtained for standard size droplets [see, for example, Akagawa (1974)]. For such mixtures the gaseous wave motion is decoupled from the droplet motion, in which case, detonation as discussed in this analysis is not to be expected.

Let us seek a condition under which the droplets move with gaseous wave motion with tolerable delay: Provided that the size of droplets is small enough to obey the Stokes law of friction, the time τ_1 required for droplets to recover its delay in motion relative to the surrounding gas motion is (Marble 1970)

$$\tau_1 = \frac{2}{9} \frac{v_g}{v_l} \frac{r^2}{\nu} \tag{24}$$

where ν is the kinematic viscosity. Since the recovery should be accomplished within time τ_2 elapsed by the droplet in passing the shock front through the CJ plane we have

$$\tau_1 \lesssim \tau_2 \tag{25}$$

When τ_2 is defined by l/u_0, where l is the distance between the shock front and the CJ plane and u_0 is the shock velocity, inequality (25) solved for the droplet radius r is

$$r \lesssim \left(\frac{9}{2} \frac{1}{u_0} \frac{v_1}{v_g} \nu\right)^{1/2} \tag{26}$$

For the realistic values of $v_g/v_1 \sim 10^3$, $\nu \sim 10^{-1}$ cm^2/s, $l \sim$ 1 cm, and $u_0 \sim 10^4$ cm/s, we have $r \lesssim 2\mu$. This value is considered to be the lower limit of the size of a droplet that can be produced in laboratories using a conventional apparatus, for example, the ultrasonic mist generator. Most probably, the average size of droplets present in industrial hazards such as spillage of cryogenic liquids from a tank or collision accidents during its transportation is much greater than the value obtained here. The actual nature of detonations of this type can be checked only through careful laboratory experiments.

References

Akagawa, K. (1974) Two phase flow (in Japanese) (Corona Sha Book Co.) p. 239.

Board, S. J., Hall, R. W., and Hall, R. S. (1975) Detonation of fuel coolant explosions. Nature 254, 319-321.

Fickett, W. and Davis, W. C. (1979) Detonation, p. 228. University of California Press, Berkeley, Calif.

Fujitsuna, Y. and Tsuge, S. (1973) On detonation waves supported by diffusion flames I. Fourteenth Symposium on Combustion, pp. 1265-1275. The Combustion Institute, Pittsburgh, Pa.

Fujitsuna, Y. (1979) On detonation waves supported by diffusion flames II. Acta Astron. 6, 785-793.

Hasegawa, T. and Fujiwara, T. (1980) Shock waves through bubbles (in Japanese) pp. 275-277. Preprint, Combustion Symposium (Japanese Section).

Hayes, W. D. and Probstein, R. F. (1966) Hypersonic Flow Theory, p. 129. Academic Press, New York.

Katsuyama, K., Shiota, K., Matsuda, T., and Kato, S. (1975) LPG-water vapor explosion. Saiko To Hoan 21, 169-176.

Marble, F. E. (1970) Dynamics of dusty gases. Annu. Rev. Fluid Mech. 2, 397-446.

Nakanishi, E. and Reid, R. C. (1971) Liquid natural gas-water reactions. Chem. Eng. Prog. 67, 36-41.

Rabie, R. L., Fowles, G. R., and Fickett, W. (1979) The polymorphic detonation. Phys. Fluids 22, 422-435.

Rausch, A. H. and Levine, A. D. (1973) Rapid phase transformations caused by thermodynamic instability in cryogens. Cryogenics 13, 224-229.

Rausch, A. H. and Levine, A. D. (1974) Shock wave overpressure due to metasable phase transformation in single component cryogens. Cryogenics 14, 139-146.

Reid, R. C. (1976) Superheated liquids. Am. Sci. 64, 146-156.

Sato, K. (1970) Evaluations of material constants, pp. 235-263. Maruzen Book Co., Tokyo, Japan.

Schneider, A. L. (1979) Liquefied natural gas safety research overview. Notes for Seminar and Study Tour on LNP Peak Shaving. U.S. Coast Guard Headquarters.

Tsugé, S. (1971) The effect of boundaries on the velocity deficit and the limit of gaseous detonations. Combust. Sci. Technol. 3, 195-205.

Tsugé, S. and Fujiwara, T. (1974) On the propagation velocity of a detonation-shock combined wave. Z. Angew. Math. Mech. 54, 157-164.

Urano, Y., Hashiguchi, Y., Ogawara, T., and Iwasaki, M. (1978) Explosions caused by liquefied propane spillage on water. Tokyo Kogyo Shikenjo Hokoku 73, 187-192.

Effect of Liquid Films on Detonation in a Gaseous Mixture

J.P. Saint-Cloud,* C. Guerraud,† and N. Manson‡
Université de Poitiers, Poitiers, France

Abstract

Previous investigations indicate that polyhedral aqueous foams of constant gaseous density but different foam structure exhibit two different detonation modes. The most significant foam parameter which distinguishes these two regimes is the specific area of the foam. The authors report a study of the propagation of detonations in a rectangular tube (20 x 23 x 492 mm) filled with foam whose gas is a stoichiometric propane oxygen mixture at room temperature and pressure. In a first set of experiments the gaseous mixture was divided by means of liquid films in equal volumes from 20 to 35 mm in length. In a second set, the tube was filled with foams of polyhedral bubbles with a diameter greater than 5 mm. The phenomena have been recorded by means of streakphotography and the velocity and pressure have been measured with piezoelectric gages. For the two set of experiments, the liquid film is instantaneously broken in droplets by the detonation front. The droplet velocity just behind the detonation front is 100-200 m/s, decreases downstream, and finally increases again due to the effects of reflected pressure waves created by the interaction of the detonation with the liquid films.

Introduction

In a previous study (Saint-Cloud et al. 1976) on the propagation of detonation in an aqueous foam filled with a

---
Paper presented at the 8th ICOGER, Minsk, USSR, Aug. 23-26, 1981. Copyright © American Institute of Aeronautics and Astronautics, Inc. 1982. All rights reserved.

*Assistant Professor, Laboratoire d'Energétique et de Détonique.

†Research Engineer, Laboratorie de'Energétique et de Détonique.

‡Professor, Laboratorie de'Energétique et de Détonique.

reactive gaseous mixture, two different regimes of
detonation were observed for a given density but different
foam structure. For a foam for which the mean bubble
diameter is greater than 3-4 mm, the detonation is very
similar to the one observed in gaseous mixtures; and for
smaller diameters (less than 1-2 mm), the shock and
combustion fronts are clearly separated. Their velocity and
pressure behind the combustion front are markedly lower than
those in the gaseous mixture alone, as in "galloping"
detonations (Saint-Cloud et al. 1972).

The most important parameter was found to be the
specific area of the foam, i.e., the number and thickness of
liquid walls of the bubbles (Saint-Cloud et al. 1979).

The foams studied were of the "dry kind" (Ross 1969) in
which the structure was composed of polyhedric planar liquid
films. Because of the large number of bubbles it was
impossible to determine the influence of each liquid film on
the propagation of the detonation. This paper reports an
investigation of the influence of foam systems on the
propagation of detonations in a rectangular tube. The
reactive mixture was a stoichiometric mixture of propane and
oxygen, and the foam was either regular interval planar
liquid films or large diameter polyhedral bubbles (> 5 mm).

Experimental Device

The Detonation Tube

The experimental tube (Fig. 1a) had a removable top and
was 20 mm x 23 mm in cross section and 492 mm in length.
The two lateral walls were glass windows. Streak photograph
records were taken through a lateral slit (1 mm width and
164 mm length) parallel to the axis of the tube. Onset of
the detonation was detected by an optical probe SI located
on the top of the tube 2 cm from extremity A. Pressure in
the detonation products was observed with piezoelectric
gages P_1 and P_2 located 171 and 247 mm from A (Brochet et
al. 1969). The locations of the pressure transducers were
in the field of view so that pressure and streak photograph
records could be correlated. The flash source S (duration 1
ms) located at the focus of a lens L provided backlighting
of the phenomenom. The streak records were taken with a
20-cm-diam rotating drum camera at rotational speeds of 80-
120 rad/s.

Production of Liquid Films

To produce liquid films, the top of the tube was
removed, and the three internal walls were covered with a

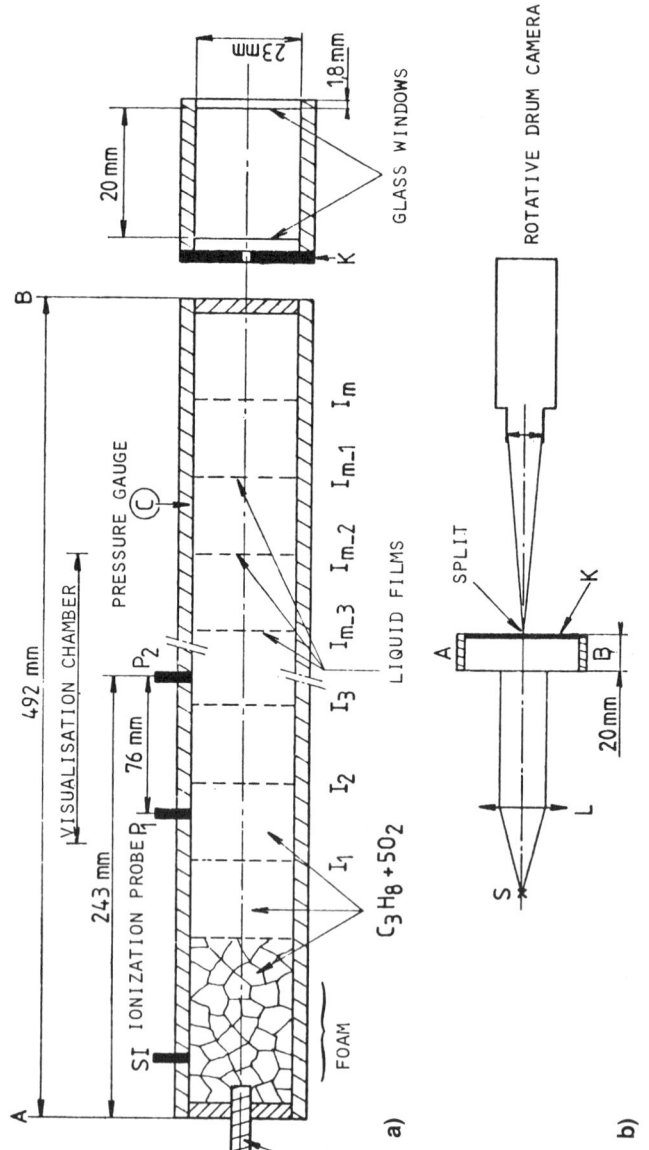

Fig. 1 Experimental apparatus: a) detonation tube; b) visualization system.

thin liquid film of water containing 5% of a detergent. The top of the tube was also wetted and replaced. The gaseous mixture ($C_3H_8 + 5\ O_2$) was injected through the bottom liquid film by means of a 2 mm i.d. tube. Parallelepipeds whose cross-section walls were the planar liquid films I_i were obtained by injection at regular intervals. Wetting of the top was necessary to stabilize the foam structure. The distance between two liquid interfaces ranged between 20 and 35 mm. The gaseous mixture flow rate was adjusted to obtain polyhedric foams with mean bubble diameters in the ranges 5-7 or 10-12 mm.

Initiation of the Detonation

Initiation was produced by an electrical ignitor located at tube extremity I (see Fig. 1a). With this experimental tube it was not possible to observe the transition between deflagration and detonation with the gaseous mixture alone or with planar liquid films or foams with large bubble diameters. To facilitate transition, a foam column (about 10 cm long) of small bubbles (2-3 mm in diameter) was required adjacent to the electrical ignitor. As previously shown (Saint-Cloud et al. 1977), the transition was accelerated by a mechanism similar to that observed by Shchelkin (1945). The stoichiometric propane-oxygen mixture was prepared as reported in a previous study (Saint-Cloud et al. 1976). The foams were ignited at room conditions ($T_0 = 292 \pm 5$ K, $P_0 = 1$ atm).

Experimental Results

Interaction Between Detonation and Planar Liquid Films

A streak photograph for a planar film detonation is shown on Fig. 2. On Fig. 3 the pressure record is superposed on the space-time diagram deduced from Fig. 2. On the space-time diagram: 1) I_1, ..., I_n represent the planar liquid films; 2) F corresponds to the detonation front; 3) Ω_i are the reflected pressure waves produced by the interaction between the detonation front F and liquid films I_i; and 4) t_i are the paths of the water droplets which result from the disintegration of the liquid film by the detonation front F.

The space-time diagrams obtained for experiments with and without liquid films indicate that the velocity of the

Fig. 2 Streak photograph of a planar film detonation (stoichiometric propane oxygen; distance between films 20 mm).

detonation front F is quite the same as that in the gaseous mixture alone and is unchanged when F interacts with a liquid film. The disintegration of liquid films is instantaneous, and the resulting small water droplets are swept along by the detonation gaseous products, whose predicted velocity is 1090 m/s. The observed maximum velocity of the droplets lies between 100 and 200 m/s (see Fig. 4), and is reached a few microseconds after passage of the detonation.

Subsequently, the droplets are drastically decelerated, stopped, and again accelerated backward to a velocity of about 20-60 m/s, and then they decelerate and move forward again.

The velocities of the reflected pressure waves Ω_i are approximately 500 + 100 m/s. Analysis of the space-time diagram indicates that the acceleration and deceleration of water droplets can be related to these pressure waves. This observation was confirmed through experiments in which only one liquid film was established in the tube and was located between the two pressure gages P_1 and P_2. The streak records showed that the droplets created by film breakup were accelerated but did not experience a deceleration.

Interaction of the Detonation with Liquid Films of Large Diameter Bubbles

Two different sets of experiments (mean diameters of bubbles: 5-7 and 10-12 mm) were conducted to show that the interaction of the detonation front with the liquid films of a bubble foam is similar to that observed in planar flims. The streak photograph records (Fig. 5) and the space-time diagrams deduced therefrom, and the pressure

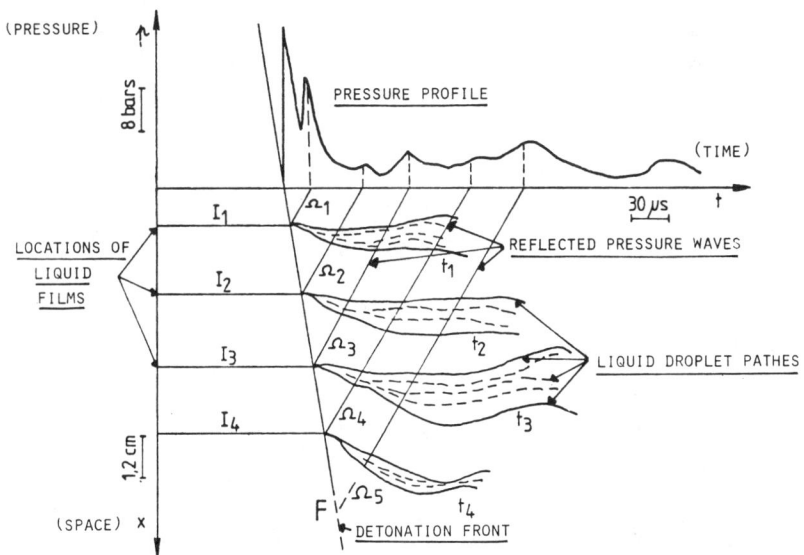

Fig. 3 Pressure record and space-time diagram, deduced from Fig. 2.

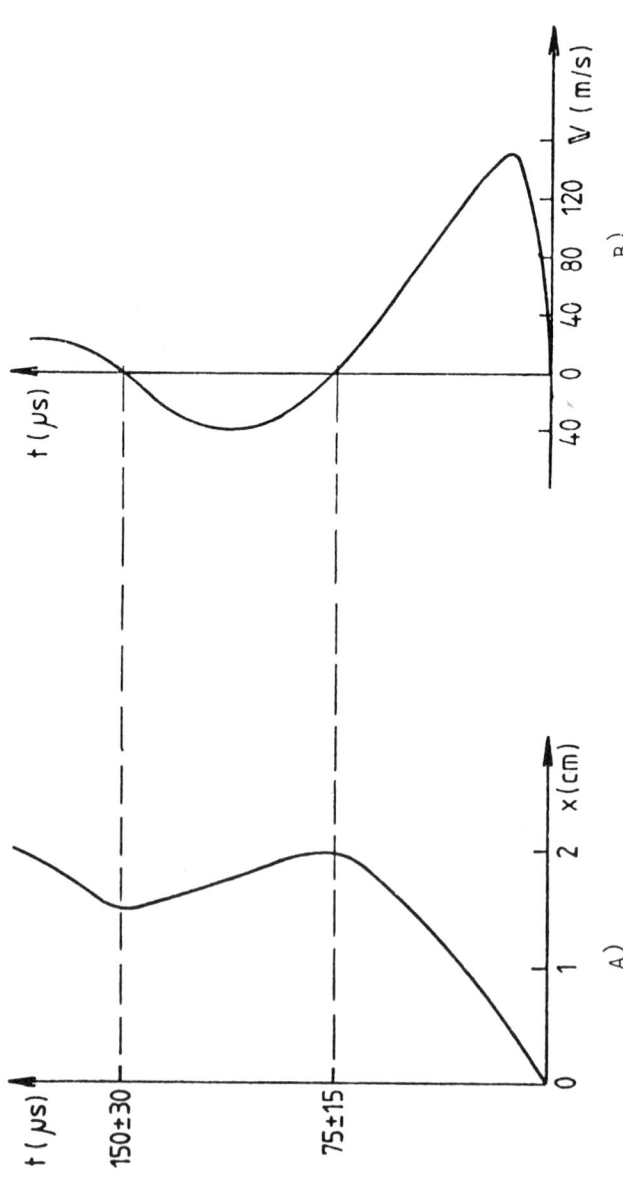

Fig. 4 Motion of water droplets produced by film break-up: a) space-time diagram; b) droplet velocity.

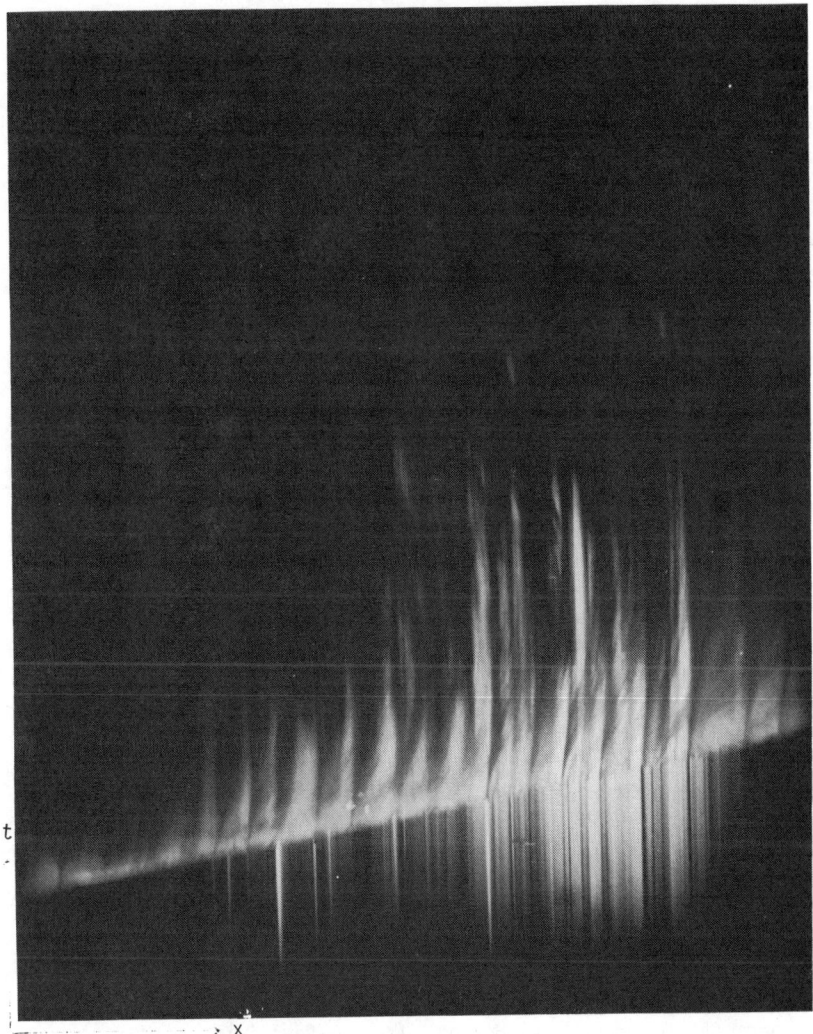

Fig. 5 Streak photography of a detonation propagating in foam with large bubble diameters (stoichiometric propane oxygen; mean bubble diameter 5-7 mm).

records (Fig. 6) are really similar. The most significant difference is that the deceleration of the droplets is faster because of the smaller distance between two liquid films in the foams.

Conclusions

The experiments show that 1) the detonation velocity for a planar film, or a large bubble diameter foam, is

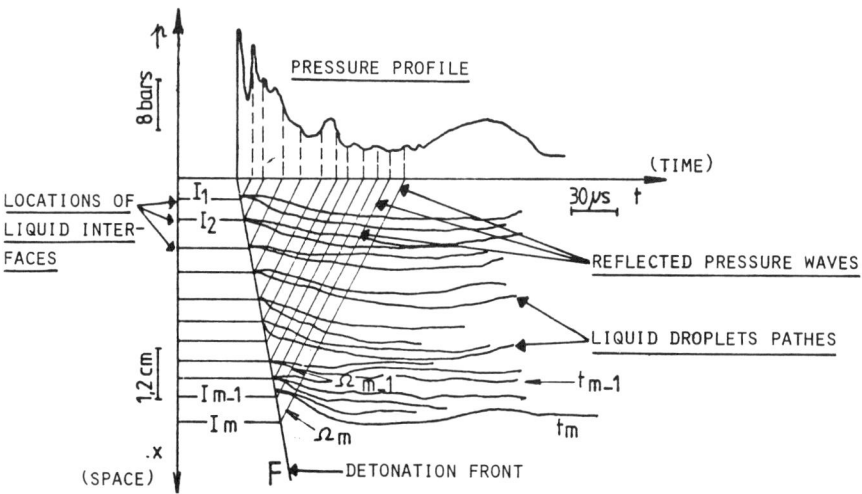

Fig. 6 Pressure record and space-time diagram deduced from Fig. 5.

nearly the same as that for the gaseous mixture alone; 2) the liquid films are broken instantaneously into droplets by the detonation front; 3) the droplet velocity a few microseconds behind the detonation front is about 100-200 m/s and then decreases downstream and finally increases again owing to the effects of reflected pressure waves which are created by the interaction of the detonation with the liquid films; and 4) the recorded pressure peak in foam detonations is 30% below that measured in the gaseous mixture alone and did not vary significantly with foam characteristics as long as the foam mean bubble diameter was greater than 2-3 mm.

References

Brochet, C., Guerraud, C., Manson, N., and Veyssière, M. (1969) Variation de la pression des gaz derrière le front des détonations dans les mélanges stricts propane-oxygène, influence du diamètre du tube et de la pression initiale. C.R. Acad. Sci. Paris 268B, 361-364.

Ross, S. (1969) Bubbles and foams. Ind. Eng. Chem. 61, 48-57.

Saint-Cloud, J. P., Guerraud, C., Moreau, M., and Manson, N. (1976) Expériences sur la propagation des détonations dans un milieu biphasique. Acta Astron. 3, 781-794.

Saint-Cloud, J. P., Guerraud, C., Brochet, C., and Manson, N. (1972) Quelques particularités des détonations très instables dans les mélanges gazeux. Acta Astron. 17, 487-498.

Saint-Cloud, J. P., Guerraud, C., and Moreau, M. (1979) Mesure de la masse volumique de l'aire interfaciale liquide gaz et de la célérité des ondes de pression dans des mousses aqueuses. Rapport de l'A.T.P. C.N.R.S. No. 3272.

Saint-Cloud, J. P., Guerraud, C., Moreau, M., and Manson, N. (1977) Formation et structure des détonations dans des mousses aqueuses renfermant un mélange de propane et d'oxygène, Communication au 6ème Colloque International sur la Dynamique des gaz en explosion et des systèmes réactifs, Stockholm, Sweden.

Shchelkin, K. I. (1945) Decrease of detonation velocity in rough tubes. Acta Phys. Chem. USSR 20, 303-306.

Ignition of Aluminum Particles in a Gaseous Detonation

B. Veyssiere[*]
Université de Poitiers, Poitiers, France

Abstract

Values of the ignition delay of an aluminum particle deduced from simple models are calculated and compared with our previous experimental results for the detonation of C_2H_4/air/Al particles mixtures. The choice of the Friedman-Macek criterion for ignition of aluminum particles is discussed. It is concluded that the results corroborate the previously proposed scheme of the detonation structure.

Nomenclature

C	= heat capacity of particle
CJ index	= Chapman-Jouguet point
c_{ox}	= oxidizer concentration
c_{xp}	= drag coefficient of particle
d	= particle diameter
E	= activation energy
h	= heat exchange coefficient
k_n	= pre-exponential factor
L	= heat of fusion of aluminum
m	= reaction order relative to oxidizer concentration
n	= reaction order relative to thickness of oxide film
Nu	= Nusselt number
Q	= $-\Delta H_R$ = heat of reaction
r	= equivalence ratio of gaseous mixture
R	= perfect gas constant
Re	= Reynolds number
S	= particle surface

Presented at the 8th ICOGER, Minsk, USSR, Aug. 23-26, 1981. Copyright © American Institute of Aeronautics and Astronautics, Inc., 1982. All rights reserved.

[*]Attaché de Recherches CNRS.

t	= time
T	= temperature
T_o	= initial temperature of particle
T_g	= gas temperature
T_i	= ignition temperature of particle
$T_{m_{Al}}$	= melting temperature of aluminum
$T_{m_{Al_2O_3}}$	= melting temperature of aluminum oxide
v	= particle volume
V_g	= gas velocity
V_p	= particle velocity
δ_p	= thickness of oxide film at the surface of particle
ε_p	= emissivity of particle
λ_g	= thermal conductivity of gas
	= dynamic viscosity of gas
ρ_g	= density of gas
ρ_p	= density of solid particle
σ_p	= Boltzmann constant
τ	= ignition delay

Introduction

Previous works (Veyssiere 1978; Veyssiere, Bouriannes, and Mason 1979), reported experimental results for the detonation of two-phase media which were composed of a gaseous explosive mixture (ethylene-air, equivalence ratio r = 1.15, at initial conditions p_f = 1 atm, T_f = 293 K) and a suspension of small metallic particles (aluminum particles; mean diameter 10 μm and mass concentration in gases σ = 30 g/m^3). Their principal findings were: 1) on the streak photographs, behind the detonation front (Fig. 1), there was a dark zone, due to the presence of the aluminum particles, the duration of which ranged from 10 to 70 μs; and 2) the characteristic reaction time of aluminum particles was more than ten times that of gaseous compounds.

The collation of these facts with velocity measurements, pressure, and brightness evolution in burnt products led to a first schematic description of the structure of the detonation in these mixtures (Veyssiere, Bouriannes, and Mason 1979), which is reported again on Fig. 2. According to this scheme, behind the detonation front, aluminum particles behave as chemically inert species as long as they have not reached their ignition temperature. This provides an explanation of the decreasing of velocity and of the pressure evolution behind the detonation front.

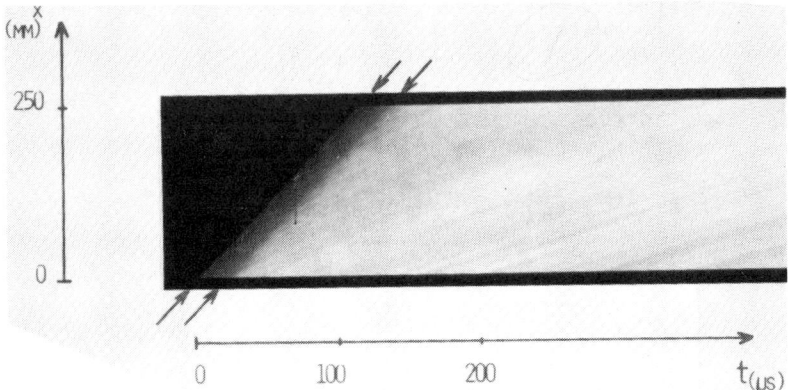

Fig. 1 Streak photography of detonation in a C_2H_4/air mixture containing a suspension of aluminum particles.

To go into further detail regarding the structure of the detonation in these mixtures, the dimensions of this zone should be connected with the ignition delay of aluminum particles. For this purpose the ignition delay of a single particle in the hot products of the gaseous detonation has been calculated. It is a quite rough approximation, since in the experiments the number of particles per unit volume was very important. Moreover, the comparison with experimental results is somewhat difficult because of the large dispersion inherent in the results of experiments in two-phase mixtures. So, one must be very cautious in arriving at definitive conclusions. The purpose in this work is to examine whether the results deduced from simple classical calculations of ignition delay of solid particle are of the same order of magnitude as experimental results and, consequently, whether they agree with the proposed scheme of the detonation or not.

An important parameter in this problem is the ignition temperature T_i of the particle. The value of the ignition delay of the particle will strongly depend upon the choice of an ignition criterion suitable to the experimental conditions. In the particular case of aluminum, the classical ignition criterion was proposed by Friedman and Macek (1962) and also by Bouriannes (1973). The criterion is based on a model assumption that an oxide layer is formed at the surface of the particle when it is heated. This oxide layer impedes the diffusion of the species, and no ignition is possible before the melting of this oxide film, which occurs at a surface temperature of 2310 K (the melting point of aluminum oxide). The validity of this criterion

IGNITION OF ALUMINUM PARTICLES IN A DETONATION

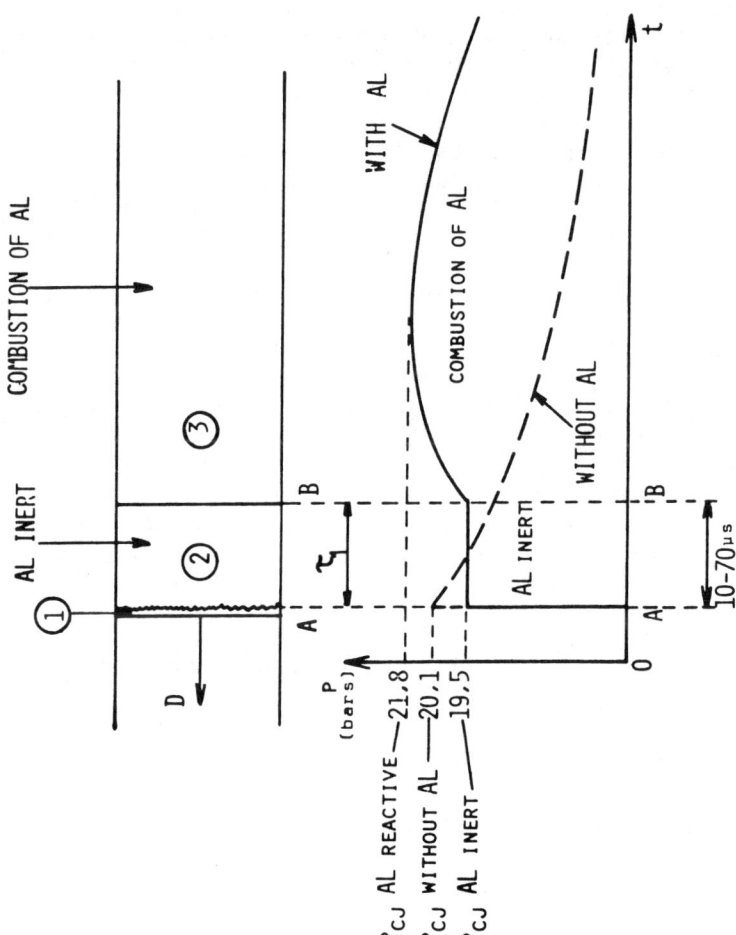

Fig. 2 Schematic structure of the detonation in a $C_2H_4/O_2/N_2$ mixture with Al particles.

may be questionable in some cases, and some important discussions were already developed about this criterion [see, for example, Khaikin et al. (1970)]. So far, it has not been possible to measure accurately the particle ignition temperature. One difficulty arises from the fact that the variation of particle emissivity during the process is unknown. The objectives of the present work are 1) to test whether the Friedman-Macek criterion provides values of the ignition delay in agreement with the proposed scheme of the detonation or not, 2) to determine the order of magnitude of the ignition delay of an aluminum particle deduced from simple, classical calculations, and 3) to discuss whether these results corroborate the previously proposed scheme of the detonation or not.

Calculation of the Ignition Delay

As mentioned above, according to Friedman and Macek, ignition of an aluminum particle occurs when the temperature of the surface of the particle exceeds the melting point of aluminum oxide. At lower surface temperatures the inert particle is heated by hot gaseous products of the ethylene detonation. Consequently, in this model it is assumed that 1) the particle is spherical; 2) the equilization of temperature inside the particle is attained instantaneously; 3) slow reactions between the particle and gases before ignition are neglected; and 4) the particle is initially at rest with a temperature T_o and is placed instantaneously in a hot gas at temperature T_g and velocity V_g.

The assumption of equilibrium in heat exchange during a lapse of time dt yields the following equation:

$$[h\ S(T_g - T) + \varepsilon_p\ S\sigma(T_g^4 - T^4)]\ dt = \rho_p\ C\ v\ dT \quad (1)$$

$$\underbrace{}_{\text{convection}} \quad \underbrace{}_{\text{radiation}}$$

with

$$h = Nu\ \lambda_g/d \quad (2)$$

The ignition delay of the particle is obtained by integrating Eq. (1) between T_o and T_i. According to Friedman and Macek, a corrective term, $\Delta\tau$, must be added to take into account the melting of aluminum at 932 K. Thus

$$h\ S(T_g - T_{m_{Al}})\Delta\tau = \rho_p\ v\ L \quad (3)$$

A rather detailed study of this kind of calculation has been previously done by Fox et al. (1977) to predict the ignition delay of magnesium particles by a shock wave.

In the same way, the effects of a detonation wave are considered to be similar to those of a shock wave (i.e., a discontinuity in temperature, velocity, and pressure). As a consequence the conclusions for shock wave ignitions should be applicable: 1) heat transfer by convection is the predominant mode, 2) heat transfer by radiation is negligible compared to that by convection, and 3) the equalization of temperature inside the particle is quasi-instantaneous for d < 100 μm.

The numerical values deduced from integration of Eqs. (1) and (3) are strongly dependent on the velocity V_p of the particle, through the heat exchange coefficient h. If the rate of change of particle momentum is due to aerodynamic drag forces, the following equation is applicable:

$$d V_p = \frac{3}{4} \frac{\rho_g (V_g - V_p)^2}{\rho_p d} c_{xp} \, dt \qquad (4)$$

The choice of the expression of c_{xp} as a function of V_p has been discussed in detail by Rudinger (1970) and also by Fox et al. (1977). The expression given by Klyachko (Fuchs 1964), which is a satisfactory approximation of the effective drag coefficient for Re ≤ 1000 has been used in this analysis:

$$c_{xp} = \frac{24}{Re} (1 + \frac{1}{6} Re^{2/3}) \qquad (5)$$

with

$$Re = [\rho_g (V_g - V_p)/\mu] \, d \qquad (6)$$

As mentioned by Fox, the choice of an expression of Nu is more difficult. For this investigation the simple expression given by McAdams (1961) has been used:

$$Nu = 0.37 \, (Re)^{0.6} \qquad (7)$$

for 17 < Re < 70,000. Obviously, this choice for the expression of c_{xp} and Nu is somewhat arbitrary but is expected to provide an order of magnitude estimate of the ignition delay of an aluminum particle.

For numerical calculations it was also assumed that 1) the gas phase was air, and its thermal conductivity λ_g was

Table 1 Ignition delay for different particle diameters
(T_g = 2980 K; V_g = 800 mm/s)

d, μm	τ, μs
5	12
10	34
100	835

dependent on temperature [for this purpose we have used and extrapolated the data given by Eckert and Drake (1972)]; 2) the heat capacity C of aluminum varied with temperature [data from JANAF (1971)]; 3) the emissivity ε_p of the aluminum particle before ignition was about 0.35 (Bouriannes 1973); and 4) the temperature T_g and velocity V_g of gases are those derived from the propagation of the gaseous detonation in our experimental conditions (Veyssiere, Bouriannes, and Manson 1979).

Several computations were made for different particle diameters. Table 1 reports the values of the ignition delay for 5-, 10-, and 100-μm-diam particles, with T_g = 2980 K and V_g = 800 m/s. Figure 3 shows the evolution behind the detonation front of the temperature T_p and the velocity V_p for the particular case of concerns: a 10-μm-diam particle. When ignition of the particle occurs, its velocity is about 500 m/s and $V_g - V_p \cong 300$ m/s and velocity equilibrium is not reached behind the detonation front at the time of particle ignition. The delay necessary for melting aluminum at 932 K is about 10% of the entire ignition delay, and the corresponding value of the ignition delay, obtained with the Friedman-Macek hypothesis Nu = 2 (stagnant heat transfer) would be about three times greater.

The variation of the ignition delay of a 10-μm particle with temperature T_g of the burnt gases (in the case of V_g = 800 m/s) is represented on Fig. 4. When the temperature T_g of burnt gases is increased, the ignition delay decreased. In spite of the dispersion of the experimental results, they seem to be in a good agreement with theoretical calculations.

Whether the assumption of constant T_g behind the detonation front is satisfactory may be questioned in view of the order of magnitude (several tens of microseconds) of the ignition delay calculated above. Desbordes et al. (1981) have demonstrated that for a CJ detonation, the

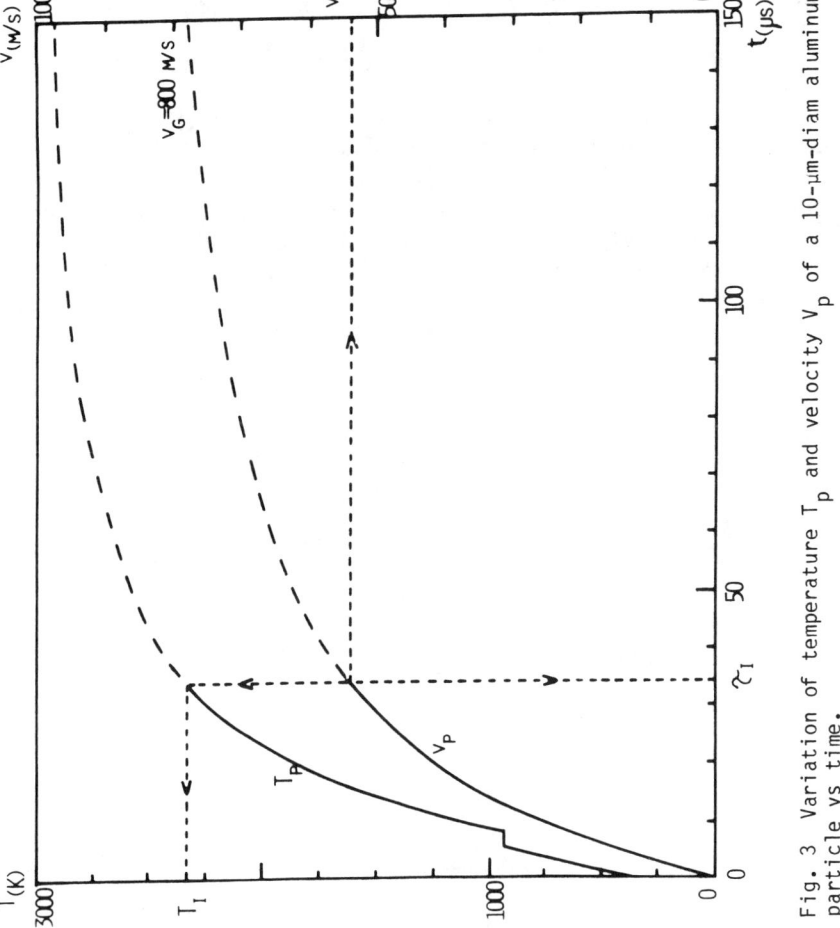

Fig. 3 Variation of temperature T_p and velocity V_p of a 10-μm-diam aluminum particle vs time.

evolution of pressure and temperature behind the front may be rather well predicted by the Taylor-Zeldovich (TZ) model. For the present case the TZ model predicts a decrease in temperature behind the front is about 50 K/100 μs. Even though the experimentaly observed temperature decrease is slightly faster during the first 200 μs behind the front (Desbordes 1981), the temperature T_g of gases may be considered approximately constant before particle ignition.

Influence of the Ignition Temperature of the Particle

The preceding calculations were based on the assumption of an aluminum particle ignition temperature of T_i = 2310 K. This assumption rests on 1) the ignition temperature of the particle not depending on its diameter, and 2) there being no chemical reactions between the particle and the gas before the particle reaches the temperature T_i = 2310 K.

As mentioned in the Introduction, this ignition criterion had been proposed in the analysis of previous

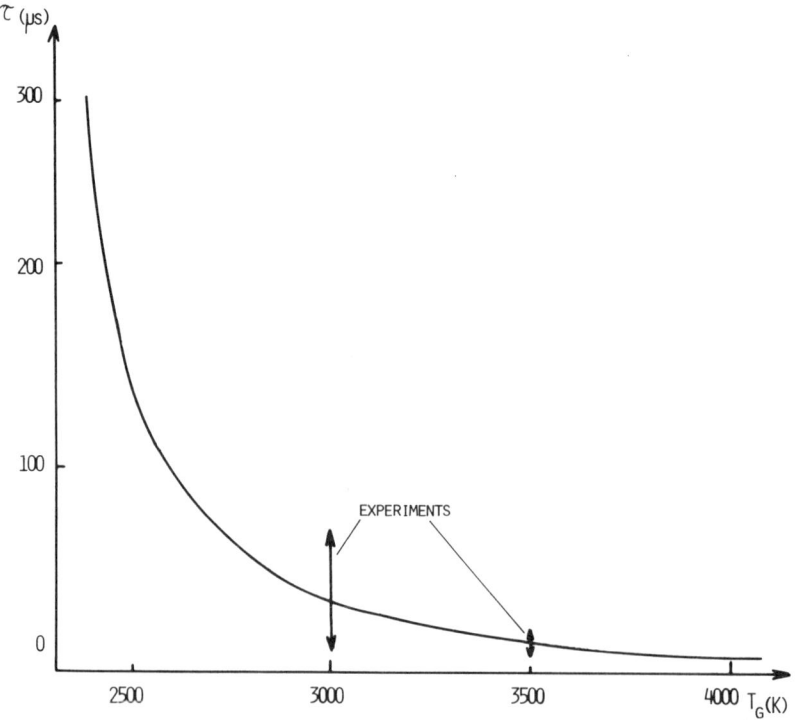

Fig. 4 Variation of the ignition delay of a 10-μm particle vs the temperature T_g of the gas.

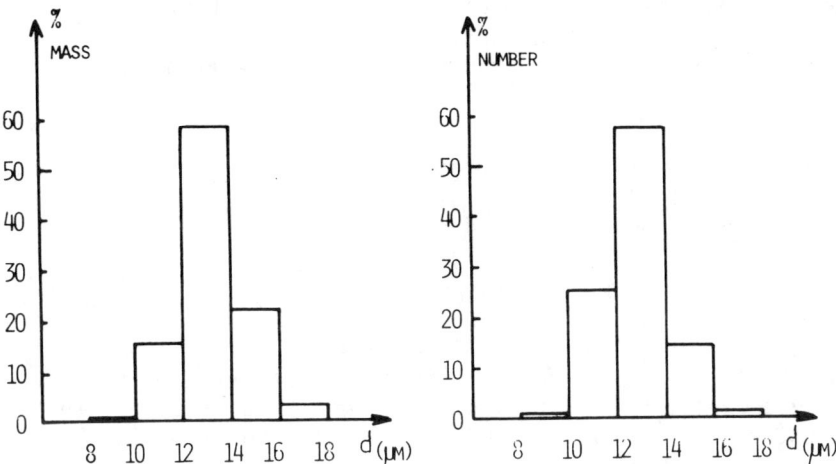

Fig. 5 Size distribution of aluminum particles used in experiments.

experimental results (Friedman and Macek 1962; Bouriannes 1973). However, those experiments were made at somewhat different conditions; in particular, the time during which the particles stayed in hot gases was much longer (several milliseconds) than in the present case. Thus an oxide layer might be formed at the surface of the particles which became a barrier to diffusion between oxygen and aluminum. In those conditions, the melting of oxide film must be a necessary condition for igniting the particle.

Behind a detonation front, ignition of aluminum particles occurs after less than 100 µs for a 10-µm particle. So the role played by the surface oxide film is questionable in the ignition process. The critical conditions for the ignition of a solid particle in an oxidant gas may be deduced from the solution of the equations of heat balance and kinetics of oxidation (Khaikin et al. 1970):

$$\rho_p \frac{Cd}{6} \frac{dT}{dt} = Q \rho_p \frac{d}{dt} + h(T_g - T) + \sigma \varepsilon_p (T_g^4 - T^4) \quad (8)$$

$$\frac{d\delta}{dt} = \frac{k_n \, c_{ox}^m}{\delta^n} \exp\left(-\frac{E}{RT}\right) \quad (9)$$

where δ is the thickness of the oxide film at the surface of the particle at the instant t.

The critical temperature of ignition will be given by

$$\left[\frac{3\ QE}{C\ RT_i^2}\right]^n \frac{C\rho_s}{3\lambda} \frac{k_n\ c_{ox}^n}{(d/2)^{n-1}} \exp\left(-\frac{E}{RT_i}\right) = Cte \qquad (10)$$

It is obvious, according to Eq. (10), that the ignition temperature generally depends on the diameter of the particle, except for n = 1. In this case (the parabolic oxidation law), the critical ignition temperature is independent of particle diameter; and it is precisely this hypothesis that was assumed by Friedman and Macek (1962) to explain their results.

However, some experimental results show that ignition of aluminum particles can be obtained at temperatures lower than 2310 K; for example, Gurevich et al. (1970) showed that for aluminum particles of diameter < 40 μm, the ignition temperature was dependent on particle diameter. They proposed an empirical relation to correlate the ignition temperature with controlled variables:

$$T_i = T_{m_{Al_2O_3}} - 0.6\ (c_{ox}^{0.3}/\lambda_g) \exp(-0.85\ \sqrt{d}) \qquad (11)$$

where λ_g is the thermal conductivity of gases calculated at the temperature $T = \tfrac{1}{2}T_i + \tfrac{1}{2}(T_o + T_{m_{Al_2O_3}})$.

Calculations based this criterion predict lower values of ignition temperature of the particle. For example, if Nu = 2 and T_g = 2980 K, Eq. (11) yields an ignition temperature of T_i = 1463 K and Eqs. (8) and (9) yield an ignition delay of τ = 56 μs instead of τ = 110 μs if $T_i = T_{m_{Al_2O_3}}$ = 2310 K. The ignition delay is divided by 2.

Unfortunately, it is very awkward to choose a suitable ignition criterion in the present work. While the ignition delay is very dependent on the ignition temperature of the particle, the dispersion of the experimental results does not permit a posteriori deduction of the ignition temperature of the aluminum particles. This dispersion may arise, among other reasons, from the particles size distribution (as shown in Fig. 5). This leads to different ignition delays between the smallest and the highest particles (τ increases with the diameter of the particle).

Khaikin et al. (1970) suggest that experimental observations about the ignition of aluminum particles are strongly influenced by the aluminum samples used in experiments and speculated that lower ignition temperatures observed by Gurevich et al. (1970) may be due to some impurities in the samples which would modify the protective properties of the oxide film.

An attempt was made to obtain a direct estimate of the ignition temperature of particles in the experiments by brightness measurements with a short-rise time pyrometer (Veyssiere et al. 1979). Although those measurements indicated that particle ignition occurred at temperatures higher than 2000 K, the accuracy of this observation is questionable because of 1) the difficulty in estimation of the variation of particle emissivity during the process, and 2) the fact that at this temperature the pyrometer was at its response limit.

Nevertheless, all previous experimental observations indicate that ignition temperatures of aluminum particles are much higher than that of melting of aluminum (932 K) (even in the Gurevich experiments). So, even if the exact role played by the oxide film at the surface of aluminum particles in the mechanism of ignition is not yet well identified, it seems acceptable to follow the Friedman-Macek criterion for estimation of the ignition delay of aluminum particles.

Conclusions

A simple model involving elementary calculations has been used to produce an order of magnitude estimate of the ignition delay of a single aluminum particle by a gaseous detonation. These calculated values corroborate the experimental results for detonation of C_2H_4/air mixtures which contain suspended aluminum particles. The effect of the number and size distribution of particles in the suspension may markedly modify the observed ignition delay (Fox et al. 1977). In the absence of more precise information the Friedman-Macek criterion for ignition temperature of aluminum particles was used, even though this assumption may be questionable. In spite of these reservations, the results confirm the assumption that the solid aluminum particles do not react in the gaseous detonation front. The characteristic time for ignition of particles is several tens of microseconds or about five times the characteristic induction time of gaseous chemical reactions in the detonation. The zone immediately behind the gaseous detonation front and prior to the postdetonation zone of increasing pressure is a zone in which aluminum particles are preheated. To define precisely the structure of the detonation in such media, pressure wave generation due to the combustion of the aluminum must now be interpreted. This task is obviously related to a better knowledge of the exact mechanism of the combustion of aluminum. The modeling of the zone in which particles were heated before ignition and that in which heat release occurs because of combustion is difficult because of the lack of more precise information on the heterogeneous kinetics of aluminum reaction.

References

Bouriannes, R. (1973) Etude expérimentale de la combustion de l'aluminium dans les mélanges oxygène-argon, dans l-azote et dans l'air. Rev. Int. Hautes Temp. Refract. 10, 113-124.

Desbordes, D., Souletis, J., and Manson, N. (1981) Evolution de la pression dans les produits des détonations autonomes "cylindriques divergentes" et "planes" des mélanges gazeux. Proceedings of the First Specialists Meeeting (International) of the Combustion Institute, pp. 443-559. The Combustion Institute, Pittsburgh, Pa.

Eckert, E. R. G. and Drake, R. M. Jr. (1972) Analysis of Heat and Mass Transfer. McGraw-Hill, New York.

Fox, T. W., Rackett, T. W., and Nicholls, J. W. (1977) Shock wave ignition of magnesium powders. Proceedings of the 11th International Symposium on Shock Tubes and Waves (edited by Ahlborn, Hertzberg, and Russell), pp. 262-268. Univ. of Washington Press, Seattle, Wash.

Friedman, R. and Macek, A. (1962) Ignition and combustion of aluminum particles in hot ambient gases. Combust. Flame 6, 9.

Fuchs, N. A. (1964) The Mechanics of Aerosols. MacMillan Co., New York.

Gurevich, M. A., Lapkina, K. I., and Ozerov E. S. (1970) Ignition limits of aluminum particles. Fiz. Goreniya Vzryva 6, 172.

IGNITION OF ALUMINUM PARTICLES IN A DETONATION

JANAF Thermochemical Tables (1971). U. S. Department of Commerce, National Bureau of Standards, Washington, D.C.

Khaikin, B. I., Bloshenko, V. N. and Merzhanov, A. G. (1970) On the ignition of metal particles. Fiz. Goreniya Vzryva 6, 474-488.

McAdams, W. H. (1961) Transmission de la Chaleur. Dunod, Paris.

Rudinger, G. (1970) Effective drag coefficient for gas particle flow in shock tubes. J. Basic Eng. 92, 165-172.

Veyssiere, B. (1978) Caractéristiques de la detonation dans des mélanges gazeux éthylène-oxygène-azote contenant des particules fines d'aluminum en suspension. Thèse de docteur-engénieur, Université de Poitiers, Poitiers, France.

Veyssiere, B., Bouriannes, R., and Manson, N. (1979) Detonation characteristics of two ethylene-oxygen-nitrogen mixtures containing aluminum particles in suspension. Gasdynamics of Detonations and Explosions: AIAA Progress in Astronautics and Aeronautics (edited by Bowen, Manson, Oppenheim, and Soloukhin), Vol. 78, pp. 423-438. AIAA, New York.

Veyssiere, B., Kato, Y., Brochet, C., Bouriannes, R., and Manson, N. (1979) Pyrometric studies of Al combustion in the wake of two-phase detonations. Paper presented at the Sixth International Symposium on Combustion Processes, Poland (to be published).

A Model of Blast Waves Propagating in Coal Mines

V.P. Korobeinikov* and I.S. Men'shov†
Academy of Sciences, Moscow, USSR

Abstract

The present paper suggests theoretical models for the motion of a two-phase mixture of air and methane with solid particles. The processes of dust-gas mixture combustion and the possibility of detonating such a mixture are analyzed. The equations of motion for one- and two-fluid models are presented. The problem of explosion development in a dusty tunnel for planar flow geometry is formulated. Variations in the initial dust density distribution with height were taken into account. A two-dimensional problem of dusty gas flow is then analyzed by use of a one-fluid model. Finally, self-similar solutions of the point explosion problem in a two-phase medium are discussed.

Introduction

The study of blast wave propagation in a two-phase medium consisting of a mixture of gas and solid particles has significant practical and theoretical interest.

An important example of such processes is that of blast wave propagation in mine tunnels filled by a mixture of air, methane, and coal particles [methane-dusty-air mixture (MDAM)]. The propagation of shock waves in MDAM is accompanied by several effects. Some of the unique two-phase characteristics of the medium are the following: evaporation (sublimation, melting) of the coal particles; shock shattering of the particles; outflow of volatiles from

---

Presented at the 8th ICOGER, Minsk, USSR, Aug. 23-26, 1981. Copyright © 1982 by V. P. Korobeinikov. Published by the American Institute of Aeronautics and Astronautics, Inc., with permission.
*Professor, Steklov Mathematical Institute.
†Senior Scientist, Steklov Mathematical Institute.

the coal; combustion of the coal particles and gases (methane and volatiles); motion of the combustion products; and heat transfer from hot regions by radiation and heat conduction. In principle, it is possible to include all the above processes in theoretical models of this problem; however, in practice, such an approach becomes very complicated.

In the present paper simple models describing shock propagation in MDAM are proposed, and problems which can clarify some important aspects of the phenomenon under consideration are solved.

Basic Equations

The following physical assumptions are made: 1) all particles are of the same size, shape, and mass; 2) particles do not fracture, divide, or collide; 3) the burning process is schematically described as an induction zone followed by a heat release zone (a two-front model with a distributed heat release); 4) thermal radiation is negligible; and 5) viscosity and heat conductivity of the medium are taken into account only in the mechanical and thermal interactions between the solid and gas phases. For the problems under consideration, two models of the mixture flow are used: the two-fluid model (I), and the one-fluid model (II) (Korobeinikov 1977, 1979).

Model I

The mixture of a gas with solid particles is modeled as two interacting continua. The volume occupied by the solid phase is considered as negligibly small. Conservation of mass is given by:

$$\frac{\partial \rho_g}{\partial t} + \text{div } \rho_g \vec{v}_g = \dot{m}$$

$$\frac{\partial \rho_s}{\partial t} + \text{div } \rho_s \vec{v}_s = -\dot{m}$$

(1)

where subscript s denotes the solid phase and g the gas phase; $\dot{m}$ is the rate of mass addition per unit volume resulting from solid/gas transformations.

The momentum equations have the form

$$\rho_g \left[\frac{\partial \vec{v}_g}{\partial t} + (\vec{v}_g \nabla)\vec{v}_g \right] = -\nabla p - \frac{\rho_s(\vec{v}_g - \vec{v}_s)}{\tau_v} + \dot{m}(\vec{v}_g - \vec{v}_s)$$

$$\rho_s \left[\frac{\partial \vec{v}_s}{\partial t} + (\vec{v}_s \nabla)\vec{v}_s \right] = \rho_s \frac{\vec{v}_g - \vec{v}_s}{\tau_v} - \dot{m}(\vec{v}_g - \vec{v}_s)$$

(2)

where τ_v is the velocity relaxation time, which is determined by a law expressing the mechanical (force) interactions between phases. If the internal energy ε and density ρ for the mixtures are expressed as:

$$\rho\varepsilon = \rho_s \varepsilon_s + \rho_g \varepsilon_g \qquad \rho = \rho_g + \rho_s$$

then the energy equations are:

$$\frac{\partial \rho\varepsilon}{\partial t} = -\text{div}[\rho_g \vec{v}_g (\varepsilon_g + \frac{p}{\rho_g} + \frac{v_g^2}{2})] - \text{div}[\rho_s \vec{v}_s (\varepsilon_s + \frac{v_s^2}{2})] \quad (3)$$

$$\rho_s \frac{\partial \varepsilon_s}{\partial t} + \rho_s(\vec{v}_s \nabla)\varepsilon_s = -\dot{m}Q_v + \frac{\rho_s C_s (T_g - T_s)}{\tau_T} \quad (4)$$

$$\varepsilon_g = p/\rho_v(\gamma - 1) + \beta Q \qquad \varepsilon_s = C_s T_s$$

(5)

$$p = \rho_g R T_g \qquad \dot{m} = \psi(p \, \rho_g T_g)$$

Here, τ_T is the temperature relaxation time determined by thermal interaction between phases, C_s the specific heat for the particles, Q_v the heat of vaporization; and the mass concentration of unburned combustible mixture.

For simplicity, the equation describing the kinetics of chemical reactions is assumed to be

$$\frac{d\beta}{et} = -k_1 \beta^m p^{n_1} \rho_g^{\ell_1} \exp\left(-\frac{E_1 \rho_g}{p}\right) \quad (6)$$

where k_1, m, n_1, ℓ_1, and E_1 are empirical constants.

In addition, combustion behind a shock front is assumed to occur after an induction period which can be determined (as in the gas case) by the following equation:

$$dc = -\frac{dt}{t_i} \qquad (7)$$

Here t_i is the empirical function of the mixture parameters behind the shock. At the shock front $C = 1$; reaction (6) begins when C approaches zero. The ignition front is determined by the empirical relation (Petrukhin et al. 1974):

$$U_f = A\, p_{sh} + B \qquad (8)$$

where A and B are constants determined from experiments; U_f denotes the flame front velocity; and p_{sh} represents the pressure immediately behind the shock.

Alternatively, the flame front position can be determined by the following equation:

$$U_f = -\frac{r_f}{t} / |\nabla r_f| \qquad (9)$$

where $r_f = r_f'(X,Y,Z,t)$ is flame front equation. The following initial conditions are prescribed as

$$r_f(X,Y,Z,0) = \phi(X,Y,Z) \qquad (10)$$

The shock jump conditions can be determined from the conservation laws across the shock. If there is insufficient time for phase transitions to occur in the shock region, the shock jump conditions in the shock coordinate frame are

$$(\rho_s v_s^n)_1 = (\rho_s v_s^n)_2 = J_s$$

$$(\rho_g v_g^n)_1 = (\rho_g v_g^n)_2 = J_g$$

(equation continued on next page)

$$J_j[\vec{v}_s] + J_g[\vec{v}_g] = \vec{n}[p]$$

$$J_s[\varepsilon_1 + \tfrac{1}{2}v_s^2] + J_g[\varepsilon_g + \tfrac{1}{2}v_g^2 + p/\rho_g] = 0 \tag{11}$$

where square brackets denote jumps in values of the parameters at the shock front.

This system of equations is not closed. It is necessary to prescribe the exchange of momentum and energy between phases at the shock front. Janenko et al. (1980), Nigmatulin and Vestnik (1969), and Kraiko and Sternin (1969) used the following relations are used to close the system (11):

$$[v_s] = 0 \qquad [\varepsilon_s] = 0 \tag{12}$$

However, in some cases additional conditions of type (12) are not necessary. The jump of solid phase parameters can be found when solving a specific problem.

Model II

In this model the relaxation times τ_v, τ_T required to achieve velocity and temperature equilibrium are assumed to be negligibly small compared to other characteristic times of the problem. In such a case velocities and temperatures of the phases do not differ: $\vec{v}_g = \vec{v}_s = \vec{v}$, $T_g = T_s = T$.

Equations (1-5) reduce to the form

$$\frac{d\vec{v}}{dt} = -\frac{1}{\rho}\nabla p \qquad \frac{\partial \rho}{\partial t} + \text{div}\,\rho\vec{v} = 0$$

$$\frac{d\varepsilon}{dt} = \frac{p}{\rho^2}\frac{d\rho}{dt} \qquad \varepsilon = \frac{p}{\rho(\gamma-1)} + \frac{\rho_g}{\rho}\beta Q \tag{13}$$

These model equations reduce to those for a perfect gas with an effective adiabatic exponent:

$$\gamma = (c_p + xc_s)/(c_v + xc_s) \tag{14}$$

The density ratio $x = \rho_s/\rho_g$ is given by

$$\frac{dx}{dt} = -\frac{(x+1)^2}{\rho}\dot{m} \qquad (15)$$

Equation (6) defines β for the reaction which is assumed to occur after an induction period.

Formulation of the Problem of Blast Propagation in Mines

The analysis is performed for a plane tube filled by a combustible mixture of dust and gas. The particle dimensions as well as particle composition, amount of volatile matter, water content of the mixture, and composition of carrying gas phase are of significance for initiation of detonation by combustion or by shock heating. The gas usually consists of a methane-air mixture. Experimental data show that a flame starts at the end of the shock tube; it is rapidly accelerated and drives a shock wave ahead of it. The transition to detonation is possible if the tube is sufficiently long and the mixture composition is suitable.

During initiation of combustion by shock waves, burning is assumed to occur first in the vicinities of the particles and then spreads over the entire volume. Details of the process are not clear now, and hence they are omitted in the analysis. Combustion of MDAM can be caused by explosion of an initiating charge, by electrical discharge, and by other impulsive energy addition sources.

Equilibration of the velocity and temperature of coal particles with the use of the gas phase does not occur immediately behind the shock wave. However, if the characteristic size of the coal particles is small (e.g., less than 20 μ), nonequilibrium effects are negligible.

The initating shock is assumed to result from a rapid energy release (explosion) at the closed end of a plane tube. The plane tube (tunnel) is filled by MDAM with coal particles having a diameter of order 10 μ. In addition, there are dust layers near the upper and lower walls of the tunnel. The layer thickness is much less than tunnel height, and varies from 1 to 10 mm. This situation is typical for coal mines (Petrukhin et al. 1974). Thus the density distribution in the tunnel varies with the height, $\rho_1 = \rho_s(Y)$, where Y is the coordinate perpendicular to the tunnel wall. The density of the particles increases sharply near the walls. Initiation is assumed to proceed uniformly along the Y coordinate. Experiments and theory show that

the particles on a bottom wall are accelerated with the
passage of the shock front and they are lofted, leading to
increases in densities behind the shock.

From the mathematical point of view, one has to solve
the equations of two-fluid hydrodynamics; [Eqs. (1-5)] must
be solved with the above initial and boundary conditions
(e.g., the normal velocity components vanish at the wall).
This model of blast wave propagation in coal mine tunnels is
so complex that it is difficult to solve completely.
Therefore more simplified models, which demonstrate the
influence of the different effects, are formulated.

The following are simplified problems:
1) a plane explosion at the end of a tunnel (one-fluid approach);
2) plane piston motion in a dusty, nonuniform medium (one-fluid approach);
3) a self-similar plane explosion in a two-phase medium.
Numerical and analytical methods can be employed to solve such problems.

Two-Dimensional Problems with a One-Fluid Approach
(Problems A and B)

For problem A, the following method of solution can be
adopted. First, a full numerical calculation (with initial
data on self-similar explosion or compressed gas expansion)
can be performed. An example (see Fig. 1) is the problem of
expansion of a compressed gas volume with initial pressure
p_1 and density ρ_0 placed at the end of a plane channel,
filled by dusty gas with pressure $p_1 \gg p_0$ and a step layer
distribution of density

$$\rho = \rho_2 \qquad 0 \leq y \leq h$$

$$\rho = \rho_1 \qquad h < y \leq H$$

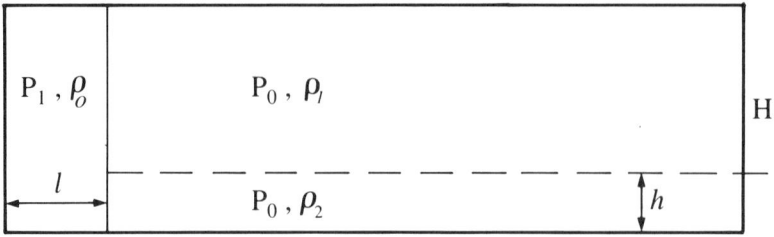

Fig. 1 Initial conditions for the problem of shock propagation in a two-layer medium ($h \ll H$, $p_1 \gg p_0$, $\rho_2 \gg \rho_1$).

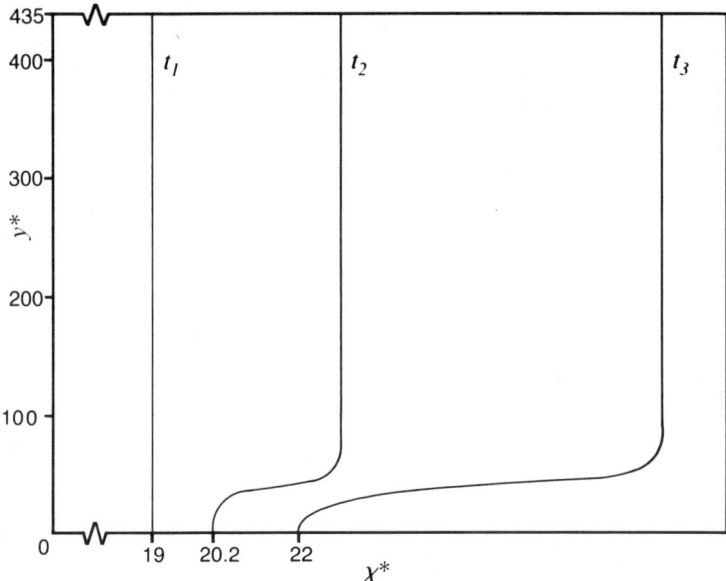

Fig. 2 Shock wave configurations at different times. Parameters of calculation: $y^* = Y/y_0$, $y_0 = 0.023$ cm, $H = 10$ cm, $h = 0.8$ cm, $\ell = 0.4$ cm, $p_0 = 1$ atm, $p_1 = 2.03.1$ atm, $\rho_0 = \rho_1 = 10^{-3}$ g/cm^3, $\rho_2 = 10^{-2}$ g/cm^3, $\gamma = 1.2$. Elapsed times are $t_1 = 0$, $t_2 = 12$ μs, $t_3 = 41$ μs.

Here, $\rho_1 \ll \rho_2$ and Y is a coordinate perpendicular to the tunnel axis. Some results of the calculations are shown in Figs. 2 and 3. The calculations were carried out by using the two-dimensional code of Markov (1981).

Clearly, the shock is curved, and its velocity is largest near the upper channel boundary. Calculated pressure at the shock front (shown in Fig. 3) indicates that a pressure gradient exists along the shock front, with the pressure being maximum in the vicinity of the lower boundary. Analysis of the velocity field indicates that a vortex flow is generated near the wall. This rotational flow distorts the shape of the contact surface (initially a plane boundary between the dense dusty layer and the upper gas volume).

A linearized solution of the model is possible for a shock wave which moves into a dusty gas with pressure p_a, density ρ_{go}, and variable density of solid phase:

$$\rho_{s0} = \rho_{s0}^* + \varepsilon^* \rho_{go} \phi(\delta Y) \qquad (16)$$

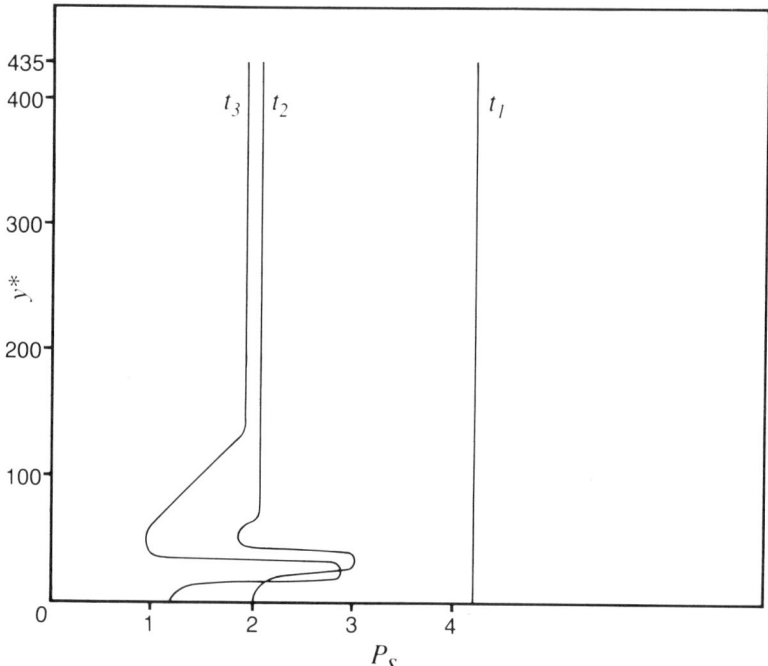

Fig. 3 Pressure distribution at the shock front for the parameters and conditions specified for Fig. 2.

where ε^* is a small parameter, and δ and ρ_{s0}^* are constant parameters. As a rough approximation, the flow is locally one dimensional along parallel layers, and for the self-similar strong explosion along the layers the shock wave motion is given by

$$X_2 = [e_0/\alpha\rho(Y)]^{1/3} t^{2/3} \qquad E_0 = \text{const}$$

The parameter α is assumed to be constant; and $\rho(Y)$ decreases with increasing Y, which implies that X_2 increases with Y at fixed t. Thus the shock is curved and decays as Y increases.

Problem B is solved by the linearization method for a plane piston which moves with a constant velocity in a weakly nonhomogeneous dusty gas with initial density given by formula (16). QX is the axis along the direction of piston motion. The mixture is assumed to be in equilibrium (the one-fluid model), and phase transitions and combustion are neglected ($\dot{m} = Q = 0$).

If the system of equations is linearized and transformed to nondimensional variables (i.e., velocities divided by the piston velocity C; coordinate divided by parameter δ; time divided by δ/a_1, with a_1 being the velocity of sound in the undisturbed medium; and pressure divided by $\rho_1 a_1 C$), the following hyperbolic system results:

$$\frac{\partial \omega}{\partial \tau} + \frac{\partial u}{\partial x} + \frac{v}{y} = 0$$

$$\frac{\partial u}{\partial \tau} + \frac{\partial \omega}{\partial x} = 0$$

$$\frac{\partial v}{\partial \tau} + \frac{\partial \omega}{\partial y} = 0$$

The boundary and initial conditions are:

$$\frac{\partial v}{\partial x} = A \frac{\partial \omega}{\partial y} + a\psi'(y) \qquad u = B\omega + B\psi(y)$$

at $x = \beta\tau$, $u = 0$ at $x = 0$, $u = 0$; $v = 0$; $\omega = \omega_0 \psi(y)$, $\omega_0 = -b/B$ at $x = \tau = 0$. Here τ, x, y, u, v, ω are nondimensional time, coordinate, velocity components, and pressure, respectively. $\psi(y)$ is a function, proportional to $\phi(y)$, A, B, a, B, β, and ω_0 are constants. After transformation of the system to new coordinates r, θ, $x = rsh\theta$, $\tau = rch\theta$, $y = y$, a solution is sought in the form of a series expansion:

$$\omega(r,\theta,y) = \sum_{k=0}^{\infty} \omega_{2k}(\theta) \psi^{(2k)}(y)^{(2k)}$$

$$u(\theta,y,r) = \sum_{k=0}^{\infty} u_{2k}(\theta) \psi^{(2k)}(y) r^{2k} \qquad (17)$$

$$v(r,\theta,y) = \sum_{k=0}^{\infty} v_{2k+1}(\theta) \psi^{(2k+1)}(y) r^{2k+1}$$

The functions ω_k, u_k, v_k are related by recursion equations; they satisfy the following inequalities:

$$\max|\omega_{2k}| \leq q_1^k |\omega_0|/(2k\ k!)^2$$

$$\max|u_{2k}| \leq q_1^k |\omega_0|/(2k\ k!)^2 \qquad (18)$$

$$\max|v_{2k+1}| \leq q_1^k |\omega_0|/(2k\ k!)^2$$

The inequalities (18) are used to determine the convergence of series (17) for a specific function $\psi(y)$ (q_1 is a constant here).

If $\phi(\delta Y) \sim \exp[-(\delta Y)^2]$, distribution is characteristic of coal dust in mines, and series convergence can be easily proved. If dominate terms are retained in Eq. (17), a simple formula for the initial stage of piston movement ($<<1$) is obtained:

$$u = 0 \qquad v = [\omega_0 ch\theta - (A\omega_0 + a)sh\theta]2ye^{-y^2}$$

$$\omega = \omega_0 e^{-y^2} \qquad g(y,\tau) = \frac{2b(\beta A - 1) - \beta a B}{2\beta B} \tau e^{-y^2}$$

These results predict that the fluid particles move toward the lower wall immediately after the shock front passage. After some time the particles stop and begin to move up. The trajectories of the particles $y(\tau)$ are shown in Fig. 4. Thus dust lofting begins at a time interval $\Delta\tau$ after the shock arrival. From an experimental explosion in a mine, Petrukhin et al. (1974) have shown that the dusty concentration near the wall begins to move upward at 300 µs after shock passage. This process creates dangerous explosive concentrations of the dust in the mine. The cause of the delay for dust lifting is thus explained by the linearized analysis.

Planar Explosion in a Two-Phase Medium
(Problem C)

In the problems A and B, velocity and temperature nonequilibrium effects between phases have been neglected. The treatment of nonequilibrium effects is illustrated by

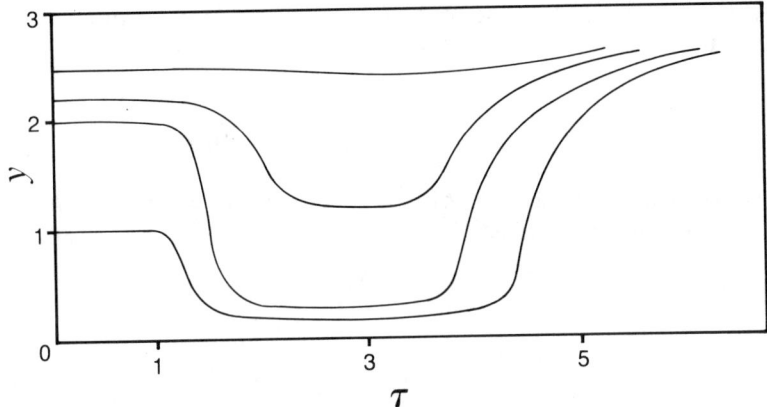

Fig. 4 Particle trajectories.

the problem of a plane explosion in the mixture of a gas with solid spherical dust particles of radius a. Combustion and phase transition processes are neglected. The mixture motion is described by the one-dimensional Eqs. (2-5) with $\dot{m} = Q = 0$. The time relaxation may be determined by relations (Soo 1967):

$$\frac{1}{\tau_v} = \frac{3}{8} \frac{\rho_g}{\rho_s^0 a} C_D |\vec{v}_g - \vec{v}_s|$$

$$\frac{1}{\tau_T} = \frac{3}{2} \frac{x_T}{\rho_s^0 a^2} Nu$$

where C_D is the drag coefficient for sphere; Nu represents the Nusselt number; x_T is the gas thermal conductivity; and ρ_1^0 denotes the solid density of the particle. The following interaction relations (Soo 1967) are assumed:

$$C_D = \frac{24}{Re} \quad Nu = 2 \quad Re = \frac{2a\rho_g |\vec{v}_g - \vec{v}_s|}{\mu}$$

Also, the viscosity μ and conductivity x_T are assumed to be power law functions of temperature:

$$\mu = \mu_0(T_g/T_0)^n \qquad x_T = x_0(T_g/T_0)^n$$

For these interaction relations, the system of equations (1-5) includes the parameter

$$A = \frac{g}{2} \frac{\mu_0}{\rho_1^0 a^2 \varepsilon_0^n} \qquad \varepsilon_0 = c_v T_0$$

This parameter has dimensions $[A] = T^{2n-1}L^{2n}$.

Thus if energy E_0 per unit area is released in a quiescent mixture of gas and solid particles (with an initial pressure, p_0, and densities ρ_{go}, and ρ_{so}), the following known parameters characterize the problem: E_0, A, p_0, ρ_{so}, ρ_{go}.

For negligible counterpressure, $p_0 = 0$ and $n = 3/2$, the problem becomes self-similar. For this case, the new dimensionless functions are

$$V_i = v_i/D \qquad R_i = \rho_i/\rho_{go} \qquad \xi_i = c_v T_i/D^2$$

(i=1 for the gas and 2 for the particles, respectively). These functions depend only on the similarity variable $\lambda = r/r_{sh}$. Here

$$r_{sh} = (E_0/\alpha \rho_{go})^{1/3} t^{2/3}$$

is the shock wave coordinate; $D = dr_{sh}/dt$ is the shock wave velocity; the parameter α is determined by the global energy conservation equation for the mixture (Sedov 1967). In this case, Eqs. (2-5) reduce to the following:

$$R_1'(V_1-\lambda) + R_1 V_1' = 0$$

$$R_2'(V_2-\lambda) + R_2 V_2' = 0$$

(equation continued on next page)

$$R_2(V_1-\lambda)V_1' + (\gamma-1)(R_1\xi_1)' - \tfrac{1}{2}R_1V_1 =$$

$$- BR_2\xi_1^{3/2}(V_1-V_2)(V_2-\lambda)V_2' - \tfrac{1}{2}V_2 = B\xi_1^{3/2}(V_1-V_2)$$

$$R_1(V_1-\lambda)\xi_1' - (\gamma-1)(V_1-\lambda)\xi_1 R_1' - R_1\xi =$$

$$BR_2\xi_1^{3/2}(V_1-V_2)^2 \quad -B\sigma_1 R_2\xi_1^{3/2}(\xi_1-\xi_2)$$

$$(V_2-\lambda)\xi_2' - \xi_2 = \sigma_2 B\xi^{3/2}(\xi_1-\xi_2) \tag{19}$$

where the prime denotes differentiation with respect to λ, and B, σ_1, and σ_2 are nondimensional constants. The solution of the above equations is sought in the domain $0 \le \lambda \le 1$. The velocities of the two phases must satisfy the conditions

$$V_1 = 0 \quad V_2 = 0 \tag{20}$$

at the explosion center ($\lambda = 0$), while the conditions (11) must be satisfied at the shock front ($\lambda = 1$). In nondimensional form, these conditions are

$$R_1(1-V_1) = 1 \quad R_2(1-V_2) = x \quad x = \rho_{so}/\rho_{go}$$

$$V_1 + V_2 = (\gamma-1)R_2\xi_1$$

$$\frac{V_1^2}{2} + \xi_1 + x\frac{V_2^2}{2} + \frac{\sigma_1}{\sigma_2}\xi_2 - (\gamma-1)R_1\xi_1 V_1 = 0 \tag{21}$$

With six ordinary differential equations and six boundary conditions and there is no need to use an additional condition of the type (12). Jumps in velocities and temperatures at the shock are governed by the character of interactions between phases. Conditions (12) are constraints which are satisfied in special cases. The self-similar equations contains the interaction parameter B. If B = 0, there are no interactions between phases; this corresponds to the case of an explosion in a pure gas (we can adopt $V_2 = \xi_2 = 0$). For this case, jumps in particle velocity and temperature are zero. For large B ($B \to \infty$), the model reduces to the case of an explosion in a equilibrium mixture which was considered by Sedov (1967). The jump in particle velocity and temperature is not zero in this case. The system of equations (19) has an analytic integral (obtained from global energy conservation of the following form):

$$(\lambda-V_1)R_1[V_1^2/2 + \xi_1] + (\lambda-V_2)R_2[(\sigma_1/\sigma_2)\xi_2 + V_2^2/2]$$
$$- (\gamma-1)R_1\xi_1V_1 = 0$$

The self-similar solution for a planar dusty gas explosion has been obtained by the above method for the case of full temperature nonequilibrium ($\tau_T = \infty$). The density, velocity, and temperature distributions are presented in Figs. 5, 6 and 7, respectively.

The problem considered above can be generalized.

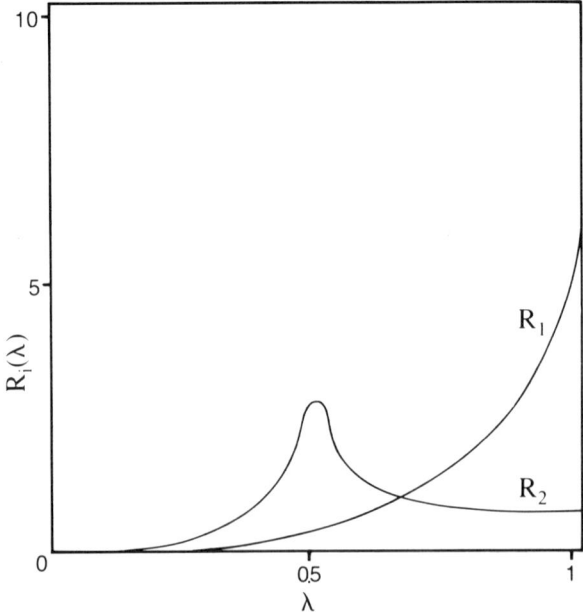

Fig. 5 Self-similar density distribution for a planar dusty gas explosion assuming temperature non-equilibrium. $\tau_T = \infty$, $B = 1$, $\sigma_1 = 0.025$, $\sigma_2 = 0.05$.

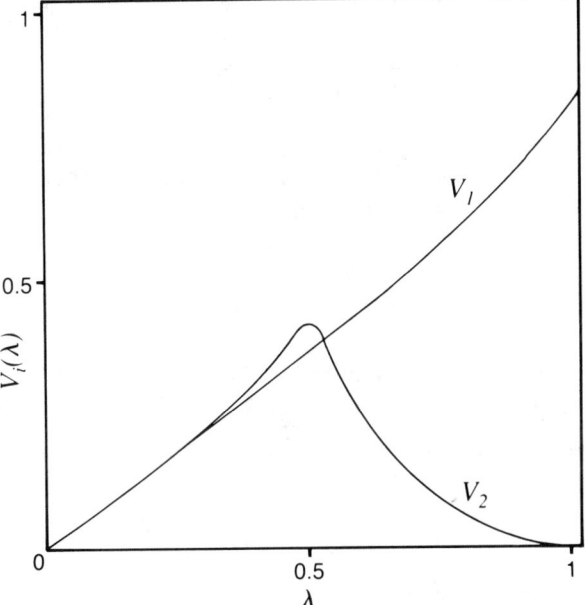

Fig. 6 Self-similar velocity distribution corresponding to Fig. 5.

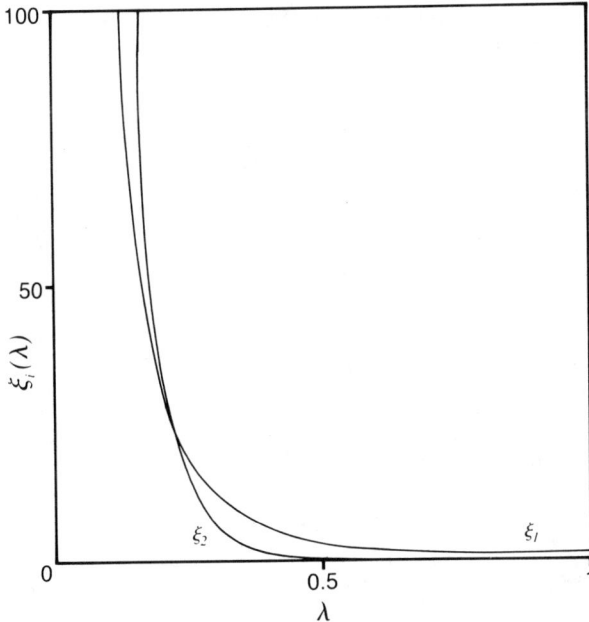

Fig. 7 Self-similar temperature distribution corresponding to Fig. 5.

First, it can be formulated and investigated for the cases of cylindrical and spherical symmetry. The self-similarity condition for these cases is $n = (2+v)/2v$ where $n = 2$ or 3 for cylindrical or spherical symmetry. Second, it is possible to study this problem in self-similar variables and include chemical reactions effects. If a chemical kinetic equation of the form (6) is involved and an induction law of heat addition of the following form is assumed:

$$t_i = k_2 \, p^{n_1} \, \rho_g^{\ell_2} \exp[-E_2 \rho/p]$$

then the point explosion problem becomes self-similar, and $n_1 = n_2 = (2+v)/2v$; $\ell_1 = \ell_2 = \tfrac{1}{2}$; $E_1 = \delta_1 p$; $E_2 = \delta_2 p$; δ_1, δ_2.

However, the method of solution for this case, as well as the existence of the solution remains open.

Conclusions

Simplified models of two-phase flows here have been considered. Some idealized problems were solved which can be useful for the study of processes of MDAM explosions in mines. A complete investigation of the problem requires considerable effort. It must include two-dimensional, two-phase flow with phase transitions, chemical reactions, and heat transfer by radiation.

Acknowledgment

The authors express their gratitude to V. V. Markov for his help in completing the numerical solutions, and for useful discussions with him.

References

Janenko, N. N., Soloukhin, R. I., Papirin, A. N., and Fomin, V. M. (1980) <u>Supersonic Nonequalibrium Two-Phase Flow</u>. Nauka, Novosibirsk, USSR.

Korobeinikov, V. P. (1977) <u>Archivum Termodynomiki Spalania</u>, 8, Polish Academy of Sciences.

Korobeinikov, V. P. (1979) <u>Acta Astron.</u> 6, 931-941.

Kraiko, A. N., and Sternin, L. E. (1969) <u>Priklady Matemotyki i Mechaniki</u>, USSR Academy of Sciences.

Markov, V. V. (1981) <u>Dokl. Acad. Nauk SSSR</u> 258.

Nigmatulin, R. I. (1969) Vestn. Mosk. Univ. Ser. Mat. Mekh.

Petrukhin, P. M., Netseplajaev, M. I., Katchan, V. N., and Sergeev, V. S. (1974) Prevention of Dust Explosions in Coal and Slate Mines. Nedra, Moscow, USSR.

Sedov, L. I. (1967) Methods of Similarity and Dimensions in Mechanics, 5th ed. Nauka, Moscow, USSR.

Soo, S. L. (1967) Fluid Dynamics of Multiphase System. Blaisdell Publishing Co., Waltham, Mass.

Multifront Combustion of Two-Phase Media

L.A. Afanasieva,* V.A. Levin,† and Y.V. Tunik‡
Moscow State University, Moscow, USSR

Abstract

This work shows that the combustion of easily ignitable fuel and solid combustible particles mixture may be realized in several detonation or flame fronts following each other. For infinitely thin fronts of detonation waves and flames and a perfect gas behavior it is shown that in cylindrical and spherical cases the first detonation wave may be followed by the second one and that a critical value of heat release exists for the second wave, beyond which the double detonation is not stable. The combustion of a mixture also results as a flame front separated from the leading detonation wave by the shock wave. In the planar case with momentum and heat transfer from the products of combustion to the walls of a constant area channel, the possibility of the constant speed propagation of double detonation was demonstrated. This study was extended to real systems in an investigation of coal dust combustion in a mixture of acetylene with oxygen or air. Because the heat of combustion of coal dust exceeds the critical value for double detonations, the multifront flow structure is not stable. Analysis of ignition of coal dust suspensions in oxygen indicates the existence of critical ignition parameters.

Nomenclature

c_S = specific heat for coke
c_L^S = specific heat for solid volatile substances

Presented at the 8th ICOGER, Minsk, USSR, Aug. 23-26, 1981. Copyright © American Institute of Aeronautics and Astronautics, Inc., 1983. All rights reserved.
*Post Graduate Student, Institute of Mechanics.
+Professor, Chief of Laboratory, Institute of Mechanics.
‡Senior Research Worker, Institute of Mechanics.

d	=	particle diameter
Nu	=	Nusselt number
Pr	=	Prandtl number
$Q^{(1)}$	=	thermal effect of the heterogeneous combustion of the coke per unit mass of the coke
$Q^{(2)}$	=	thermal effect of the homogeneous combustion of the volatile substances per unit mass of fuel
Re	=	Reynolds number
T_s	=	temperature of the solid phase
V_S	=	coke volume for the particle
V_L	=	volume of the volatile substances in a solid phase for the particle
β_{O_2}	=	mass concentration of the oxygen in gas
β_2	=	mass concentration of the volatile substances in gas

Introduction

Combustion of an easily ignitable fuel mixed with combustible solid or liquid particles can occur as a series of reaction waves (e.g., detonation and flame fronts). For example, a gaseous fuel can burn in the first detonation wave; or liquid particles can be heated in the first wave and burned in the second wave. This phenomenon has been experimentally confirmed by Veyssiere et al. (1979). Two-front combustion has also been observed in detonations of heterogeneous fuels in gaseous oxidizer by Barr-Or et al. (1981) and Pierce and Nicholls (1973).

The following analysis of the phenomenon utilizes the perfect gas with constant specific heat ratio and relies on the assumptions that the reaction fronts are infinitely thin, the heat release per unit mass in each reaction front is constant Q_i, and the ignition energy is negligible.

Multifront Combustion

Spherical and Cylindrical Gases

Ignition of a uniform combustible mixture is assumed to occur instantaneously at t = 0 and to be symmetric, i.e., a straight line ignition source in the cylindrical case or a point ignition source in the spherical case. The resulting waves are also assumed to be symmetric, i.e., a cylindrical detonation wave for the former case and spherical for the latter. The analysis shows that the detonation wave may be followed by a flame front for both case.

From the above approximations, self-similar solutions are developed if the reaction wave velocities are constant.

The continuous adiabatic flow behind each wave is described by the following equations:

$$\frac{dV}{d\xi} = (\nu-1) \frac{V}{\xi} \left[\left(\frac{V-\xi}{A}\right)^2 - 1\right]^{-1} \quad (1)$$

$$\frac{dA}{d\xi} = \frac{\gamma-1}{2} \frac{dV}{d\xi} \frac{(\xi-V)}{A}$$

where $\xi = r/D_J t$; $V = v/D_J$; $A = a/D_J$; a is the sound velocity; D_J the velocity of the first Chapman-Jouguet (CJ) detonation wave; v the velocity; and $\nu = 2, 3$ for cylindrical and spherical waves, respectively.

Sedov (1957) presents an analysis of the integral curves in the phase (Z,V) plane (where $V = tv/r$, $Z = t^2 a^2/r^2$) for single discontinuities.

The phase-plane diagram for the double discontinuity propagation is shown in Fig. 1. The supersonic rarefaction wave (see curve AB) following the first detonation wave (see point A) is bounded by the second CJ detonation wave (see point D). In the rarefaction wave, following the second detonation wave, the velocity decreases to zero, and a stagnation zone follows. The nondimensional flowfield

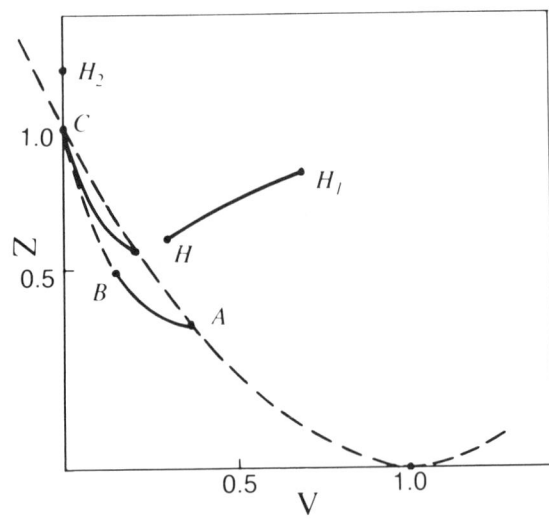

Fig. 1 Phase plane representation of similarity solution for the propagation of double detonation wave.

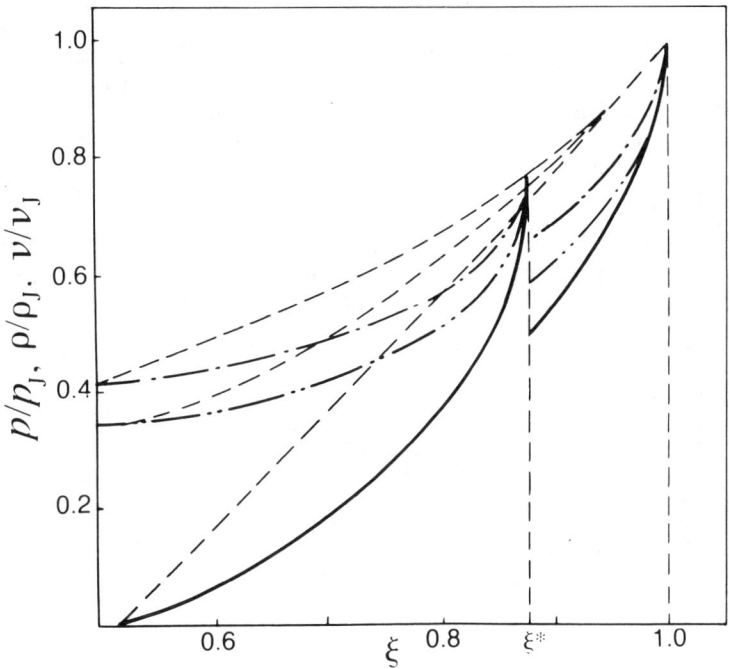

Fig. 2 Flowfield for spherically symmetric double detonation wave. $\nu = 3$; $\gamma = 1.3$; $Q_1/D_J^2 = 0.637$; (—··—) P/P_J; (—·—) ρ/ρ_J; (——) v/v_J.

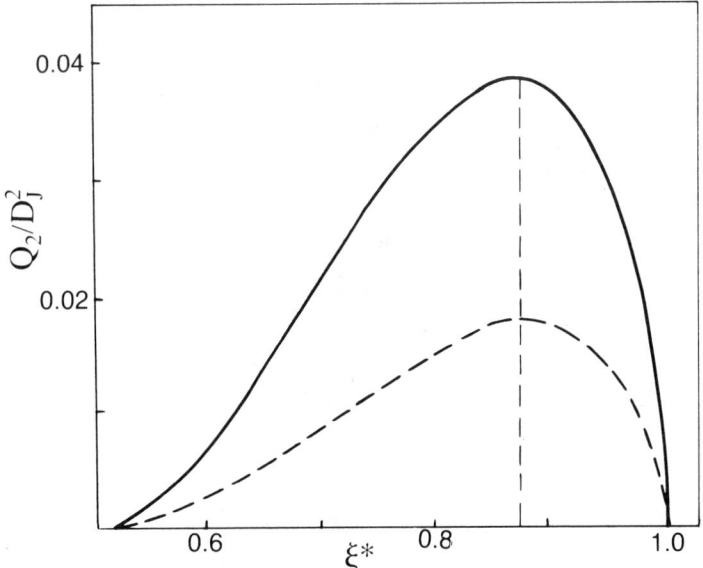

Fig. 3 Influence of heat release on the position of the second detonation wave. (——) spherical symmetry $\nu = 3$; (----) cylindrical symmetry $\nu = 2$.

determined by a numerical solution of Eqs. (1) for the spherical case is presented in Fig. 2. The distribution of parameters behind the second CJ wave is shown as a function of its own position; this solution is denoted by the broken lines.

For each value of Q_1 there exists a maximum value of Q_2^* beyond which the double detonation solution fails to exist. The relationship between the heat of combustion value for the second wave and its own position is shown in Fig. 3. When $Q_2 < Q_2^*$ there are two possible propagation velocities for the second wave.

In addition to the solution regime with double detonations, there is a second regime for which the flame front H_1H_2 (see Fig. 1) is separated from the leading detonation wave (point A) by the shock wave BH. A compression wave HH_1 links this shock with the flame front. Behind the flame front, the stagnation zone is formed as indicated in Fig. 4.

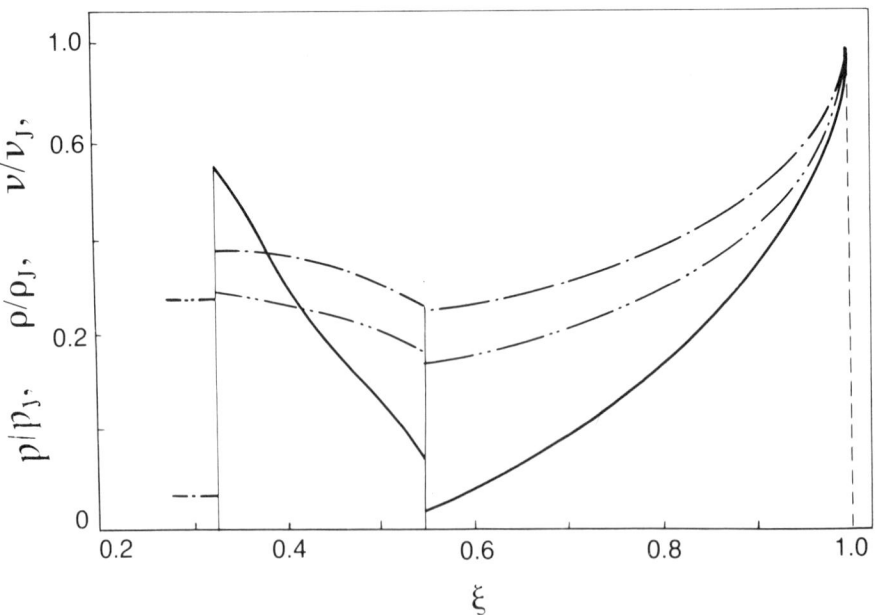

Fig. 4 Flowfield for a detonation shock wave flame front. (—··—) P/P_J; (—·—) ρ/ρ_J; (——) v/v_J; $\nu = 3$; $\gamma_1 = \gamma_2 = 1.3$; $\nu_F/a_0 = 0.4$; $Q_1/D_J^2 = 0.637$; $Q_J/Q_F = 3$; $\lambda_S = 0.55$.

One-Dimensional Plane Case

It is well known that in the one-dimensional planar case, the propagation of a multifront detonation wave is impossible. The leading detonation wave may be followed by one or several flame fronts. For a single flame front propagating in a closed-end tube, the gas motion behind the detonation wave is modeled by the following. When $\xi_1 = \xi \leq 1$,

$$v = v_J \left(\frac{2}{1-q_J} \cdot \frac{x}{D_J t} - \frac{1+q_J}{1-q_J} \right)$$

$$p = p_J \left[\left(\frac{\gamma-1}{\gamma+q_J} \right) \left(\frac{x}{D_J t} + \frac{1+q_J}{\gamma-1} \right) \right]^{2\gamma/(\gamma-1)}$$

$$\rho = \rho_J \left[\left(\frac{\gamma-1}{\gamma+q_J} \right) \left(\frac{x}{D_J t} + \frac{1+q_J}{\gamma-1} \right) \right]^{2/(\gamma-1)}$$

$$\xi_1 = \frac{1-q_J}{2} \left[(\lambda-1) \frac{V}{v_J} + \frac{1+q_J}{1-q_J} \right]$$

When $\xi_2 \leq \xi \leq \xi_1$,

$$v = (\lambda-1)V$$

$$p = p_J \left[\left(\frac{\gamma-1}{\gamma+q_J} \right) \left(\xi_1 + \frac{1+q_J}{\gamma-1} \right) \right]^{2\gamma/(\gamma-1)}$$

$$\rho = \rho_J \left[\left(\frac{\gamma-1}{\gamma+q_J} \right) \left(\xi_1 + \frac{1+q_J}{\gamma-1} \right) \right]^{2/(\gamma-1)}$$

$$\xi_2 = \lambda V/D_J$$

When $0 \leq \xi \leq \xi_2$, $v = 0$.

The value $q_J = a_1^2/D_J^2$ is obtained from relation [see Levin and Cherniy (1967)]:

$$\frac{a_1}{\gamma+1} \frac{1-q_J}{\sqrt{q_J}} = \sqrt{2 \frac{\gamma-1}{\gamma+1} Q_1}$$

a_1 is the sound velocity before the detonation wave, v, the flame front velocity with respect to the particles, being the physical-chemical constant; and λ, the density ratio, determined by the value of heat released in the flame front. The state created by the lead CJ detonation is denoted by the subscript J. The flowfield exhibits the following structure: the CJ detonation wave is followed by a Riemann rarefaction wave, in which the particle velocity is reduced to that compatible with the flame propagation speed. The flow parameters are constant between the trailing edge of the rarefaction wave and the flame front. A stagnation zone with a zero particle velocity is formed behind the flame front. This solution is valid for $(\lambda-1)V < v_J$, i.e., gas velocities, immediately ahead of the flame front, less than those immediately after the CJ detonation wave, and for gas velocities behind the flame front with respect to the front velocity less than the sound speed. While the latter condition is usually satisfied, the former may not because of a large difference in densities, i.e., for $\lambda \gg 1$. In this case the solution is simple. The overdriven detonation wave propagates with the velocity sufficient to create a particular gas velocity ahead of the flame front. This velocity is determined in terms of the particle flame velocity. The region with constant flow parameters is confined to the region between the detonation wave and the flame front. Subsequent to the flame front is the stagnation zone.

Propagation of Double Detonation in a Tube

When wall friction and heat-transfer effects are included in the analysis for the one-dimensional planar case, it will be shown in the following that the double detonation wave still exists and propagates with constant front velocities. In this analysis the energy losses to the channel walls are averaged over the entire section, although gas cooling and deceleration actually occur in the wall boundary layers. The Prandtl number is assumed to be unity, and the Reynolds analogy between a heat and momentum transfer is used. The wall shear stress τ_w and heat transfer q_w are described by relations [see Abramovich (1969) or Skinner (1967)]:

$$\tau_w = -\frac{C_f}{2}\rho u^2 \qquad q_w = \frac{\lambda(C_p T - C_p T_w + u^2/2)}{C_p \mu u}$$

where λ is the thermal conductivity; μ the viscosity; C_p the gas-specific heat at constant pressure; T_w the wall temperature (assumed to be a known constant value); and C_f, the friction coefficient.

With the above assumptions, the continuity, momentum, and energy equations for the mean flow are

$$\frac{\partial \rho}{\partial t} + \frac{\partial \rho u}{\partial \xi} = 0$$

$$\rho\left(\frac{\partial u}{\partial t} + u\frac{\partial u}{\partial \xi}\right) + \frac{\partial p}{\partial \xi} = -\frac{C_f}{2L^*}\rho u^2$$

$$\frac{\partial p}{\partial t} + u\frac{\partial p}{\partial \xi} + \gamma p \frac{\partial u}{\partial \xi} = (\gamma-1)\frac{C_f}{2L^*}\rho u\left(-\frac{u^2}{2} - \frac{\gamma}{\gamma-1}\frac{p}{\rho} + C_p T_w\right)$$

(2)

where $L^* = F/P$, F is the cross-sectional area of the channel and P its perimeter.

The gas flow is assumed to be stationary in a coordinate frame relative to the detonation wave propagating at the constant velocity D_J. With this assumption, the partial differential equations (2) are reducible to a set of ordinary differential equations which are solved by standard methods.

Under these assumptions, the constant-rate double detonation is described in terms of the CJ values. The first detonation is followed by a rarefaction wave whose

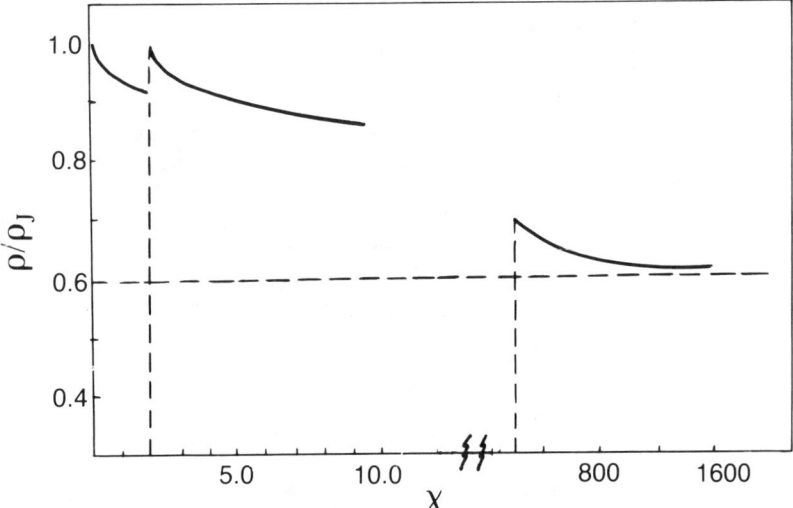

Fig. 5 Double detonation in tube with wall friction and heat transfer. $D_J/v_0 = 1.667$; $C_p T_w/v_0^2 = 0.78$; $c_f/L^* = 0.01$; $Q_1/D_J^2 = 0.54$; $Q_2/D_J^2 = 0.07$.

trailing edge is immediately ahead of the second detonation wave. The second detonation wave is followed by a second rarefaction wave whose trailing edge particle velocity is zero.

The solution of Eqs. (2) for constant detonation front velocity is shown in Fig. 5. Here, $x = (D_J t - \xi)/L^*$.

Heterogeneous Combustion of Coal Dust

Mathematical Model

As an example of the analysis of real flows of two-phase mixtures, the combustion of a suspension of coal dust in an oxygen-acetylene mixture is considered in the following. The system of equations describing the phenomenon is based on a widely used multiphase model (Nigmatulin 1978), and on a coal dust combustion model with constant reaction values (Hitrin 1957). The conservation equations in standard notation for a mixture with n particles per unit volume in flow and chemical reaction in a variable cross-sectional channel of area F are as follows.

The conservation equation for the number of particles:

$$\frac{\partial}{\partial t}(nF) + \frac{\partial}{\partial x}(nvF) = 0$$

for the gas phase mass:

$$\frac{\partial}{\partial t}(\rho F) + \frac{\partial}{\partial x}(\rho u F) = F(W_C^{(1)} + W^{(3)})$$

for oxygen:

$$\frac{\partial}{\partial t}(\rho \beta_{O_2} F) + \frac{\partial}{\partial x}(\rho \beta_{O_2} u F) = -F(W_{O_2}^{(1)} + W_{O_2}^{(2)})$$

for volatile substances in the gas phase:

$$\frac{\partial}{\partial t}(\rho \beta_L F) + \frac{\partial}{\partial x}(\rho \beta_L u F) = F(W^{(3)} - W_L^{(2)})$$

for coke:

$$\frac{\partial}{\partial t}(F n \rho_S V_S) + \frac{\partial}{\partial x}(F n \rho_S v) = -F W_C^{(1)}$$

for volatile substances in the solid phase:

$$\frac{\partial}{\partial t}(F n \rho_L V_L) + \frac{\partial}{\partial x}(F n \rho_L V_L v) = -F W^{(3)}$$

The momentum equations for the gas phase:

$$\frac{\partial}{\partial t}(\rho u F) + \frac{\partial}{\partial x}[F(\rho u^2 + p)] = F_S' p - F\phi_n + F(W_C^{(1)} + W^{(3)})v$$

and for the particles:

$$\frac{\partial}{\partial t}[Fnv(\rho_S V_S + \rho_L V_L)] + \frac{\partial}{\partial x}[Fnv^2(\rho_S V_S + \rho_L V_L)]$$

$$= F\phi_n - F(W_C^{(1)} + W^{(3)})v$$

The energy conservation equations for the mixture:

$$\frac{\partial}{\partial t}F[\rho H + n(\rho_S V_S E_S + \rho_L V_L E_L)]$$

$$+ \frac{\partial}{\partial x}F[\rho u H + nv(\rho_S V_S E_S + \rho_L V_L E_L)] = 0$$

for the particles:

$$\frac{\partial}{\partial t}[Fn(\rho_S V_S e_S + \rho_L V_L e_L)] + \frac{\partial}{\partial x}[Fnv(\rho_S V_S e_S + \rho_L V_L e_L)]$$

$$= F[\frac{n\,Nu\,F_S'}{d}(T-T_S) - W^{(1)}e_S^0 - W^{(3)}e_L]$$

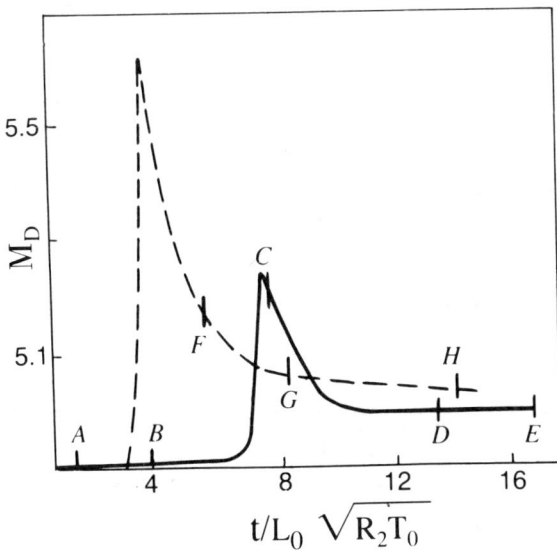

Fig. 6 Mach number of the leading detonation wave in a suspension of coal dust in mixture of acetylene. (----) planar symmetry; (——) spherical symmetry. $L_0 = 1$ m; $T_0 = 293$ K; $d = 2 \times 10^{-5}$ m; $Pr = 0.75$; $R_2 = 0.28 \times 10^3$ J/kg-K.

The equations of state:

$$H = \frac{k}{k-1} \frac{p}{\rho} + \beta_L Q^{(2)} + \frac{u^2}{2}$$

$$E_S = e_S + \frac{v^2}{2} = c_S T_S + Q^{(1)} + \frac{v^2}{2}$$

$$E_L = e_L + \frac{v^2}{2} = c_L T_L + Q^{(2)} + \frac{v^2}{2}$$

$$e_S^0 = c_S T_S$$

The supplemental relations:

$$\phi_n = n\xi(Re) S \rho \frac{(u-v)|u-v|}{2}$$

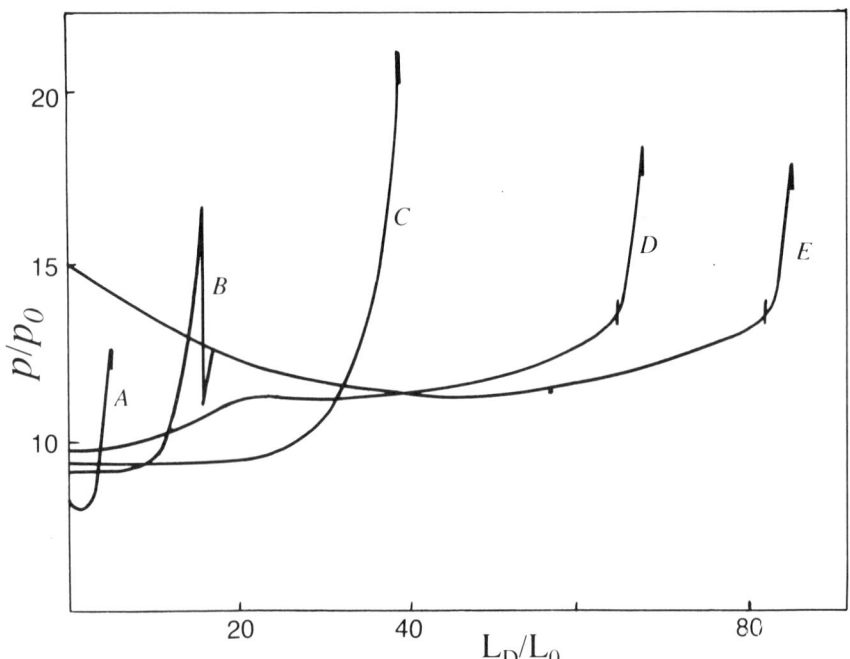

Fig. 7 Pressure distribution for a spherically symmetric detonation of a coal dust suspension in a mixture of acetylene-oxygen for parameters shown in Fig. 6.

$$Re = \frac{\rho |u-v|}{\mu} d \qquad Nu = 2 + 0.6 \, Pr^{1/3} Re^{1/3}$$

$$d = \left(\frac{6V}{\pi}\right)^{1/3} \qquad S = \frac{\pi d^2}{4}; \qquad V = V_L + V_S$$

$$F_S' = \begin{cases} \pi d^2, & \text{when } T_S < 3500 \text{ K} \\ 0, & \text{when } T_S \geq 3500 \text{ K and } T > T_S \end{cases}$$

The reactions:

$$x_1 C + x_2 O_2 \xrightarrow{W^{(1)}} x_3 CO + x_4 CO_2 + Q^{(1)}$$

$$L_g + O_2 \xrightarrow{W^{(2)}} M + Q^{(2)}$$

$$L_S \xrightarrow{W^{(3)}} L_g$$

$$W_C^{(1)} = n\rho_S V_S \rho \beta_{O_2} K_1' F_S$$

$$W_{O_2}^{(2)} = \rho \beta_L \rho \beta_{O_2} K_2 \sqrt{T} \exp(-E_2/R_o T)$$

$$W^{(3)} = n\rho_L V_L \, n\rho_S V_S K_3 \exp(-E_3/R_o T_S)$$

$$(K_1')^{-1} = [K_1 \exp(-E_1/R_o T_S)]^{-1} + \frac{d}{NuD}$$

$$D = D_{O_2}^o \left(\frac{P_o}{P}\right)\left(\frac{T}{T_o}\right)^2 \qquad F_S = \pi d^2$$

where $Q^{(1)}$ is the heat of combustion of coke per unit mass of coke; $Q^{(2)}$ the heat of combustion of volatile substance per unit mass of fuel; V_S the volume of coke in a particle; V_2 the volume of volatile substances in the solid phase per particle; T_S the temperature of the solid phase; β_{O_2} the mass concentration of oxygen in the gas phase; and β_L the mass concentration of volatile substances in the gas phase.

Results

After initiation, the detonation wave propagates through the gas phase of oxygen-acetylene mixture. After the passage of the detonation wave, coal dust particles are accelerated, heated, and ignited. Particle combustion produces a second reaction wave in the flow.

Antonov and Gladilin (1972) have shown that when the energy released in particle combustion is small the combustion zone did not overtake the preceding detonation wave.

This work reports the results of a calculation for the propagation of detonation waves in stoichiometric mixtures of coal dust with a particle radius of 10^{-5} m. The model equations mixture were solved by a numerical method due to Godunov et al. (1961). The Mach number M_D for the leading detonation wave is shown in Fig. 6. Here L_0 is the characteristic length, R_2 the gas constant in the disturbance zone, T_0 the mixture temperature at a rest.

In Figs. 7 and 8 the gas pressure and velocity distributions as a function of L_D, the leading wave

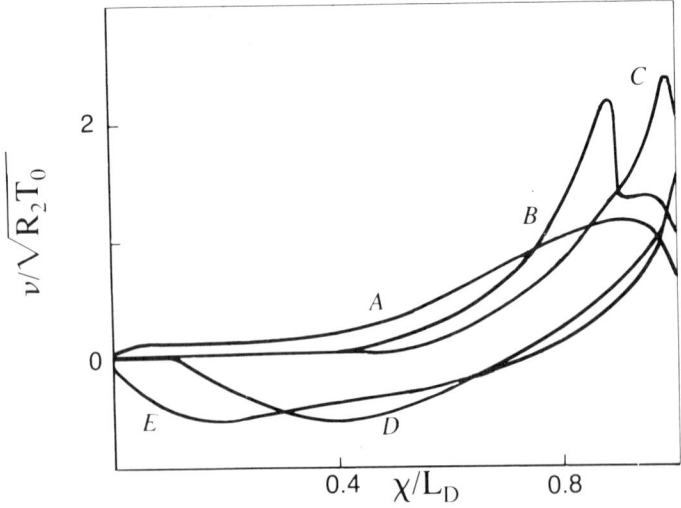

Fig. 8 Velocity distribution behind the leading wave for a spherically symmetric detonation of a coal dust suspension in a mixture of acetylene-oxygen for parameters shown in Fig. 6.

MULTIFRONT COMBUSTION OF TWO-PHASE MEDIA 407

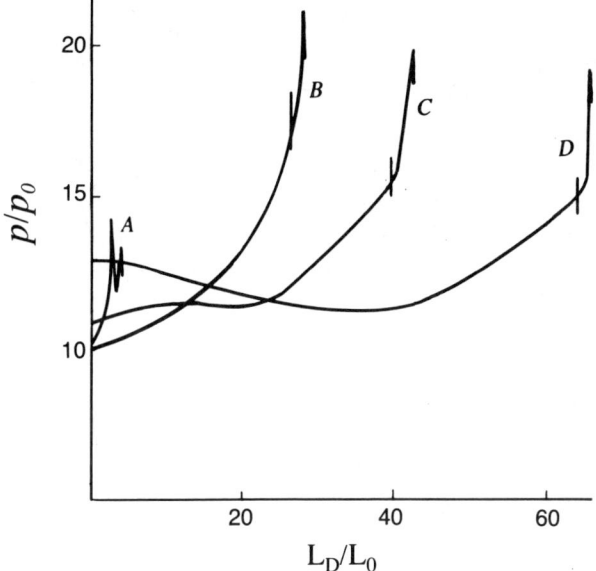

Fig. 9 Pressure distribution for a planar detonation of a coal dust suspension in a mixture of acetylene-oxygen for parameters shown in Fig. 6.

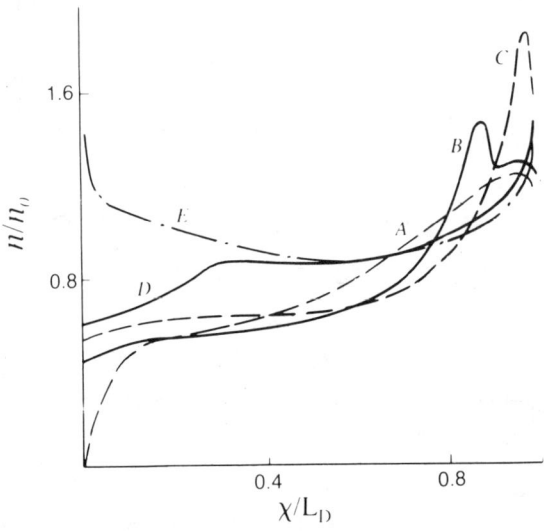

Fig. 10 Particle density disbribution behind the leading wave for a spherically symmetric detonation of a coal dust suspension in a mixture of acetylene-oxygen for parameters shown in Fig. 6.

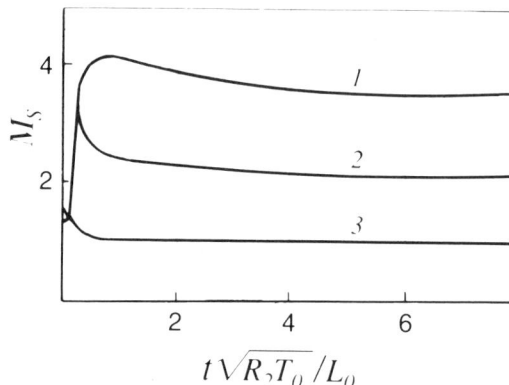

Fig. 11 Damping of shock waves supported by deflagration of coal dust in oxygen. Planar geometry. d = 2 x 10^{-5}m. Curve 1) v_f = 0.2 m/x, L* = 1.0 m. Curve 2) v_f = 0.1 m/s, L* = 0.25 m. Curve 3) v_f = 0.01 m/s, L* = 0.05 m.

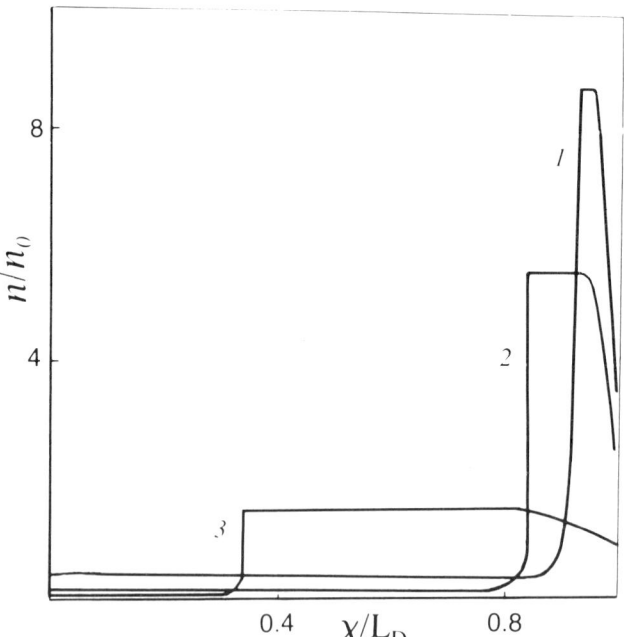

Fig. 12 Particle density distribution for initiation of deflagration in suspensions of coal dust in oxygen. Conditions of initiation shown on Fig. 11.

coordinates are shown for the spherical case, while in Fig. 9 the pressure distributions for the plane case are given. The maximum pressure observed for the spherical case is 24.5 and for the planar case 34.5 atm. The velocity and particle density distributions behind the leading wave for spherical symmetry are shown in Fig. 10 for $t = (t/L_0)\sqrt{R_2 T_0}$: 0.9, 3.5, 7.6., 13.7, 16.8.

These results show that as a result of coal dust combustion in front of the intense heat-release zone a compression wave is developed. This wave runs after the detonation front and accelerates the ignition zone and coal dust burning. This process leads to the formation of a detonation wave which propagates with a greater velocity together and to the heat-release zone being adjoined to the shock wave. In the limit a self-similar flow is formed with only one detonation wave. This outcome is to be expected since Q_2 for combustion of coal dust exceeds Q^*, the maximum heat release for multifront propagation.

The development of a deflagration in a coal dust suspension in oxygen as air and the transition from deflagrative to detonative combustion is also of interest. Ignition is modeled as a momentary release of energy in a finite volume; this energy pulse produces a momentary increase of the gas and particles temperature to T_*, but does not change the gas density. The energy release is sufficient to cause chemical reactions with the ignition volume, and a shock wave is formed. The shock wave, initially damped, precedes the combustion zone.

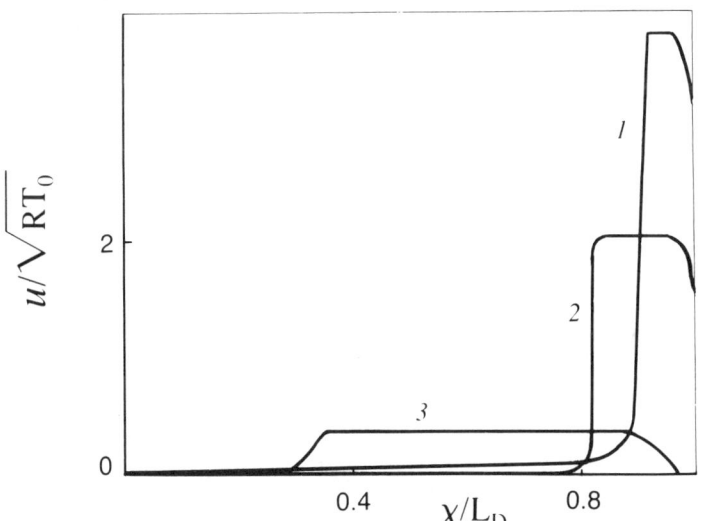

Fig. 13 Velocity distribution behind lead shock wave for initiation of deflagration in suspensions of coal dust in oxygen. Conditions of initiation shown on Fig. 11.

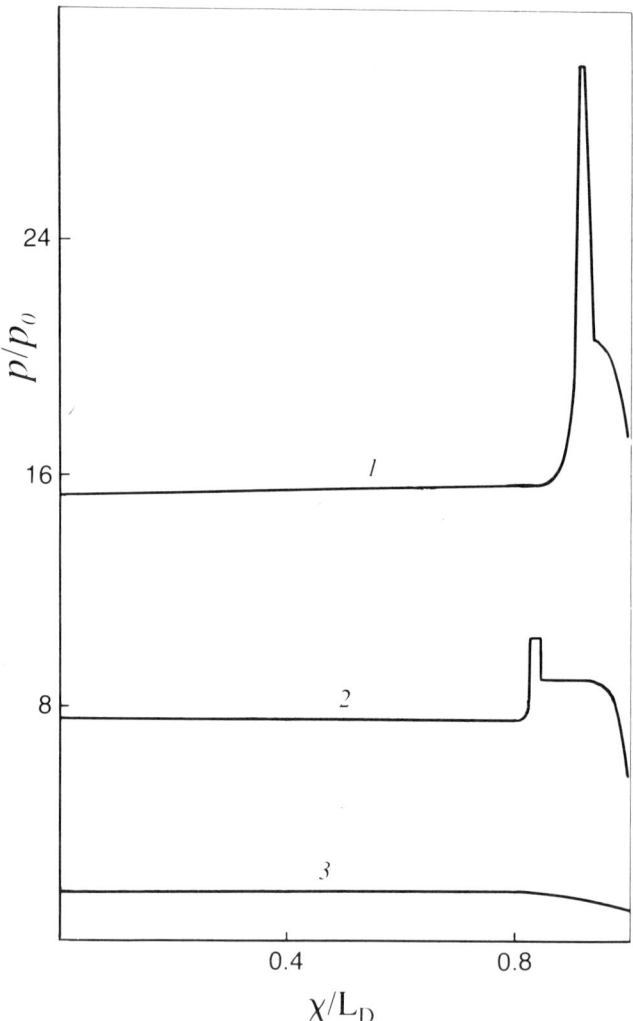

Fig. 14 Pressure distribution behind lead shock wave for initiation of deflagration in suspensions of coal dust in oxygen. Conditions of initiation shown in Fig. 11.

The results of a numerical investigation which used the model developed for this work and the above referenced numerical method are reported in the following. The combustion, in oxygen, of anthracite coal dust with a particle radius of 10^{-5} m is analyzed. In the planar case the ignition source is uniform over the cross-sectional area of the tube and the volume is defined by the distance L_* from the end of the closed tube.

The results for the stoichiometric mixture at $T_* = 1500$ K are presented in Figs. 11-15. The shock wave produced within the ignition volume will be either damped to a sonic wave (e.g., $L_* = 0.05$ m) or will be intensified by the perturbations produced by particulate combustion (e.g., $L_* = 0.25$ and 1.0 m). In Fig. 12 v_f is the velocity of the flame relative to the gas in front of the combustion zone.

In all cases considered, the transition to the self-similar regime of combustion occurs, but the propagation velocities and mixture parameters behind the lead shock wave differ according the ignition volume. The flow pattern behind the lead shock wave is characterized by the existence of a region in which the gas is near the state of rest or in which the gas before the flame front is moving. The mixture parameters in each area are practically constant. Decreasing L_* to 0.5×10^{-2} m or T_* to 1100 K leads to damping the shock wave to a sonic wave.

The calculations suggest the existence of critical ignition parameters which determine whether the flame can propagate in a self-similar regime at detonative or deflagrative velocity.

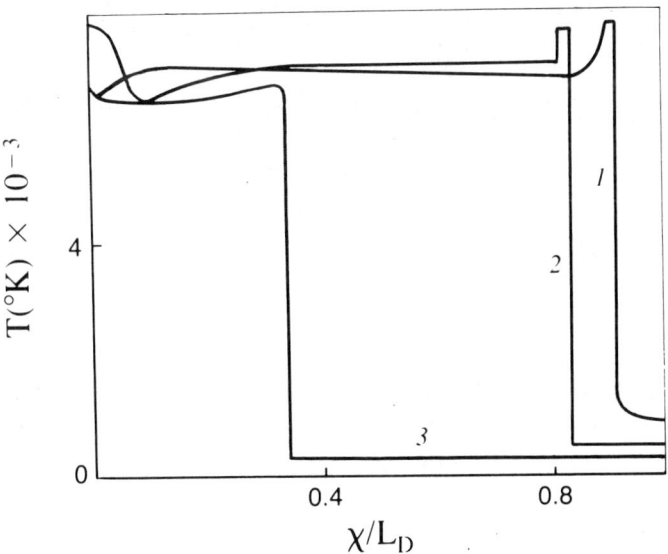

Fig. 15 Temperature distribution behind lead shock wave.

Conclusions

This analysis suggests the possibility of multifront combustion of the two-phase media. In cylindrical and spherical cases, the first detonation wave may be followed by the second one. For each value of heat of combustion in the first wave, there is a maximum value of heat to release in the second one beyond which multifront combustion is not possible. In addition to the double detonation regime, combustion may also be realized in a flame front separated from the leading detonation wave by the shock wave.

As an example of calculation of real two-phase mixture flows, coal dust combustion in oxygen with acetylene was analyzed. Multifront combustion was not possible since the heat release in coal combustion exceeded the critical heat release for multifront combustion. Analysis of the initiation of combustion in suspension of coal dust in oxygen showed that there are critical values of ignition parameters which determine the velocity of flame propagation in a self-similar regime.

References

Abramovich, G. N. (1969) <u>Prikladnaja Gazovaja Dinamika</u>. Nauka, Moscow, USSR.

Antonov, E. A. and Gladilin, A. M. (1972) Usilenie detonacionnoj volny zonoj vtorichnyh reakcij v dvuhfazoj srede. <u>Ivz. Akad. Nauk. SSSR Mekh. Zhidk. Gaza.</u> 5, 92-96.

Bar-Orr, R., Sichel, M., and Nicholls, J. A. (1981) The propagation of cylindrical detonations in monodisperse sprays. <u>Eighteenth Symposium (International) on Combustion</u>, pp. 1599-1606. The Combustion Institute, Pittsburgh, Pa.

Godunov, S. K., Zabrodin, A. V., and Prokopov, G. P. (1961) Raznostnaja shema dlja dvuhmernyh zadach gazovoj dinamiki i raschet obtekanija s otoshedshej udarnoj volnoj. <u>Zh. Vychisl. Mat. Mat. Fiz.</u> 1, 1020-1050.

Hitrin, L. N. (1957) <u>Fiz. Goreniya Vzryva.</u>

Levin, V. A. and Chernij, G. G. (1967) Asimptoticheskie zakony povedenija detonacionnyh voln. <u>Prikl. Mat. Mekh.</u> 31, 393-405.

Nigmatulin, R. I. (1978) <u>Osnovy Mechaniki Geterogennyh Sred</u>. Nauka, Moscow, USSR.

Pierce, T. H. and Nicholls, J. A. (1973) Time variation in the reaction zone structure of two phase spray detonations. <u>Fourteenth Symposium (International) on Combustion</u>, pp. 1277-1284. The Combustion Institute, Pittsburgh, Pa.

Sedov, L. I. (1957) Metody podobija i razmernosti v mehanike. M., Gostehizdat, Moskva, SSSR.

Skinner, T. H. (1967) Friction and heat-transfer effects on the nonsteady flow behind a detonation. <u>AIAA J.</u> 5, 2069-1284.

Veyssiere, B., Bouriannes, R., and Manson, N. (1979) Detonation characteristics of two ethylene-oxygen, nitrogen mixtures containing aluminum particles in suspension. <u>Gasdynamics of Detonations and Explosions: AIAA Progress in Astronautics and Aeronautics</u> (edited by Bowen, Manson, Oppenheim, and Soloukhin), Vol. 75, pp. 423-438, AIAA, New York.

Flame Propagation in Dust-Air Mixtures at Minimum Explosive Concentration

Wojciech Buksowicz*
Fire Protection Research and Development Center, Józefów, Poland
and
Piotr Wolański†
Technical University of Warsaw, Warsaw, Poland

Introduction

The lean flammability limit is expressed in terms of the minimum concentration of dust dispersed uniformly in air at which flame can be self-sustained. This value is indispensable for the appraisal of fire and explosion hazards. It is consequently called minimum explosive concentration (MEC). Studies of lean flammability limits have been carried out for a long time. They dealt with several kinds of dust and were conducted on many different test stands using different criteria for flammability. Consequently, existing data present a widespread range of results even for the same kind of dust. For example, MEC for coal dust varies from 0.005 to 0.31 kg m^{-3}, while that for linen dust extends from 0.01 to 0.25 kg m^{-3}. It is clear that the data are affected by different physical and chemical properties, but the large spread of data must arise mainly as a consequence of the differences in testing conditions. Previous studies yielded precise values for MEC. However, the mechanism governing the flame propagation under these conditions were not fully explored. It is for this reason, then the present studies have been undertaken.

Presented at the 8th ICOGER, Minsk, USSR, Aug. 23-26, 1981. Copyright © American Institute of Aeronautics and Astronautics, Inc., 1982. All rights reserved.
*Research Engineer.
+Associate Professor.

Experimental Technique

Three types of chambers have been widely used for the evaluation of minimum explosive concentration (MEC). Of these the spherical chambers have the best volume to surface ratio (minimum cooling effect at the walls), the subicoidal chambers have the largest volume (a few cubic meters), and the cylindrical chambers (also called Hartmann tubes) are very simple in construction and design. Long cylindrical tubes are also used to observe flame propagation through dust-air mixtures.

The present studies were conducted with the use of a cylindrical chamber, made out of glass, of a volume equal to about 5.5 liters and a diameter to length ratio $D/H = 0.5$. The dust dispersion unit was located at the bottom plate. It consisted of a dispersion cup and a supply system using compressed air. Ignition was performed by an electric spark generated between two electrodes located at the center of the chamber. Before dispersion of the dust, the chamber was partially evacuated. The dust was dispersed by pressurized air delivered from a tank reservoir in such a way that the pressure in the chamber reached about 1 atm at the end of the pulse. The top of the chamber was closed by a paper membrane to protect it from destruction that could be caused by pressure rise due to the combustion process.

Experiments were also performed using cylindrical steel chambers of the same dimensions as the glass chamber. The wall temperature was controlled by cooling fluid. The pressure was measured by a transducer located at the top of the chamber. In both glass and steel chambers the same concentration of the dust was initially provided. The electric discharge, ignition, and the flame propagation process was observed by the use of the Schlieren technique. This required the use of a special type of chamber with flat windows ($\phi = 80$ mm) to observe the area near the electrodes. A detailed description of the chambers and test stands was provided by Buksowicz et al. (1978).

Preparation of the Dust-Air Mixture

Evaluation of the quality of dust dispersion was made on the basis of data obtained by the visualization technique. The process of dispersion was recorded directly by a cinecamera (Fig. 1). A special system for illumination of the chamber was used for this purpose. The photographs show the successive phases of dispersion. After about 0.25 s, the mixture is evidently quite uniform. In the next period of time there are only small changes in the concentration. However, there are small local fluctuations that are not visibly in evidence because the photographs provide only information averaged over the full depth of the chamber. In order to determine such nonuniformities, local changes in the concentration of the dust were measured by the use of a laser forward scattering technique (using a He-Ne laser). The apparatus was aligned in such a way [see Wolanski (1979) and Bucksowicz et al. (1978)] that the recording photo-multiplier detected the scattering signal from a volume of 0.5 cm^3. Thus the signal of dust concentration was proportional to light scattered within this volume.

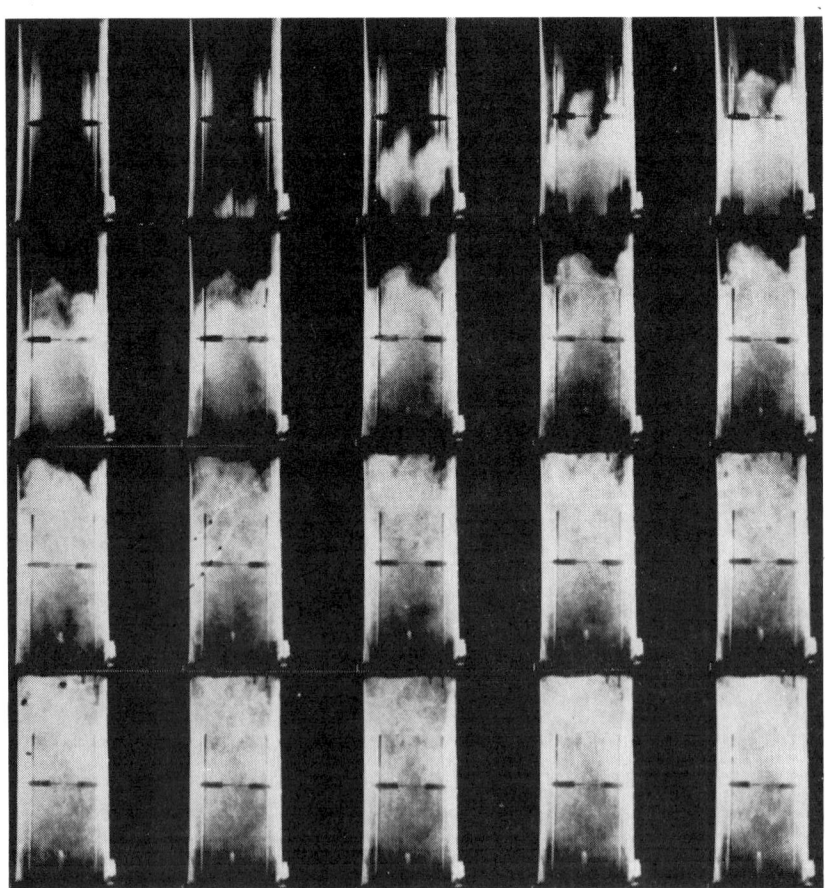

Fig. 1 Photographs of the dust dispersion process in a cylindrical chamber. Film speed: 48 frames per second.

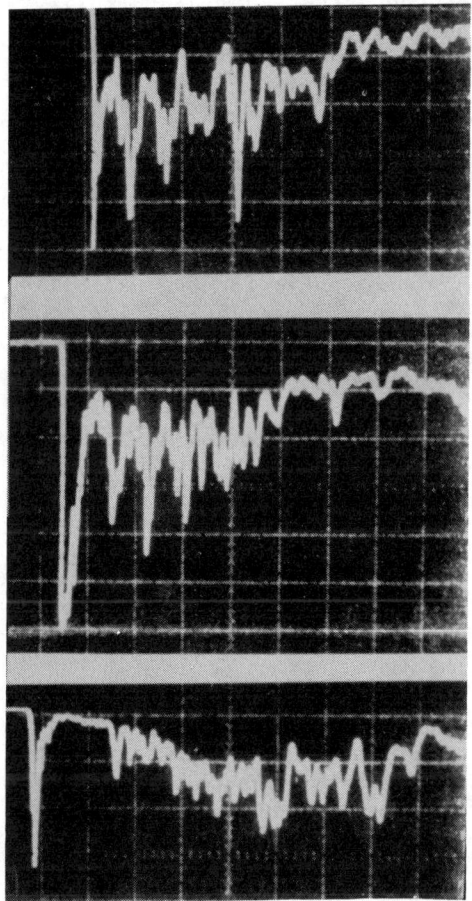

Fig. 2 Oscilloscope records showing fluctuations in dust concentration at the center of the test chamber. a = 3/4 H; b = 1/2 H; c = 1/4 H, where H is the height of the chamber. Sweep: 0.05 s per division.

Typical records of concentration fluctuations recorded by the oscilloscope are presented in Fig. 2. As the concentration is increased, the beam is deflected downward owing to negative polarization of the photomultiplier signal. As can be seen in Fig. 2, the first sharp rise in concentration is followed by a steady state, with fluctuations reaching 50% of the mean. The concentration decreases after about 0.3-0.4 s because the dust starts to fall down.

Similar changes occur throughout the whole extent of the chamber. Laser concentration measurements and direct photography are very helpful in measuring dust dispersion and choosing the right moment for ignition.

Ignition

Studies of ignition and flame propagation require suitable ignition sources. The appearance of the flame cannot be used as the basis for the evaluation of MEC in small chambers because a strong ignition source affects the flame propagation process. Literature data indicate that inappropriate ignition energy and ill-timed moment of ignition produce inconsistent data (Klemens 1977). Ignition energy must be strong enough to ignite the mixture, but it should not affect the ensuing process of flame propagation. This is especially important at concentration limits where too strong ignition can support a flame which otherwise would have been extinguished. Furthermore, ignition must be activated at a proper time to match local concentration fluctuations in the dust. It has been found that the best ignition was provided by an electrical discharge of low inductance. The frequency of the discharge was about 100 Hz (two sparks over one period in an alternating current of 50 Hz), while individual spark energy was equal to 1.4 J.

Typical Schlieren photographs of electric discharge in air and ignition process in a dust-air mixture are presented in Fig. 3. The discharge energy was evidently too high, enhancing the flame propagation in mixtures where dust concentrations were below the MEC. Figure 4 displays Schlieren records of the flame propagation process in a lean dust-air mixture (twice the MEC). One can see that after ignition the flame propagates mainly in the upward direction and is associated with large-scale fluctuations. These fluctuations are due to the nonuniformity of dust dispersion in the mixture. Ignition delay varied from a few to several milliseconds, the latter occurring in mixtures near MEC.

It has been found that for the majority of organic dust-air mixtures a successful ignition source is produced by a number of oscillating electric discharges. Ignition energies were in the range of hundreds of joules and the duration time was from 0.2 to 0.3 s. Dusts made of materials that are difficult to ignite required the use of individual ignition sources providing an adequate amount of energy.

FLAME PROPAGATION IN DUST-AIR MIXTURES

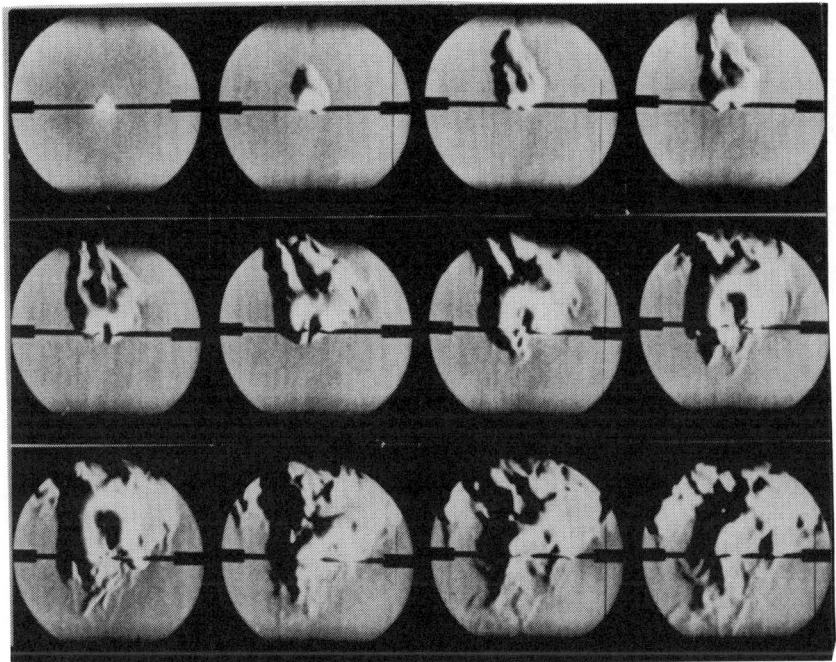

Fig. 3 High-speed Schlieren photography of electric discharge in air. Time between frames: 0.01 s.

Determination of the Minimum Explosive Concentration

Flame propagation in a dust-air mixture was observed in the glass chamber. A sequence of photographic records of the flame propagation process in a mixture below MEC is presented in Fig. 5. In this case the flame propagates only in the upward direction. Its intensity decreases with distance from the ignition source. This means that flame propagation decays in long vertical tubes. Figure 5 demonstrates that the combustion process occurs only near the electrodes, while the flame is sustained solely by the energy of the electric spark.

For mixtures above MEC, the flame propagates freely up to the chamber top. In all cases, flame propagates only in the upward direction owing to the bouyancy effects.

The concentration at which the flame begin to propagate freely in the upward direction is equal to MEC. For such limiting conditions the pressure rise due to combustion was measured in a cylindrical steel chamber. A typical dependence of the pressure rise in this chamber on dust concentration near MEC is depicted in Fig. 6. The

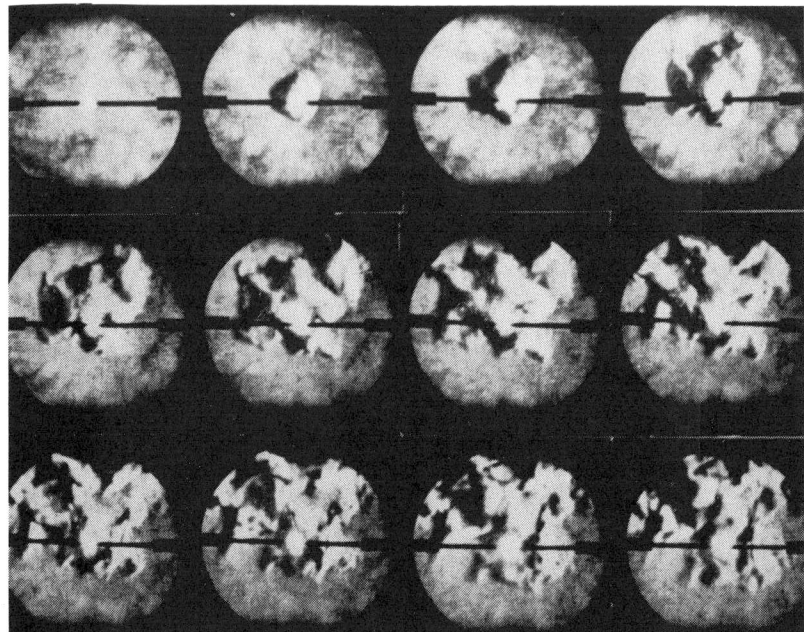

Fig. 4 High-speed Schlieren photographs of the ignition of grain dust-air mixture by electric spark. Time between frames: 0.01 s.

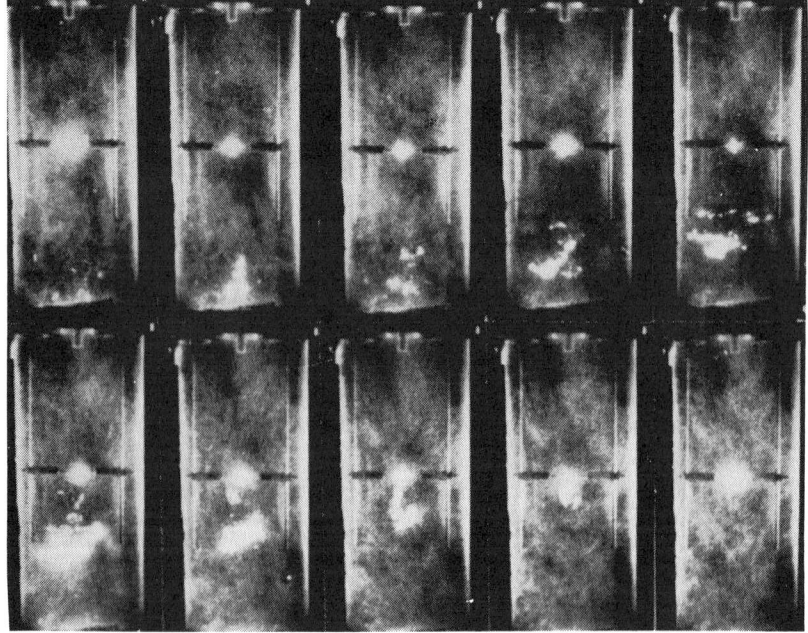

Fig. 5 Flame propagation in linen dust-air mixture. Dust size below 200 µ.

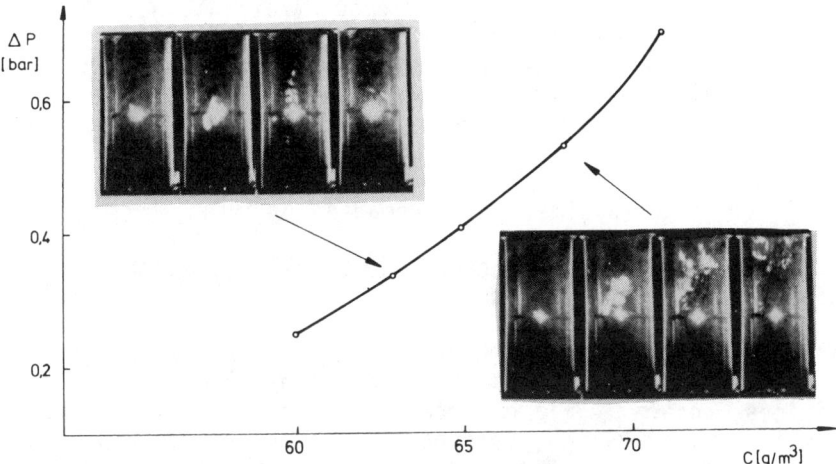

Fig. 6 Pressure rise in the test chamber containing linen dust-air mixture near MEC.

results were obtained for linen dust smaller than 200 μm and heat of combustion of 20,870 kJ/kg. A pressure rise of 0.05 MPa was slected as the criterion for limiting concentration (for the particular volume of the chamber used in the test, while the igniting electrodes were located at its center). It has been noted that at the limiting conditions the flame propagates only in the upward direction and the pressure rise is dependent on the location of the igniting electrodes in the chamber. MEC evaluated in such a way was in good agreement with minimum concentrations at which flame propagated in long tubes 100 mm in diameter. For example, MEC of dust from shoveboard grating determined in the cylindrical steel chamber was equal to 0.106 kg m^{-3}, while that measured in a long tube varied from 0.08 to 0.103 kg m^{-3}, with an accuracy of up to 30% (Gorecki 1978). This means that the cylindrical steel chamber can be used with success for the determination of MEC in a dust-air mixture. However, to obtain reliable results, the dispersion of the moment of ignition, the type of ignition source, and the pressure must be carefully controlled. Taking all this properly into account, MEC for organic dust, whose heat of combustion was in the range of 14,200-24,000 kJ/kg, was determined using the cylindrical chamber.

Experimental data determined in this manner are presented in Fig. 7 as a function of the higher heating value (HHV) of the dust. The results are very close to those of gaseous hydrocarbon flames a lower explosive limits.

This means that near lean-limit concentrations, gaseous hydrocarbons are produced by the thermal decomposition of dust particles accompanying the combustion process. Moreover, similar adiabatic flame temperatures of 1600 K (Hertzberg et al. 1979) were obtained also for coal dust at MEC.

Physical Interpretation of MEC

Flame propagation in combustible mixtures at the lean limit can be explained on the basis of a thermal model of combustion. It has been found that adiabatic flame termperature of the majority of hydrocarbon-air mixtures at the lean limit is 1400-1600 K (Fig. 7). Thus, under such conditions, adiabatic flame temperatures for dust-air mixtures attain similar values, irrespective of the exact nature of the dust material. The heat produced during the combustion process is then just sufficient to maintain a positive balance of energy in the flame. Under such circumstances it appears that MEC depends only on the heating value of the dust. One should be able, therefore, to deduce the value of MEC from the heating value of the dust and the adiabatic flame temperature.

In Fig. 7, displaying the relationship between the heating value of the dust and its concentration, two theoretical lines of the adiabatic flame temperature for hydrocarbon-air mixtures at lean limit are shown. Experimental points are evidently quite close to these lines. By the use of a line regression technique the results can be expressed approximately in terms of a relatively simple theoretical relation, namely,

$$MEC = 15,500 \, Q^{-1.21} \text{ kg m}^{-3}$$

where Q is the heating value of the dust in kilojoules per kilogram.

This correlation was plotted in Fig. 7 as the dotted line. It can be seen that the calculated theoretical value of MEC does not differ much from experimental data. Mean departure of experimental values from the correlation is 15%. Hence for quick evaluation of MEC, the above formula can be considered to be quite satisfactory. It should be noted that this relationship was established only for a special kind of dust (i.e., grain and linen) whose size was below 200 μm. It was based, moreover, on a vastly simplified thermal model of the process. Consequently, it should not be used without particular verification for other types of dust.

FLAME PROPAGATION IN DUST-AIR MIXTURES

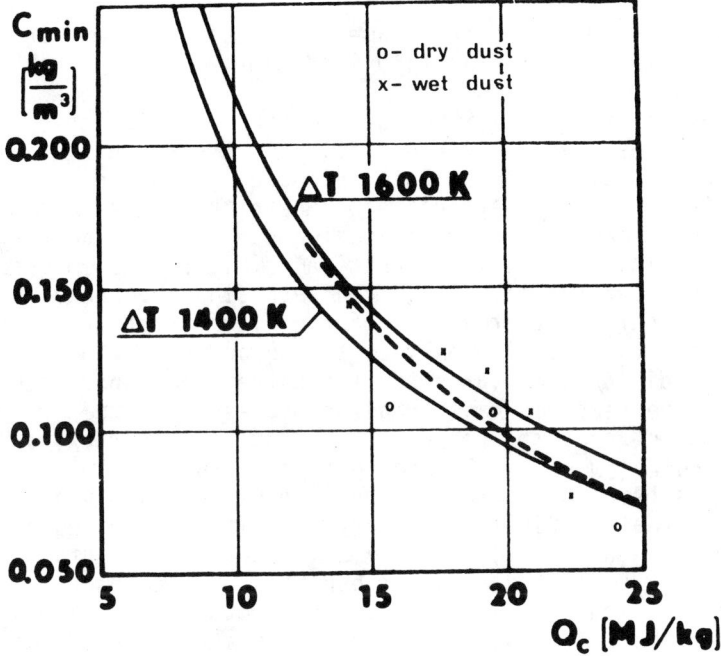

Fig. 7 Influence of heat of combustion on MEC.

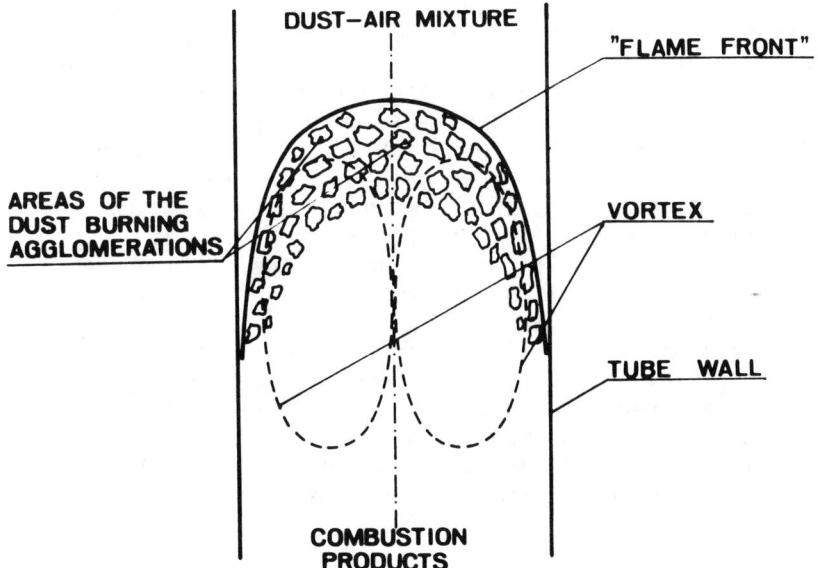

Fig. 8 Schematic diagram of flame structure in a dust-air mixture.

Physical Model of the Flame Propagation Process Under Limiting Conditions

As it is generally known, the burning velocity decreases as the dust concentration in the combustible mixtures is diminished. This decrease in the burning velocity is especially noticeable near the lean-limit concentration where free convection due to buoyance plays a critical role in flame propagation, similarly as in gaseous mixtures. At the lean-limit concentration, buoyancy produces strong deformations of the flame front and causes the flame to propagate only in the upward direction. Under such conditions, buoyancy stretches the flame and intensifies the heat and mass transfer processes in the combustion zone (the region of recirculation vortex) and may contribute toward the quenching of the flame.

A schematic diagram of these processes is given in Fig. 8. For real mixtures with small nonuniformity of dust dispersion (near MEC), the flame is partitioned into distinctly separated areas of burning dust agglomerates. These are clearly evident in direct photographs. Schlieren photographs show that the deformed areas consist of hot combustion products moving in an upward direction, while the concomitant vortex motion of the dust particles has been recorded by streak photography. Flame thickness in the dust-air mixture is much larger than in a gaseous medium, varying from a few to several centimeters. The extent of the flame zone depends on the burning velocity of the dust. For example, in a grain dust-air mixture the flame thickness is of an order of a few centimeters, while in a coal dust-air mixture containing a solid residue it can be as large as 1 m. In the latter case, at the limiting condition, the first period of the combustion process (the combustion of volatiles) affects mainly the flame propagation rate, while combustion of the solid phase, being relatively slow, occurs far behind the flame front and does not have any effect on its motion.

Hertzberg et al. (1979) demonstrated that the combustion process in the coal dust mixtures near MEC depends on the rate at which the volatiles are released and burned. However, for the same kind of dusts which do not produce volatiles (e.g., metallic dust, graphite, diamond) flame propagation is controlled only by heterogeneous processes.

Conclusions

1) At limiting conditions, free convection plays a significant role in the flame propagation process occurring

in dust-air mixtures. It causes the flame to be strongly stretched and it creates vortices that intensify heat and mass transfer in the combustion zone. This may eventually upset the heat balance in the combustion zone and cause the flame to be quenched.

2) For dusts whose burning velocity is sufficiently high (e.g., organic dusts), critical conditions of flame propagation at the lower combustion limit can be expressed in terms of just one parameter: the heat of combustion. Thus, for all practical purposes, this limit can be specified by a simple algebraic expression.

3) For dusts whose burning process consists of a number of stages, the lower combustion limit is controlled by the combustion rate of just one phase (e.g., the combustion of volatiles). Combustion of other phases occurs far behind the flame front and therefore it cannot effect the flame propagation rate or its lower combustion limit.

References

Buksowicz, W., Lizut-Skwarek, M., and Wolanski, P. (1978) Dust air mixture explosivity study. Technical University of Warsaw, Warsaw, Poland (in Polish).

Gorecki, J. (1978) Report I-20/SPR-12/79. Technical University of Warsaw, Warsaw, Poland (in Polish).

Hertzberg, M., Cashdollar, K. L. and Opferman, J. J. (1979) The flammability of coal dust-air mixtures. RI 8360, U.S. Bureau of Mines, Pittsburgh, Pa.

Klemens, R. (1977) Heterogeneous mixture ignition by electrical spark. Ph.D. Thesis, Technical University of Warsaw, Warsaw, Poland (in Polish).

Wolanski, P. (1979) Explosion hazards of agricultural dust. Proc. International Symposium on Grain Dust, pp. 422-446. Manhattan, Kansas.

Chemical Kinetics of Detonation in Some Organic Liquids

B.N. Kondrikov*
Mendeleev Institute for Chemical Technology, Moscow, USSR

Abstract

The failure diameter of liquid O and C-nitrocompounds, of their mixtures with organic solvents, as well as of the mixtures of some liquid fuels with concentrated nitric acid was measured and used as a tool for approximate determination of leading chemical reaction rates at high-velocity detonation of the liquids. Good correlation between the failure diameter d_f and adiabatic temperature of explosion T_v is obtained for nitric esters containing C-C, C-H, C-O, and O-N bonds and for the mixtures of the esters with methanol, chloroform, and bromoform: $d_f = A \exp(B/T_v)$, where $A = 0.059$ mm and $B = 7300$ K. The corresponding values of A and B for solutions of nitric acid in dichloroethane are 0.053 mm and 6400 K, and for solutions of nitric acid in acetic anhydride are 0.097 mm and 4600 K. The addition of olefins to nitrocompounds reduces the value of d_f. The failure diameter of nitromethane containing 2% of allylalcohol or acrylic acid is two times less than that of the pure substance. The addition of strong organic bases to the substance is much more effective. A very small quantity (0.02%) of diethylamine reduces d_f from 13 to 6 mm. Nitromethane containing 2% of the additive has a d_f as small as 1.8 mm. The most effective catalytic additive is allylamine $CH_2 = CHCH_2NH_2$. The effect of amines on the reaction rate at detonation of di- and trinitrotoluenes and nitropropane is also determined. Strong inorganic acids also have the

---
Presented at the 8th ICOGER, Minsk, USSR, Aug. 23-26, 1981. Copyright © American Institute of Aeronautics and Astronautics, Inc., 1983. All rights reserved.
*Professor.

catalytic activity but a more weak one than organic bases.
A chain mechanism of acceleration of nitromethane decomposition in a detonation wave under the influence of bases
and acids is proposed. The chain length in the case of
amines may reach 100-150 of elementary steps. At the same
time amines do not accelerate the nitromethane decomposition
reactions at low-velocity detonation or at laminar burning
in the Andreev-Crawford strand burner in the pressure range
from 10 to 200 atm. The definite relations between the
kinetics of thermal decomposition of some alkylnitrites and
their reaction rates with burning and detonation are
observed, and some explanations are proposed.

Introduction

J. B. Khariton (1947) proposed that chemical reaction
parameters of liquid explosives be determined from the
detonation failure diameter of an explosive solution in an
inert solvent. He postulated that the failure diameter d_f
is related to a shock compression temperature of the
solution T_s by the Arrhenius-type expression as

$$d_f = A \exp(E/RT_s)$$

If T_s is a function of the explosive concentration, the
kinetic constants in Eq. (1) could be easily estimated.
Kurbangalina (1947) performed detailed experiments on the
failure diameters of liquid explosives (mainly nitric
esters) diluted with inert organic solvents, but did not
observe direct dependence of d_f on T_s. Apin and Velina
(1967) in analyzing Kurbangalina's data suggested that
organic solvent molecules do not react in a detonation wave
and that the failure diameter of the nitric ester solutions
depends exclusively on the relation between the number of
molecules of a solvent and molecules of an explosive (a
molar part of the solvent, α_m).

Kusakabe and Fujuwara (1970, 1976) proposed that the
dipole moment of a solvent strongly influenced the
phenomenon. A small dipole moment favors detonation; a
large moment hinders detonation. For low-velocity
detonations, they showed that the influence of a solvent on
d_f could be related to the solvent density. Dremin (1962)
developed a new hydrodynamic theory of the detonation
failure diameter, in which the value of d_f is determined by
the reaction rate at temperature T_3 slightly less than T_s.

Enig and Petrone (1970) and Tarver et al. (1976) have confirmed the validity of Dremin's theory, although Tarver and co-workers noted that different equations of state can make a difference in the d_f values up to 670 times.

The aim of this work was to demonstrate some of the ideas concerning the nature of the failure diameter of organic liquids and to accumulate some new experimental data.

Experimental Technique

Methylnitrate and nitroglycol were prepared and purified by the routine technique and mixed with a solvent. The increments in solvent concentration changed by 5% mass

Table 1 Characteristics of solvents and "critical" solutions containing methylnitrate and nitroglycol

Solvents	Characteristics of solvents			Characteristics of critical solvents[b]			
	Molecular weight	Density g/cm^3	Dipole moment, D	Explosive solute[a]	Solvent conc., %	Density g/cm^3	Solvent mole fraction
Toluene	92	0.867	0.37	MN	23	1.11	0.20
				NG	28	1.24	0.39
Dichloroethane	99	1.256	1.27	MN	35	1.23	0.30
Methanole	32	0.791	1.71	MN	38	1.01	0.60
				NG	38	1.11	0.74
Nitrobenzene	123	1.201	4.23	MN	48	1.21	0.37
				NG	43(?)	1.35	0.48
Methylene chloride	85	1.336	1.62	MN	65	1.39	0.63
Chloroform	119	1.489	1.06	MN	78	1.42	0.70
				NG	78	1.49	0.82
Carbon tetrachloride	154	1.594	0	MN	73	1.47	0.57
Bromoform	254	2.89	0.99	MN	88	2.48	0.69
				NG	88	2.60	0.79

[a] MN is methylnitrate, NG, nitroglycol.
[b] Experiments conducted in steel tubes.

from one test to the next. The critical solution was defined as the most dilute solution to detonate. The mixture was poured into steel or glass tubes which were wider at the top for setting a booster. The booster mainly consisted of two pressed pellets of RDX (2 g each) and in some experiments of one pellet of tetryl (1.5 g). The booster, protected with a thin polyethylene film, was immersed in a liquid. It was initiated by the standard electric detonator No. 8. The steel tubes were 10 mm i.d., 36 mm o.d., and 150-160 mm long. Small (2-mm-diam) orifices were drilled in the tube walls to measure the detonation velocity by a streak camera. The orifices were sealed up from the outside by a transparent film. The glass tubes were 150-160 mm long. Wall thickness was about 1-2 mm.

Results

Nitric Esters: The Critical Solvent Concentration

The experimental results are shown in Table 1. The density, dipole moment, or molar fraction of a solvent in a mixture appear to be of no importance in the determination of the detonation limits. Tables 1 and 2 show also that the difference between chloroform and carbon tetrachloride observed earlier (Kusakabe and Fujiwara 1970) can be attributed to use of a weak initiator (tetryl), explosive liquid of small density (methylnitrate), and weak confinement (glass tube). Beyond these circumstances no significant difference between two diluents were observed (Kondrikov and Raykova 1974, 1977).

Table 2 Critical concentrations of solvents in mixtures with methylnitrate and nitroglycol[a]

Solvent[b]	Methylnitrate			Nitroglycol	
	P-30[c]	Tetryl	RDX	Tetryl	RDX
Methanol	...	...	23	...	...
Toluene	...	...	...	18	18
Nitrobenzene	...	...	23	25	25
Chloroform	24	23	63	53	63
Carbon tetrachloride	57	53	68	...	...
Bromoform	...	68	68	73	73

[a] Experiments made in glass tubes with different ignitors.
[b] Solvent concentrations are given in mass percent.
[c] Data of Kusakabe and Fujiwara (1970).

Nitric Esters: The Failure Diameter

The dependence of the failure diameter of methylnitrate solutions in chloroform and carbon tetrachloride on the solvent concentration is shown in Fig. 1. The dependence d_f (C) for methylnitrate solutions in methanol and bromoform reported by Kurbangalina (1947) is approximately the same as that found in this work (see Fig. 1). A very good correlation was obtained between the failure diameter of detonation of nitric esters and their solutions in organic solvents and explosion temperature T_v (Fig. 2).

$$\log d_f = \log d_0 + B/T_v$$

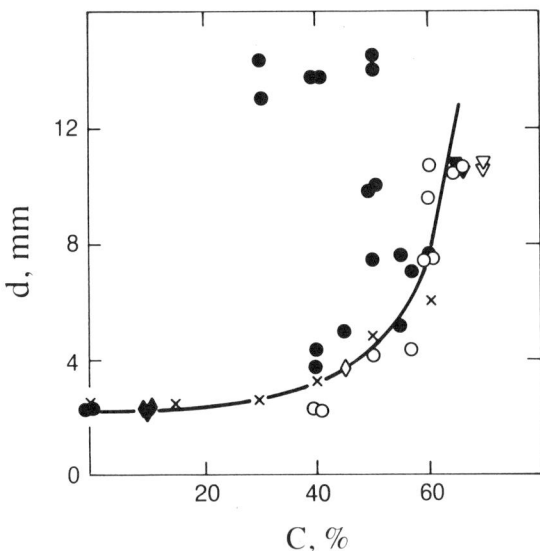

Fig. 1 Detonation failure diameters of solutions of methylnitrate in chloroform and carbon tetrachloride (see Table 3).

Table 3 Legend for Figure 1

Result Solvent	Detonation	Detonation failure	Symbol Detonation extinguishment	Low-velocity detonation
$CHCl_3$	●	○	◆	◇
CCl_4	◀	▽	—	—

[a]Data of Kurbangalina (1947) on nitroglycol in chloroform.

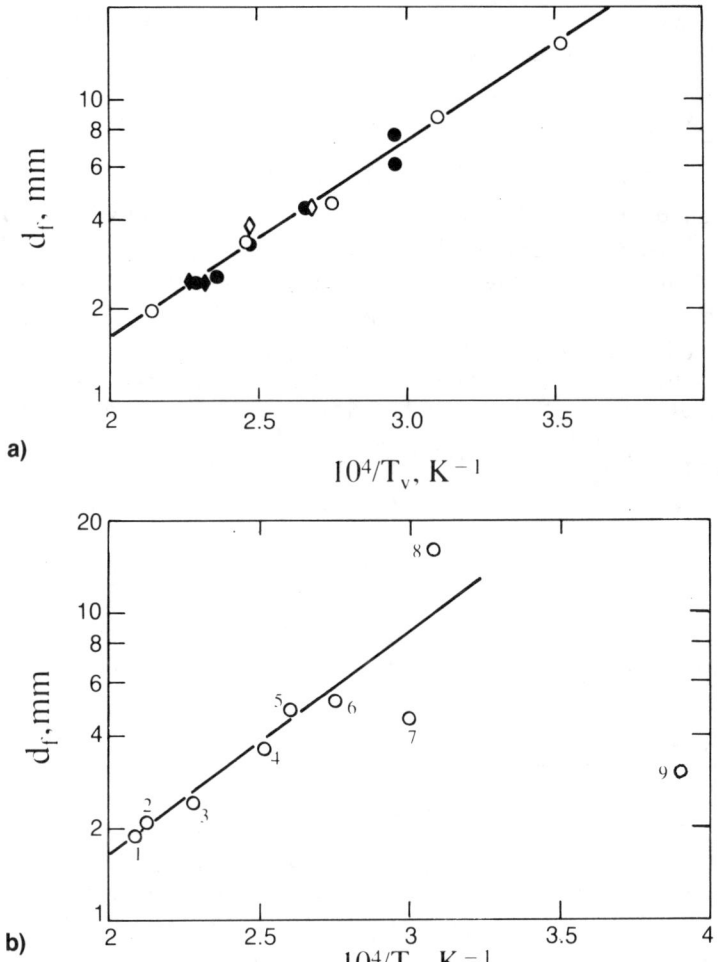

Fig. 2 Correlation of detonation failure diameters with the explosion temperature. a) Data of Kurbangalina (1947) for solutions of nitroglycol. ○ = CH_3OH; ● = $CHCl_3$; ■ = $CHBr_3$. b) Present results solid line identical to that shown in Fig. 2a. 1) nitroglycol, 2) nitroglycerine, 3) methylnitrate, 4) glycerine dinitrate, 5) dinitrochlorohydrin, 6) propylenedinitrate, 7) glycidolnitrate, 8) diethylene glycoldinitrate, 9) allynitrate.

Five nitric esters of simple chemical structure (only C-C, C-H, and $C-ONO_2$ bonds) and 20 mixtures of methylnitrate and nitroglycol with organic solvents tend to form the same straight line (Fig. 2) with constants d_o = 0.059 mm and B = 7300 K. If the T_s values calculated by the conventional

Nitric Acid Solutions

The dependence of d_f on concentration of fuel for mixtures of HNO_3 with dichloroethane, acetic anhydride, di- or trinitrotoluenes are very similar to each other. In Fig. 3 the results for HNO_3-dichlotoeyhane solutions are shown. The experimental points for both lean and rich mixtures can be fitted with an Arrhenius correlation with the explosion temperature. The Arrhenius line for dichloroethane in the region of interest is over the line for acetic anhydride, but both are under the line for nitric esters. The constants of Eq. (2) follow for dichlorethene, $d_o = 0.053$ mm, $B = 6200$ K; and for acetic anhydride, $d_o = 0.097$ mm, $B = 4600$ K.

The data for di- and trinitrotoluenes do not conform to a simple correlation. Rich and lean mixtures give different straight lines probably because of the large nitrocompounds concentration (up to 70%) in the rich mixture (Raykova 1980).

Unsaturated Compounds

The detonation failure diameters of chemical compounds (nitric esters and organic fuels or diluents) containing double or triple chemical bonds or epoxy group cannot be correlated by an Arrhenius expression. Allylnitrate $CH_2 = CH-CH_2ONO_2$ and glycidol nitrate $O-CH_2-CH-CH_2ONO_2$ have respective detonation failure diameters 1/13 and 1/2 those of nitric esters of the usual chemical structure which have identical temperatures of explosion (see Fig. 2b). Double bonds and epoxy group, which are susceptible to addition reactions, evolve heat quickly and accelerate the explosion process as a whole.

The failure diameters of allylnitrate and nitroglycol are 3 and 1.9 mm, respectively. A solution of allynitrate and nitroglycol has a smaller d_f. At allylnitrate concentrations from 10 to 50% d_f is 1.5 mm, and even at 60% it is still less than d_f of pure nitroglycol (Kozak et al. 1979; Kondrikov et al. 1980). Allylalcohol addition leads to a more pronounced reduction of d_f. The mixture (90/10 NG/alcohol) has $d_f = 1.3$ mm. The effect of allylchloride

KINETICS OF DETONATION IN ORGANIC LIQUIDS 433

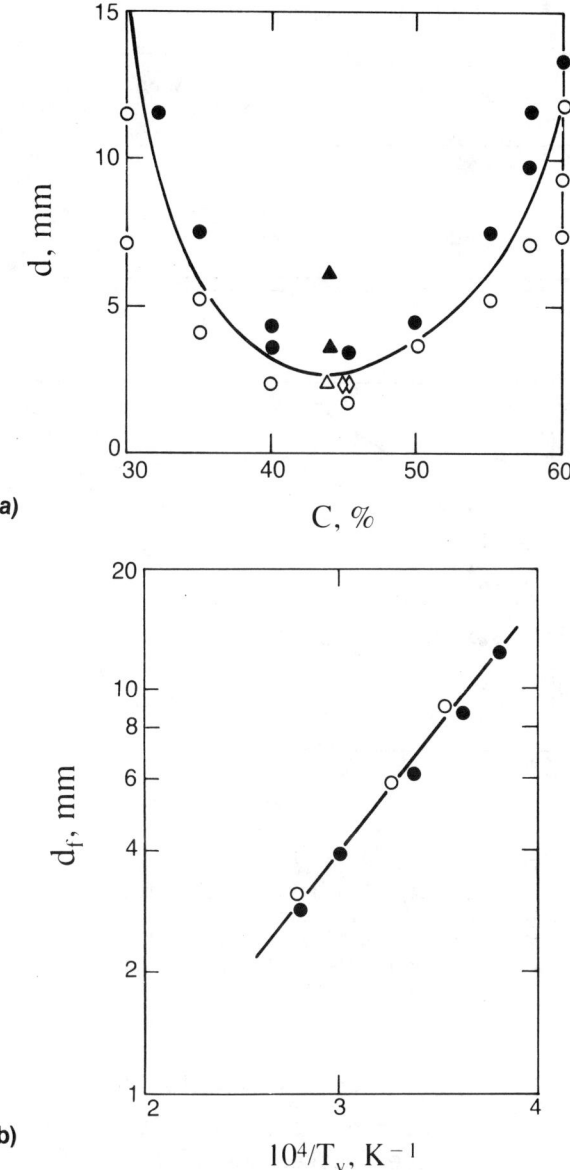

Fig. 3 Detonation failure diameters of nitric acid dichloroethane solutions.
a) Effect of dichloroethane concentration.
b) Arrhenius correlation with explosion temperature. ○ = lean mixtures, left hand branch of Fig. 3a). ● = rich mixtures, right hand branch of Fig. 3a) (see Table 4).

Table 4 Legend for Figure 3

Result	Detonation	Symbol Detonation failure	Detonation extinguishment
This work	●	○	◇
Data of Kurbangalina (1947)	▼	▽	—

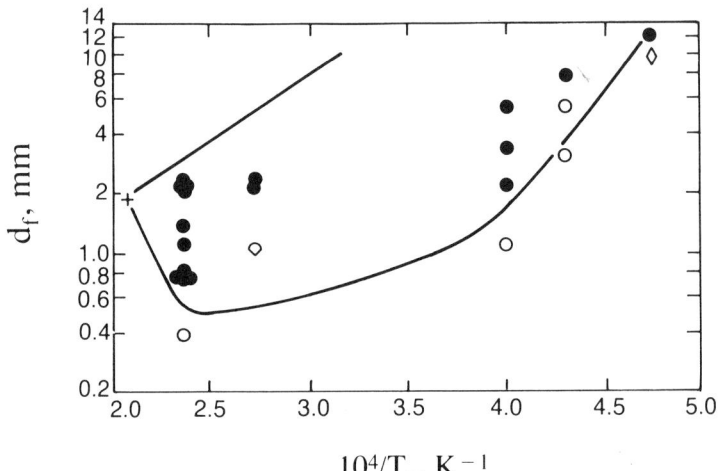

Fig. 4 Detonation failure diameters of nitroglycol/propargyl alcohol solutions. ● = detonation; ○ = detonation failure; ◐ = low-speed detonation; ◆ = detonation extinguishment. (———): correlation for nitric esters in Fig. 3b.

$CH_2=CH-CH_2Cl$ is weaker. Presumably chlorine at least partly blockades the double-bond effect. Epichlorhydrin $O-CH_2-CH-CH_2Cl$ is an even weaker additive.

The most pronounced effect was obtained with addition of propargyl alcohol to nitroglycol. Propargyl alcohol $CH-C-CH_2OH$ has a triple bond. A mixture containing 10% of propargyl alcohol steadily detonates in a glass capillary of an internal diameter as small as 0.7-0.8 mm (four runs). Detonation failure was observed only at d = 0.5 mm (Fig. 4).

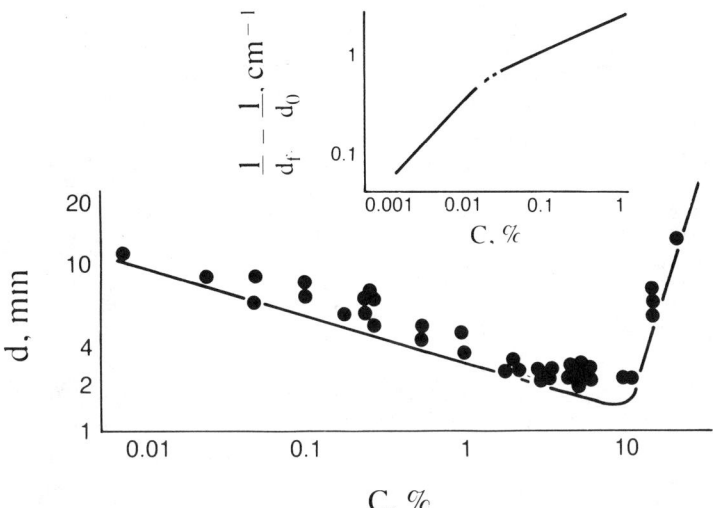

Fig. 5 Detonation failure diameters of solutions of nitromethane and triethylamine. ● = detonation. Failure points all under the line are not shown.

Nitromethane

Van Dolah et al. (1958) described the strong effect of additives (amines, acids, and some other substances) in small concentrations (1-5%) on the gap test results for nitromethane. The thickness of a plastic attenuator necessary for prevention of the detonation transmission in the gap test experiment more than doubled for only 1% of some of the additives. The failure diameter of nitromethane appeared to be much more sensitive to the additives than the gap test (Kondrikov et al. 1977). Just 0.02% di- or triethylamine or pyridine reduces d_f by a factor of 2; 2% reduces it from 13 to 1.5 mm. The dependence of d_f on the amines concentration C_a (Fig. 5) has a form

$$\log d_f = \log d_1 - b \log C_a$$

where d_1 and b are constants (see Table 2). The effect of strong mineral acids is less pronounced, although even distilled water diminishes d_f at only 2% concentration by almost a factor of 2.

The addition of amines also strongly diminishes the detonation front instability in nitromethane. Streak camera photographs of detonation propagating in nitromethane

Table 5 Parameters for the equation $d_f/d_1 = C_a^b$

Amine	b	d_1, mm	Δ, mole/l [a]
Diethylamine	0.263	1.35	0.005-1.0
Triethylamine	0.295	1.35	0.001-0.5
Pyridine	0.285	1.4	0.005-1.0
Allylamine	0.407	1.0	0.02-1.0
Dimethylaniline	0.149	3.0	0.008-0.2
Formamide	0.157	4.0	0.03-1.0
Dimethylformamide	0.175	5.8	0.01-1.0
Nitroaniline	0.291	2.4	0.005-0.2

[a] C_a is the amine concentration; Δ is the concentration range over which the correlation is valid.

indicate that, in the limits of our recorder sensitivity, the instability disappears if 0.1% of diethylamine is added to the nitromethane. HNO_3 acts far more weakly: even 1.5% of HNO_3 does not completely eliminate the instability. A mixture of nitromethane with 2.2% of nitroglycerine gives the same striped streak photographs as does the pure nitromethane itself. The addition of usual solvents to nitromethane sensitized by triethylamine (3%) increases d_f. Acetone, chloroform, and carbon tetrachloride give the Arrhenius dependence of Eq. (2) with constants $d_0 = 1.7 \times 10^{-4}$ mm, B = 14,400 K.

Unsaturated compounds (allyl alcohol $CH_2=CH-CH_2OH$, butyndiol $HOCH_2-C\ C-CH_2OH$) react at detonation in the mixture with nitromethane, as well as with nitric esters, and as a result causes d_f to decrease. Benzoyl peroxide $(C_6H_5CO)_2O_2$ acts almost identically to unsaturated compounds. Acrylic acid $CH_2=CH-COOH$ (unsaturated compound and acid simultaneously) exerts a stronger influence on d_f than allyl alcohol and acetic acid separately do. The most powerful additive in this respect is allylamine $CH_2=CH-CH_2NH_2$, which is both a strong organic base and an unsaturated compound (Starshinov et al. 1977).

A beautiful red color appears when small amounts of some amines are mixed with di- or trinitrotoluenes. The failure diameter decreases with an increase in the amine concentration until the failure diameter reaches a minimum.

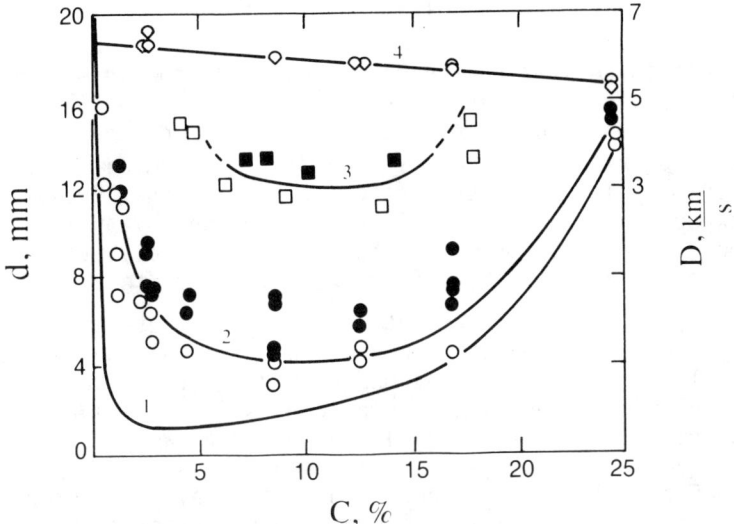

Fig. 6 Detonation failure diameters of solutions pyridine with nitromethane, trinitrotoluene, or mixtures of the two (see Table 6).

Table 6 Legend for Figure 6

Curve	Solution		Symbol Detonation	Detonation failure
1	Detonation limits	Nitromethane	–	–
2	Detonation limits	Trinitrotoluene Nitromethane	(50/50) ●	○
3	Detonation limits	Trinitrotoluene	■	□
4	Detonation velocity ◆	Trinitrotoluene Nitromethane	(50/50) –	–

Subsequently it increases with a further increase of amine concentration. The concentration of amine at which the minimum is located, for trinitrotoluene/pyridine mixture is about 12% compared with approximately 2% for mixtures nitromethane/pyridine (Fig. 6). The relation d_f^o/d_f^{min} for the first mixture is 5, and for the second is 8 (Kondrikov et al. 1979).

Discussion

This investigation and others cited in the foregoing clearly demonstrate that the effect of usual, so-called "inert" diluents on the critical diameter of detonation of liquid explosives is strongly dependent on chemical structure and physical characteristics of the diluent. Presumably the experimental results could be interpreted by application of the d_f theory of Dremin (1962). Unfortunately, this theory requires values: (T_3), u_3, c_3, D_3, v, that are very difficult to define experimentally and to compute theoretically. As a consequence, it would be desirable to develop a correlation between d_f and one or more of the usual physio-chemical or explosive characteristics of solvent or solutions. Correlations involving molecular weight, dipole moment, density of diluents, and detonation shock temperature of solutions have been proposed in previous works.

However, as shown in this work, a correlation involving one of the factors mentioned above can explain the dependence of d_f on the nature and the concentration of the inert solvents used. The only parameter which gives a good quantitative correlation is the explosion temperature of the solutions (see Figs. 2 and 3). The value of T_v was calculated by means of three well-known computational methods (Andreev and Beljajev 1960; Sorkin 1967; Stanjukovich 1975). The absolute value of T_v was dependent on the method of its calculation, but the temperature variation which results from dilution of the explosive was always about the same. All of the calculations led to the straight lines in the coordinates $1/T_v$-log d_f. The values of T_v on Figs. 2-4 were calculated by the method developed by Sorkin (1967) at $p = 100$ kbars.

The relative success of this correlation probably results from the fact that the explosion temperature is related to many of the other parameters of explosive, i.e., detonation, temperature, detonation velocity, or Dremin's T_3 temperature. A more direct connection between d_f and T_v may be possible, i.e., fast small-scale turbulent mixing in the reaction zone or a significant increase of heat conductivity in the shock compressed media. Both mechanisms lead to an increased significance of the final reaction temperature (T_{CJ} or T_v) in contrast to that of the initial temperature of the reaction zone, T_s. If such a direct connection between d_f and T_v were to exist, the activation energy E_a

and pre-exponential factor Z_o of a leading chemical reaction in the detonation wave can be calculated from the constants of Eq. (2). For instance, Kondrikov (1980) has proposed

$$d_f = 3/2 \, e \, D_i \, a \, (C_o Z_o)^{-I} e^a$$

where $a = E_a/RT_v$, D_i is the ideal detonation velocity, and C_o us the concentration of a reactant. We obtain from Eq.(4) at $d_0 = 0.059$ mm, $B = 7300K$, and $C_0 = 10$ mole/l for nitric esters: $E_a = 33.4$ kcal/mole and $Z_0 = 10^8$ 1/mole-s. These data are close to the results obtained earlier (Kondrikov et al. 1973) for a leading reaction of the nitrocompounds burning ($E_a = 28$ kcal/mole, $Z_0 = 10^{10.3}$ 1/mole-s).

Explanations of the active dilunts (unsaturated compounds, epoxygroup, amines, acids) influence on the d_f of liquid nitrocompounds has been given elsewhere (Kondrikov et al. 1979, 1980; Kozak at al. 1979; Starshiniv et al. 1977).

Acknowledgments

The author gratefully acknowledges the assistance and contributions of Dr. G. D. Kozak and Dr. V. M. Raykova of the Mendeleev Institute Laboratory.

References

Andreev, K. K. and Beljajev, A. F. (1960) Theory of Explosives, Oborongiz, Moscow, USSR.

Apin, A. Y. and Velina, N. F. (1967) On the failure diameter and detonation velocity of Hexogen. Vzryvnoje Delo, 63, 5-21.

Dremin, A. N. (1962) On failure diameter of detonation. Dokl. Akad. Nauk SSSR 147, 870-873.

Enig, J. W. and Petrone, F. (1970) The failure diameter theory of Dremin. 5th Symposium on Detonation, pp. 99-104, ONR Rept. ACR 184, Arlington, Vir.

Khariton, J. B. (1947) On detonation ability of explosives. Vopr. Teor. Vzry. Vesch. Izd. Akad. Nauk SSSR, 1, 7-28.

Kondrikov, B. N., Raykova, V. M., and Samsonov, B. S. (1973) On kinetics of chemical reactions of nitrocompounds burning at high pressures. Fiz. Goreniya Vzryva 9, 84-90.

Kondrikov, B. N. and Raykova, V. M. (1974) Detonation of nitric esters solutions. Trans. Mendeleev Inst. Chem. Technol. 83, 147-153.

Kondrikov, B. N. and Raykova, V. M. (1977) The detonation limits of explosive solutions. Fiz. Goreniya Vzryva 13, 55-62.

Kondrikov, B. N., Kozak, G. D., Raykova, V. M., and Starshinov, A. V. (1977) On detonation of nitromethane. Dokl. Akad. Nauk. SSSR 233, 402-405.

Kondrikov, B. N., Kozak, G. D. and Starshinov, A. V. (1979) On influence of amines on the failure diameter of nitrocompounds. Trans. Mendeleev Chem. Technol. 104, 91-95.

Kondrikov, B. N. (1980) Detonation, Mendeleev Institute of Chemcial Technology, Moscow, USSR, 80 pp.

Kondrikov, B. N., Kozak, G. D., Starshinov, A. V., and Polikarpov, V. A.(1980) On chemical kinetics at detonation of nitric esters. 6th Symposium on Combustion and Explosion, Detonation, p. 18-21. Alma Alta, USSR.

Kozak, G. D., Kondrikov, B. N., and Starshinov, A. V. (1979) The failure diameter of detonation of nitric esters. Trans. Mendeleev Inst. Chem. Technol. 104, 87-91.

Kurbangalina, R. K. (1947) Ph.D. Thesis, Institute of Chemical Physics, Akademii Nauk SSSR, Moscow, USSR.

Kusakabe, M. and Fujiwara, S. (1970) Explosive behavior of methylnitrate and its mixtures with liquid diluents. 5th Symposium on Detonation, pp. 267-273. ONR Rept. ACR 184, Arlington, Vir.

Kusakabe, M. and Fujiwara, S.(1976) Effects of liquid diluents on detonation propagation in nitromethane. 6th Symposium on Detonation, pp. 60-70. ONR Rept. ACR 221, Arlington, Vir.

Raykova, V. M. (1980) The detonation of mixtures on the base of nitric acid. Trans. Mendeleev Inst. Chem. Technol. 112, 97-102.

Sorkin, R. E. (1967) Gas- and Thermodynamics of Solid Propellant Rocket Engines, Nauka, Moscow, USSR.

Stanjukovich, K. P., ed. (1975) Physics of Explosion, Nauka, Moscow, USSR.

Starshinov, A. V., Kondrikov, B. N., Kozak, G. D., and Raykova, V. M. (1977) The homogeneous catalysis of detonation of nitromethane. 5th Symposium on Combustion and Explosion, Detonation, pp. 73-76. Chernoglovka, USSR.

Tarver, C. M., Shaw, R., and Cowperthwaite, M. (1976) Detonation failure diameter studies of four liquid nitroalkanes. J. Chem. Phys, 64, 2666-2673.

Van Dolah, R. W., Herickes, J. A., Ribovich, J., and Damon, G. H. (1958) Shock sensitivity studies of nitromethane systems, Communications of the XXXI Congress (International) Industrial Chemistry, pp. 121-126. Liege, Belgium.

Chapter V. Explosions in Solids

Shock Induced Hot-Spot Formation and Subsequent Decomposition in Granular, Porous HNS Explosive

D.B. Hayes*
Sandia National Laboratories, Albuquerque, N. Mex.

Abstract

Experimental and theoretical studies on granular porous hexanitrostilbene (HNS) explosive have yielded an increased understanding of microstructural processes occurring during initiation by shock loading. Experiments give time-resolved pressure, hence chemical decomposition history at the impact interface following planar impact of HNS specimens onto fused-silica targets. The data are interpreted in terms of a quantitative two-temperature model which considers hot spots to be formed at pore sites as a result of the irreversible work accompanying the shock. Subsequently, decomposition completion is achieved by burn fronts which propagate radially out from each hot spot at a velocity which can be inferred from the bulk decomposition rate. Analysis of the data in the context of the model leads to several observations: 1) The delay times corresponding to hot-spot decomposition are shorter than expected, based on extrapolated low-temperature kinetics. 2) Contrary to interpretation of data on other explosives, the velocity of the burn front which propagates radially from each hot spot does not have a strong dependence upon the prevailing pressure but is more nearly constant. 3) The inferred burn velocity has a very strong dependence upon the <u>initial</u> shock pressure.

Introduction

It is generally accepted that the passage of a shock wave through a granular porous explosive leads to

---

Presented at the 8th ICOGER, Minsk, USSR, Aug. 23-26, 1981. Copyright © 1982 by the U.S. Dept. of Energy. Published by the American Institute of Aeronautics and Astronautics, Inc., with permission.

*Manager, Fluid and Thermal Science Department.

heterogeneous heating and to the production of hot spots which chemically decompose more rapidly than if the material had been shock heated homogeneously (Bowden and Yoffe 1952). Hot spots, however, constitute only a small mass fraction of the shocked explosive material. Subsequent decomposition is thought to take place by the propagation of burn fronts which propagate radially from each hot spot, ultimately consuming the material (Eyring et al. 1949). This general picture has recently provided a basis for a variety of mathematical descriptions to model both the decomposition process and mechanical behavior of the decomposing mixture (Lee and Tarver 1980; Hayes and Mitchell 1978; Kipp et al. 1981). Most descriptions of the decomposition process are empirical or semiempirical, requiring the use of constants which are adjusted to obtain a good match between predicted and experimental behavior (Forest 1979; Kanel 1978; Wackerle et al. 1978; Hayes and Mitchell 1978; Lee and Tarver 1980, Kipp et al. 1981). It is desirable to develop models which are more physically based and attempt to relate the parameters which appear in the decomposition function to independently measurable material properties.

A variety of experiments have been performed to measure the decomposition history in a high explosive after the passage of a shock. In one common technique, arrival of the shock and subsequent wave shape are observed at a witness plate or window on the rear surface of the explosive material (Kennedy and Nunziato 1976; Kipp et al. 1981). Through a series of experiments on explosive specimens of different length, evolution of a low-amplitude plane shock can be measured as it grows to a detonation. In principle, a superior technique is the use of multiple embedded magnetic or piezoresistive gages which yield the velocity or pressure history at a number of stations, and thus make a measure of the entire flowfield possible. From these whole-field data and from the equations of motion, the pressure, volume, and energy histories can be deduced at each material point, and hence the decomposition history inferred (Cowperthwaite 1973; Wackerle et al. 1978; Anderson et al. 1981; Kanel and Dremin 1977). Such an analysis has the advantage that it directly measure the decomposition history without requiring assumption of a specific analytical form for the decomposition function which must be reconciled with indirect shock data. However, the benefits have been limited by some difficulties which this experimental technique. A third technique, the one considered here, involves the measurement and use of pressure transients at the impact interface after planar loading by an elastic impactor (Kennedy 1970; Hayes and Mitchell 1978; Mitchell and Hayes 1981). This method has the advantage of simplicity of use and ease of interpretation up to a point; a disadvantage is the requirement for an extra assumption to extract pressure-volume-energy (P-V-E) response which may only be warranted under special circumstances.

In this paper we examine a specific microstructural model for the shock initiation process in the porous granular explosive hexanitrostilbene (HNS) and test the results against available data. These data are from impact experiments in which pressure histories at the impact interface were obtained by an interferometric technique which views through a transparent impactor. Pressure excursions are used to imply the extent of the decomposition reaction. The model assumes that the majority of each grain of explosive is reversibly compressed along an isentrope and, hence, heated only to the isentropic compression temperature (Hayes and Mitchell 1978). The substantial irreversible work, which must take place in the shock front, takes the form of heat which is deposited preferentially in the vicinities of pore sites, and thus a small fraction of the material achieves a considerably elevated temperature. The hot spots thus formed must chemically decompose, leading to a measurable delay in the onset of the major pressure excursion. Subsequently, the material is supposed to decompose through the radial propagation of burn fronts from each ignited hot spot into grain interiors. The dependence of velocity of the burn front upon the thermodynamic state and the shock history of the unburned explosive grain interior into which it propagates are of prime interest in this study.

This paper is organized in the following way: a brief review of the experimental technique is followed by a description and discussion of the experimental results. In the next section the specific model is developed and macroscopic observables are related to microstructural behavior. Finally, there is a discussion of the seven major findings in this study which come from comparing the model with data from experiments upon porous HNS explosive, initially shock-loaded to a pressure in the range from 1.2 to 5.3 GPa.

With regard to the hot spot formation and decomposition, the measured delay (decomposition) times: 1) are shorter than expected, based on extrapolated low-temperature kinetics; 2) do not depend strongly upon initial HNS particle size, indicating that hot spots are formed by dissipative processes with small-scale length/time and that

hot spots remain adiabatic during their decomposition; and
3) do not depend strongly upon initial explosive pressing
density, which is consistent with our assumption that the
mass fraction of hot spots approximates the initial volume
fraction of pores within the solid prior to shock loading.

With regard to the propagation of intragranular burn
fronts:

1) The inferred burn velocity does not display a
strong dependence upon the prevailing pressure, even though
in any given experiment that pressure rises in excess of a
factor of 2 between reaction initiation and completion.
This apparent pressure independence differs from results on
several other explosives in which the decomposition rate is
observed to depend strongly upon prevailing pressure (Lee
and Tarver 1980; Mader 1979).

2) The burn velocity is observed to have a very strong
dependence upon the initial shock pressure above a threshold
of 2 GPa; both Wackerle et al. (1978) and Kanel (1978),
using direct analyses, have developed decomposition-rate
laws which depend to some extent on the initial shock
pressure rather than through a direct dependence on
prevailing pressure.

3) The same burn velocity is inferred for specimens
pressed from HNS of two different initial particle sizes,
and shocked in otherwise identical experiments.

4) The inferred grain burning in velocities of a few
tens of meters per second, are greatly in excess (by a
factor of 10^2) of the classical laminar burn velocity.
Although the velocities generally agree in magnitude with
extrapolation of burn rate vs pressure for a variety of high
explosives and propellants, they lack a dependence on
subsequent pressure rise.

The purpose of this paper is to interpret experimental
results within a consistent framework and to give guidance
to further theoretical and experimental work. Our present
data do not warrant a more sophisticated and precise
mathematical description than given here. Of course,
consistency of experimental results with any model is not
sufficient reason to assert that the model accurately
captures those processes which are occurring; such a
comparison does, however, provide a basis for the design of
future experiments which can test various hypotheses
offered.

Experimental

The experiments were performed by impacting an
HNS/fused-silica assembly onto a stationary fused-silica

target material and observing the velocity history near the impact interface. A schematic of the experimental configuration is shown in Fig. 1. The details of the experimental configuration, measurement technique, and data interpretation methods are described in detail elsewhere (Hayes and Mitchell 1978; Mitchell 1981). Therefore only a brief description is given here. The symmetric impact of fused-silica on fused-silica produces a particle velocity exactly one-half of the independently measured projectile velocity. Soon after impact the HNS specimen is shock-loaded and the amplitude of the shock as well as all subsequent pressure changes are transmitted to a viewing plane back in the fused-silica target. An interferometric technique (Barker and Hollenbach 1972) is used to measure particle velocity. Because the fused-silica target assembly shown in Fig. 1 only experiences simple, right-going waves, there exists a unique one-to-one relation between pressure and particle velocity. Thus we can determine the pressure history experienced at the HNS/fused-silica interface. These pressure histories are used to infer the decomposition history of HNS near the impact interface.

Figure 2 shows the results for experiments which were performed on a high-explosive powder designated HNS-I (Hayes

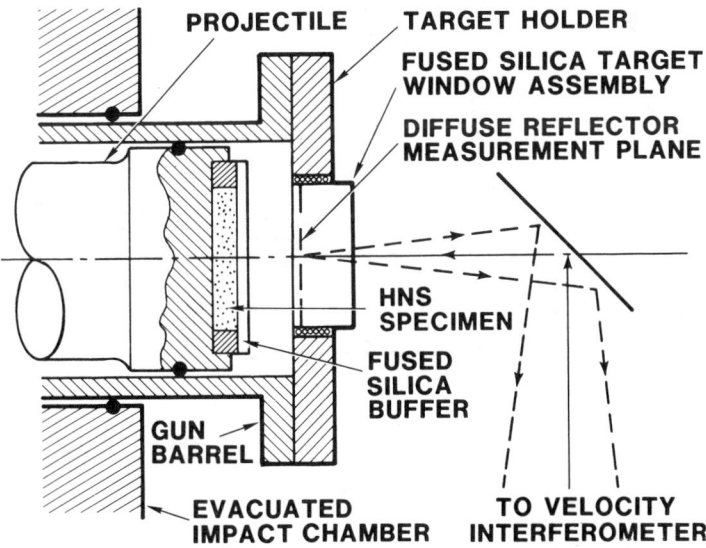

Fig. 1 Schematic of the experimental configuration. An HNS specimen (with fused-silica cover plate) impacts a fused-silica target with a thin embedded opaque metallic foil. Velocity interferometry measures the particle velocity history, which is used to infer the pressure history at the HNS/fused-silica interface.

and Mitchell 1978). HNS-I bulk powder has a distribution of grain sizes which average a 21-μm equivalent spherical diameter (Mitchell 1981). Experimental conditions are elaborated in Table 1. At the lowest impact pressure (1.25 GPa in experiment G1), the shocked state is quiescent with no observable reaction. At the highest impact pressure (5.28 GPa in experiment G6), a prompt decomposition is noted immediately after the HNS is shock-loaded. At intermediate pressures, the records are characterized by a pronounced delay after the HNS is shocked, which we have tentatively identified as the period of time while the hot spots decompose. This delay is followed by a pressure excursion which is produced by the major decomposition reaction and ends at a value which we assume represents reaction completion. Delay time is a strong decreasing function of increasing initial impact pressure, and the subsequent major decomposition reaction is noted to become more rapid for increasing impact pressure. Figure 3 shows results for experiments on HNS-I material which was pressed to initial

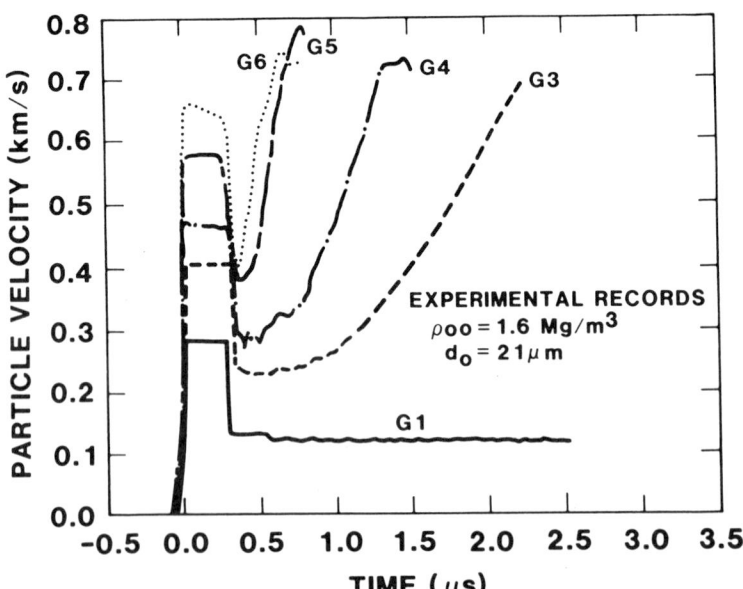

Fig. 2 Experimental records for experiments on HNS-I (equivalent spherical diameter of the grains of the bulk powder is 21 μm) pressed to initial density 1.60 Mg/m^3 (8.7% porosity) and shock-loaded. See Table 1 for details. The particle velocity increases after t = 0.35 μs is identified with the pressure rise accompanying chemical decomposition.

densities of 1.39 and 1.48 Mg/m^3. These records display the same qualitative features as the other results. Finally, Fig. 4 shows the records for specimens fabricated from HNS-II. This is a more coarse material with a more uniform particle size distribution; individual grains have an average equivalent spherical diameter[†] of 37 μm. These specimens were also pressed to an initial density of 1.60 Mg/m^3. There are the same qualitative features in these experimental records as others. It should be noted that fused-silica is not linearly elastic. Therefore compressive waves steepen slightly as they propagate from the high-explosive/fused-silica interface, through fused-silica, to the interferometer viewing plane, a distance of 2.31 mm in each experiment. For particle velocities in a simple wave of 0.3, 0.5, and 0.7 km/s, the Lagrangian sound speeds are 4.98, 5.10, and 5.15 km/s, respectively. Therefore there is a slight uncertainty introduced into the arrival times of various portions of the records shown in Figs 2-4. The uncertainty in time intervals deduced is, at most, 0.05 μs and in most cases far less than that. However, in all cases the compression waves were broad enough so that no shocks were formed during periods subjected to analysis. Before describing the manner in which these pressure histories are analyzed to obtain the decomposition rates, it is necessary to give a fairly detailed account of the microstructural model construction.

Model

The model for the shock compaction and subsequent decomposition of HNS considers three distinct regimes which will be discussed separately: the shock compaction process which forms the hot spots; the period of decomposition of the hot spots; and the intragranular burning which originates from decomposed hot spots.

During the shock compaction process the HNS material is assumed to be elevated to one of two temperatures. The bulk of the material in the interior of grains is considered to be compressed slowly and to experience only the isentropic compression thermal excursion. This is reasonable because the rise time for a shock propagating through a porous

---

[†]The quoted equivalent spherical diameters are based on a Ziess analysis, which uses visual inspection of a photomicrograph to obtain particle size distribution. Other methods yielded different results, but the ratio of diameters for HNS-II:HNS-I always remained at about a factor of 2.

material is necessarily significantly longer than that of a shock propagating through the same material with no heterogeneities (Butcher et al. 1974; Dunin and Surkov 1978). The irreversible work which is performed by the shock is assumed to be perferentially concentrated in a small fraction of the material in the vicinity of pore sites. The magnitude of the irreversible work can be quantified by using the Hugoniot energy jump condition, reconciled with the partitioning of energy into the two regions:

$$\frac{P + P_0}{2}(V_{00} - V) = W_H E(P, T_H) + (1 - W_H) E_S(P) - E(P_0, T_0)$$
(1)

The left-hand side is the energy rise which accompanies the shock process assuming that the shock is steady, while the right-hand side partitions that energy rise between the cold and hot materials. The functions E and E_S are known (Hayes and Mitchell 1978; Sheffield et al. 1976), as are the initial and final pressures ($P_0 = 0$ and P) and volumes (V_{00}

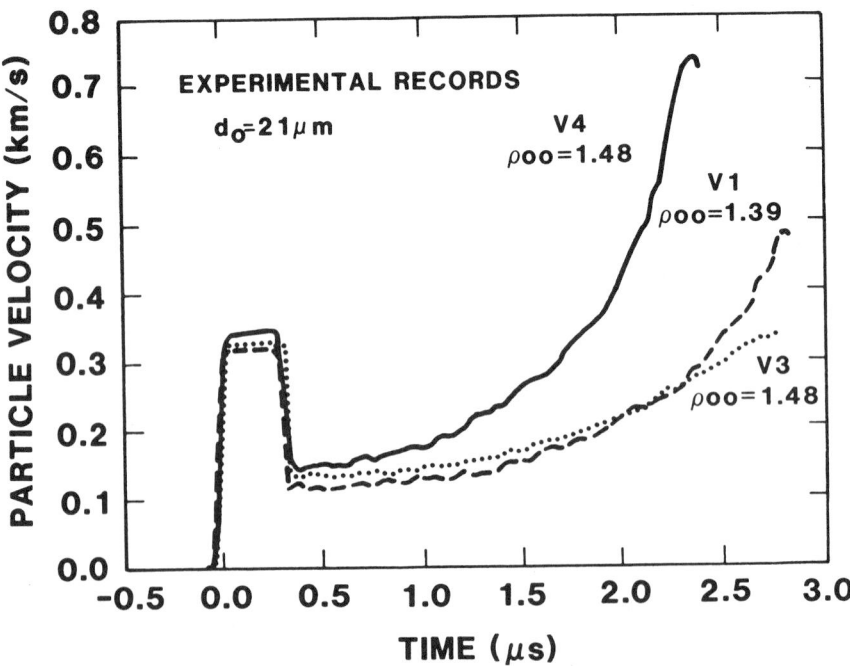

Fig. 3 Experimental records for HNS-I pressed to initial densities of 1.39 and 1.48 Mg/m^3. See Table 1 for details.

SHOCK INDUCED DECOMPOSITION IN HNS   453

Table 1 Conditions for the experiments and results

Shot	Specimen density, Mg/m^3	HNS powder equivalent spherical diameter, μm	Impact velocity, km/s	Impact pressure, GPa	Calculated hot-spot temperature, K	τ_1 hot-spot ignition time, μs	τ_2 characteristic burn time, μs	Particle burn velocity, m/s
G1[a]	1.60	21	0.503	1.25	561.	b	b	b
G3	1.60	21	0.814	2.72	920.	0.68	1.17	9.0
G4	1.60	21	0.951	3.30	1079.	0.39	0.58	18.0
G5	1.60	21	1.164	4.33	1394.	0.155	0.31	33.9
G6	1.60	21	1.346	5.28	1718.	0.076	0.21	50.0
G8	1.60	37	1.321	4.96	1607.	c	0.54	34.2
G9	1.60	37	0.932	3.34	1093.	0.42	1.03	18.0
G11	1.60	37	0.800	2.68	909.	0.72	1.68	11.0
V1	1.39	21	0.638	1.55	619.	1.56	2.10	5.0
V3	1.48	21	0.647	1.78	672.	1.7	2.92	3.6
V4	1.48	21	0.720	1.92	703.	1.01	1.00	10.5

[a]Experiment G1 has an impactor facing material of X-cut crystalline quartz. All other experiments used fused-silica.
[b]No decomposition observed.
[c]Early portion of record noisy.

and V) through experiment. The mass fraction of hot material W_H and its temperature T_H are unknown.

Mader (1965) first quantified the extent of the disrupted region in the vicinity of a hot spot in a classical set of numerical simulations of a shock-induced pore closure in the liquid explosive nitromethane. He also made the qualitative observation that the volume of disrupted material was approximately equal to the volume of the pore from which the hot spot was formed. We have also performed hot-spot closure simulations using a two-dimensional hydrodynamic code (Thompson 1979) for a variety of pore number densities and shock rise times. Our results generally agree with those of Mader (1965). Figure 5b shows the two-dimensional temperature distribution after the collapse of an 8.4-µm-diam isolated spherical pore (initial pore shape is given in Fig. 5a) in HNS which was loaded to 3.3 GPa by a shock with rise time of 2 ns. In spite of the complex spatial temperature distribution, the extent of the highly disrupted region is approximately equal in volume to the initial pore volume. Since the temperature contour plotting interval is coarse, and the transition to background temperature gradual, this is more graphically illustrated in Fig. 5c, where the volume (expressed in terms

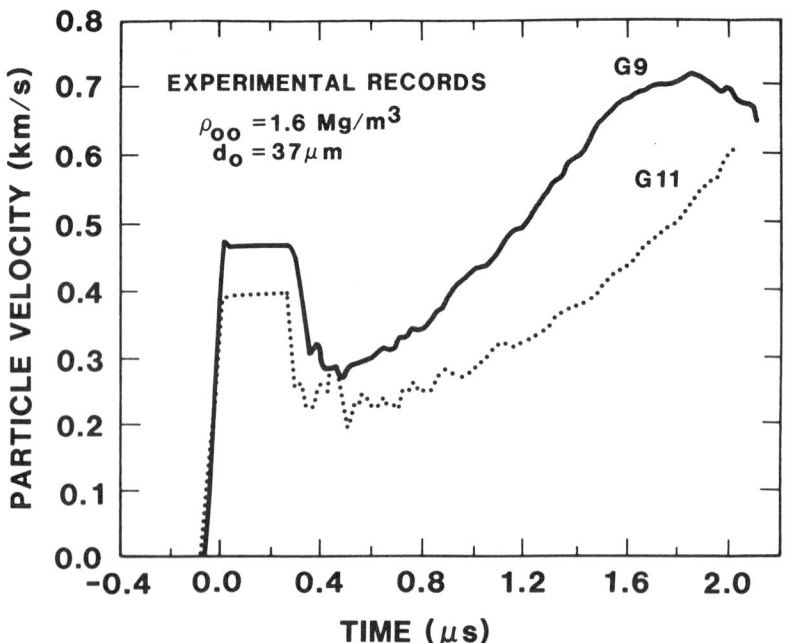

Fig. 4 Experimental records for HNS-II (equivalent spherical diameter of powder grains is 37 µm) pressed to 1.60 Mg/m^3. See Table 1 for details.

of an equivalent spherical radius) of material shock heated above a given temperature is graphed against that temperature.

Using such numerical studies for justification, we have, as an integral part of the model developed here, assumed equality between intitial volume fraction of pores prior to shock arrival, and the mass fraction of hot material just after shock passage.

$$W_H = (\rho_0/\rho_{00}) - 1 \qquad (2)$$

The double-zero subscript denotes the porous state, while the single zero pertains to the fully dense solid. The temperature distribution obtained analytically using Eqs.

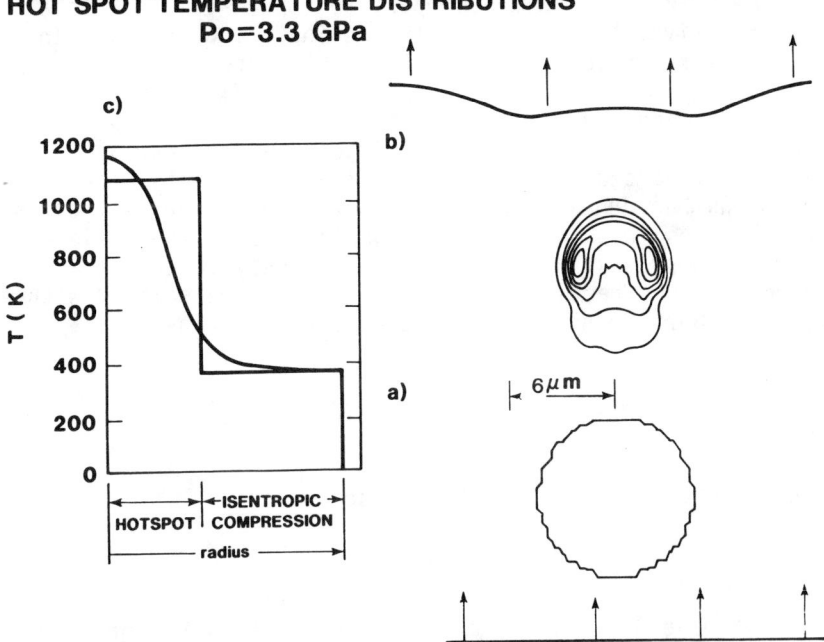

Fig. 5 Typical results from numerical simulation of disruption caused by shock-induced pore collapse. a) Initial pore shape at t = 2 ns. Note the isotherm T = 342 K is on the approaching shock. b) Temperature distribution after shock passage at t = 8.5 ns. The pore is entirely collapsed. Contours start with T = 342 K (0.03 eV), which marks the shock front which has passed. All contours are separated by δT = 116 K (0.01 eV). c) Comparison between actual distribution of temperature (expressed in terms of an equivalent spherical radius) and that obtained with a two-temperature model, using $W_H = \rho_0/\rho_{00} - 1$.

(1) and (2), is shown in Fig. 5c, and is in good agreement with results from the more complex method.

The second regime of the decomposition process is the interval of time during which the hot spots decompose. In the experimental records, that instant when the observed mass fraction decomposed just becomes equal to W_H is identified as the point when decomposition of hot spots is complete and the burning mode of decomposition begins. Of course, this transition from one mode of decomposition to another is not in actuality expected to be as abrupt as we choose to calculate. This partitioning does, however, provide a convenient means of quantifying the experimental hot-spot decomposition time.

The final stage of decomposition entails burn fronts which propagate intragranularly from each hot spot. Experiments measure bulk decomposition; burn velocity is a more primitive, hence physically meaningful, quantity. In order to extract that velocity from bulk rate it is necessary to make a correction of the specific area of contact between the enlarging hot gas pockets and the unburned isentropically compressed solid. Consider a randomly dispersed distribution of ignition points with average number density, N, equal to the number density of initial pores (with one pore per grain) within the solid. Ignoring compressibility effects, it is easy to show that the decomposed fraction x is related to the velocity V with which the burn fronts have propagated radially as

$$x = 1 - \exp\left(-(4\pi N/3)\int_0^t V dt\right)^3 \quad (3)$$

which can be conveniently reexpressed

$$\dot{x} = (3V/r_0)(1-x)\log 1/(1-x)^{2/3} \quad (4)$$

where we have denoted the average radius of an HNS grain as r_0. Thus the decomposition rate $\dot{x}$ is proportional to V, the velocity of the burn fronts, with a geometric correction which accounts for the specific area of contact between the gas pockets and the isentropically compressed, unburned material. This geometric correction is very similar in its effect to those used elsewhere (see, for instance, Lee and Tarver 1980) differing only in algebraic detail. Thus we can use the measured bulk decomposition rate in conjunction with Eq. (4) to extract the velocity of the burn front from the decomposition history.

Data Reduction Using Model

The difficulty with the front-surface-impact technique is that insufficient information is obtained to extract P-V-E histories without invoking an additional assumption. To estimate the decomposition rate from experimental results we have made the assumption here that the extent of reaction is proportional to the pressure rise at the impact interface above the initial shock pressure P_s:

$$x = (P - P_s)/(P_f - P_s) \tag{5}$$

P_f is the final pressure. This assumption has some justification: Firstly, we have performed numerical simulations of the experiments.[‡] For a range of assumed decomposition functions in these numercial simulations, a nearly linear relationship between pressure rise and extent of reaction is computed. Prototypic P-x histories calculated during numerical simulations for experiment G9 are shown in Fig. 6. Secondly, we find this process reversible in the sense that we can use the computed pressure profiles and extract the initial kinetics (to an accuracy of about 10-20%). Thus this ad hoc assumption of linearity produces self-consistent results but leaves open the question of uniqueness.

Records were reduced using Eq. (5). Given the resultant decomposition histories, it remains to interpret them within the framework of the proposed model. Examine Fig. 7, where the decomposition history is shown for experiment G9. The interval $0 < t < 0.42$ μs is the period of time required for the decomposed mass fraction to achieve the mass fraction supposed to reside in the hot spots: $W_H = 0.087$. This is the delay time which is denoted τ_1 in Table 1. The subsequent portion of the history, corresponding to decomposition through burning, is fit with a numerical integration of Eq. (4), with the parameter $\tau_2 = r_0/V$ chosen as a best-fit value of 1.03 μs. Each experimental record was treated in a similar way, with results tabulated in Table 1.

It should be noted that the present interpretation of the experimental records require postulating some delay time

---

‡In these calculations, at any given extent of reaction, our model of a pressure-equilibrated mixture of gaseous decomposition products and isentropically compressed solid was assumed. Each phase is assumed adiabatic.

(time for hot spots to react), prior to the onset of the burning portion of the decomposition history. If the analytical fit [Eq. (4)] for the burning portion of the decomposition is extrapolated to earlier times and hence smaller values of x, say, to less than 0.01, there still remains an unaccounted-for interval of time between initial shocking and decomposition beginning.

Results and Discussion

Figure 8 shows the calculated hot-spot temperature as a function of initial porosity. Since there is no scale length or time assumed in Eqs. (1) or (2), results are independent of particle size. This was also the case in numerical simulations of hot-spot closure, where artifical viscous stress excited by the gross hole closure was very small; the majority of the dissipation came from the shocks produced by the collisions of the imploding surfaces. Neither do the calculated hot-spot temperatures depend significantly upon the initial porosity. This result can be interpreted as indicating that the effect of increased irreversible work, which is performed on the lower initial density material, must be distributed over a proportionately larger mass fraction [see Eq. (2)].

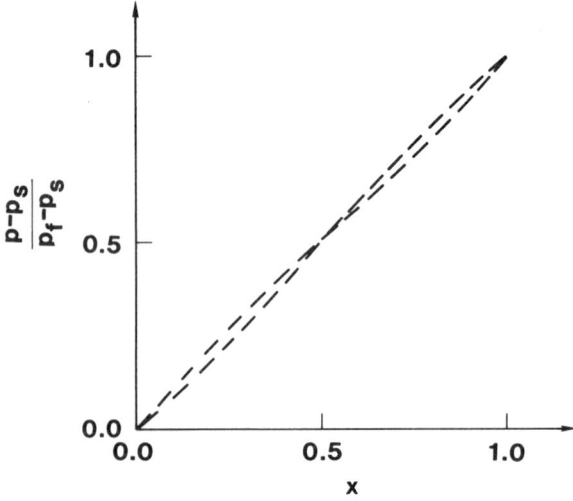

Fig. 6 Results of two different numerical calculations on experiment G9 showing emperical verification of the assumed proportionally between pressure rise after shock loading and extent of decomposition. These two calculations are made using two different forms for the equation of state of the gaseous decomposition products.

The delay time data can provide several critical tests of these model results. Experimental results on Fig. 9 show that all inferred delay times fall on a single smooth curve. Notice the independence of delay time on grain, hence pore size (HNS-I vs HNS-II at ρ_{00} = 1.60 Mg/m^3). Therefore data do not contradict the assumption made in the calculation of hot-spot temperatures: that the dissipative processes responsible for hot-spot formation do not contain a scale length or scale time which is large compared with the pore radius or collapse time, respectively.

For several reasons, the best side-by-side comparison of this independence of grain size comes from experiments G4 and G9. First the experiments were performed at almost exactly the same pressure, making comparison easy. Also the records are very high quality, being among the best obtained. It was seen in Fig. 9 and in Table 1 that the measured delay times for these two experiments were almost equal. Figure 10 graphically demonstrates that fact in that the early portion of the two decomposition histories are nearly identical.

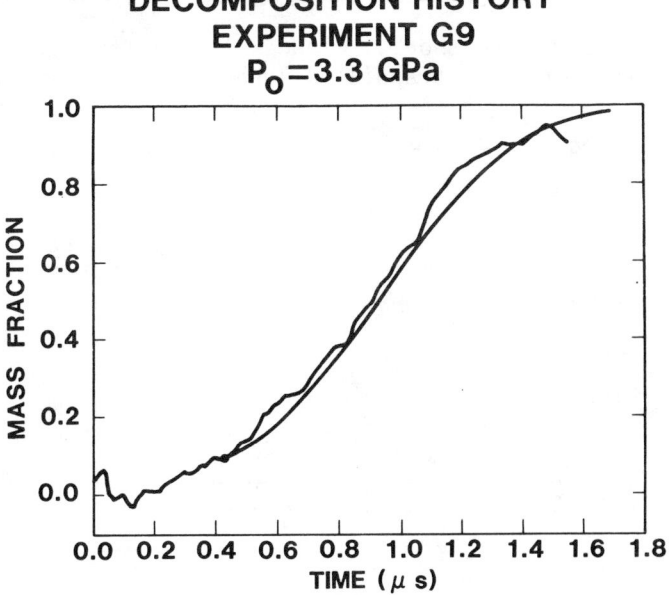

Fig. 7 Experiment G9 decomposition history showing two time regimes: t < 0.045 µs, x < 0.087 = W_H (hot-spots decompose); 2) t > 0.45 µs, x > 0.87 (grain burning consumes the material, with characteristic time τ_2 = 1.03 µs). The orgin of the slight early dip in pressure, which leads to small (but aphysical) negative mass fractions is unknown.

The observed particle size independence of hot-spot decomposition time is also evidence that those hot-spots remain adiabatic during that period. Based on thermal diffusivity measured at standard conditions, this is expected by a rather wide margin. There have, however, been some decreases in shock sensitivity in extremely fine-grained HNS material at low pressure and long pulse width (Schwarz 1981). Unfortunately, if these prove to be due to lack of hot-spot ignition, the effect of dissipation mechanism length or time scale, which can affect the spatial extent of the hot spot during its formation, may be confounded with subsequent thermal diffusion time scale, which can affect the persistence time of hot spots once produced, to give them time to decompose.

Figure 11 shows the measured delay time τ_1 as a function of the calculated hot-spot temperature T_H. As expected, there again is a consistent dependence of delay time on T_H which appears to be independent of grain size. Note that delay time might also be independent of initial pressing density [G-series (circles on Fig. 11) vs V-series (non-circles)], although this is somewhat speculative since the calculated hot-spot temperature ranges for each series do not overlap.

Delay times are short. Figure 11 shows the calculated delay time assuming extrapolated Arrhenius kinetics, A =

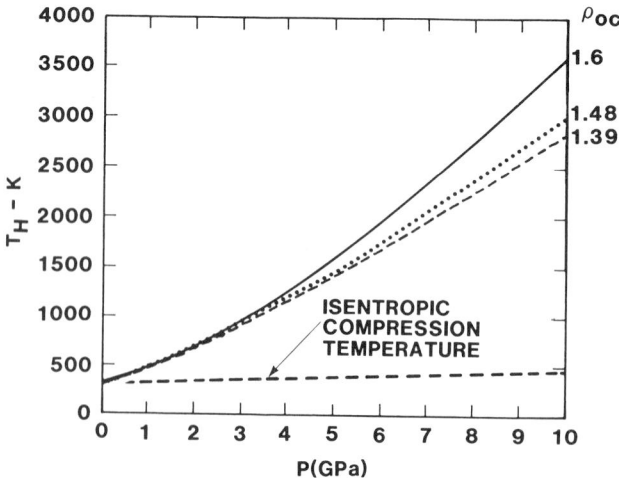

Fig. 8 Calculated hot-spot temperatures vs pressure for different initial pressing densities.

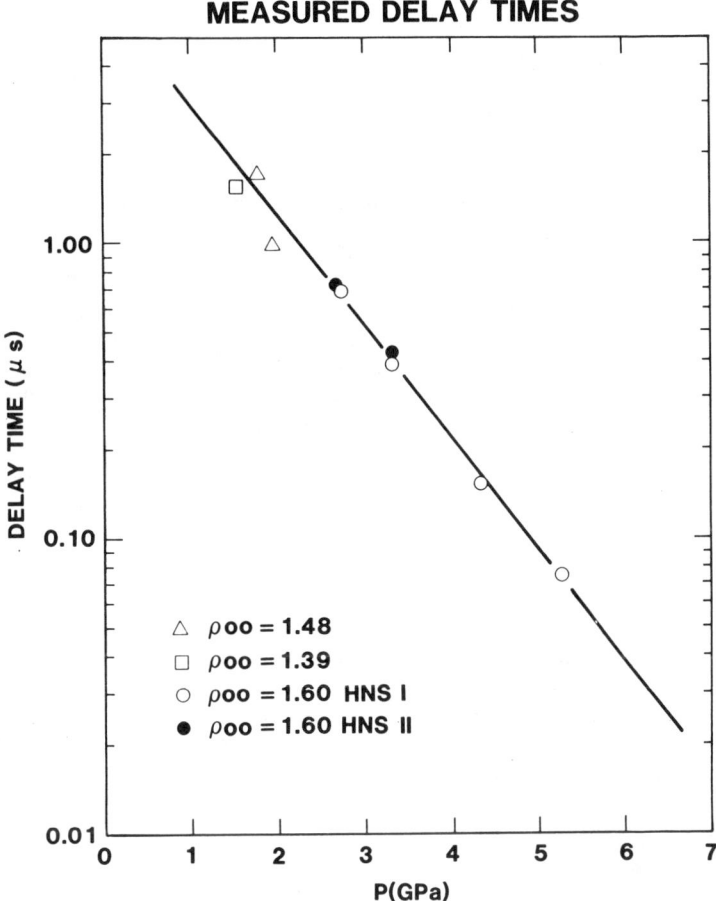

Fig. 9 Inferred delay times τ_1 identified with the hot-spot decomposition time vs impact pressure.

1530 μs^{-1}, $E^{\ddagger}/k$ = 15,300 K [after Rogers 1975], which have been measured at lower temperature. It is obvious that the measured delay times are considerably less than those that would be anticipated by extrapolation of these low-temperature kinetics to these temperature regimes. In fact, if an attempt is made to fit the observed delay times with a thermal explosion model, using the low-temperature kinetics, hot-spot temperatures would have to be substantially larger than those supposed here.

Several important insights follow from the inferred burn velocities. An inspection of the particle velocity histories, which were measured in the experiments on HNS-I pressed to an initial density of 1.60 Mg/m^3, makes it obvious that the prevailing pressure is not the dominant variable in determining decomposition rate. Examine, for instance, Fig. 2 at the particle velocity amplitude of 0.5 km/s, which corresponds to a prevailing pressure of 5.8 GPa. At this value of pressure the decomposition rates are markedly different, being about 0.8 μs^{-1} at this instant for experiment G2 and 5 μs^{-1} for experiment G6 - a factor of 6 difference in rate. A comparable spread in rates is observed in comparisons with these and other records at other amplitudes. Even during an individual experiment, pressure rise does not seem to markedly affect rate. The best example is experiment G9, where pressure rises from 3.3 to 8.2 GPa during decomposition, yet the history is very well fit by a constant velocity burn (see Fig. 7). We have chosen to reduce each record assuming a constant velocity, because the accuracy of the data do not warrant a more detailed treatment. It is possible in some of the records that the velocity could vary by 30% and we could not detect that variation. However, the possibility of a strong pressure dependence can be eliminated.

Figure 12 shows that the inferred burn velocity is an extremely strong function of the initial shock pressure. Again, to within the scatter in the data, it appears that very similar burn velocities are obtained, independent of particle size and initial pressing density.

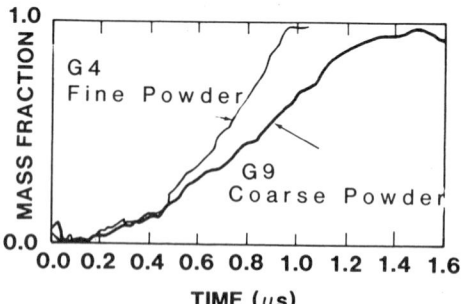

Fig. 10 Comparison of decomposition histories for specimens pressed from powder of different particle sizes. In experiments G4 and G9, HNS powders have grains with equivalent diameters of 21 and 37 μm, respectively. Similar decomposition histories for the first 0.4 μs are evidence that the dissipation process is self-similar on this scale. Later decomposition proceeds more rapidly in the "fine" material, which supports the observation that burn velocities are comparable.

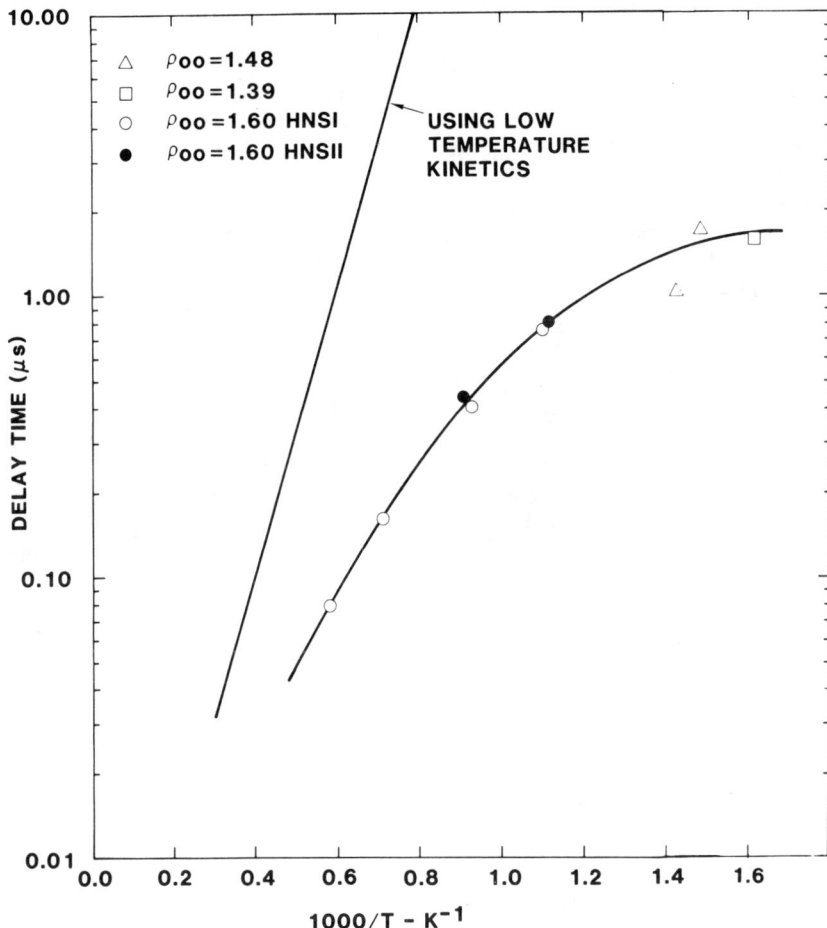

Fig. 11 Inferred delay times vs hot-spot temperature. Extrapolated low-temperature kinetics lead to calculations of much longer delays than those observed.

Summary and General Remarks

We have compared a specific microstructural model with experimental observations on the decomposition history of shock-loaded specimens of porous HNS explosive. Hot-spot mass fraction is determined by the pore volume from which the hot spot was formed. Velocities of subsequent burn fronts which propagate radially from each hot spot are observed to weakly depend upon the prevailing pressure, in contradiction with some current interpretations of data on other explosives, and are seen to be strongly dependent upon the initial shock pressure above a 2-GPa threshold. Velocities are very large compared with classical burn velocities and are independent of particle size and initial pressing density, within experimental resolution.

If the interpretation presented here is correct, results would suggest a general increase of reactivity of material which has been subjected to the shock environment. The reactivity change would have to be dramatic within the highly disrupted hot spots and less so in that material which is in the grain interior. These general observations would then be in agreement with results of studies of chemical alteration produced by the shock process (Dodson and Graham 1981). Several interpretations of the increase in reactivity produced by the shock process center on the high-frequency content in the shock front and the attainment of a nonequilibrium temperature distribution which can be responsible for sission of particular chemical bonds (for a recent example, see Dremin et al. (1981). Because the initially porous material under study here is very heterogeneous, shock rise times are relatively long and such

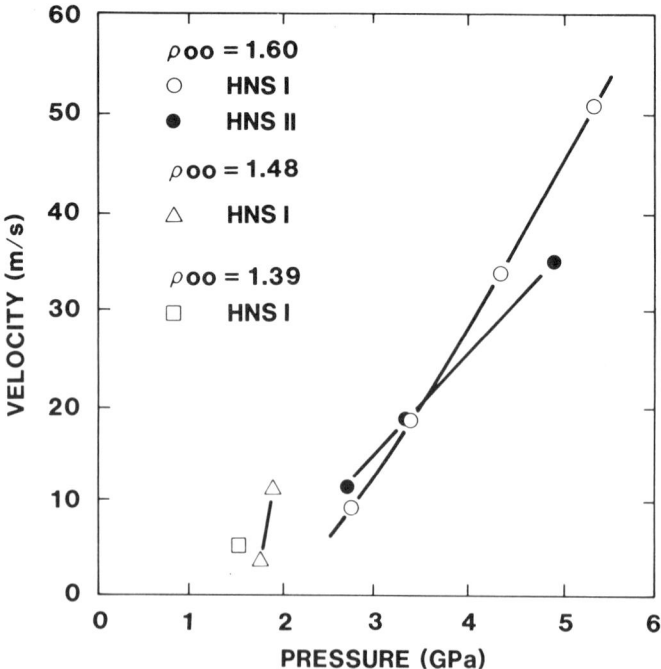

Fig. 12 Intragranular burn velocities as a function of impact pressure. Neither particle size nor initial pressing density greatly affect the velocity.

high-velocity gradients are not present on the molecular scale. Therefore an alternative mechanism would be required to interpret the present experimental results.

Acknowledgments

This article was supported by the U.S. Department of Energy under Contract DE-AC04-76-DP00789. This work has benefited from fruitful discussions with several colleagues; D. E. Mitchell, J. E. Kennedy, P. L. Stanton, and J. W. Nunziato have been particularly helpful. Thanks are also due to C. M. Korbin, who performed almost all of the numerous calculations cited here.

References

Anderson, A. B., Ginsberg, M. J., Seitz, W. L., and Wackerle, J. (1981) Shock initiation of porous TATB. 7th Symposium (International) on Detonation, Annapolis, Md., pp. 385-393.

Barker, L. M. and Hollenbach, R. E. (1972) Laser interferometer for measuring high velocities of any reflecting surface. J. Appl. Phys. 43, 4669.

Bowden, F. P. and Yoffe, A. D. (1952) Initiation and Growth of Explosion in Liquids and Solids. Cambridge University Press, Cambridge, Mass.

Butcher, B. M., Carroll, M. M., and Holt, A. C. (1974) Shock-wave compaction of porous aluminum. J. Appl. Phys. 45, 3864-3875.

Cowperthwaite, M. (1973) Determination of energy-release rate with the hydrodynamic properties of detonation waves. Proceedings of the Fourteenth Symposium (International) on Combustion, pp. 1259-1264. The Combustion Institute, Pittsburgh, Pa.

Dodson, B. W. and Graham, R. A. (1981) Shock-induced organi chemisty. AIP Conference Proceedings, 78, Shock Waves in Condensed Matter--1981, Menlo Park, Calif.

Dremin, A. N., Klimenko, V. Y., and Michailjuk, K. M. (1981) On decomposition reaction kinetics in shock wave front. 7th Symposium (International) on Detonation, Annapolis, Md., pp. 789-794. (1981) Shock initiation of porous TATB. 7th Symposium (International) on Detonation, Annapolis, Md., 385-393.

Dunin, S. Z. and Surkov, V. V. (1978) Structure of a shock wave front in a porous solid. Zh. Prik. Mekh. Tekh. Fiz. (5) 106-114.

Dunin, S. Z. and Surkov, V. V. (1979) Structure of a shock wave front in a porous solid. J. Appl. Mech. & Tech. Phys., 20, 612-618.

Eyring, H., Powell, R. E., Duffey, G. H., and Parlin, R. B. (1949) The stability of detonation. Chem. Rev. 45, 69.

Hayes, D. B. and Mitchell, D. E. (1978) A constitutive equation for the shock response of porous hexanitrostilbene (HNS) explosive. Proc. of the Symposium HDP, Commissariant a L'Energie Atomique, Paris, France, pp. 161-172.

Kanel, G. I. (1978) Kinetics of the decomposition of cast TNT in shock waves. Fiz. Goreniya Vzryva 14, 113-117.

Kanel, G. I. and Dremin, A. N. (1977) Decomposition of cast trotyl in shock waves. Fiz. Goreniya Vzryva 13, pp. 85-92.

Kennedy, J. E. (1970) Quartz gauge study of upstream reaction in a shocked explosive. Proceedings of the Fifth Symposium International on Detonation, Pasadena, Calif., pp. 435-445.

Kennedy, J. E. and Nunziato, J. W. (1976) Shock-wave evolution in a chemically reacting solid. J. Mech. Phys. Solids 24, 107-124.

Kipp, M. E., Nunziato, J. W., Setchell, R. E., and Walsh, E. K. (1981) Hot spot initiation of heterogeneous explosives. 7th Symposium (International) on Detonation, Annapolis, Md., pp. 394-406.

Lee, E. L. and Tarver, C. M. (1980) Phenomenological model of shock initiation in heterogeneous explosives. Phys. Fluids 23, 2362-2372.

Mader, C. L. (1965) Initiation of detonation by the interaction of shocks with density discontinuities. Phys. Fluids 10, 1811-1816.

Mader, C. L. (1979) Numerical Modeling of Detonations,(edited by D. H. Sharp and L. M. Simmons Jr.),pp.208-272. University of California Press, Berkeley, Calif.

Mitchell, D. E. (1981) Private communication. Sandia National Laboratories, Albuquerque, New Mex.

Rogers, R. N. (1975) private communication. Los Alamos National Laboratory, Los Alamos, N. Mex.

Schwarz, A. C. (1981) Study of factors which influence the shock-initiation sensitivity of hexanitrostilbene (HNS). Report SAND80-2372, Sandia National Laboratories, Albuquerque, N. Mex.

Sheffield, S. A., Mitchell, D. E., and Hayes, D. B. (1976) The equation of state and chemical kinetics for hexanitrostilbene (HNS) explosive. Sixth Symp. International on Detonation, p. 748.

Thompson, S. A. (1979) CSQII -- An eulerian finite difference program for two-dimensional material response - Part 1. Material Sections. Report SAND77-1339, Sandia Laboratories, Albuquerque, N. Mex.

Wackerle, J., Rabie, R. L., Ginsberg, M. J., and Anderson, A. B. (1978) A shock initiation study of PBX-9404. Proc. of the Symposium HDP, Commissariant a L'Energie Atomique, Paris, France, pp. 127-138.

Initiation of Detonations

Charles L. Mader*
Los Alamos National Laboratory, Los Alamos, N. Mex.

Abstract

The initiation of propagating detonation in PBX 9404, PBX 9502, and X0219 by hemispheric initiators of PBX 9404, 1.8-g/cm^3 TATB, and X0351 is described numerically, using a two-dimensional Lagrangian code and the Forest fire explosive decomposition rate to describe the heterogeneous explosive shock initiation process. The initiation of propagating detonation in the insensitive PBX 9502 by triple-shock-wave interaction from three initiators has been modeled using a three-dimensional, reactive, Eulerian hydrodynamic code.

Introduction

The initiation of propagating, diverging detonation is usually accomplished by small conventional initiators; however, as the explosive to be initiated becomes more shock insensitive, the initiators must have larger diameters (>2.5 cm) to be effective.

The image intensifier camera (I^2C) (Winslow et al. 1976) was used to examine the nature of the diverging detonation waves formed in the shock sensitive explosive PBX 9404 (94/3/3 HMX/nitrocellulose/Tris-β-chlorethyl phosphate) and in the shock insensitive explosives X0290 or PBX 9502 (95/5 TATB/Kel-F at 1.894 g/cm^3) and X0219 (90/10 TATB/Kel-F at 1.914 g/cm^3) by hemispheric initiators. (These data were supplied by James R. Travis, Group M-3, Los Alamos National

---

Presented at the 8th ICOGER, Minsk, USSR, Aug. 23-26, 1981. Copyright © American Institute of Aeronautics and Astronautics, Inc., 1982. All rights reserved.
*Fellow, Theoretical Division.

Laboratory, Los Alamos, N. Mex.) The geometrics of the initiators were 1) a 6.35-mm-radius hemisphere of PBX 9407 (94/6 RDX/Exon at 1.61 g/cm^3) surrounded by a 6.35-mm-thick hemisphere of PBX 9404; 2) a 6.35-mm-radius hemisphere of 1.7-g/cm^3 TATB surrounded by a 19.05-mm-thick hemisphere of 1.8-g/cm^3 TATB; or 3) a 16-mm-radius hemisphere of X0351 (15/5/80 HMX/Kel-F/TATB at 1.89 g/cm^3).

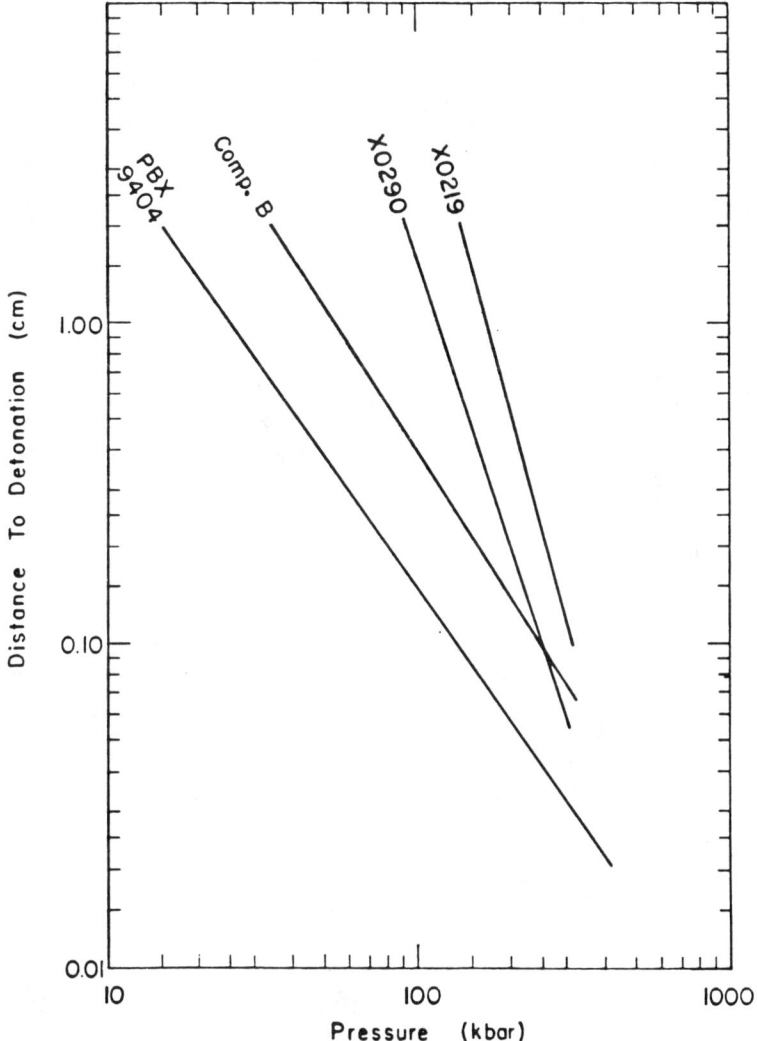

Fig. 1 The distance of run to detonation as a function of the shock pressure.

We have numerically examined systems with similar geometries by use of a hydrodynamic, two-dimensional Lagrangian code (Mader 1979) and the Forest fire explosive decomposition rate (Mader 1979) to describe the shock initiation process. As the explosive to be initiated becomes more shock insensitive, the initiators must have larger diameters or some other method must be used to achieve the required high pressures of adequate duration. High pressures are achieved if two or more shock waves interact to form regular or Mach shock reflections. We will investigate propagating detonation initiation in the insensitive high explosive PBX 9502 by the double- and triple-

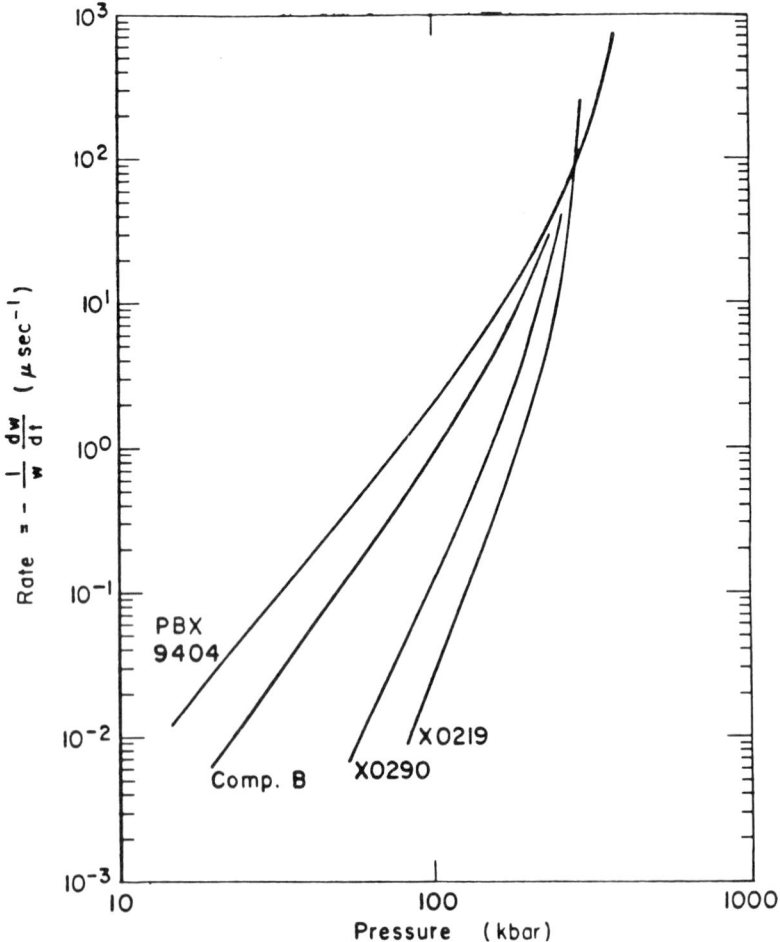

Fig. 2 The Forest fire decomposition rates as a function of shock pressure.

INITIATION OF DETONATIONS 471

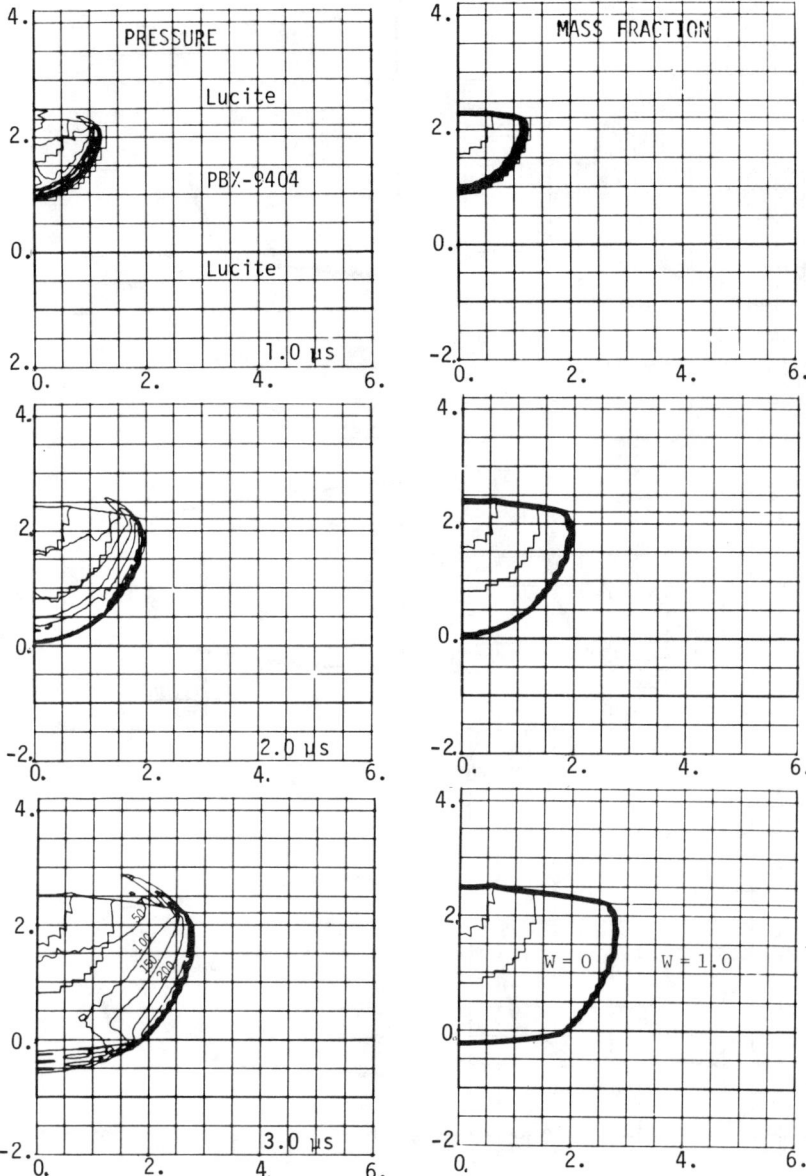

Fig. 3 The pressure and mass fraction contours at various times for a hemispheric initiator of 6.35-mm-radius PBX 9407 surrounded by 6.35 mm of PBX 9404 initiating PBX 9404. The pressure contour interval is 50 kbars, and the mass fraction contour is 0.1. All of the explosive is decomposed. The outermost pressure contour is 50 kbars and increases to 250 kbars.

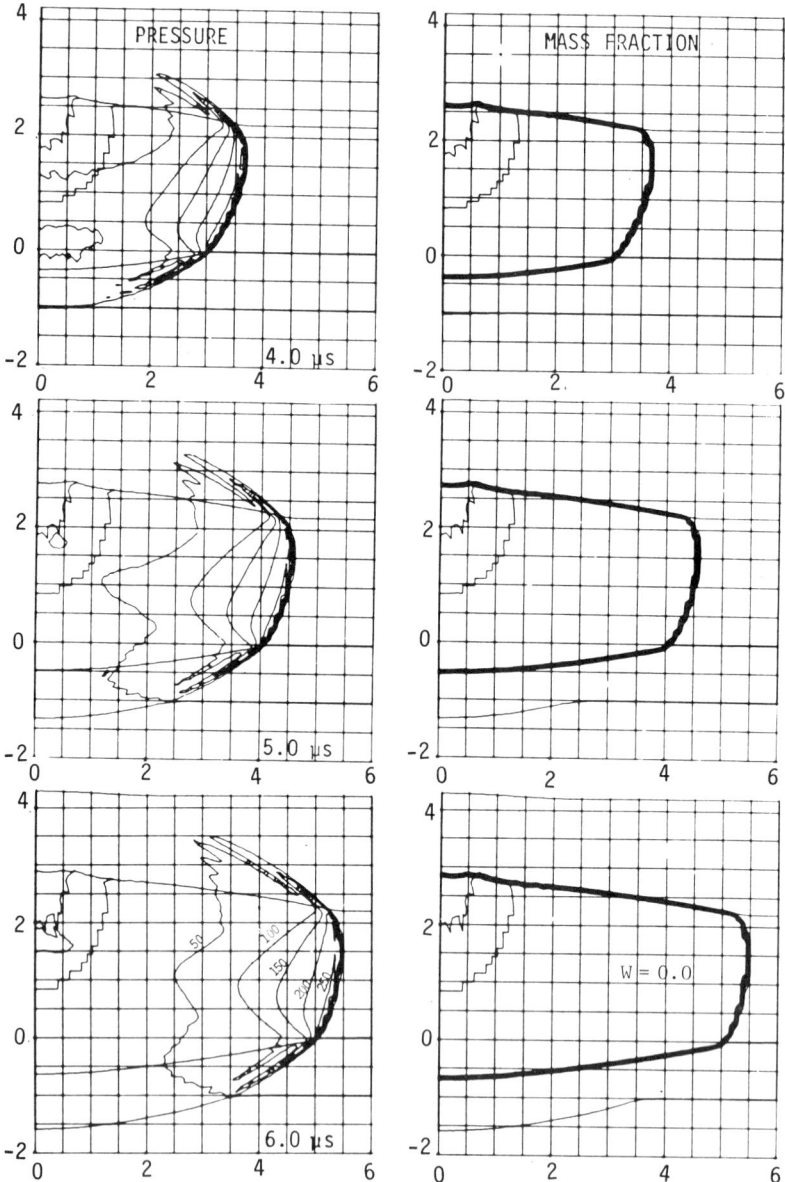

Fig. 3 (cont.) The pressure and mass fraction contours at various times for a himispheric initiator of 6.35-mm-radius PBX 9407 surrounded by 6.35 mm of PBX 9404 initiating PBX 9404. The pressure contour interval is 50 kbars, and the mass fraction contour is 0.1. All of the explosive is decomposed. The outermost pressure contour is 50 kbars and increases to 250 kbars.

INITIATION OF DETONATIONS 473

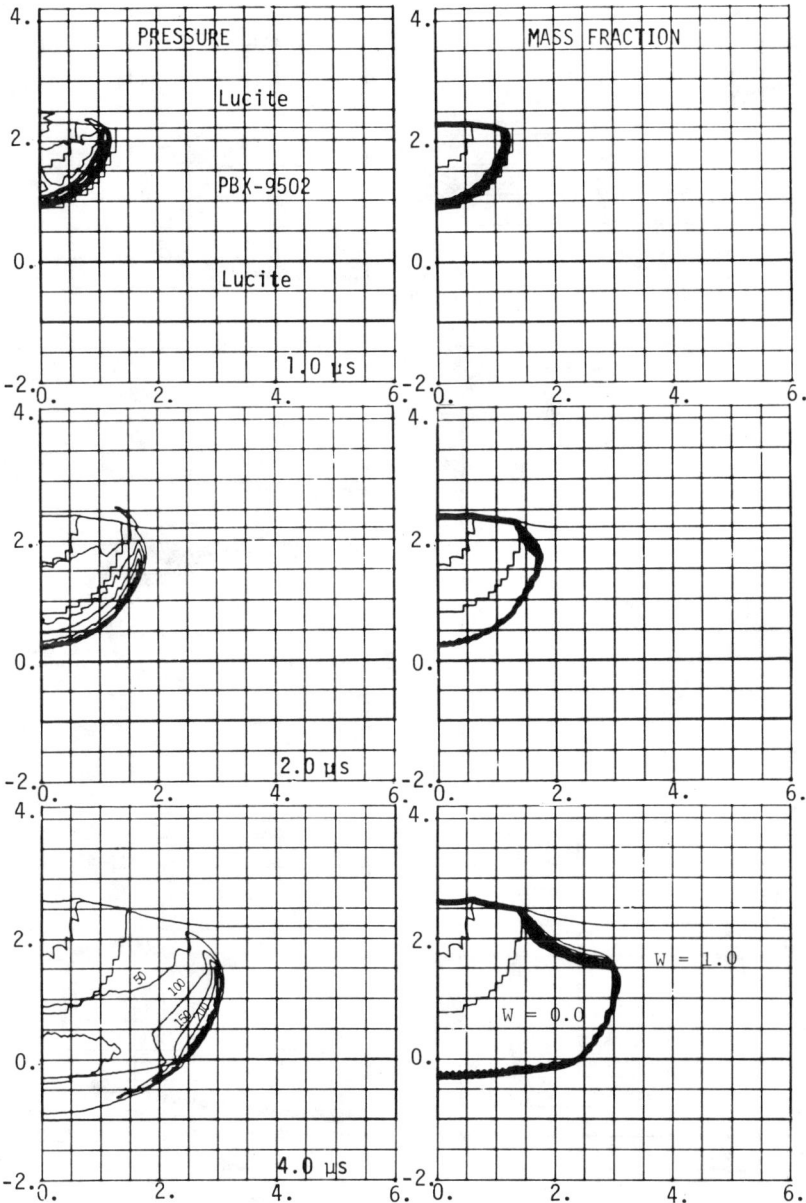

Fig. 4 The pressure and mass fraction contours at various times for a hemispheric initiator of 6.35-mm-radius PBX 9407 surrounded by 6.35 mm of PBX 9404 initiating PBX 9502 (X0290). The pressure contour interval is 50 kbars, and the mass fraction contour is 0.1. A region of undecomposed explosive occurs near the top of the PBX 9502.

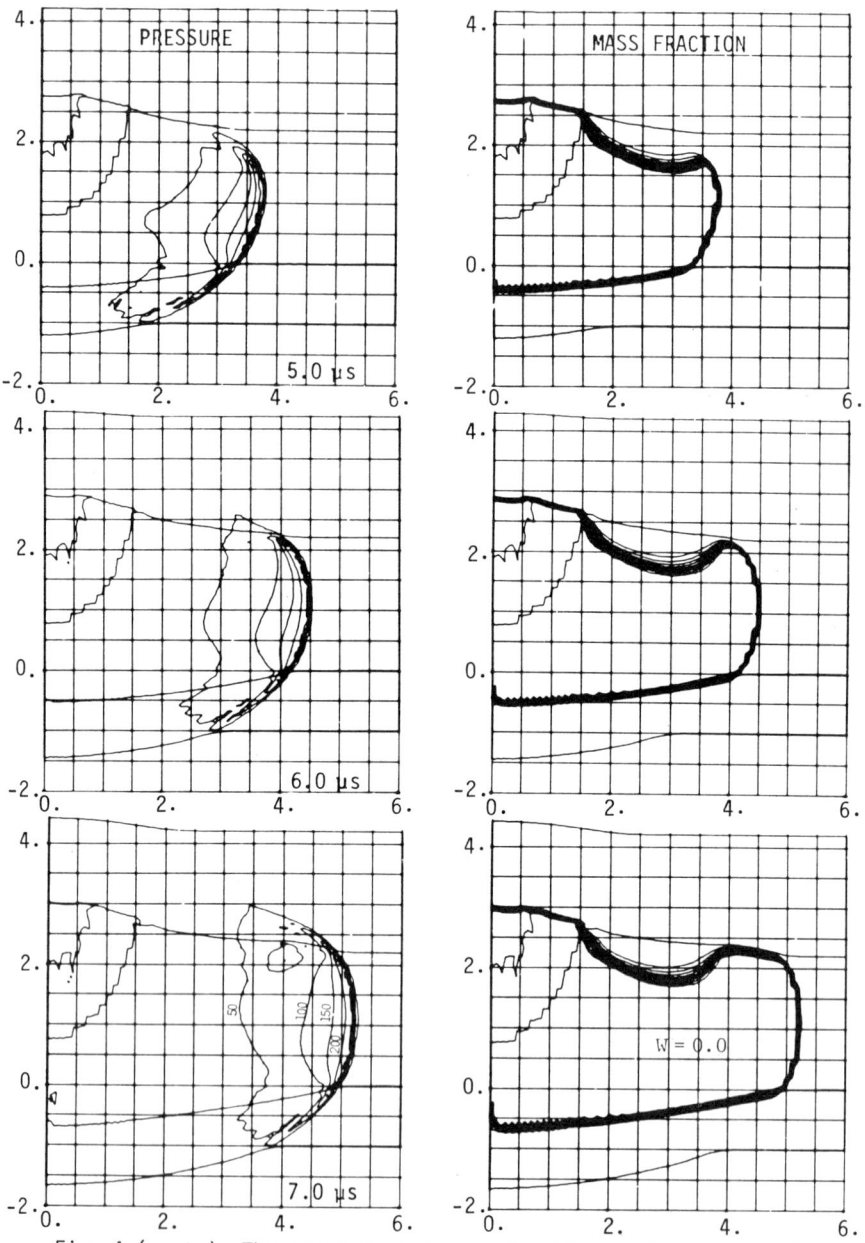

Fig. 4 (cont.) The pressure and mass fraction contours at various times for a hemispheric initiator of 6.35-mm-radius PBX 9407 surrounded by 6.35 mm of PBX 9404 initiating PBX 9502 (X0290). The pressure contour interval is 50 kbars, and the mass fraction contour is 0.1. A region of undecomposed explosive occurs near the top of the PBX 9502.

Table 1 Calculational parameters

Calculation Initiator	Acceptor	Mesh size, cm	Time step, µs	Viscosity coefficient Mbar
PBX 9407/PBX 9404	PBX 9404	0.05	0.02	4.0
PBX 9407/PBX 9404	PBX 9502	0.05	0.02	5.0
PBX 9407/PBX 9404	X0219	0.05	0.02	4.2
1.7 TATB/1.8 TATB	PBX 9502	0.1	0.02	5.0
X0351	PBX 9502	0.1	0.02	5.0

wave interaction of shock waves formed by initiators that are too weak to initiate propagating detonation individually.

Numerical Modeling of Initiation by Single Initiators

A two-dimensional, reactive, Lagrangian hydrodynamic code was used to describe the reactive fluid dynamics. The Forest fire explosive decomposition description of heterogeneous shock initiation was used to describe explosive burn. The HOM equation of state for undecomposed explosive, detonation products, and mixtures of the two (Mader 1979), and the Forest fire explosive decomposition rate constants for PBX 9502, PBX 9404, and X0219 were identical to those described by Mader (1979). The Pop plots (pressure as a function of distance of run to detonation) are shown in Fig. 1 and the Forest fire explosive decomposition rates in Fig. 2. The Becker-Kistiakowsky-Wilson (BKW) detonation product equation-of-state (Mader 1979) constants for X0351 and for 1.7- and 1.8-g/cm^3 TATB are given by Mader (1980).

The calculations were done in cylindrical geometry with Lucite confinement rather than the air confinement present in the experimental study. The Lucite confinement prevents the mesh distortion that can be fatal to Lagrangian calculations, but does not affect the initiation of propagating detonation process being studied.

The central 6.35-mm region of the detonator is initially exploded. This initiates the remaining explosive in the detonator using a Chapman-Jouguet (CJ) volume burn. For any given mesh size and time step, the viscosity (Mader 1979) must be adjusted to give a peak pressure at the detonation front near the effective CJ pressure to compensate for inadequate numerical resolution of the reaction zone. The resulting viscous pressure at the shock front is about a tenth of the total pressure. The parameters used are given in Table 1.

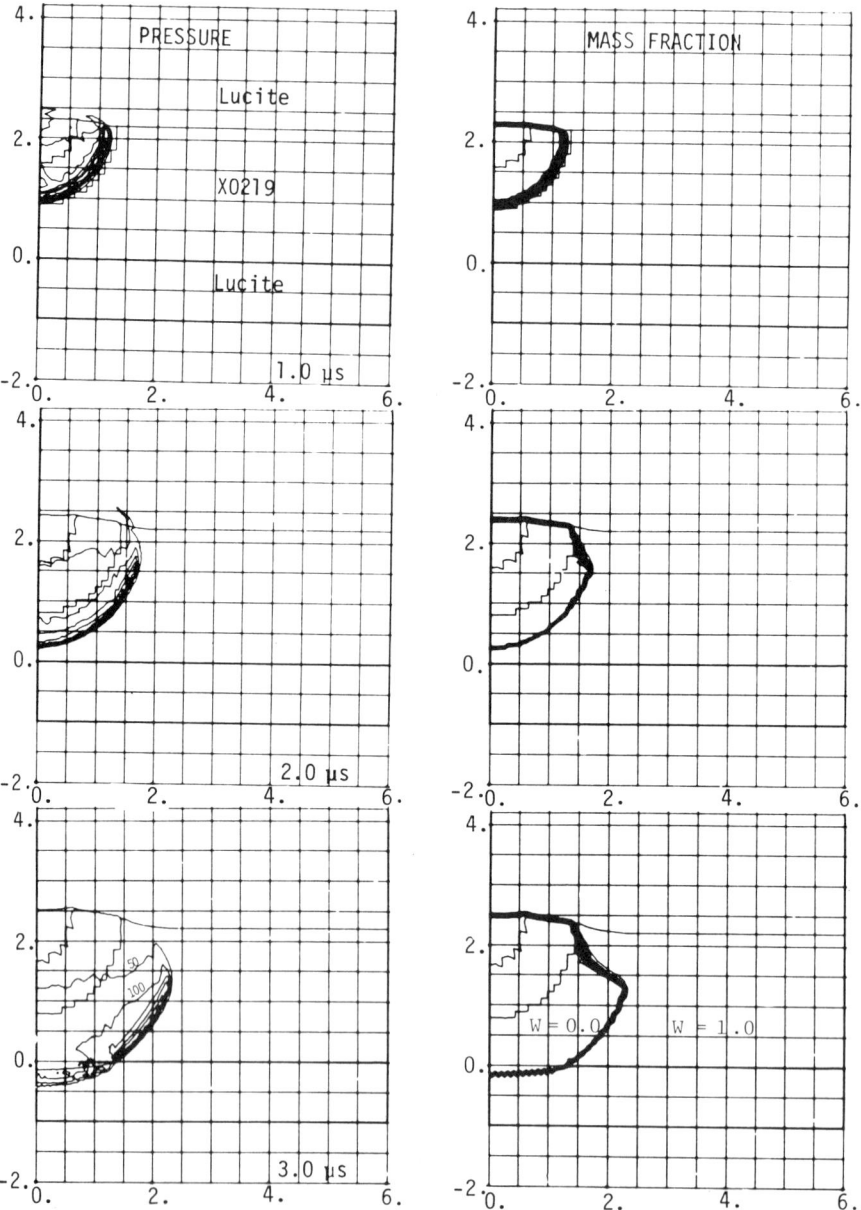

Fig. 5 The pressure and mass fraction contours at various times for a hemispheric initiator of 6.35-mm-radius PBX 9407 surrounded by 6.35 mm of PBX 9404 initiating X0219. The pressure contour interval is 50 kbars, and the mass fraction contour is 0.1. The region of undecomposed explosive initially occurring near the top of the X0219 expands, causing failure of propagating detonation.

INITIATION OF DETONATIONS

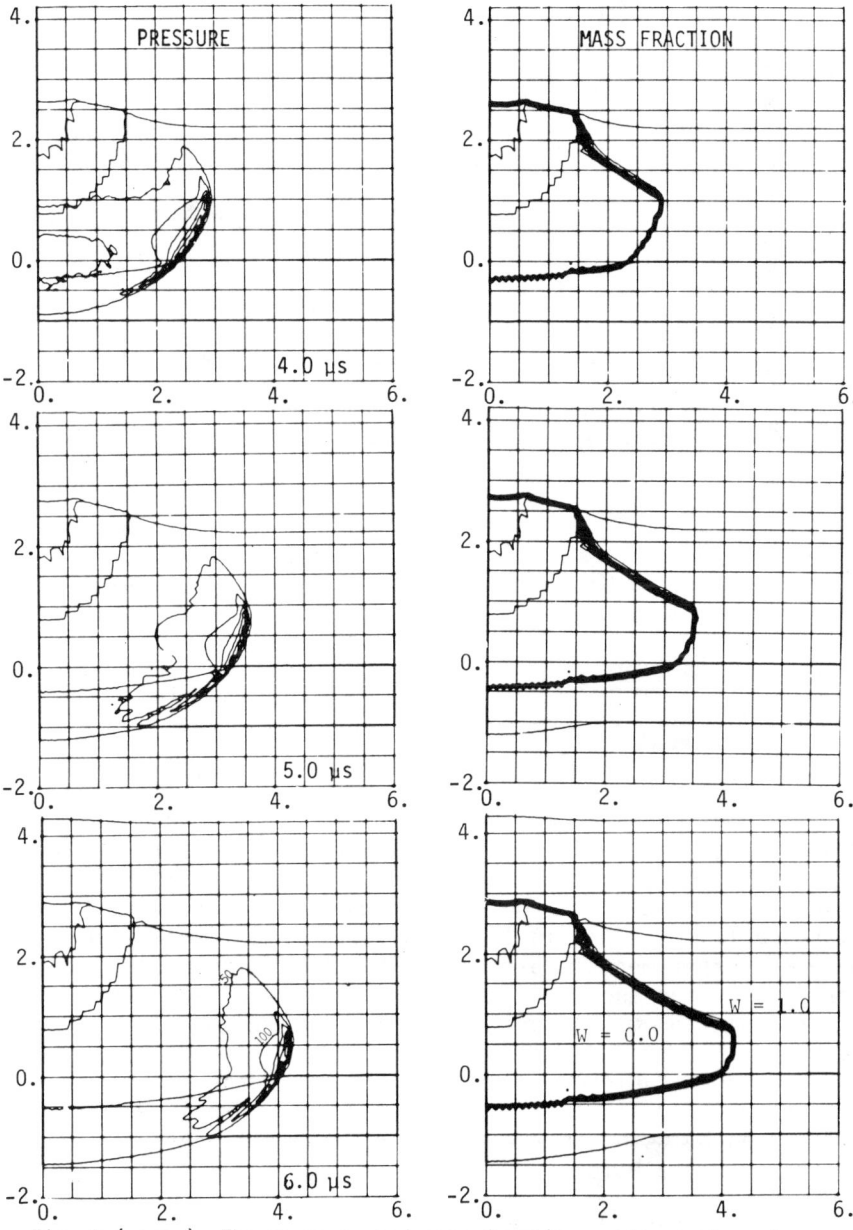

Fig. 5 (cont.) The pressure and mass fraction contours at various times for a hemispheric initiator of 6.35-mm-radius PBX 9407 surrounded by 6.35 mm of PBX 9404 initiating X0219. The pressure contour interval is 50 kbars, and the mass fraction contour is 0.1. The region of undecomposed explosive initially occurring near the tip of the X0219 expands, causing failure of propagating detonation.

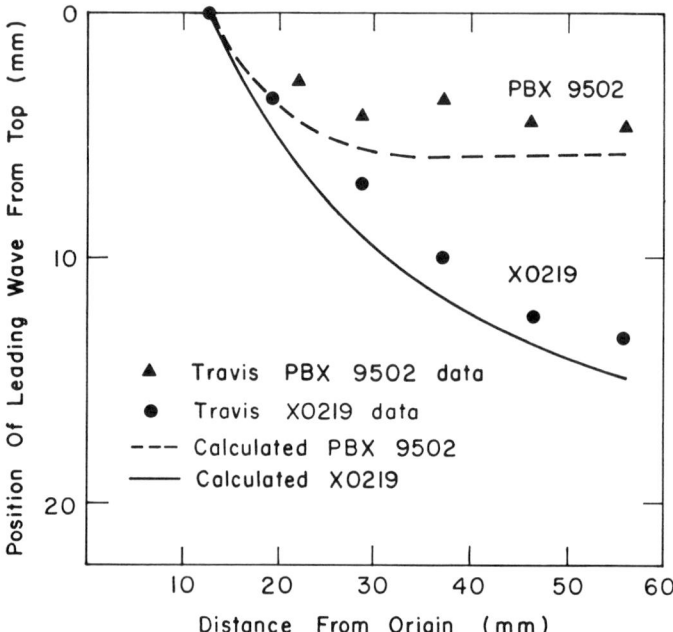

Fig. 6 The experimental and calculated position of the leading wave from the top of the explosive block as a function of the distance of the leading front of the wave from the origin.

The pressure and mass fraction contours are shown for a PBX 9404 hemisphere initiating PBX 9404, PBX 9502 (X0290), and X0219 in Figs. 3-5, respectively. The experimental and calculated position of the leading wave as a function of distance from the origin is shown in Fig. 6. Comparison with the experimental data shows that the calculated position of the shock and the detonation wave is correct and the calculation may be used to define the region of partially decomposed explosive.

The burn can become unstable when it turns a corner. The instability is apparently numerical because it was eliminated by using an average of nearby cell pressures for the Forest fire burn rather than the individual cell pressure.

The pressure and mass fraction contours are shown in Fig. 7 for the 1.8-g/cm^3 TATB hemisphere initiating PBX 9502. Very little undecomposed explosive was observed experimentally, in agreement with the calculated results. The contours are shown in Fig. 8 for an X0351 hemisphere

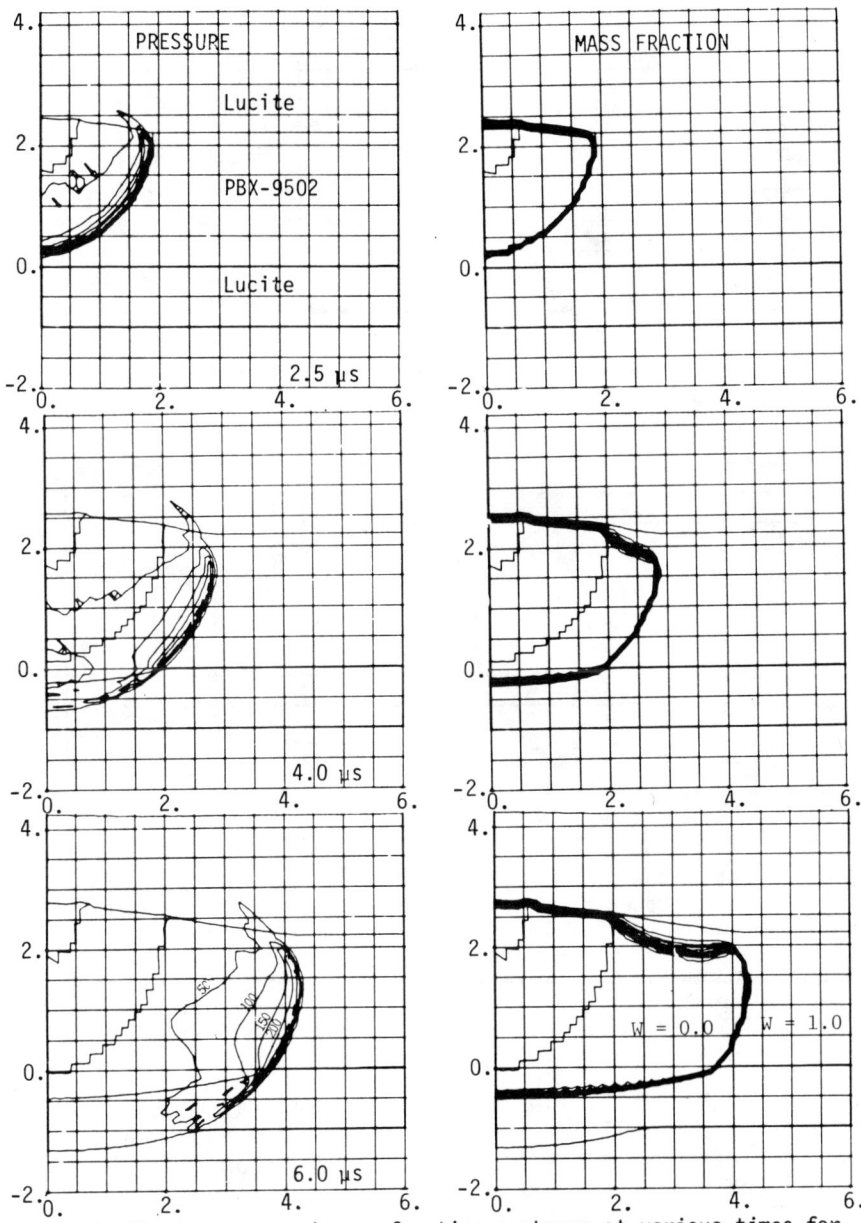

Fig. 7 The pressure and mass fraction contours at various times for a hemispheric initiator of 6.35-mm-radius TATB at 1.7 g/cm^3 surrounded by 19.05 mm of TATB at 1.8 g/cm^3 initiating PBX 9502. The pressure contour interval is 50 kbars, and the mass fraction contour is 0.1. A region of undecomposed explosive occurs near the top of the PBX 9502, showing that even large initiators can result in regions of undecomposed insensitive explosive.

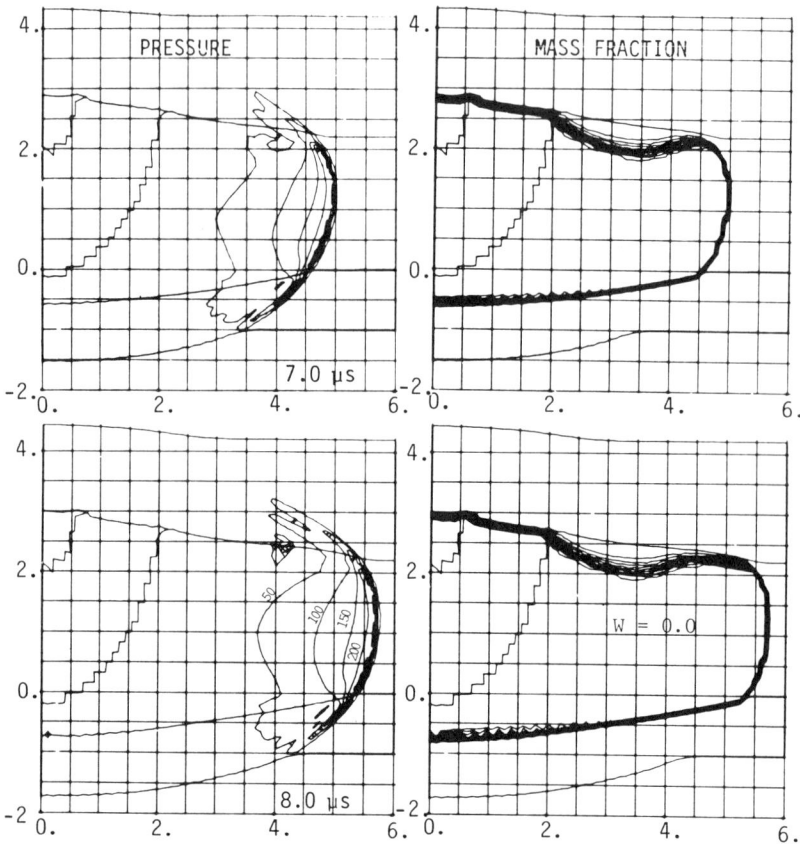

Fig. 7 (cont.) The pressure and mass fraction contours at various times for a hemispheric initiator of 6.35-mm-radius TATB at 1.7 g/cm^3 surrounded by 19.05 mm of TATB at 1.8 g/cm^3 initiating PBX 9502. The pressure contour interval is 50 kbars, and the mass fraction contour is 0.1. A region of undecomposed explosive occurs near the top of the PBX 9502, showing that even large initiators can result in regions of undecomposed insensitive explosive.

initiating PBX 9502. The experimental and calculated regions of partially decomposed PBX 9502 are shown in Fig. 9. Comparison with the experimental data shows that the calculated region of partially decomposed explosive is consistent with the experimental observations.

Numerical Modeling of Initiation by Multiple Initiators

A three-dimensional Eulerian hydrodynamic computer code (Mader and Kershner 1980) was used to model numerically the

INITIATION OF DETONATIONS

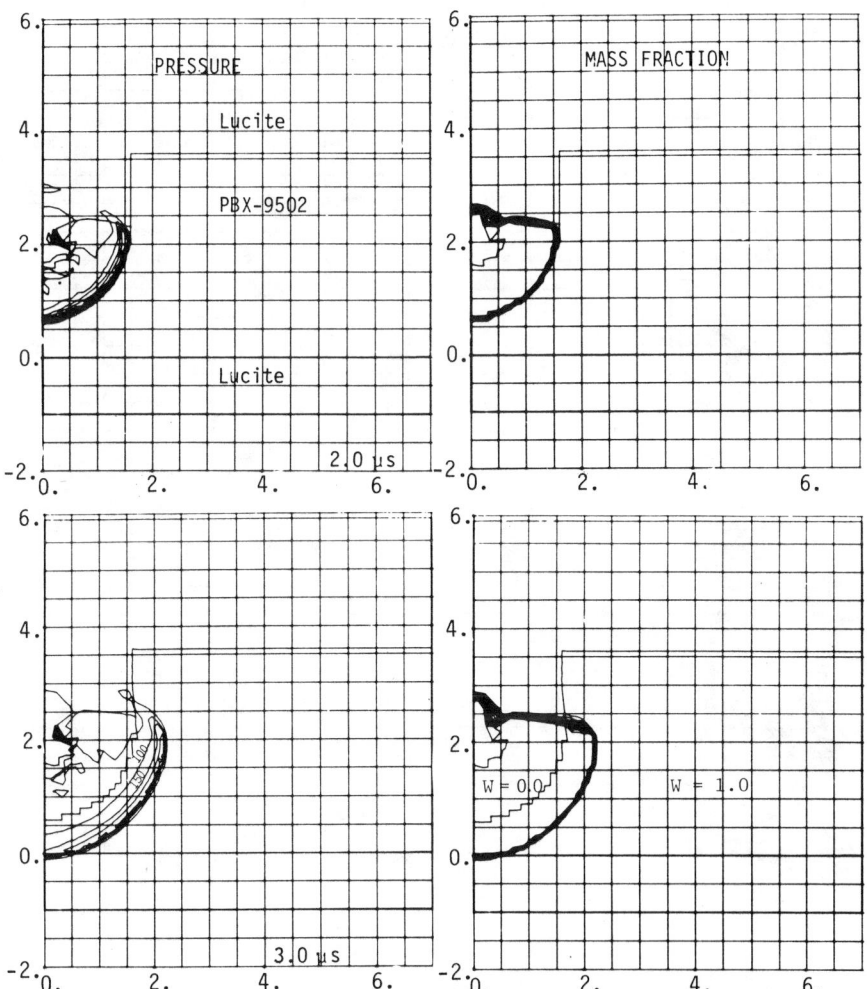

Fig. 8 The pressure mass fraction contours at various times for a hemispheric initiator of 16-mm-radius X0351 initiating PBX 9502. The pressure contour interval is 50 kbars, and the mass fraction contour is 0.1. A region of undecomposed explosive occurs when a detonation wave turns a corner is show.

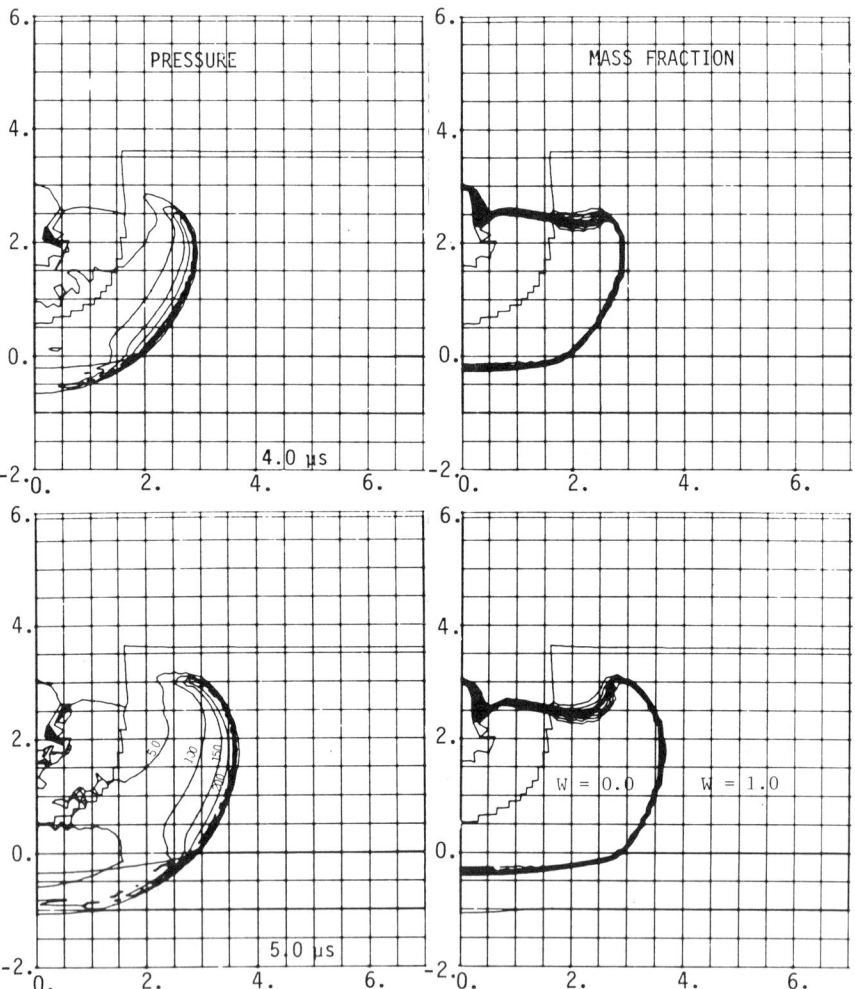

Fig. 8 (cont.) The pressure mass fraction contours at various times for a hemispheric initiator of 16-mm-radius X0351 initiating PBX 9502. The pressure contour interval is 50 kbars, and the mass fraction contour is 0.1. The region of undecomposed explosive that occurs when a detonation wave turns a corner is shown.

INITIATION OF DETONATIONS 483

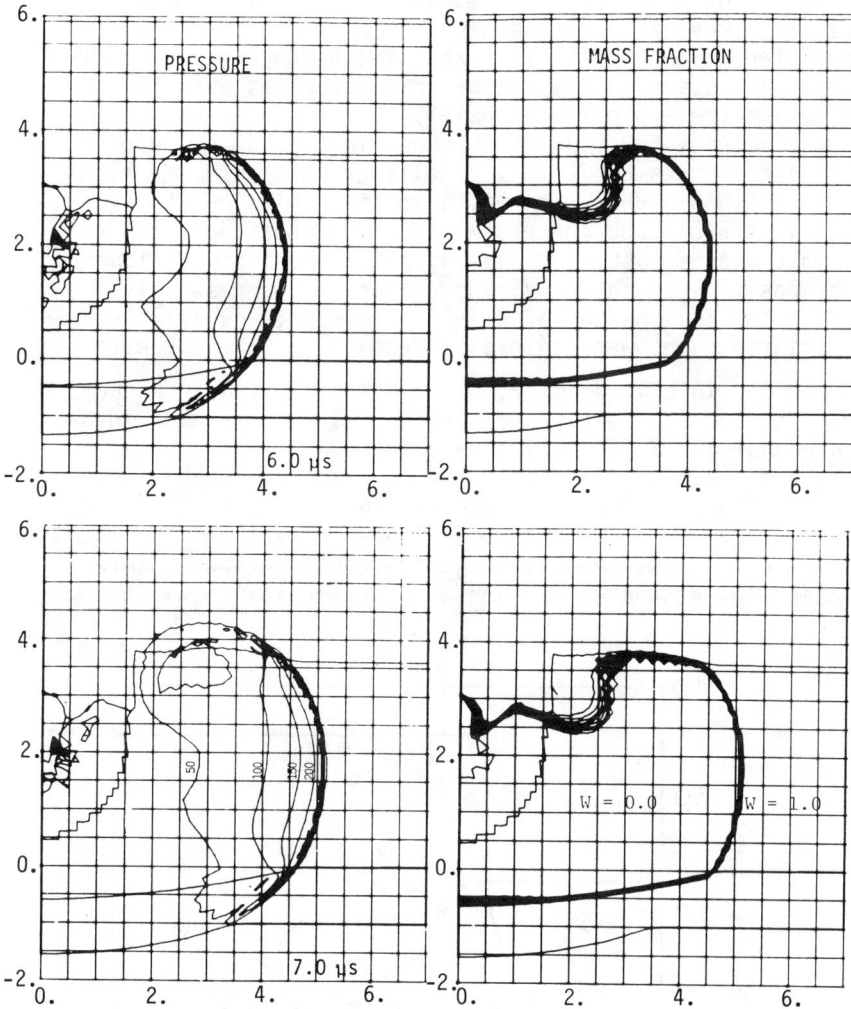

Fig. 8 (cont.) The pressure mass fraction contours at various times for a hemispheric initator of 16-mm-radius X0351 initiating PBX 9502. The pressure contour interval is 50 kbars, and the mass fraction contour is 0.1. The region of undecomposed explosive that occurs when a detonation wave turns a corner is shown.

interaction of shock waves in PBX 9502 formed by initiators that are too small to initiate propagating detonation. Multiple shock-wave interactions to initiate propagating detonations have been studied experimentally by Goforth (1978). The calculations were performed on the CRAY computer. The Forest fire explosive decomposition model of heterogeneous explosive shock initiation was used to describe the explosive burn.

The geometry studied is shown in Fig. 10. Two or three initiator cubes of 7 x 7 x 7 cells are placed symmetrically in a PBX 9502 cube with continuum boundaries on its sides. The initiator cube centers were 1.6 cm apart and 1.09 cm from the cube bottoms. The indices i, j, and k designate the position of the x, y, and z coordinates, respectively. The total cube height of k is 31, i is 29, and j is 25. The initiator cubes were initially decomposed PBX 9502 with a 2.5-g/cm^3 initial density, which has an initial pressure of 245 kbars. This sends a diverging ~100-kbar shock into the surrounding PBX 9502. The computational cell size was 0.114 cm, and the time step was 0.022 µs. The computer time for the 22,475 cells was about 50 min for 150 cycles.

The expected wave interactions are sketched in Fig. 11. The sketch shows the waves, just after double-wave interaction, as dashed lines, and the dark region shows the

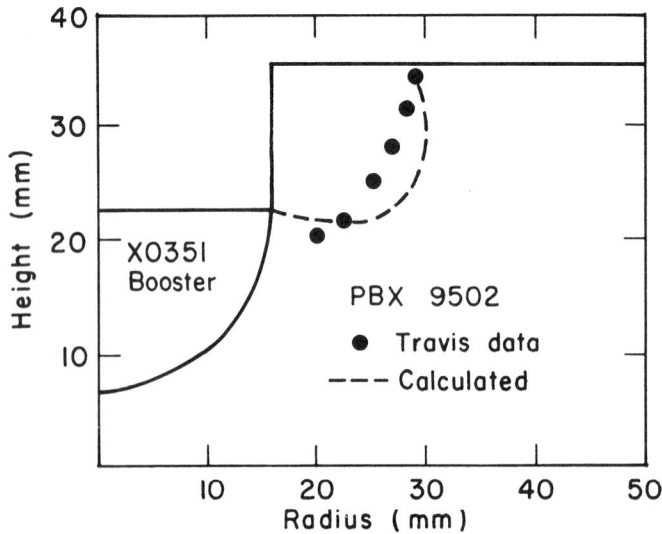

Fig. 9 The calculated and experimental region of partially decomposed PBX 9502 when initiated by an X0351 initiator.

INITIATION OF DETONATIONS 485

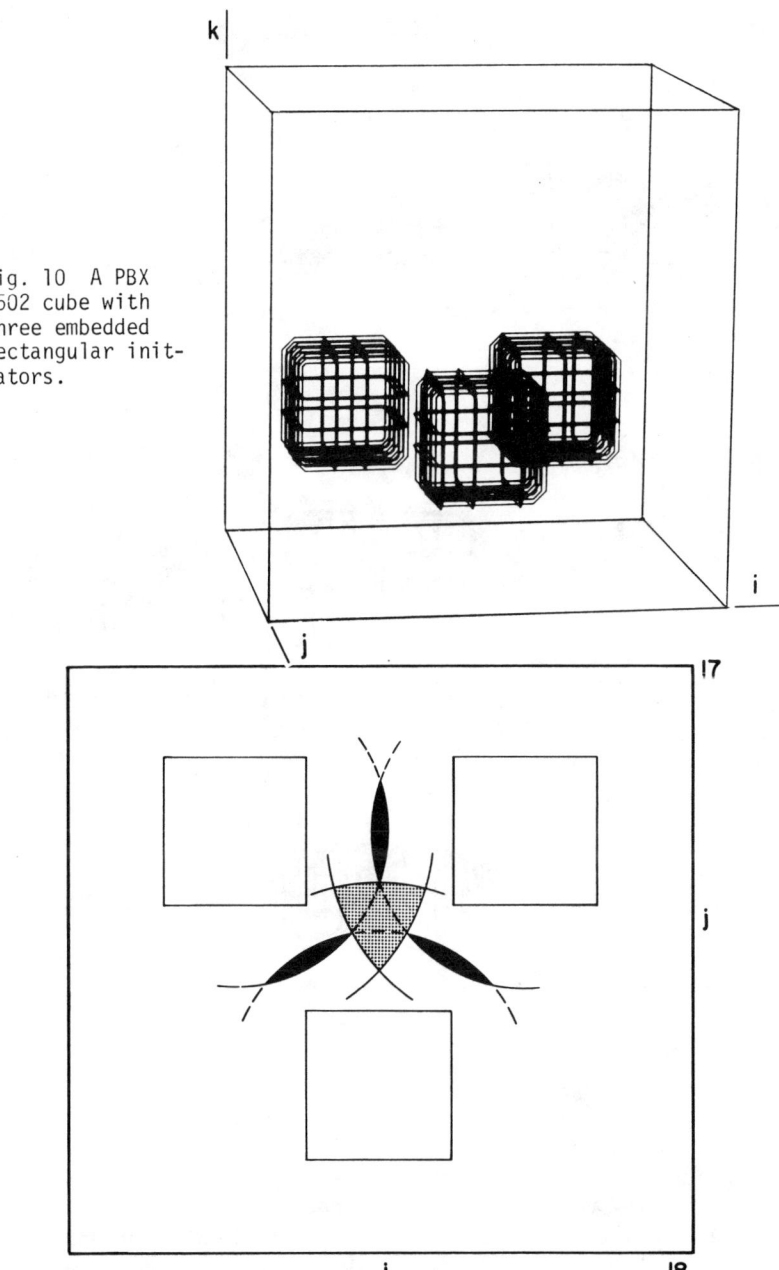

Fig. 10 A PBX 9502 cube with three embedded rectangular initiators.

Fig. 11 The expected double- and triple-wave interactions from three initiators. The dashed lines and dark regions show the double-wave interaction. The solid lines and dotted regions shown the triple-wave interaction.

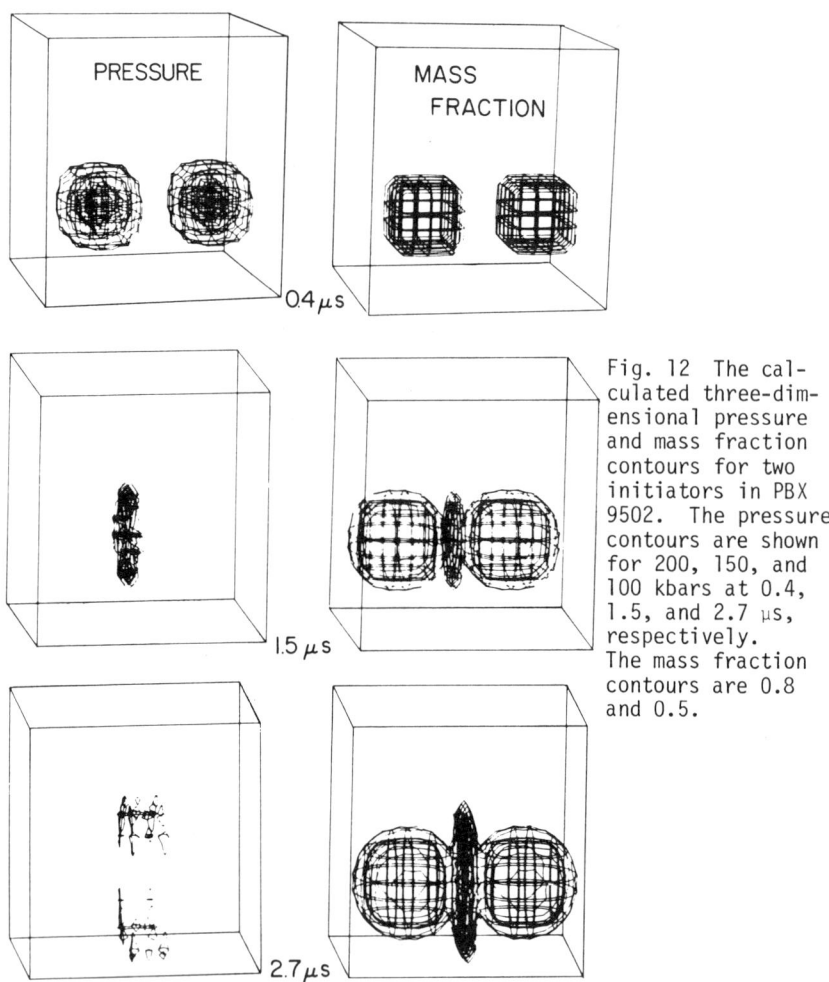

Fig. 12 The calculated three-dimensional pressure and mass fraction contours for two initiators in PBX 9502. The pressure contours are shown for 200, 150, and 100 kbars at 0.4, 1.5, and 2.7 μs, respectively. The mass fraction contours are 0.8 and 0.5.

double-wave interactions. The solid lines and dotted regions show the waves after triple-wave interaction.

The pressures from the diverging double-wave interaction in inert PBX 9502 are about 200 kbars, and those from the triple-wave interaction are about 300 kbars.

The calculated three-dimensional pressure and mass fraction contours for two initiators are shown in Fig. 12 and for three intiators in Fig. 13. The isobar and mass fraction cross sections for layer j of 9 (across the detonator centers) are shown for two initiators in Fig. 14 at 1.34 μs, in Fig. 15 at 1.78 μs, and in Fig. 16 at 2.66

INITIATION OF DETONATIONS

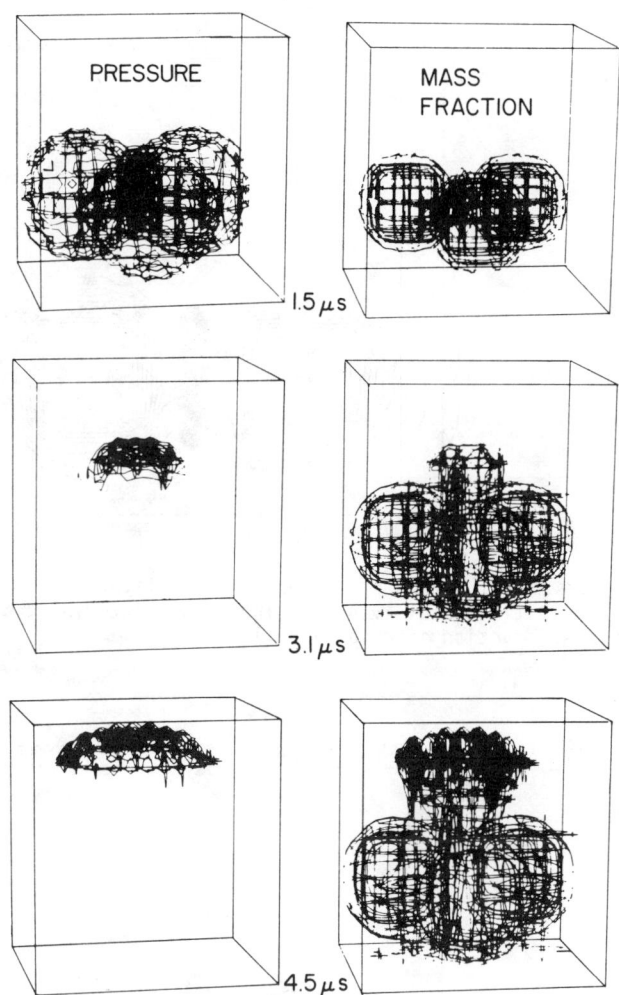

Fig. 13 The calculated three-dimensional pressure and mass fraction contours for three initiators in PBX 9502. The pressure contours are shown for 200, 150, and 100 kbars at 1.5, 3.1, and 4.5 μs, respectively. The mass fraction contours are 0.8, and 0.5.

μs. The isobar and mass fraction cross sections for layer j of 11 (across the edge of the detonators) for three initiators are shown in Fig. 17 at 1.78 μs, in Fig. 18 at 3.10 μs, and in Fig. 19 at 4.42 μs.

Although two initiators cause double-wave interaction that results in considerable decomposition, propagating detonation does not result.

Three initiators fail to initate propagating detonation at the double-wave interaction points but do at the triple-wave interaction region. The higher triple-wave interaction

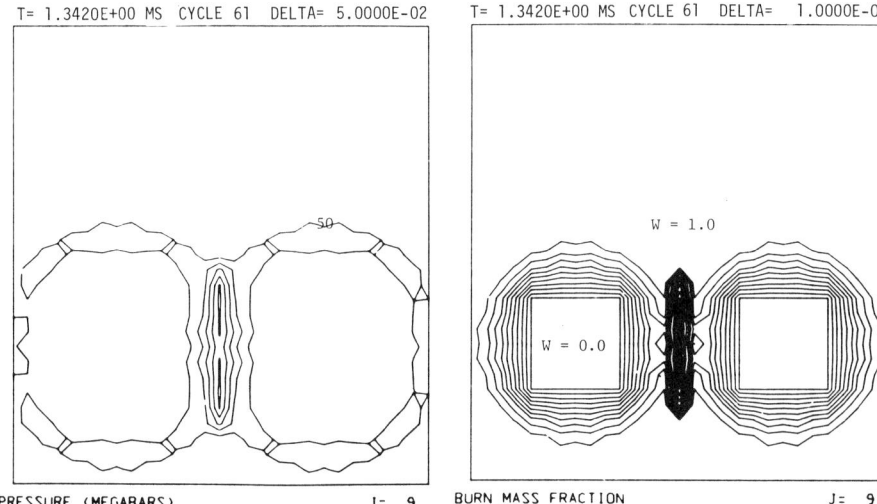

Fig. 14 The isobar and mass fraction cross sections for layer j of 9 are shown for two initiators at 1.34 μs. The isobar interval is 50 kbars and the mass fraction interval is 0.1.

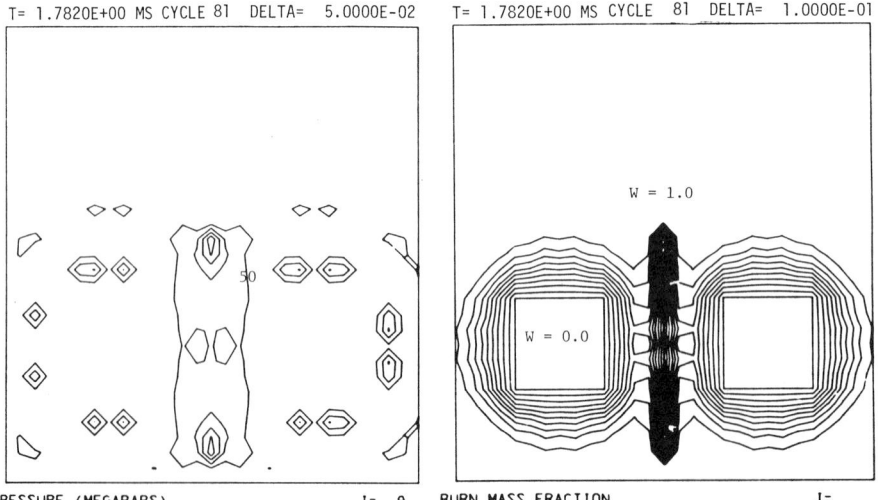

Fig. 15 The isobar and mass fraction cross sections for layer j of 9 are shown for two initiators at 1.78 μs. The isobar interval is 50 kbars, and the mass fraction interval is 0.1.

INITIATION OF DETONATIONS 489

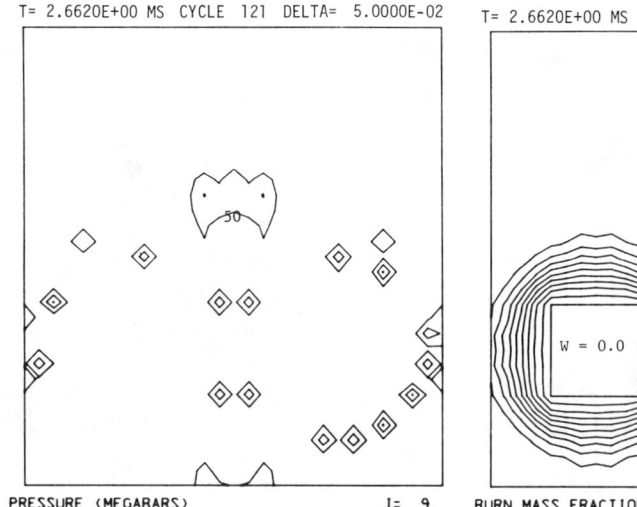

Fig. 16 The isobar and mass fraction cross sections for layer j of 9 are shown for two initiators at 2.66 µs. The isobar interval is 50 kbars and the mass fraction interval is 0.1.

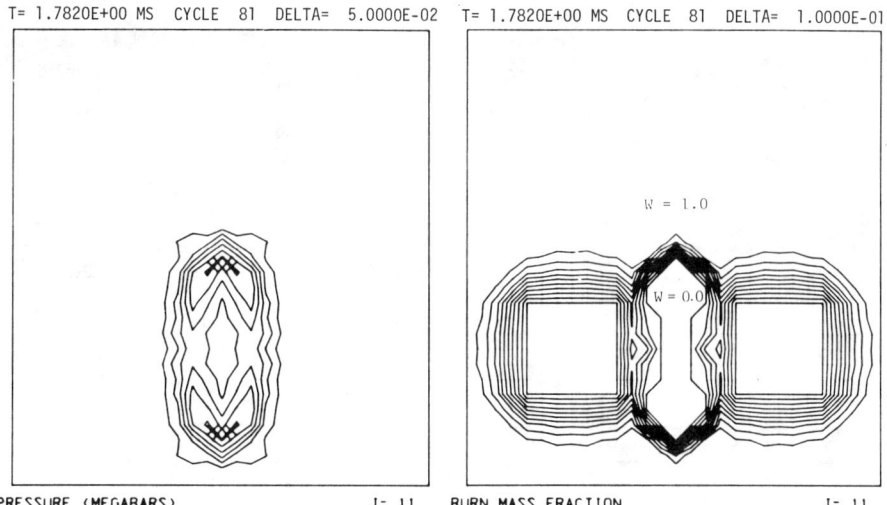

Fig. 17 The isobar and mass fraction cross sections for layer j of 11 are shown for three initiators at 1.78 µs. The isobar interval is 50 kbars and the mass fraction interval is 0.1.

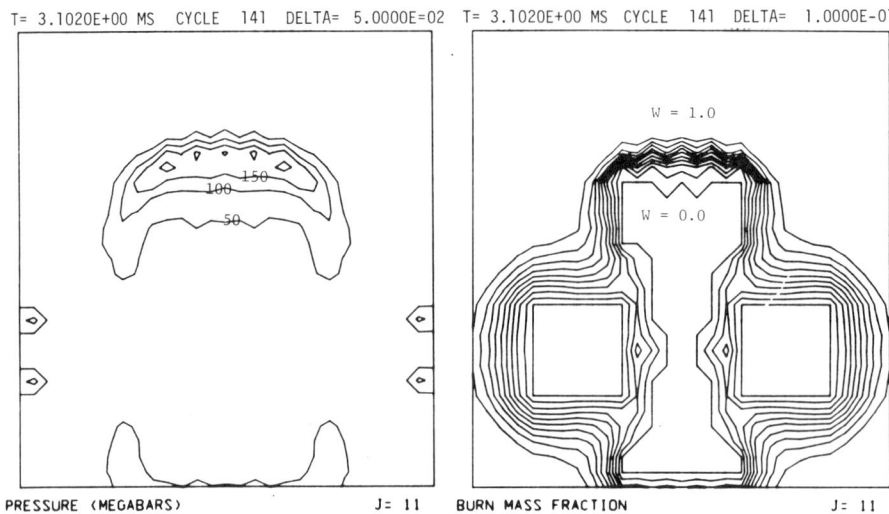

Fig. 18 The isobar and mass fraction cross sections for layer j of 11 are shown for three initiators at 3.10 µs. The isobar interval is 50 kbars and the mass fraction interval is 0.1.

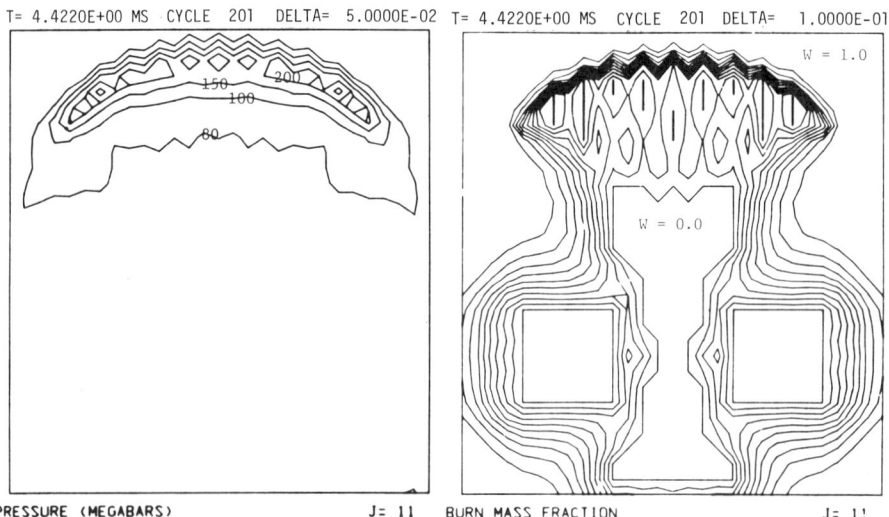

Fig. 19 The isobar and mass fraction cross sections for layers j of 11 are shown for three initiators at 4.42 µs. The isobar interval is 50 kbars and the mass fraction interval is 0.1.

pressure results in a shorter run to detonation. The
detonation can be maintained long enough to become a
propagating, diverging detonation. These computed results
reproduce the experimentally observed failure of two initiators to initiate propagating detonation and of three
initiators to result in propagating detonation.

Conclusions

The initiation of propagating detonation in sensitive
(PBX 9404) and insensitive (PBX 9502 and X0219) explosives
by hemispheric initiators can be described numerically using
a two-dimensional Lagrangian code and the Forest fire
explosive decomposition rate. Large regions of partially
decomposed explosive occur even when insensitive explosives
are initiated by large initiators. For a given initiator,
the regions of undercomposed explosive become large for the
more shock insensitive explosives and can result in failure
to initiate propagating detonation if the explosive is
sufficiently insensitive.

A three-dimensional Eulerian hydrodynamic computer code
has been used to examine the interaction of two and three
shock waves from initiators in PBX 9502. The dynamics of
initiating propagating detonation in an insensitive
explosive by multiple shock-wave interactions has been
modeled numerically. The ability to numerically model the
initiation of propagating detonation permits the study of
devices designed to enhance or mitigate the effects of
shock-wave interactions on initiation of insensitive
explosives.

References

Goforth, J. H. (1978) Safe-stationary detonation train for army
 ordnance. Los Alamos Scientific Laboratory, Los Alamos, N. Mex.,
 LA-7123-MS.

Mader, C. L. (1979) Numerical Modeling of Detonations.
 University of California Press, Berkeley, Calif.

Mader, C. L. (1980) Numerical modeling of insensitive
 high-explosive initiators. Los Alamos Scientific Laboratory,
 Los Alamos, N. Mex., LA-8437-MS.

Mader, C. L. and Kershner, J. D. (1980) Three-dimensional
 Eulerian calculations of triple-initiated PBX 9404. Los
 Alamos Scientific Laboratory, Los Alamos, N. Mex., LA-8206.

Winslow, O. G., Davis, W. C., and Chiles, W. C. (1976)
 Multiple-exposure image intensifier camera. Sixth Symposium
 (International) on Detonation, p. 664. ACR-221. Office of
 Naval Research, Arlington, Va.

Shock Wave Predetonation Processes in Porous High Explosives

B.A. Khasainov,* A.A. Borisov,† B.S. Ermolayev‡
Academy of Sciences, Moscow, USSR

Abstract

The existing models of the reaction center formation caused by the compression of high explosives (HE) in shock waves, for example, Mader's mechanism of "hydrodynamic hot spots," do not allow any explanation of the relatively high sensitivity of solid HE containing 0.1-10-μ voids to the shock waves with amplitudes of the order of 1 GPa. In the present paper, the dynamics of void collapse is analyzed. It is shown that the latter occurs in solid HE in a substantially viscous regime, and almost all energy to be evolved in the course of collapse is converted by viscous and plastic forces into heat in the vicinity of a void. Unlike Mader's mechanism (which is typical for large pores with sizes $\sim$ 1-10 mm), the temperature rise near a collapsing void in this case is not accompanied by a local pressure increase in the hot spot. Thus the "lifetime" of a hot spot is controlled by thermal conductivity of HE rather than by diverging hydrodynamic waves (as in the case of Mader's mechanisms), and is long enough to provide the conditions for the initiation of a reaction even if the voids are as small as 1 μ. The values of a pore surface temperature rise, ignition delays, and critical pressures required to generate reaction centers behind a shock are estimated. The calculated critical pressures agree by an order of magnitude with the available experimental ones. The calculations demonstrate a high effectiveness of the considered mechanism

---

Presented at the 8th ICOGER, Minsk, USSR, Aug. 23-26, 1981. Copyright © American Institute of Aeronautics and Astronautics, Inc., 1982. All rights reserved.

*Junior Researcher, Institute of Chemical Physics.
†Head of Lab., Institute of Chemcial Physics.
‡Senior Researcher, Institute of Chemical Physics.

PREDETONATION PROCESSES IN POROUS EXPLOSIVES   493

of the hot spot formation, shed light on the origin of the reaction centers, and support the surface (hole) burning concept in the treatment of the initiation of detonation in HE by weak shock waves.

Nomenclature

a_0, b_0 = initial inner and outer radii of an equivalent spherical pore, respectively
C = constant in Eq. (2)
C_s = heat capacity of HE
D = shock wave velocity
P_{ch} = minimum initiating shock wave pressure that can cause ignition of a pore surface
P_s = shock amplitude
P_s^{scr} = critical pressure for the initiation of detonation
P_{yo} = plastic yield strength of a pore = $(2Y/3) \ln(1/\phi_0)$
Re = Reynolds number of a radial flow
  = $2a_0 \sqrt{\rho(P_s - P_{yo})}/\mu$
T_0 = initial bulk temperature of HE (behind a shock)
T_+^{ig} = pore surface temperature at the moment of ignition
T_+ = variable void surface temperature = $T_+(t)$
T_{+max} = pore surface temperature without heat losses due to heat conduction = $T_{+max}(t)$
t_μ = characteristic pore collapse time when $2a_0 \ll \delta_\mu$
  = $4\mu/(P_s - P_{yo})$
t_χ = characteristic cooling time = a_0^2/χ
Y = material yield strength
Z = ratio of inner to outer radius of an equivalent pore = a/b
δ_μ = critical value of a pore diameter below which viscosity influence becomes very important
  = $8.4\mu/\sqrt{\rho(P_s - P_{yo})}$
δ_χ = critical value of a pore diameter at which heat conduction influence becomes important
  = $2\sqrt{\mu\chi/(P_s - P_{yo})}$
δ_{min} = minimum pore diameter at which self-sustained ignition can still occur
μ = viscosity of HE
ρ_{max} = theoretical maximum density of HE
ρ = density of porous explosive
σ = yield strength parameter = Y/P_s
τ = reduced time in Eq. (2) = t/t_μ

ϕ_0 = initial porosity of HE = $1 - \rho/\rho_{max}$
χ^0 = thermal diffusivity of HE

Introduction

It is commonly accepted that in the course of shock initiation of detonation in porous high explosives (HE), chemical reactions behind the shock front (when the wave has an amplitude $P_s \simeq 1-3$ GPa) are initiated in so-called hot spots, or reaction centers. Among the possible mechanisms of hot spot generation in HE with relatively small pores, only the model of "hydrodynamic reaction centers" (Mader 1965) has a reliable experimental and theoretical substantiation. According to this model, local ignition occurs owing to the overcompression of HE in the vicinity of a collapsed void. The minimum initial size of a void (δ_{min}) which could be responsible for the local self-sustained ignition of HE in this case is rather large, for instance, in melted TNT, $\delta_{min} = 1$ mm at $P_s \simeq 10$ GPa (Enig and Petrone 1966). This is due to the fact that the expansion of overcompressed material near the collapsed void decreases the hot spot temperature. This expansion results in lesser effectiveness of smaller pores in the initiation of a reaction. However, there is experimental evidence (Belyaev et al. 1973; Howe et al. 1976) that a critical pressure P_{scr} for the initiation of detonation in porous TNT ranges from 0.5 to 3 GPa depending on charge density. Here the medium and maximum sizes of voids in TNT do not exceed 5 and 50 μ, respectively, for porosity $\phi_0 = 1 - \rho/\rho_{max} = 0.21$, and 0.5 and 1 μ, respectively, for $\phi_0 = 0.05$ (Belyaev et al. 1973).

Thus Mader's mechanism, which probably prevails in the case of shock initiation of liquid HE with relatively large bubbles, does not explain the high impact sensitivity of solid high-density heterogeneous HE.

In the previous paper (Khasainov et al. 1981) the authors proposed a viscoplastic mechanism of reaction centers formation in high-density HE which under certain conditions is more effective than the Mader mechanism. This mechanism accounts for the heating of the material around a void by viscous forces and the essentially viscous character of void collapse in solid HE. The latter implies that the collapse occurs without pressure overshoot and without generation of compression waves emanating from the site of collapse; it also implies that the slow heat conduction process is a rate-determining one in cooling hot spots. This in turn results in rather small values of δ_{min} (0.1-1 μ).

The previous analysis (Khasainov et al. 1981) was carried out for high-density HE with rarely spaced voids; the yield strength of voids is assumed to be constant. In the present paper an analysis of the dynamics of essentially viscous deformation of voids is given which approximately accounts for 1) the mutual influence of adjacent voids (what makes the analysis applicable in the case of relatively large porosity), and 2) the heating of the material around voids due to both viscosity and plastic strain (in the previous paper only the work of viscous forces was considered). Estimated values of P_{ch} and of ignition delays at the walls of collapsing voids demonstrate the very high effectiveness of the viscoplastic mechanism of the reaction centers generation and agree with the concept of surface burning[§] of HE after it is ignited in hot spots behind an initiating shock wave (Howe et al. 1976).

The Dynamics of Void Collapse

The dynamics of void collapse changes drastically when the equivalent pore size $2a_0$ becomes less than $\delta_\mu = 8.4\,\mu/\sqrt{\rho(P_s - P_{yo})}$ (Zababakhin 1970). Here μ and ρ are respectively the viscosity and density of the material surrounding a void; $P_{yo} = (2Y/3)\ln(1/\phi_0)$ is the plastic yield strength of the pore (Butcher et al. 1974); and Y is the plastic yeild strength of the solid material. Unlike the case of the collapse of bubbles in regular liquids, the essentially viscous regime of collapse ($2a_0 \ll \delta_\mu$) does not lead to a pressure rise in the vicinity of a pore and results in a very low radial velocity of void walls: $v_\mu = a_0(P_s - P_{yo})/4\mu$. For μ = 10-100 Pa-s, indirectly estimated by Khasainov et al. (1981) and for P_s = 1-3 GPa and P_{yo} = 0.1-0.3 GPa, the value of δ_μ exceeds 30-300 μ. These values of δ_μ exceed substantially the mean (and sometimes the maximum) size of pores in pressed HE (Belyaev et al. 1973), hence the collapse of voids in pressed HE behind relatively low-amplitude shock waves should occur in an essentially viscous regime. The large characteristic collapse time $t_\mu = 4\mu/(P_s$

[§]HE in this case burns from inside of a pore and thus this kind of burning will be referred to as hole burning.

$- P_{yo}$) (in comparison with $2a_0/D$, where D is a shock wave velocity) and the extremely low Mach number of the flow around the pore make quite reasonable the assumption that the solid material around the pore is incompressible. For convenience the collapse is assumed to be spherical to characterize the process quantitatively. Qualitatively the results for voids of an arbitrary shape are believed to be analogous. There are two arguments in favor of this: 1) It is generally believed that the thickness of the shock front is much larger than the size of pores (heterogeneities). In addition, in the considered viscous collapse regime, the shock front passes by the pore for a time interval which is much less than the characteristic time of collapse. Therefore the pressure drop at pore diameter is always less than the shock pressure P_s. 2) The boundary condition in the vicinity of the impact face point A of the pore interacting with a shock is

$$(-P + q)_{HE} = -P_{gas}$$

where the viscous stress $q = 2\mu \partial U/\partial r$. For considered pores [$2a_0 \ll 8.4\mu/\sqrt{\rho(P_s - P_{yo})}$ and $P_{gas} = 0$], the estimated value of the static pressure is

$$P_A = |q_A| = 2\mu \frac{\partial U}{\partial r}\bigg|_A \approx \frac{2\mu U_s}{a_0} \gg \frac{U_s \sqrt{\rho P_s}}{2.1} \approx 0.1 \text{ GPa}$$

Here U_s is a characteristic particle velocity in a shock with $P_s \approx 1-3$ GPa. The high value of viscous stress at the pore surface tends to make the pressure more evenly distributed along the pore surface than in the absence of viscosity. Nearly isotropic deformation of the small pores generates high particle velocity gradients (and respectively high energy dissipation due to the viscosity) in sites where the pore surface is curved. In that sense the heating process of material around the void for the considered mechanisms of collapse is similar for any void having a curved surface (including spherically symmetric pores).

For further simplification the effects caused by melting of HE are neglected, μ and Y are considered as constants, consider voids as spheres having the same initial radius a_0, and neglect the gas pressure in voids. Under these assumptions (which are estimated to be realistic ones) the dynamics of void collapse in a shock wave with $P_s > P_{yo}$ can be described by the following equation (Butcher et al.[yo])

1974):

$$-\rho a \frac{d^2 a}{dt^2} = \frac{P_s + 2Y\ln Z}{1-Z} + \frac{4\mu}{a} \frac{da}{dt} \frac{1-Z^3}{1-Z} + \rho\left(\frac{da}{dt}\right)^2 \cdot \left(2 - \frac{1}{2}\frac{1-Z^4}{1-Z}\right) \quad (1)$$

At $t = 0$: $a = a_0$ and $da/dt = 0$. Here $Z = a/b$ $= a/\sqrt[3]{a^3 + b_0^3 - a_0^3}$, and $b_0 = a_0/\sqrt[3]{\phi_0}$ is the initial external radius of an equivalent sphere; it defines the amount of the solid material which can be influenced by the flow around a single void.

For $2a_0 \ll \delta_\mu$ [or $Re = 2a_0 \sqrt{\rho(P_s - P_{yo})}/\mu \ll 8$], an approximate solution for Eq. (1) obtained by matching two asymptotic expansions (for small and large times) is

$$\frac{a}{a_0} = \sqrt[3]{\frac{1-\phi_0}{\phi_0} \frac{1}{\exp[R(\tau)]-1} + \left(\frac{Re}{8}\right)^2 \cdot C \cdot \frac{1-\sqrt[3]{\phi_0}}{(1-\phi_0)^2}} \times \left\{1 - \exp\left[-\frac{64}{Re^2} \frac{1-\phi_0}{1-\sqrt[3]{\phi_0}} \tau\right]\right\}$$

(2)

where

$$R(\tau) = \frac{3P_s}{2Y}\left[1 - C \exp\left(\frac{-2Y}{P_s}\right)\tau\right] \quad C = 1 + \frac{2Y}{3P_s}\ln\phi_0 \quad \tau = \frac{P_s t}{4\mu}$$

Since $(Re/8)^2 \ll 1$, the second term in Eq. (2) is important only at the very beginning of the process, when $a/a_0 \simeq 1$. It contributes substantially to the initial acceleration of the void walls. Thus the main part of the $a(t)$ curve is practically independent on Re. Figure 1 shows $a(t/t_\mu)$ curves for different values of the porosity and of the yield strength parameter $\sigma = Y/P_s$. It can be seen that the characteristic collapse time is approximately the same (equal to $\sim t_\mu$) for all of the curves and that the more rigid the pore (i.e., the larger the Y and the smaller the ϕ_0), the larger is its final radius. The absence of pressure overshoot and of rarefaction waves mentioned above follows from the approximation

$$[P(r,t) - P_s]/P_s \sim (Re/8)^2 \ll 1$$

The Dynamics of Pore Surface Heating

The heat gain due to the work of dissipating forces (viscosity and plastic strain) competes with the heat loss due to heat conduction to give the surface temperature vs time curve with a maximum. The thermal balance is expressed by the following equation:

$$\frac{\partial T}{\partial t} + u\frac{\partial T}{\partial r} = \frac{\chi}{r^2}\frac{\partial}{\partial r}\left(r^2\frac{\partial T}{\partial r}\right) + \frac{12\mu \dot{a}^2 a^4}{\rho C_s r^6} - \frac{2Y \dot{a} a^2}{\rho C_s r^3} \quad (3)$$

At $r = a$: $\partial T/\partial r = 0$; at $r = b$: $\partial T/\partial r = 0$; at $t = 0$: $T = T_0$. Here χ is the thermal diffusivity, $u = \dot{a}a^2/r^2$, C_s is the specific heat capacity of HE, and $a(t)$ and $\dot{a}(t) = da/dt$ are defined by Eq. (2). The voids are assumed to be evacuated; therefore the temperature gradient at the surface of a void is equal to zero. Thus for the temperature at the void surface T_+, one has from Eq. (3):

$$\frac{dT_+}{dt} = \chi \left.\frac{\partial^2 T}{\partial r^2}\right|_+ + \frac{12\mu \dot{a}^2}{\rho C_s a^2} - \frac{2Y \dot{a}}{\rho C_s a} \quad (4)$$

If $2a_0 \ll \delta_\chi = 2\sqrt{\mu\chi/(P_s - P_{yo})}$, the characteristic cooling time $t \cong a_0^2/\chi$ is much less than t_μ, which means that T_+ in this case remains at the level of initial temperature T_0. On the contrary, if $2a_0 \gg \delta_\chi$, all the heat dissipated during the collapse at first should be localized in a surface layer of a void [according to theoretical results (Khasainov et al. 1981) the thickness of heated layer is approximately $a_0/4$]. The surface temperature can be expected to reach the magnitude T_{+max}, which is estimated from Eq. (4) with $\chi = 0$; that is,

$$\frac{\rho C_s (T_{+max} - T_0)}{P_s} = \left[\ln X - \left(\frac{1}{X} + \frac{2\sigma}{3}\frac{1}{X}\right) + 1 \ln(1 + X)\right]\bigg|_{w_0}^{w}$$

$$- \frac{2\sigma}{3}\int_{w_0}^{w}\frac{\ln(1 + X)}{X}dx + O\left(\frac{Re^2}{64}\right) \quad (5)$$

Here $w = w(\tau) = \exp[R(\tau)] - 1$, $w_0 = w(0)$, and $R(\tau)$ is defined in Eq. (2). This temperature will last until the time elapsed from the beginning of collapse becomes comparable with t_χ, then the temperature will drop due to the heat conduction.

Figure 2 shows T_{+max} as a function of time. Increasing the yield strength of voids results in decreasing the surface temperature rise. At early stages of deformation,

T_+ increases linearly with time:

$$T_+ \cong T_0 + \frac{3(P_s - P_{yo})^2 t}{4\mu \rho C_s} \qquad (6)$$

If $2a_0 \gg \delta_\chi$, the surface temperature T_+ exceeds 1000 K even when $P_s - P_{yo} \geq 0.5$ GPa only. Khasainov et al. (1981) showed that ignition of a void surface layer sets in when $T_+ = T_{ig} \cong 1000$ K; this ignition temperature decreases slightly with a decrease in the amplitude of a shock wave. Thus the considered viscoplastic mechanism undoubtedly can be responsible for the fast initiation of the reaction behind shock waves with amplitudes of 1-3 GPa around pores as small as 0.1-1 μ.

In the real case, density changes arise owing to temperature and pressure changes during the pore collapse process. However, it is believed that these density changes do not alter the final results drastically at least at moderate pressures.

If for an approximate estimation one assumes that T_{ig} is constant, the ignition delay t_{ig} appears to be proportional to the effective viscosity of HE and to $1/(P_s - P_{yo})^2$; its value for PETN is in the range 0.1-0.01 μs when pressure changes are in the range 1-3 GPa. Equation (6) shows also that

$$(P_s - P_{yo})^2 t_{ig} \cong \text{const}$$

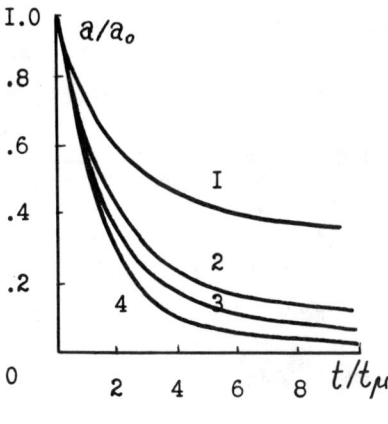

Fig. 1 Radius of a void as a function of time. Curve 1 corresponds to $\sigma = Y/P_s = 0.2$ and $\phi_0 = 0.01$; curves 2, 3, and 4 correspond to $\sigma = Y/P_s = 0.1$ and $\phi_0 = 0.001$, 0.01, and 0.1, respectively.

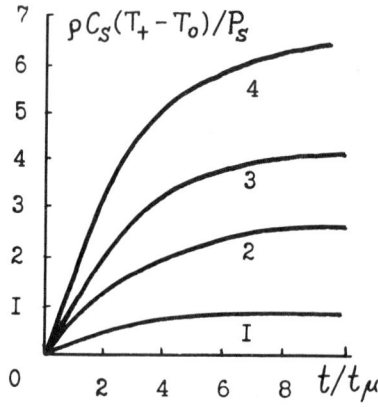

Fig. 2 Maximum possible temperature rise $\rho C_s(T_+ - T_0)/P_s$ vs time. $\bar{\sigma}$ and ϕ are the same as in Fig. 1.

which is similar to the well-known critical shock energy criterium of Walker and Wasley (1969).

The Critical Pressure Evaluation

The exact value of critical pressure for the initiation of chemical reaction (P_{ch}) should be derived from the solution of the heat balance equation (3) with a chemical release source added and should correspond to the condition of $t_{ig} \to \infty$. However, it requires the numerical solution of an equation with partial derivatives. Since the experimental data on P_{ch} are very scarce, we have made an attempt here to derive a simplified analytical expression for P_{ch} and to compare experimental and calculated values of P_{ch}.

The case of sufficiently large pores is considered, that is, $2a_0 \gg \delta_\chi$ or $t_\mu \ll t_\chi$ but still $2a_0 \ll \delta_\mu$. In this case at $t \sim t_\mu$ the pore surface temperature will be practically equal to T_{+max} defined by Eq. (5). (Note that even for $2a_0 \sim \delta_\chi$, heat conduction cannot quickly cool the heated layer around the pore.) The moment of time $t \cong t_\mu$ corresponds to the most intense contraction of voids and the pore surface temperature rise. The heat conduction will lead to an essential decrease of pore surface temperature only when t becomes comparable with t_μ. Ignition of HE should not be expected if T_{+max} does not reach the value of T_{ig} for the time period of the order of t_μ. Thus the condition

$$T_{+max}(t \cong t_\mu) = T_{ig}$$

is an approximate equation for the estimation of the critical pressure P_{ch}. The solid line of Fig. 3 shows P_{ch} calculated from Eq. (5) as a function of ϕ_o; and the dashed line represents the approximate estimation, which is written as follows:

$$P_{ch} = P_{yo} + \rho C_s (T_{ig} - T_o)/3$$

Experimental data of Howe et al. (1976) are also shown in Fig. 3. The latter represent the minimum pressure amplitude of shock waves which cause immediately observable chemical reactions in TNT of various densities. Calculations are made for T_{ig} = 1000 K and Y = 0.05 GPa. Calculated values of P_{ch} at various ϕ_o agree at least by an order of magnitude and by the trend with experimental ones. The discrepancy in the slope of experimental and calculated curves should not be considered as a serious argument against the model because of the approximate character of calculations and also the uncertainty of the definition of P_{ch} in the experiment. Further, both experimental and theoretical verification is needed.

Shock to Detonation Transition Modeling

Thus ignition occurs at an early stage of deformation of voids with sizes in the range $\delta_\mu \gg 2a_o > \delta_\chi$ in a layer having a thickness of $\sim a_o/4$ (Khasainov et al. 1981). At later stages the process proceeds to the formation of propagating surface flames inside of those of the voids around which the heat flux from the reacted HE material (either conductive or convective) is capable of creating the necessary (for the steady flame) preheated layer.

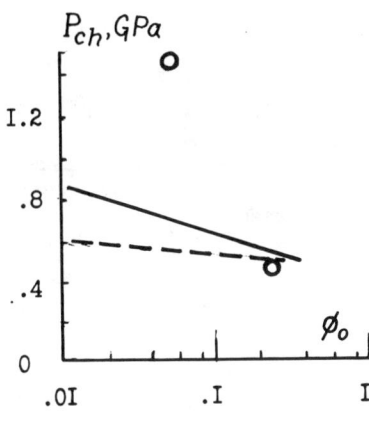

Fig. 3 Critical pressure for the initiation of chemical reaction P_{ch} vs porosity. Lines are calculated values; points represent experimental data (Howe et al. 1976).

The lower the shock wave amplitude, the smaller the pore surface temperature is at the moment of ignition (Khasainov et al. 1981) and consequently the larger the ignition delays and the critical sizes of hot spots (Merzanov et al. 1963). In accordance with the experimental observations, the latter means that ignition becomes more heterogeneous when the shock wave amplitude decreases. To support this conclusion the above mechanism is used in numerical modeling of the initiation of the reaction in PETN with a thick aluminum plate impacting the HE at various velocities. The density of PETN was 1750 kg/m^3. Figure 4 demonstrates the results of the calculations (dashed lines) together with the experimental pressure profiles (solid lines) obtained by Wackerle et al. (1976). The best fit of the experimental and calculated curves is obtained when the effective diameter of voids which are to become the reaction centers is increased with decreasing P_s. The calculated curves 1-5 in Fig. 4 are obtained for fixed density of HE (1750 kg/m^3) and monomodal distributions of voids with $2a_0$ = 2.4, 4.0, 5.0, 5.5, and 8.0 μ, respectively. Since the initial porosity of HE is fixed, the increase of the initial pore radius a_0 leads to a decrease in the burning surface area and consequently to a decrease in the heat release rate. That is why the larger the a_0, the lower the curve is in Fig. 4. Ignition delays in all the considered cases do not exceed 0.05 μs. Hence the time to detonation is determined by the burning rate of HE (that is, by a specific surface area of ignited voids and by the deflagration velocity in bulk HE) rather than by ignition delays. However, this will not be the case when P_s is approaching P_{ch}, since P_{ch} corresponds to $t_{ig} \to \infty$.

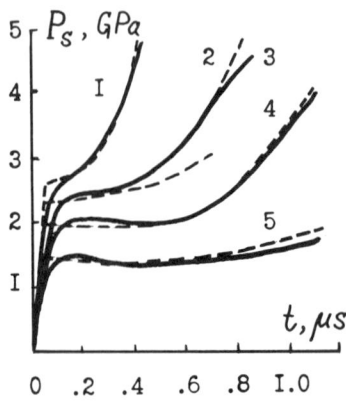

Fig. 4 Impact-face pressure-time histories: Solid lines represent experiments (Wackerle et al. 1976) and dashed ones represent calculations. $2a_0$ = 2.4, 4.0, 5.0, 5.5, and 8.0 μ for the curves 1-5, respectively; μ = 30.0 Pa-s for all curves.

It can be concluded from the previous analysis that the principal parameter which defines the limits of applicability of the viscoplastic model is the equivalent size of ignited voids. This size must be large enough ($2a_0 > \delta_\chi$) not to allow a decrease in T_+ by heat conduction, and on the other hand it must be small enough ($2a_0 \ll \delta_\mu$) for a sufficient portion of the potential energy of compressed HE to be converted into heat by viscosity and plastic stress. Note that for usually considered large pores (of the order of 1-10 mm) the proposed viscoplastic mechanism is not effective and other mechanisms will predominate. The following estimations characterize the region of the applicability of the considered model. If $P_s - P_{yo} = 1.5$ GPa and $\mu = 1$ Pa-s, spherical voids with sizes between 0.3 and 5 μ can be ignited according to above model, and if $\mu = 1000$ Pa-s, this interval of pore sizes is much wider: $1 < 2a_0 \ll 5000$ μ. These intervals cover at least partly the range of possible void size distributions in real HE; that is, the above model has wide limits of applicability.

Concluding Remarks

The considered viscoplastic mechanisms of hot spot initiation of reaction in high explosives should be regarded as one of the possible mechanisms. In reality there may exist many causes of reaction center formation, in particular pure hydrodynamic collapse (without viscosity effects), brittle fragmentation, adiabatic gas compression, and friction. Depending on the conditions, some of these factors can be prevailing. The foregoing calculations suggest there are such conditions (very small pores characteristics for high-density solid HE and relatively low-pressure amplitude of initiating shock waves) under which the viscoplastic mechanism is responsible for the reaction centers generation. Unfortunately the experimental data on the viscoplastic behavior of solid HE under pressures of the order of 1-3 GPa and high temperatures are not available. Therefore all the estimations are awaiting experimental confirmation.

Acknowledgment

Helpful discussions with Dr. A. V. Dubovik from the Institute of Chemical Physics, USSR Academy of Sciences, Moscow, are acknowledged.

References

Belyaev, A. F., Bobolev, V. K., Korotkov, A. I., Sulimov, A. A., and Chuiko, S. V. (1973) Transition from Deflagration to Explosion in Condensed Systems, 292 pp. Nauka, Moscow.

Butcher, B. M., Carrol, M. M., and Holt, A. C. (1974) Shock-wave compaction of porous aluminum. J. Appl. Phys. 45, 3864-3876.

Enig, J. W. and Petrone, F. J. (1966) Equation of state and derived shock initiation criticality conditions for liquid explosives. Phys. Fluids 9, 398-408.

Howe, P., Frey, R., Taylor, B., and Boyle, V. (1976) Shock initiation and the critical energy concept. Proceedings of Sixth Symposium (International) on Detonation, pp. 11-19. ONR ACR-221.

Khasainov, B. A., Borisov, A. A., and Ermolayev, B. S. (1981) Two phase viscoplastic model of shock initiation of shocks with density discontinuities. Paper submitted to Seventh Symposium (International) on Detonation.

Mader, C. L. (1965) Initiation of detonation by the interaction of shocks with density discontinuities. Phys. Fluids 8, 1811-1816.

Merzanov, A. G., Barzykin, V. V., and Gontkovskaya, V. T. (1963) A problem of local thermal explosion. Dokl. Akad. Nauk SSSR 148, 380-387.

Wackerle, J., Johnson, J., and Halleck, P. (1976) Shock initiation of high-density PETN, Proceedings of Sixth Symposium (International) on Detonation, pp. 20-28. ONR ACR-221.

Walker, F. E. and Wasley, R. J. (1969) Critical Energy for shock initiation of heterogeneous explosives. Explosivstoffe, 17, 9.

Zababakhin, E. I. (1970) Unlimited cumulation phenomena, Mechanics in the USSR for 50 Years, pp. 313-342. Nauka, Moscow.

Author Index for Volume 87

Afanasiva, L.A. 394
Anisimov, S.I. 218
Atkinson, R. 318
Bauer, P. 231
Boiko, V.M. 71
Borisov, A.A. 88, 492
Branch, M.C. 22
Brochet, C.. 231
Brossard, J. 302
Buksowicz, W. 414
Bull, D.C. 318
Cherashin, A.V. 96
Chistyakov, V.P. 205
Cunningham, W.G. 9
Desbordes, D. 302
Ermolayev, B.S. 492
Fedorov, A.V. 71
Fomin, V.M. 71
Gavrilenko, T.P. 244
Gel'fand, B.E. 88
Geurraud, C. 352
Gilinsky, M.M. 251
Guirguis, R.H. 121
Hardy, J.R. 9
Hayashi, K. 22
Hayes, D.B. 445
Inogamov, N.A. 218
Ivanov, M.F. 218
Kadowaki, S. 335
Kamel, M.M. 121
Karo, A.M. 9
Khasainov, B.A. 492
Klimkin, V.F. 196
Kondrikov, B.N. 426
Korobeinikov, V.P. 376
Kudinov, V.M. 96
Kuhl, A.L. 41, 175

Kurlyo, J. 262
Lebed, S.G. 96
Levchenko, S.A. 64
Levin, V.A. 394
Mader, C.L. 468
Malakhov, A.T. 96
Manson, N. 302, 352
Martynenko, O.G. 64
Men'shov, I.S. 376
Nigmatulin, R.I. 88
Novikova, T.S. 251
Ohsawa, A. 157
Ohyagi, S. 157
Oppenheim, A.K. 121
Palamarchuk, B.I. 96
Papyrin, A.N. 71
Pickalov, V.V. 196
Prokhorov, E.S. 244
Rakhmatulin, K.A. 88
Rudin, G.I. 64
Saint-Cloud, J.P. 352
Sauer, F.M. 262
Shvetsov, G.A. 205
Soloukhin, R.I. 71, 196
Stadnichenko, I.A. 205
Stolovich, N.N. 64
Thomsen, J.M. 262
Timofeev, E.I. 88
Titov, V.M. 205
Tsugé, S. 335
Tunik, Y.V. 394
Vakhnenko, V.A. 96
Veyssiere, B. 362
Walker, F.E. 9
Wolański, P. 414
Zak, L.I. 251
Zel'dovich, Y.B. 1, 218